ADVANCES IN CHEMICAL PHYSICS

VOLUME 108

EDITORIAL BOARD

Advances in
CHEMICAL PHYSICS

Global and Accurate Vibration
Hamiltonians from High-Resolution
Molecular Spectroscopy

Edited by

MICHEL HERMAN, JACQUES LIEVIN
AND JEAN VANDER AUWERA*

Laboratoire de Chimie Physique Moléculaire CP 160/09
Université Libre de Bruxelles, Belgium

and

ALAIN CAMPARGUE

Laboratoire de Spectrométrie Physique (UMR5588) [†]
Université Joseph Fourier de Grenoble, BP 87, France

*Research Associate with the FNRS (Belgium)
[†]Associated with the CNRS

Series Editors

I. PRIGOGINE

Center for Studies in Statistical
Mechanics and Complex Systems
The University of Texas
Austin, Texas
and
International Solvay Institutes
Université Libre de Bruxelles
Brussels, Belgium

STUART A. RICE

Department of Chemistry
and
The James Franck Institute
The University of Chicago
Chicago, Illinois

VOLUME 108

AN INTERSCIENCE® PUBLICATION
JOHN WILEY & SONS, INC.
NEW YORK · CHICHESTER · WEINHEIM · BRISBANE · SINGAPORE · TORONTO

To the memory of Ikhlef HADJ BACHIR

CONTENTS

SERIES INTRODUCTION

Few of us can any longer keep up with the flood of scientific literature, even in specialized subfields. Any attempt to do more and be broadly educated with respect to a large domain of science has the appearance of tilting at windmills. Yet the synthesis of ideas drawn from different subjects into new, powerful, general concepts is as valuable as ever, and the desire to remain educated persists in all scientists. This series, Advances in Chemical Physics, is devoted to helping the reader obtain general information about a wide variety of topics in chemical physics, a field that we interpret very broadly. Our intent is to have experts present comprehensive analyses of subjects of interest and to encourage the expression of individual points of view. We hope that this approach to the presentation of an overview of a subject will both stimulate new research and serve as a personalized learning text for beginners in a field.

I. Prigogine
Stuart A. Rice

I. GENERAL INTRODUCTION

Over the last 50 years (i.e., since the late 1940s or early 1950s) or so, high-resolution molecular spectroscopy has provided more and more precise, accurate, and reliable information about the internal energy-level structure in gas-phase molecules, with unmatched quality. High-resolution molecular spectroscopy contributed as a major driving force to the advances of chemical physics, and significantly promoted and stimulated connected fields such as quantum chemistry, astrophysics, atmospheric sciences, metrology, and intramolecular dynamics.

By the end of this millennium, among the hot research topics in high-resolution molecular spectroscopy are those considering (1) *unstable species*, including van der Waals complexes, atmospheric radicals, and astrophysical ions; (2) *larger species*, such as organic species and clusters; and (3) *highly excited molecules*, in vibrational overtones and in Rydberg states. They all confirm the fascinating complexity and richness of the pattern of electronic vibration–rotation energy levels in polyatomic species. Interestingly, they simultaneously seem to demonstrate the emergence of some global, universal trends. These are of essential importance in various stimulating perspectives, including predicting the intramolecular dynamics and mastering the related chemistry, at the microscopic level.

In the present review, we focus on a very specific aspect in this general context. We deal with molecular vibration as revealed by *high-resolution* vibration–rotation spectra in the gas phase. We are concerned with energy and intensity features related to molecules in low and (mainly) highly excited vibrational levels, of particular interest to bond selected chemistry. To avoid disappointing some readers, let us clearly emphasize that our strategy is of spectroscopic rather than dynamical nature, that our interest is restricted to ground electronic states, for asymmetric and linear top molecules, in particular, and that electron and nuclear spins, the influence of external fields, and the role of large amplitude motions are almost completely neglected. Keywords of the present review are coordinate systems, overtone experimental and theoretical spectroscopy, and acetylene.

The concept of global fitting of the pattern of vibration–rotation energy levels was more often raised in the recent (at the time of writing) literature. This trend truly supports the present review. Global fits aim at unraveling the *complete* vibration–rotation energy pattern in a molecule, by bridging the Hamiltonian and the spectrum. The Hamiltonian, on one side, is a compact mathematical object describing precisely all kinetic and potential-energy couplings in the isolated quantum system considered, specifically, a vibrating–rotating molecule containing electrons and nuclei. The resolution of the Hamiltonian, using standard quantum-mechanical techniques,

1

generates a complete theoretical spectrum. The experimental spectrum, on the other side, is a complicated object built from the superposition of many pieces, specifically, the spectral lines whose position, intensity, and profile carry information on the eigenstates and on the intramolecular couplings. Their analysis ought to reveal the full Hamiltonian. Two opposite procedures, the forward and the backward trips, can therefore be followed to bridge the Hamiltonian and the spectrum. Both still suffer from many drawbacks and bottlenecks.

Extensive work was devoted over the years to push the investigation of both trips as far as possible. Refinement procedures were developed, sequentially acting on the forward and backward trips to make the experimental and theoretical information converge. Nevertheless, many authors agree (see general discussion hold at the XXth Solvay Conference on Chemistry, 1995, in particular in Ref. 1) that deeper investigation is required to provide a description of the full pattern of the vibration–rotation energy levels allowing reliable predictions to be performed, in particular in the field of intramolecular dynamics.

In one procedure, the exact Hamiltonian expressed in the coordinate or momentum space is the starting point. The aim is to provide a *theoretical spectrum* through the resolution of the corresponding Schrödinger equation and the calculation of the matrix elements of appropriate transition operators defining the theoretical transitions intensities. This is a direct process, which connects the basic quantum description of the system to its properties. It is furthermore an exact process because it strictly follows the prescriptions of quantum mechanics: the determination of observables from appropriate quantum operators. This forward trip, achieved by quantum chemists, faces problems with the numerical applications that, despite the tremendous developments in computational methods, are not trivial to achieve, by far. These limitations are linked to the well-known bottlenecks occurring in the resolution of n-body quantum systems, facing the management of infinite basis set expansions. It is actually still impossible to derive an exact vibration–rotation Hamiltonian from *ab initio* calculations and to cope with the huge dimension of the Hamiltonian matrices needed for converging toward solutions of the Schrödinger equation for nontrivial polyatomic systems. This trip is detailed in Section II, with particular emphasis on the possible selection of coordinate systems. Basic theoretical features are introduced in that section. They are supported by a comprehensive list of symbols provided in Appendix A, actually covering the whole review. Appendix A also lists all abbreviations used throughout the text. Units in Section II are atomic units.

In the other procedure, the observed spectrum is the starting point. The aim is, by performing the global treatment of all the experimental information available, to recompose the *exact Hamiltonian*. This is an indirect process as it goes from the properties to the mathematical object generating them. This backward trip, attempted by spectroscopists, is very ambiguous in nature. It relies on a set of information that is restricted in number and quality by experimental factors, and furthermore biased by the interaction of the molecule with the radiation used to produce the spectrum. The full treatment of the data also suffers the inherent ambiguity of the least-squares procedure applied to the inverse secular problem. As a consequence, approximate Hamiltonians must be designed, adapted to the limited quantum information available. They are the so-called effective, algebraic, or spectroscopic Hamiltonians. In Section III, we build a matrix image of the molecule allowing for global treatments, using well-known theoretical developments. The aim is to provide a reliable way to perform the backward trip, *ultimately accounting for the full experimental accuracy and precision now available in high-resolution spectroscopy*. Conventional spectroscopic units, with energies in reciprocal centimeters (cm^{-1}), are used in that section. Energy and intensity features are concerned.

The present research subject was dramatically promoted by formidable developments of experimental nature, to which state-of-the-art Fourier transform spectroscopy and laser techniques extensively contributed. Measurement precision and accuracy have increased significantly, the detection sensitivity has been magnified, and broader spectral coverage and spectral decongestion means have become available. A spectacular wealth of new vibration–rotation data have appeared in the literature, shedding light on previously inaccessible regions of the potential energy surface, and stimulating theoretical developments, therefore opening new research strategies. We felt that it was most critical to highlight some of these recent instrumental achievements, as achieved in the final section (IV).

Our strategy throughtout this review is as follows. *Basic features* are extensively developed for didactic purposes, although they have already been presented in various review papers or books in the literature. *Published research results* are extensively reviewed, sometimes presented differently than in the original papers, to illustrate the actual trends and perspectives in the subject. *Still unpublished information* is also included in some places. The consistency between all parts in the review is ensured by focusing on one target molecule, *acetylene*, which is used as a benchmark of today's status of vibration–rotation high-resolution spectroscopy within the global approach. Its vibration–rotation energy levels indeed constitute a laboratory of thus far unmatched quality, because of its combined richness and accessibility, to merge the backward and forward trips discussed above. The huge literature concerning the ground electronic state of acetylene, which is provided in Appendix B, illustrates the hopefully premonitory evolution of the present research topics, namely, the merging of models unraveling the global vibrational energy pattern, at experimental high-resolution precision, allowing the intramolecular dynamics to be mapped out and bond-selected chemistry to be achieved. As an output of the treatment developed around acetylene, Appendix C provides a comprehensive list of calculated energies of the vibration energy levels in $^{12}C_2H_2$, up to 10,000 cm^{-1}, as well as additional information on those levels.

In the present review, we thus aim at building on global concepts a general and didactic introduction to vibration and vibration–rotation spectroscopy, addressing in a systematic way the entire pattern of energy levels. We simultaneously focus on actual research trends in overtone spectroscopy, exemplified with published and sometimes unpublished results on acetylene, eventually illustrated by most recent instrumental advances.

II. THE FORWARD TRIP: FROM THE HAMILTONIAN TO THE VIBRATION–ROTATION SPECTRUM

A. Introduction

The aim of this section is to give a general theoretical background to the review by detailing how to bridge the full molecular Hamiltonian and the corresponding vibration–rotation spectrum. We consider the viewpoint of the theoretician and highlight the bottlenecks to be faced and the approximations to be introduced. As pointed out in the general introduction, the *Hamiltonian* is a compact mathematical object describing precisely all the kinetic and potential-energy couplings in the isolated quantum system: a vibrating-rotating molecule containing n_e electrons and n nuclei. Our aim is to detail the steps leading to an hypothetical *theoretical spectrum*, involving the resolution of the corresponding Schrödinger equation and the calculation of the matrix elements of appropriate transition operators defining the theoretical transition intensities.

An overview of the forward, direct trip is presented in Section II.B, starting in Section II.B.1 with a general formulation of the *exact* quantum-mechanical approach. All basic equations leading from an exact Hamiltonian to an exact vibration–rotation spectrum are provided, ignoring at this stage the severe approximations practically required for computational reasons. This exact formulation provides a reference level toward which practical applications have to converge. Departure from the exact treatment in the forward trip is discussed in Section II.B.2. Of main concern in this analysis is the choice of vibration–rotation coordinates, the definition of approximate Hamiltonians, the finite basis sets to be used, and the alternative methods for getting approximate solutions of the Schrödinger equation. All such features are applied in Section II.C to acetylene, the target system adopted throughout this work.

B. The Forward Trip: An *Ab Initio* Approach

1. Exact Quantum Mechanical Formulation

Let us assume that it is possible to know, from a purely *ab initio* source, the exact vibration–rotation Hamiltonian. Let us dream some more and assume that this Hamiltonian is numerically tractable. It is then possible to solve the corresponding vibration–rotation Schrödinger equation and to obtain the exact calculated energy spectrum. One can further dream of accessing to exact expressions of transition properties and, therefore, to exact transition intensities. It is the story of such an "exact" forward trip that is detailed, step by step, in this section.

The notations adopted below obey well defined rules. Hamiltonian terms, eigenvalues and eigenfunctions have descriptive subscripts such as "e," "n," "t," "r," and "v" for electron, nuclear, translation, rotation, and vibration, respectively. Superscripts give additional information on quantum numbers or state labeling, nuclear geometry, and components in a given axis system.

a. Full Molecular Hamiltonian. We consider a molecule made of n_e electrons and n nuclei and label these particles from 1 to n_{tot} ($n_{tot} = n_e + n$). Particle i is then defined by its mass m_i, its charge Z_1 and its coordinates (X_i, Y_i, Z_i) in a Cartesian system of coordinates rigidly attached to the laboratory [laboratory axis system (LAS)]. Let $\bar{R}_i$ be the corresponding vector position for particle i. As far as the Coulomb interactions between these particles are dominant, and relativistic contributions can be neglected, one uses the full nonrelativistic Hamiltonian, which can be written in the following compact form:

$$\hat{H} = -\sum_{i=1}^{n_{tot}} \frac{\nabla^2(\bar{R}_i)}{2m_i} + \sum_{i,j=1}^{n_{tot}} \frac{Z_i Z_j}{|\bar{R}_j - \bar{R}_i|} \qquad (2.1)$$

We adopt in this section an usual unit system in theory—atomic units—in which the units of action, mass and charge are $\hbar$, the electron mass m_e, and the electron charge e, respectively. This means that for electron i the values of Z and m_i are -1 and 1, respectively and for nucleus j, Z_j and m_j are the corresponding atomic number and mass, respectively.

The nuclear and electron spins are not explicitly described in the above-mentioned electrostatic Hamiltonian. They are nevertheless present and determine, among other properties, the Pauli statistics, whose application to the relative vibration–rotation line intensities is detailed in Section III.E.3.(*e*).

b. Born–Oppenheimer Separation. A first simplification of the preceding Hamiltonian is provided by the usual separation of the electronic and nuclear motions, dictated by the so-called Born–Oppenheimer approximation [2]. This approximation seems indeed fully justified in the present review dedicated to the vibration–rotation spectroscopy of small molecules within an isolated, nondegenerate electronic potential-energy surface. We thus exclude here electronic-vibration–rotation interactions, requiring degenerated potential-energy surfaces (Renner–Teller, Jahn–Teller, or Herzberg–Teller effects) or two crossing electronic surfaces (conical intersections) [3]. Such cases are dealt with in many papers, including Refs. 4–26. They are exemplified in the study of target molecules such as NO_2, presenting a conical intersection between its two low-lying electronic states [27–30], and in some neutral and ionic first-row hydrides such as BH_2 [31], CH_2 [32,33], SiH_2 [34], NH_2^+ [35,36], BH_2^- [37] and CH_2^+ [38,39] exhibiting a Renner–Teller effect.

We thus investigate the vibration–rotation motion within an isolated adiabatic potential-energy surface arising from the resolution of the electronic Schrödinger equation:

$$[\hat{T}_e + \hat{V}_e(\bar{R}_n^{(k)}, \bar{R}_e)]\Psi_e^{(E)}(\bar{R}_n^{(k)}, \bar{R}_e) = E_e^{(E)}(\bar{R}_n^{(k)})\Psi_e^{(E)}(\bar{R}_n^{(k)}, \bar{R}_e) \qquad (2.2)$$

We now use distinct notations for electrons and nuclei by means of subscripts "e" and "n," respectively. The compact notation $\bar{R}_e$ and $\bar{R}_n$ refers to the whole set of electron and nuclear vector positions, respectively. $\bar{R}_{ej}$ refers to the vector position of the jth electron. The electronic Hamiltonian in (2.2) is a so-called *clamped nucleus Hamiltonian*, which means that all nuclei have fixed coordinates, which is denoted by $\bar{R}_n^{(k)}$, with the superscript k referring to a given geometry of the nuclear skeleton. The operators $\hat{T}_e$ and $\hat{V}_e$ represent the kinetic-energy operator of the electrons and the Coulomb potential-energy operator, respectively; the latter term accounts for the attraction between nuclei and electrons and the repulsion between couples of electrons. In atomic units; this can be expressed as

$$
\begin{aligned}
\hat{T}_e &= -\frac{1}{2}\sum_{j=1}^{n_e} \nabla^2(\bar{R}_{ej}) \\
\hat{V}_e &= -\sum_{i=1}^{n}\sum_{j=1}^{n_e} \frac{Z_i}{|\bar{R}_{ej} - \bar{R}_{ni}^{(k)}|} + \sum_{i<j}^{n_e} \frac{1}{|\bar{R}_{ej} - \bar{R}_{ei}|}
\end{aligned}
\qquad (2.3)
$$

Equation (2.2) has in principle an infinite number of eigensolutions labeled by the superscript (E). All these states are bound states, in the sense defined by quantum mechanics; that is, they all have square integrable wavefunctions and negative eigenvalues [40]. Note that this condition is compatible with the fact that some of these states may be unbound with respect to a given dissociation channel of energy E_{diss}, which means that at all possible nuclear geometries $\bar{R}_n^{(k)}$ the inequality $E(\bar{R}_n^{(k)}) \geq E_{\text{diss}}$ holds. To avoid confusion we qualify such states as dissociative rather than unbound.

The total energy of a given electronic state (E) at the clamped nucleus geometry $\bar{R}_n^{(k)}$ is given by the sum of the eigenvalue of (2.2) and the nuclear repulsion potential energy:

$$E^{(E)}(\bar{R}_n^{(k)}) = E_e^{(E)}(\bar{R}_n^{(k)}) + \sum_{i<j}^{n} \frac{Z_i Z_j}{|\bar{R}_{nj}^{(k)} - \bar{R}_{ni}^{(k)}|} \qquad (2.4)$$

Focusing on eigensolution (E) of (2.2), one defines the corresponding potential-energy surface (PES) by the multidimensional energy function $E^{(E)}(\bar{R}_n)$, in which $\bar{R}_n$ holds for the collective variable defining all possible nuclear geometries. Mathematically, all possible $\bar{R}_n^{(k)}$ geometries form a $3n$-dimensional vector space $\mathscr{R}^{3n}$ associated with an Euclidean space E^{3n} equipped with an inner product [41]. In this picture, each vector position $\bar{R}_n^{(k)}$, defining a given molecular configuration, is a point of E^{3n}. In practical applications, one is usually restricted to a given domain $\mathscr{D}$ of $\mathscr{R}^{3n}$, matching nuclear arrangements actually sampled in the application. The PES is the graph of the energy functional $E^{(E)}(\bar{R}_n)$, that is, it is mathematically defined as $\{\bar{R}_n, E^{(E)}(\bar{R}_n) | \bar{R}_n \in \mathscr{D}\} \subseteq \mathscr{R}^{3n} \times \mathscr{R}$.

In the second step of the Born–Oppenheimer scheme, one has to solve the Schrödinger equation for nuclei moving in the potential created by the electrons:

$$[\hat{T}_n + \hat{V}_n(\bar{R}_n)]\Psi_n^{(E)}(\bar{R}_n) = E_{\text{en}}^{(E)}\Psi_n^{(E)}(\bar{R}_n) \qquad (2.5)$$

in which the kinetic-energy operator is:

$$\hat{T}_n = -\sum_{i=1}^{n} \frac{\nabla^2(\bar{R}_{ni})}{2m_i} \qquad (2.6)$$

and the potential-energy operator is simply the PES for the electronic state labeled (E):

$$\hat{V}_n = E^{(E)}(\bar{R}_n) \qquad (2.7)$$

The eigenfunctions of equation (2.5) are nuclear wavefunctions describing the $3n$ degrees of freedom of the nuclei: the translation, rotation and vibration motions. The eigenvalues $E_{en}^{(E)}$ are the total molecular energies including the electronic and nuclear contributions. The corresponding total wavefunctions are simple products of the corresponding electronic and nuclear eigenfunctions:

$$\Psi_{en}^{(E)}(\bar{R}_e, \bar{R}_n) = \Psi_e^{(E)}(\bar{R}_e, \bar{R}_n)\Psi_n^{(E)}(\bar{R}_n) \tag{2.8}$$

c. *Translation-Free Hamiltonian.* The nuclear Hamiltonian in (2.5) contains, in addition to a discrete spectrum of levels, a continuum spectrum arising from the center-of-mass motion. In order to eliminate these translation nuclear degrees of freedom, we rewrite the total nuclear Hamiltonian of (2.5) in terms of translation-free coordinates. This transformation can be accomplished exactly; the translation is linearly separable from $\mathcal{R}^{3n}$ [41,42]. One usually identifies the translation coordinates to those of the nuclear center of mass $\bar{R}_0$, defined in the LAS, with the associated total nuclear mass M_n. The remaining $3n - 3$ coordinates $\{t_{\Xi i}$; with $i = 1$ to $n - 1$ and $\Xi = X, Y, Z\}$, accounting for pure rotation and vibration motions, are obtained from the $n - 1$ linear equations:

$$t_{\Xi i} = \sum_{q=1}^{n} w_{qi}\bar{R}_{\Xi q}; \qquad \Xi = X, Y, Z \tag{2.9}$$

where $\bar{R}_{\Xi q}$ are the nuclear coordinates in the LAS and the w_{qi} coefficients obey the formula

$$\sum_{q=1}^{n} w_{qi} = 0 \tag{2.10}$$

for ensuring translation invariance.

The total nuclear kinetic operator can now be factored out into pure translation $\hat{T}_t$ and vibration–rotation $\hat{T}_{rv}$ terms [43]:

$$
\begin{aligned}
\hat{T}_n &= \hat{T}_t + \hat{T}_{rv} \\
\hat{T}_t &= -\frac{\nabla^2(\bar{R}_0)}{2M_n} \\
\hat{T}_{rv} &= -\frac{1}{2}\sum_{\Xi=X,Y,Z}\sum_{i,j=1}^{n-1} G_{ij}\nabla(t_{\Xi i})\,\nabla(t_{\Xi j}) \\
G_{ij} &= \sum_{\Xi=X,Y,Z}\sum_{k=1}^{n}\frac{1}{m_k}\frac{\partial t_{\Xi i}}{\partial R_{\Xi k}}\frac{\partial t_{\Xi j}}{\partial R_{\Xi k}}
\end{aligned}
\tag{2.11}
$$

d. Vibration-Rotation Separation. A further separation of the $3n - 3$ translation-free coordinates $\{t\}$ is still needed to solve the nuclear Schrödinger equation; this is done by separating the rotation motion (uniform rotation of the molecule as a whole) from the vibration or internal motion (relative motions of the nuclei with respect to each other). The following choices have to be made:

- Definition of an axis system attached and rotating with the molecule [the molecular axis system or (MAS)].
- Definition of rotation coordinates, describing the rotation of the MAS with respect to the LAS.
- Definition of vibrational coordinates in the MAS.

In any case, complete uncoupling of rotation and vibration coordinates is impossible. The remaining coupling terms are of curvilinear nature [41]. As shown below, particular choices, however, tend to minimize the importance of such terms.

1. MOLECULAR AXIS SYSTEM (MAS). Let us define a Cartesian (x, y, z) axis system fixed to the molecule, which we refer to as the *body-fixed* or *molecular axis system* (MAS). The origin of this frame is selected to be at the nuclear center of mass. Figure 1 represents the two LAS (X, Y, Z) and MAS (x, y, z) systems. The vector position of nucleus j is thus referred to as $\bar{R}_j$ and $\bar{r}_j$ in the LAS and MAS, respectively.

2. ROTATION COORDINATES. Although other choices are possible [44], we consider here the usual Euler angles as rotation coordinates. As represented

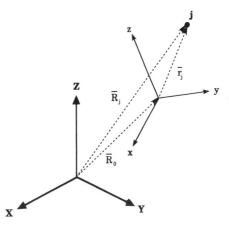

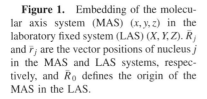

Figure 1. Embedding of the molecular axis system (MAS) (x, y, z) in the laboratory fixed system (LAS) (X, Y, Z). $\bar{R}_j$ and $\bar{r}_j$ are the vector positions of nucleus j in the MAS and LAS systems, respectively, and $\bar{R}_0$ defines the origin of the MAS in the LAS.

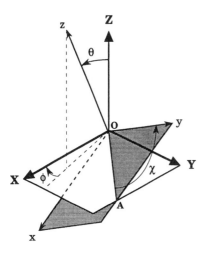

Figure 2. Definition of the Euler angles θ, ϕ, and χ, with $0 \leq \theta \leq \pi$, $0 \leq \phi \leq 2\pi$, and $0 \leq \chi \leq 2\pi$. OA denotes the intersection of the XY and xy planes.

in Fig. 2, following the convention used in [45], angles θ, ϕ and χ, define the orientation of the MAS with respect to the LAS [45–47]:

$$
\begin{bmatrix} x \\ y \\ z \end{bmatrix} = \begin{bmatrix} \lambda_{xX} & \lambda_{xY} & \lambda_{xZ} \\ \lambda_{yX} & \lambda_{yY} & \lambda_{yZ} \\ \lambda_{zX} & \lambda_{zY} & \lambda_{zZ} \end{bmatrix} \begin{bmatrix} X \\ Y \\ Z \end{bmatrix}
\tag{2.12}
$$

in which the rotation matrix λ is the so-called direction cosine matrix, given in Table I.

For a linear molecule having all nuclei lying along the z axis, only two of the Euler angles (θ, ϕ) act as rotation coordinates. The rotation about z, defined by angle χ, is an operation of the molecular symmetry point group $(D_{\infty h}$ or $C_{\infty v})$. It therefore does not correspond to an uniform rotation of all nuclei. We are thus left with a number of m_{int} internal or vibrational coordinates $\{q_1, q_2, \ldots, q_{m_{int}}\}$ with m_{int} equal to $3n - 6$ and $3n - 5$ for nonlinear and linear molecules, respectively.

TABLE I
Direction Cosine Matrix, Defining the Orientation of the MAS with respect to the LAS

	X	Y	Z
x	$\cos \theta \cos \phi \cos \chi - \sin \phi \sin \chi$	$\cos \theta \sin \phi \cos \chi + \cos \phi \sin \chi$	$-\sin \theta \cos \chi$
y	$-\cos \theta \cos \phi \sin \chi - \sin \phi \cos \chi$	$-\cos \theta \sin \phi \sin \chi + \cos \phi \cos \chi$	$\sin \theta \sin \chi$
z	$\sin \theta \cos \phi$	$\sin \theta \sin \phi$	$\cos \theta$

3. ECKART FRAME. Given the definition of the rotation coordinates, we have to decide how the molecule is attached to the MAS. We use the well known Eckart frame [42,46,48]. It presents the advantage of minimizing the vibration–rotation coupling terms and is a natural choice when using normal coordinates. More suitable choices do exist, such as when dealing with large-amplitude motion [43,49,50]. The choice of the nuclear frame does not, however, introduce approximations in the exact resolution of the vibration–rotation Schrödinger equation; rather, it, influences the convergence of the calculated vibration–rotation properties when a truncated numerical treatment is applied (see Section II.B.2.f).

Referring to Figs. 1 and 2, the position of a nucleus j can be written as

$$\bar{R}_j = \bar{R}_0 + \lambda^{-1} \cdot \bar{r}_j \qquad (2.13)$$

with λ the direction cosine matrix defined in (2.12) and $\bar{R}_0$ the origin of the MAS in the LAS. This equation depends on $3n + 6$ variables ($3n$, 3, and 3 for $\bar{r}_j$, $\bar{R}_0$, and λ, respectively). There are 6 of these free for defining unambiguously the position of the n nuclei in the LAS. One possible set of $3n$ linearly independent parameters to be used within the MAS is defined using the Eckart conditions (see, e.g., Ref. 46), for a maximal uncoupling between the different nuclear degrees of freedom. Considering a reference nuclear configuration (0) in the MAS (usually selected to be the equilibrium geometry of the molecule), one can define any other configuration (k) by means of displacement vectors $\bar{\delta}_j$ of each nucleus j with respect to the reference geometry:

$$\bar{r}_j = \bar{r}_j^{(0)} + \bar{\delta}_j \qquad (2.14)$$

Taking the origin of the MAS as the center of mass of the molecule in (0):

$$\sum_{j=1}^{n} m_j \bar{r}_j^{(0)} = 0 \qquad (2.15)$$

leading to the following two Eckart conditions:

$$\sum_{j=1}^{n} m_j \bar{\delta}_j = 0; \qquad \sum_{j=1}^{n} m_j (\bar{r}_j^{(0)} \wedge \bar{\delta}_j) = 0 \qquad (2.16)$$

The first condition indicates that the MAS follows the translation of the molecule; the second, that the components of the angular momentum cancel at the reference geometry, meaning that the separation of rotation and vibration is achieved for small-amplitude motions around equilibrium.

4. VIBRATION COORDINATES. Now that we have defined how the molecule is attached to the rotating frame, we have still to characterize the coordinates to be used for describing the vibrational motion. Two fundamental classes emerge from the literature: the *rectilinear* and *curvilinear coordinates*. The standard normal mode coordinates are falling in the former class, while all kinds of internal coordinates are in the latter class. Depending on the context in which they are used, internal coordinates are referred to in the literature as *internal bond-angle* [51–55], *curvilinear normal* [56–58], *geometrically defined* [49,59], *heliocentric-like* [60,61], and *local mode* [62–64] coordinates. The respective advantages of the different coordinate systems are discussed in Sections II.B.2.*b* and II.B.2.*c* and illustrated in Section II.C.

e. Exact Vibration–Rotation Hamiltonian

1. VIBRATION–ROTATION KINETIC-ENERGY OPERATOR. It is in principle possible at this point to derive an exact form of the kinetic part of the vibration–rotation Hamiltonian for any choice of the vibration–rotation coordinates (see, e.g., Refs. 42 and 65–70 and references cited therein). The step is straightforward in the case of rectilinear normal coordinates embedded within the Eckart frame, as detailed in Section III.A. It is more complicated to achieve when dealing with internal coordinates and non-Eckart frames. We refer to an excellent paper of Bramley et al. [43] and to the references therein, for the specific problems of singularities encountered and for the techniques available for building the operator in specific cases. It is, in any case, possible to derive an exact form of the kinetic-energy operator. This task is greatly simplified by using computer algebra programs [67].

Let us collect the $3n - 3$ vibration–rotation variables in a compact notation $\{\Theta, q\}$, in which Θ holds for the Euler angles $\{\theta, \phi, \chi\}$ and q for the whole set of m_{int} vibration coordinates $\{q_1, q_2, \ldots, q_{m_{\text{int}}}\}$, independently of their definition. The label q specifically refers to normal modes in Section III. A general form of the exact vibration–rotation kinetic-energy operator is

$$\hat{T}_{rv} = \hat{T}_v(q) + \hat{T}_r(\Theta) + \hat{T}_{\text{rovib}}(\Theta, q) \qquad (2.17)$$

in which $\hat{T}_v$ and $\hat{T}_r$ refer to the pure vibration and rotation parts of the kinetic-energy operator, respectively, and $\hat{T}_{\text{rovib}}$ gathers all vibration–rotation coupling terms.

2. POTENTIAL ENERGY SURFACE. The potential $\hat{V}_n(\bar{R}_n) = E^{(E)}(\bar{R}_n)$ appearing in the nuclear Schrödinger equation (2.5), corresponds to the PES of the considered electronic state (E). It can be expressed directly in the selected set of internal coordinates q. It is indeed invariant under translation and

under uniform rotation of the molecule. One can thus easily follow the coordinate transformation $\bar{R}_n \Rightarrow \bar{r}_n \Rightarrow q_i$, from LAS to MAS and then to translation- and rotation-free coordinates. The resulting expression $E^{(E)}(q)$ can be introduced in the final vibration–rotation Schrödinger equation as potential-energy operator:

$$\hat{V}_{rv}(q) = E^{(E)}(q) \tag{2.18}$$

If an exact expression of this potential $E^{(E)}(q)$ could ever become available from *ab initio* calculations (see Section II.B.2.*d*), it could either be an analytical function of q, sufficiently flexible for matching the exact potential in the domain $\mathscr{D}$ of interest, or a discretized form adapted to a direct numerical integration of the Schrödinger equation by collocation methods (see Sections II.B.2.*d*.5 and II.B.2.*f*.4).

3. EXACT VIBRATION–ROTATION SCHRÖDINGER EQUATION. The sum of the kinetic and potential operators defines the *exact* vibration–rotation Hamiltonian:

$$\hat{H}_{rv}^{(E)} = \hat{T}_v(q) + \hat{T}_r(\Theta) + \hat{T}_{rovib}(\Theta, q) + \hat{V}_{rv}(q) \tag{2.19}$$

and the corresponding vibration–rotation Schrödinger equation

$$\hat{H}_{rv}^{(E)}(\Theta, q)\Psi_{rv}^{(E,p)}(\Theta, q) = E_{evr}^{(E,p)}\Psi_{rv}^{(E,p)}(\Theta, q) \tag{2.20}$$

with $E_{evr}^{(E,p)}$, $\Psi_{evr}^{(E,p)}$ eigenvalues and eigenfunctions defining a vibration–rotation state (p) belonging to the electronic state (E). The importance of this equation must be emphasized. It indeed contains the whole information on the vibration–rotation spectrum. In the following sections we focus on the resolution of this eigenvalue problem.

f. Variational Resolution of the Vibration–Rotation Schrödinger Equation

1. VARIATIONAL PRINCIPLE. Among the methods provided by quantum mechanics for finding approximate solutions of the Schrödinger equation associated with a given problem, the *variational method* is most central [40]. It is widely used in atomic [71], molecular [72–76], nuclear [77], and solid-state [78] physics, and, in many cases, replaces the Rayleigh–Schrödinger many-body perturbation theory approach [72,79]. It is particularly well adapted to the description of strong interactions between particles, including the so-called electron correlation effect and, of direct interest to the present review, the vibrational anharmonicity and the Coriolis interactions, that couple vibrational and rotational degrees of freedom. One drawback of perturbation theory is that it requires an accurate zeroth-order description as starting point, which is a necessary condition for ensuring a rapid

convergence of the perturbative expansions. It moreover fails when degeneracies or quasi-degeneracies occur, which are commonplace in the vibration–rotation structure. Perturbation theory is therefore not adapted for treating the so-called resonances, which are most relevant in the present context (see Section III), and for which a specific treatment is required (see Sections II.B.2.g.2 and III).

The basic ideas of the variational method [40,72] are first summarized and then applied to the resolution of the vibration–rotation Schrödinger equation (2.20). We use the "bra-ket" notation ($< |, | >$), due to Dirac [40], in which wavefunctions are expressed in terms of "ket" vectors $| >$ and of their adjoins, the "bra" vectors $< |$. Given an Hamiltonian $\hat{H}$, one starts from a normalised trial wavefunction Ψ depending on certain parameters, and vary these parameters until the expectation value $< \Psi|\hat{H}|\Psi >$ reaches a minimum value E_{min}, which is the variational estimate of the ground-state energy. The *variational principle*, guarantees that E_{min} is an upper bound of the exact ground-state energy ε_0 ($E_{min} \geq \varepsilon_0$). The functional dependency (linear or nonlinear) of the trial wavefunction with respect to the variational parameters generates different classes of variational methods. A particular class of methods derived from the *mean-field approach* [self-consistent field (SCF) or multiconfigurational self-consistent field (MCSCF)] will be addressed in Sections II.B.2.d.1 and II.B.2.f.2 for electronic and vibrational structure calculations, respectively. We give here details on the more usual variational method in the vibration–rotation context, specifically, the linear variational method, in which Ψ is a linear function of the parameters.

2. LINEAR VARIATIONAL METHOD. The trial function is expanded on a fixed basis set of d linearly independent functions $\{\Psi_i; i = 1 \text{ to } d\}$

$$|\Psi\rangle = \sum_{i=1}^{d} c_i \Psi_i \qquad (2.21)$$

and the coefficients c_i are the only variational parameters. One obtains d variational solutions (2.21) from the diagonalization of H, the matrix representation of the Hamiltonian operator in the basis $\{\Psi_i; i = 1 \text{ to } d\}$. The matrix eigenvalue problem can be written in the following secular form:

$$Hc = SEc \qquad (2.22)$$

where H, S, c, and E are $(d \times d)$ square matrices; H is the Hamiltonian matrix and S is, in the present case, the overlap matrix, defining the orthonormality of the basis functions:

$$H_{ij} = \langle \Psi_i | \hat{H} | \Psi_j \rangle$$
$$S_{ij} = \langle \Psi_i | \Psi_j \rangle \qquad \text{for} \qquad 1 \leq i,j \leq d \tag{2.23}$$

E is a diagonal matrix containing the eigenvalues, which we can arrange in an ascending sequence as: $E_0 \leq E_1 \leq \cdots \leq E_{n_d-1}$, in which the index 0 refers to the ground state. The columns of c are the corresponding eigenvectors, the coefficients of (2.21). The variation principle implies that $E_0 \geq \varepsilon_0$, namely, that E_0 is an upper bound of the exact ground-state energy, but also that any eigenvalue E_i is also an upper bound of the exact energy ε_i corresponding to the ith excited state.

The differences $(E_i - \varepsilon_i)$ give a measure of the errors introduced by the variational method. They strictly depend on the quality of the chosen basis set, defined by its size d and by the adequacy of the basis components for reproducing the exact wavefunctions. The quality of variational results can be investigated by means of convergence tests performed in the following way. A variational calculation is considered to have converged for the set of x low-lying energy states (with $x \leq d$), if an increase of the basis set size from d to $(d + \alpha)$ keeps the energies of the x states within a defined threshold δ. The former d functions are not replaced during the procedure, and the value of δ is adapted to the desired precision on the individual energies. Therefore

$$E_i(d) - E_i(d + \alpha) \leq \delta \qquad \text{for} \qquad 0 \leq i \leq x \tag{2.24}$$

An important property of the linear variational solutions is [76] that they converge to the exact solutions of the corresponding Hamiltonian when the basis set size tends to infinity, that is, when the *completeness* of the basis set is achieved. Handy and co-workers [43] used the word *completability* to qualify a basis set becoming mathematically complete in its systematic extension toward infinite size.

The difference (2.24) is always positive in virtue [76] of the Hyleraas–Undheim theorem [80], also known as the "MacDonald theorem" [81]. This theorem guarantees that the roots of the secular problem with dimensions $(d + \alpha)$ are separated by those of the d one. When dealing with finite basis sets one generally observes that the convergence is very poor for the highest energy states. It is therefore necessary to use basis sets having a size large enough. Also, the wavefunctions obtained from (2.22) are optimized for energy considerations. They are therefore not necessarily optimal for other operators, such as, for instance, the dipole moment operator, and specific convergence tests ought to be performed when calculating matrix elements of those operators (see, for e.g., examples in atomic [82] and molecular [83] electronic structure calculations).

Last but not least, equation (2.22) can be reduced by block diagonalization, by virtue of the vanishing integral rule [46]. One uses a symmetry-adapted basis set, thus with all basis functions in (2.21) belonging to a given irreducible representation of the total symmetry group [46], to generate a secular equation (2.22) of reduced size, which provides the eigensolutions of that particular symmetry.

The variational method thus allows a set of eigensolutions of the Hamiltonian to be obtained in a single calculation, with some tests available on their convergence. This method is therefore very appealing in the context of vibration-rotation problems, as shown in the next sections.

3. CONFIGURATION INTERACTION METHOD. In most of the applications in physics cited above, the basis functions Ψ_i in (2.21) are n-body functions built as symmetry-adapted products of single body functions. In electronic structure theory, simple products of one-electron functions define *electronic configurations*, and the corresponding symmetry-adapted products are *configuration state functions* (CSF). Within this context, the linear variational method is usually referred to as the *configuration interaction* (CI) *method*. By extension, one also talk about *vibrational configuration interaction* calculations (VCI) [e.g., 84,85]. A vibrational configuration is defined as a product of m_{int} oscillator functions $\Phi(q_i)$, each characterized by a vibrational quantum number v_i

$$\Psi_{v_1, v_2, \ldots, v_{m_{int}}} = \prod_{i=1}^{m_{int}} \Phi_{v_i}(q_i) \qquad (2.25)$$

Linear combinations of such products, defining symmetry-adapted configurations for systems with degenerate modes, are the corresponding vibrational CSFs.

4. DIAGONALIZATION OF THE VIBRATION–ROTATION HAMILTONIAN. The application of the linear variation method to the resolution of the vibration–rotation problem was pioneered by Carney and Porter [86] in 1974. The availability of ever more powerful computers and more efficient computational algorithms since supported its impressive development (e.g., see Refs. 87–95, and references cited therein).

In order to variationally solve equation (2.20), one must diagonalize the matrix representation of Hamiltonian (2.19) using a suitable basis set. It is meant a symmetry-adapted basis set $\{\Psi_1(\Theta, q), \Psi_2(\Theta, q), \ldots, \Psi_d(\Theta, q)\}$, of size d, able to make a set of x vibration–rotation sates converge to a desired accuracy, with x the number of states of interest. One builds a product basis set of size $d = n_r \times n_v$, made from the direct product of a pure rotation basis set $\{\Psi_r^{(a)}(\Theta); a = 1 \text{ to } n_r\}$ of size n_r, by a pure vibration basis set $\{\Psi_v^{(b)}(q); b = 1 \text{ to } n_v\}$ of size n_v. These individual basis sets can be obtained from approximate solutions of the corresponding pure rotation and vibration problems (see Section II.B.2.c).

Standard symmetric top functions are usually adopted [49,96–98] for the rotation part. The so-called Wigner functions $D^J_{KM}(\Theta)$ are defined by three quantum numbers: J (rotational angular momentum), M (magnetic quantum number), and K (projection of J on the main symmetry axis in the MAS), all properly defined in Section III.D.2.a. Vibrational functions are usually expressed as in (2.25), that is, in terms of products of m_{int} single vibrational mode functions. Such product basis sets show better convergence properties if the couplings between the different product terms are weak and if individual basis sets are good solutions of the corresponding uncoupled problems.

Assuming the absence (still dreaming) of technical problems, the diagonalization of the huge matrix H_{rv} corresponding to the "exact" vibration–rotation Hamiltonian (2.19) built using a suitable basis set can be performed. It provides eigenvalues corresponding to the exact, or rather the converged energies $\{E_{rv}^{(E,p)}; p = 1 \text{ to } x\}$ of the x desired vibration–rotation states and eigenvectors, defining the corresponding wavefunctions $\{\Psi_P\}$.

5. ENERGY SPECTRUM OF THE VIBRATION–ROTATION HAMILTONIAN. Arranging in ascending order the eigenvalues resulting from the diagonalisation just described

$$E_0 \leq E_1 \leq \cdots E_p \cdots \leq E_d \qquad (2.26)$$

leads to the *energy spectrum* of the Hamiltonian (2.19), socalled by the theoreticians. As sketched in Fig. 3, this spectrum is split into converged and nonconverged parts; the former part is usually concerned with the x lower-lying energy states. Note that specific windowing algorithms, such as those provided within the Lanczos diagonalization scheme, can be used to make a set of eigensolutions lying in a user-defined energy window converge (see Refs. 99 and 100 and references cited therein). The energy scale of this spectrum is the one of the PES $[E^{(E)}(q)]$. If the PES in (2.20) directly comes from the *ab initio* resolution of the electronic Schrödinger equation (2.2), without further energy scaling, the vibration–rotation energies include the electronic contribution. This is why the subscript "e" was retained when labeling the eigenvalues $E_{erv}^{(E,p)}$ in (2.20). Please note that this subscript also

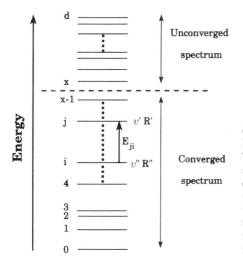

Figure 3. Schematic picture of the energy spectrum of the vibration–rotation Hamiltonian from variational calculations. The spectrum is split into a converged spectrum, consisting in the x low-lying energy levels and the remaining unconverged spectrum. The total size of the variational basis set is d. A single transition $j \leftarrow i$, or $v'R' \leftarrow v''R''$ according to the usual spectroscopic notation, is illustrated.

takes other meanings in the present review. The PES is, however, frequently normalized with respect to the equilibrium position, and the energy scale is then defined with respect to the bottom of the PES. The wavefunctions associated to (2.26) have the following form:

$$\Psi_{rv}^{(E,p)}(\Theta, q) = \sum_{a=1}^{n_r} \sum_{b=1}^{n_v} c_{ab}^{(p)} \Psi_r^{(a)}(\Theta) \Psi_v^{(b)}(q) \qquad (2.27)$$

where the square of the variational coefficients $c_{ab}^{(p)}$ gives the weights of the corresponding product basis functions $\Psi_r^{(a)}(\Theta)\Psi_v^{(b)}(q)$ contributing to state p. Those coefficients define the dominant contributions and often help in labeling the vibration–rotation states, using the rotation and vibration quantum numbers of this dominant basis function. Each couple of eigensolution $\{E_{rv}^{(p)}, \Psi_{rv}^{(p)}\}$ can then be labeled as $\{E^{(v_p R_p)}, \Psi^{(v_p R_p)}\}$, where v_p and R_p refer to the whole set of vibration and rotation quantum numbers of state p, respectively. Such unambiguous assignments, however, require the individual basis sets to contain the adequate information about the total quantum numbers. Dramatic problems may be encountered when the Hamiltonian induces strong coupling between the basis functions. It is the case for the vibration polyads and clusters discussed in Section III.C. In those cases, the constants of the vibrational motion are not the same in the zeroth-order basis set and in the cluster eigenfunctions. The normal-mode picture of the basis set then fails, and new quantum numbers need to be defined.

6. VIBRATION–ROTATION TRANSITION ENERGIES. The transition between two vibration–rotation states, such as 2 (higher energy) and 1 (lower energy), can be labeled $v_2 R_2 \leftarrow v_1 R_1$ or, according to the usual spectroscopic notation, $(v'R') \leftarrow (v''R'')$. Its energy is given by the difference between the corresponding eigenvalues (see also Fig. 3):

$$E_{21} = E_{rv}^{(E,2)} - E_{rv}^{(E,1)} = E^{(v'R')} - E^{(v''R'')} \tag{2.28}$$

We can calculate in this way the *theoretical energy spectrum*, containing the energy position of all possible transitions between the x converged energy levels. This spectrum is a theoretical object, which contains an energetical information obviously much richer than the one provided by any experimental technique. The spectroscopic information is indeed limited by different factors. A first factor is the selection rules. It comes from the fact that an experimental spectrum is revealed only from the interaction of the isolated system with a particular external electromagnetic radiation. The second one is the limited sensibility of the experimental technique making weak transitions unobserved. The third limitation comes from the initial state population, also dictated by the experimental conditions. The theoretical spectrum does not suffer from such restrictions and can in principle predict the energy positions of all $x!/(x-2)!2!$ possible transitions between the x converged energy levels. This situation is schematically depicted in Fig. 4 in the case of a four-level system. The theoretical spectrum [see part (*a*) of the figure] contains 6 lines, to which we have arbitrarily assigned an intensity of 1, as no theoretical intensity calculation has been considered at this point. We have assumed in this fictitious example that the corresponding experimental spectrum, presented in part (*b*), contains only 3 lines, considering both infrared (plain lines) and Raman (dashed lines) absorption spectra. Three transitions are supposed to be unobserved, in order to simulate a realistic situation: the very weak 3–2 and 3–1 transitions, and the 4–3 transition, because, say, level 3 is not populated under the actual experimental conditions. This means that in this example, no information on level 3 is available from the spectrum. We are here very close to the heart of the present review dedicated to the derivation of accurate vibration–rotation Hamiltonians from spectral information. Applying a backward trip to these four levels case would parameterize a Hamiltonian built from an incomplete picture of the interacting levels. It would be biased because interaction and transition selection rules may be different. It is the "least-squares fitting bottleneck," discussed by Field [1], which renders extrapola-

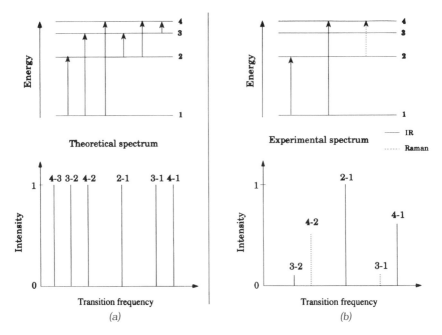

Figure 4.   Schematic comparison of theoretical (*a*) and experimental (*b*) spectra in a four-level system. In this example, only 3 transitions are experimentally observed among 6 possible, 2 of which are allowed by the IR (plain line) and 1 by the Raman (dashed line) selection rules. Experimental information on level 3 is missing and therefore also on related coupling terms in the Hamiltonian.

tions from spectroscopic Hamiltonians delicate. Let us stress once more that only an exact formulation of the forward trip can provide this full vibration–rotation information. It remains, as we also pointed out, a most inaccessible objective for nontrivial systems, and merging both forward and backward approaches must be one of the basic motivations in all theoretical investigations.

One should add at this point that the additional information obtained from the transition intensities is also worthwhile, because it brings out a rich complementary set of data on the charge distribution within the molecule, through the electric dipole moment operator (infrared spectrum) or the polarizability tensor (Raman spectrum). The theoretical calculation of intensities from the corresponding operators is a specific part of the forward trip that will be discussed in the next section.

g. Vibration–Rotation Line Intensities

1. OPTICAL VIBRATION-ROTATION TRANSITION PROBABILITIES. Using quantum electrodynamics and time-dependent perturbation theory [101,102] an expression of the *transition probability* W_{21} characterizing the transition from state $|1\rangle$ to state $|2\rangle$ can be obtained, which corresponds to the transition energy E_{21}. Except for some general comments at the end of the present subsection, only one-photon transitions involving an absorption process in an isolated molecule are considered here. Their probability is related to the corresponding Einstein absorption coefficient (see Section III.E) and is proportional to the squared modulus of the matrix element M_{21} defining the interaction with the electromagnetic radiation

$$M_{21} = \sum_{j=1}^{n_{tot}} \langle 2|e^{i\vec{k}.\vec{R}_j}\vec{\varepsilon}.\nabla(\bar{R}_j)|1\rangle \qquad (2.29)$$

where $\vec{k}$ and $\vec{\varepsilon}$ are the propagation and polarisation vectors of the radiation wave, respectively, and $\bar{R}_j$ is the vector position of the jth particle (nucleus or electron) in the LAS, as defined previously. Expanding the exponential terms in (2.29) as

$$e^{i\vec{k}.\bar{R}} = 1 + i\vec{k}.\bar{R} + \frac{1}{2!}(i\vec{k}.\bar{R})^2 + \cdots \qquad (2.30)$$

leads to a corresponding expansion of M_{21}. As $(\vec{k}.\bar{R})$ is very small compared to unity for an optical radiation, one usually keeps the first term of (2.30) only, defining the so-called electric dipole approximation. Within this approximation M_{12} becomes proportional [101,102] to the so-called electric dipole transition moment R_{21}:

$$R_{21} = \langle 2|\hat{\mu}|1\rangle \qquad (2.31)$$

in which $\hat{\mu}$ is the operator associated with the classical electric dipole moment vector of all the particles, defined in the LAS as

$$\hat{\mu} = \sum_{j=1}^{n_{tot}} Z_j \bar{R}_j \qquad (2.32)$$

The components of this vector are labeled μ_ξ and μ_Ξ, where $\xi = (x, y, \text{or } z)$ and $\Xi = (X, Y, \text{or } Z)$ refer to projections in the MAS and in the LAS, respectively. The corresponding components of R_{21} are denoted R_{21}^Ξ and R_{21}^ξ.

For completeness, we must point out that the next terms in (2.30) involve higher-order electric and magnetic multipole contributions, as, for instance, those of electric quadrupole and magnetic dipole nature arising from the linear term $(i\vec{k}.\bar{R})$. As the corresponding transition probabilities are usually several orders of magnitude weaker than the electric dipole ones, they are often neglected, except when the dipole term vanishes or when one is interested by so-called forbidden transitions. Raman- and multiphoton-type transitions, not considered here, are also of great interest in the present context as they reveal states whose access to is often forbidden by one-photon selection rules.

2. ELECTRIC DIPOLE TRANSITION PROBABILITIES. The *exact* calculation of electric dipole intensities involves the one of the components $R_{21}^{(\Xi)}$ of the corresponding transition moment (2.31) in the LAS. The bra and ket functions in this matrix element are total wavefunctions, depending not only on the electron $\bar{r}_e$ and vibration–rotation (Θ, q) coordinates but also on the electron and nuclear spin variables. As far as the coupling between spins and positions can be neglected, the total wavefunctions become

$$
|E''v''R''S''I''\rangle = |\psi_e^{(E'')}\rangle \cdot |\Psi_{rv}^{(v''R'')}\rangle \cdot |\Psi_{n_e}^{(S'')}\rangle \cdot |\Psi_{n_n}^{(I'')}\rangle
$$
$$
|E'v'R'S'I'\rangle = |\psi_e^{(E')}\rangle \cdot |\Psi_{rv}^{(v'R')}\rangle \cdot |\Psi_{n_e}^{(S')}\rangle \cdot |\Psi_{n_n}^{(I')}\rangle
$$
$$(2.33)$$

in which the two first terms of each function are Born–Oppenheimer products such as (2.8), and the two last terms refer to the nuclear and electron spin functions; E, v, R, S, and I refer to the electronic, vibration, rotation, electron spin and nuclear spin quantum numbers, respectively. Within the same electronic state—$E = E'' = E'$ and $S = S'' = S'$—the following compact notation arises:

$$
|Ev''R''SI''\rangle = |E\rangle \cdot |v''R''\rangle \cdot |S\rangle \cdot |I''\rangle
$$
$$
|Ev'R'SI'\rangle = |E\rangle \cdot |v'R'\rangle \cdot |S\rangle \cdot |I'\rangle
$$
$$(2.34)$$

and

$$
R_{21}^{\Xi} = \langle Ev'R'|\mu_\Xi|Ev''R''\rangle \cdot \langle I'|I''\rangle
$$
$$(2.35)$$

in which the unity integral involving the electron spins has been omitted. The last integral leads to the usual nuclear spin selection rule $\Delta I = 0$. Nuclear spins matter when considering the overall degeneracy of a vibration–rotation level and define the nuclear spin statistics, as explained in Section III.E.3.*e*.4 when we discuss intensity alternation features.

Using the variational solutions (2.27) within the vibration–rotation contribution of (2.35) gives

$$R_{21}^{\Xi} = \sum_{a',a''=1}^{n_r} \sum_{b',b''=1}^{n_v} c_{a'b'}^{(2)} c_{a''b''}^{(1)} \langle \Psi_e^{(E)} \Psi_r^{(a')} \Psi_v^{(b')} | \hat{\mu}_{\Xi} | \Psi_e^{(E)} \Psi_r^{(a'')} \Psi_v^{(b'')} \rangle \quad (2.36)$$

This matrix element can be expressed in terms of the dipole moment components in the MAS $\{\mu_\xi; \xi = x, y, z\}$, by means of the direction cosine matrix (2.12):

$$\mu_{\Xi} = \sum_{\xi=x,y,z} \lambda_{\xi\Xi} \mu_\xi \quad (2.37)$$

Each integral occurring in (2.36), denoted $I_{erv,\Xi}^{(a'a''b'b'')}$, is then expressed in terms of products of pure rotational $I_{r,\xi}^{(a'a'')}$ and vibronic $I_{ev,\xi}^{(b'b'')}$ contributions in the MAS:

$$I_{erv,\Xi}^{(a'a''b'b'')} = \sum_{\xi=x,y,z} \langle \Psi_r^{(a'')} | \lambda_{\xi\Xi} | \Psi_r^{(a')} \rangle \cdot \langle \Psi_e^{(E)} \Psi_v^{(b'')} | \mu_\xi | \Psi_e^{(E)} \Psi_v^{(b')} \rangle \quad (2.38)$$

in which the rotational integrals can be evaluated analytically [103,104]. The vibronic integrals must be integrated over the electronic coordinates:

$$I_{ev,\xi}^{(b'b'')} = \langle \Psi_v^{(b'')} | \langle \Psi_e^{(E)} | \mu_\xi | \Psi_e^{(E)} \rangle | \Psi_v^{(b')} \rangle = \langle \Psi_v^{(b'')} | \mu_\xi^{(E)} | \Psi_v^{(b')} \rangle \quad (2.39)$$

The operator $\mu_\xi^{(E)}$, common to all integrals (2.39), is the expectation value of the component ξ of the electric dipole moment operator for the electronic state (E). It still depends on the nuclear coordinates through both the electronic wavefunction and the dipole moment operator itself. For each clamped nuclear geometry (k), one has to calculate the following expectation value with the solutions of equation (2.2):

$$\mu_\xi^{(E)}(\bar{r}_n^{(k)}) = \langle \Psi_e^{(E)}(\bar{r}_e, \bar{r}_n^{(k)}) | \mu_\xi(\bar{r}_e, \bar{r}_n^{(k)}) | \Psi_e^{(E)}(\bar{r}_e, \bar{r}_n^{(k)}) \rangle \quad (2.40)$$

$\bar{r}_e$ and $\bar{r}_n^{(k)}$ referring to the vector positions of the electrons and nuclei at geometry (k), all coordinates expressed in the MAS. The dipole moment operator (2.32) in this geometry is, in atomic units in which the charge of the electron is -1:

$$\mu_\xi(r_{e\xi}, r_{n\xi}^{(k)}) = \sum_{j=1}^{n} Z_j r_{\xi j}^{(k)} - \sum_{k=1}^{n_e} r_{\xi k} \quad (2.41)$$

and (2.40) becomes

$$\mu_\xi^{(E)}(\bar{r}_n^{(k)}) = \sum_{j=1}^{n} Z_j r_{\xi j}^{(k)} - \left\langle \Psi_e^{(E)} \left| \sum_{k=1}^{n_e} r_{\xi k} \right| \Psi_e^{(E)} \right\rangle \qquad (2.42)$$

3. DIPOLE MOMENT SURFACE. Dreaming that the *exact* electronic wavefunction can be obtained at any geometry (k), one can calculate the exact values of the three components $\mu_\xi^{(E)}(\bar{r}_n^{(k)})(\xi = x, y, z)$ of the electric dipole moment in a given electronic state (E) at each possible nuclear geometry (k), belonging to

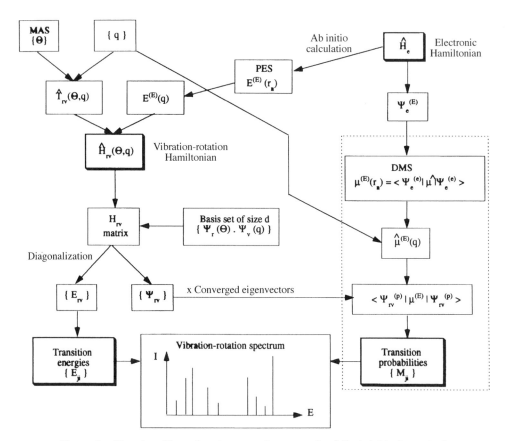

Figure 5. Flowchart illustrating the successive steps of a full *ab initio* forward trip, as detailed in Section II.B.1. The starting points (the electronic and vibration–rotation Hamiltonians) and the ending points (the transition energies and intensities) are marked by shadowed boxes. The dashed box regroups the steps in the procedure dedicated to the intensity calculation.

a restricted domain $\mathscr{D}$ of $\mathscr{R}^{3n}$. The corresponding functionals $\mu_{\xi}^{(E)}(\bar{r}_n)$ define hypersurfaces with $3n$ dimensions, the so-called components of the *dipole moment surface* (DMS) of state (E). Just as the PES, the DMS is invariant on translation and rotation of the whole system and can therefore be expressed, as $\mu_{\xi}^{(E)}(q)$, in terms of the internal set of coordinates of reduced dimension m_{int}.

Practical details on the ab initio determination of a DMS are provided in Section II.B.2.e. From the exact expression, one could perform the set of integration from (2.36) to (2.42) and obtain exact vibration–rotation transition probabilities and hence exact spectral line intensities, provided converged vibration–rotation wavefunctions (2.27) were used in (2.36).

h. Full Ab Initio Forward Trip. The successive steps we have developed so far, of full *ab initio* nature, allow a forward trip from the Hamiltonian to the vibration–rotation spectrum to be performed. The general scheme is summarized in Fig. 5. Apart from the basic Born–Oppenheimer, nonrelativistic, and electric dipole approximations, the formalism was kept as general as possible. It provides the basic theoretical background to the present review. All practical details were discarded, in particular those related to the computational feasibility of the calculations for a given molecular system. Assuming that, at each step, an *exact* treatment was possible, the guide lines of an ideal theoretical treatment were set. It is now time to see to what extent approximations need to be introduced to deal with real-life cases.

2. Towards a Converged Ab Initio Approach

The literature coverage in this section (II.B.2) is not exhaustive, but the major relevant problems are introduced and their solution illustrated. In this way, the reader is being prepared for the critical analysis of the acetylene literature, presented in Section II.C.

a. Setting the Ab Initio Approach

1. NEED FOR A PES. The computation of the PES, from the resolution of the electronic Schrödinger equation (2.2) is a critical and basic step. The analysis of the PES topology indeed allows the stationary points to be identified, and their nature (minima, saddle points, bifurcation points, higher-order critical points), their relative energy and various other static properties (nuclear geometry, harmonic and anharmonic force fields, permanent electric dipole moment, etc.) to be established. The connection to high-resolution spectroscopy is therefore most obvious. Reaction pathways connecting stationary points and leading to dissociation channels are also governed by the PES topology whose study is therefore also of interest to chemical

reactivity. In the case of unimolecular, processes, like the $AB \rightarrow A + B$ dissociation and the $AB \leftrightarrow AB'$ isomerisation reactions, the PES must then cover, in addition to the nuclear degrees of freedom of AB, those of A and B or AB'. In the case of a bimolecular reaction $A + B \rightarrow C \rightarrow D + E$, it must include the supermolecule C, which is usually an activated complex along the reaction pathway. The PES is also of interest to investigations of dynamical nature, such as classical trajectory simulations [105,106], semiclassical periodic orbits or bifurcation analysis [107–112], and propagation of quantum wavepackets [113–115]. We refer the reader to a recent special issue of *Spectrochimica Acta* [116] entitled *"Ab initio* and *ab initio* derived force fields: state of Science" for more details on some of these applications. It gathers 24 papers dealing with *ab initio* force fields applied to spectroscopy for small [93,117–129] and medium $(n \leq 30)$ [130–136] molecules. Larger systems, such as porphyrins [137], polymers [138], and soft matter [139] are also considered. In some case the *ab initio* force field has served as starting point for a refinement to spectroscopical data [93,123,128,135,137]. Applications mostly dealt with spectroscopy, but also with molecular dynamics [128,129,135,136,138].

2. GENERAL PROBLEMS WITH PES AND DMS. Depending on various questions, approximations have to be introduced in order to actually make the computations aiming at exact solutions, described in Section II.B.1, applicable for nontrivial systems. These questions concern the size of the system, the degree of vibration and rotation excitation in the problem, the presence of large-amplitude motions, in particular if they extent over more than one stationary point of the PES, the extent of coupling between the nuclear degrees of freedom; the accuracy required to bring significant information, with respect to experimental results; and the need for investigating intensity features.

It may turn out, that the system is too large for the accuracy required. This, for example, occurs for highly excited vibration–rotation levels in molecules containing more than four atoms. In such frontier cases, accurate calculations may be out of reach. In most other cases, quantum chemistry may significantly help in the problem.

It is first required to select some appropriate methodology, with respect to the general scheme presented in Fig. 5. The selection of the coordinate system is first discussed hereafter in Sections II.B.2.*b* and II.B.2.*c*. Calculation of the PES and of the DMS are detailed in sections II.B.2.*d* and II.B.2.*e*, respectively. The application of the variational method is discussed in Section II.B.2.*f*. Eventually, an alternative methodology, provided by perturbation theory, is developed in Section II.B.2.*g*.

b. Choice of a Coordinate System

1. CRITERIA. We have already pointed out that the choice of the coordinate system, in which the vibration–rotation Hamiltonian is written, determines the precision of the results and the convergence of the underlying variational or perturbative expansions. There are four related criteria helping in the selection of a coordinate system leading to the simplest form for the vibration–rotation Hamiltonian (1) the configuration space to be sampled, (2) the relative atomic masses, (3) the excitation energies, and (4) the coupling strength between the nuclear degrees of freedom.

The final choice also determines the basis sets to be used in the variational calculations. A reduction in the dimensionality of the variational problem obviously arises if one is ready to consider a partial uncoupling between some degrees of freedom, in adiabatic or effective treatments.

The most commonly used co-ordinate systems, to be illustrated with acetylene in Sections II.C.1.*e* and II.C.1.*f*, are briefly discussed below.

2. RECTILINEAR VERSUS CURVILINEAR COORDINATES. The choice between rectilinear and curvilinear coordinates is a very basic one. It is discussed in the literature since the early times of vibrational spectroscopy [45]. The exact internal coordinates are those that follow the molecular frame in its natural vibrational motion, namely, the elongation of the bonds (stretchings) and the angular variation between them (bendings). Such motions define *curvilinear coordinates. Rectilinear coordinates* are the projections of these exact internal coordinates onto a fixed Cartesian reference frame. The normal coordinates are appropriate linear combinations of these Cartesian displacements, which diagonalize the quadratic part of the kinetic- and potential-energy operators [45]. The conventional normal coordinates are thus *truly normal rectilinear coordinates.*

The major differences between both sets are illustrated by the two dimensional (2D) stretching pendulum, in parts (*a*) and (*b*) of Fig. 6, following the basic investigation by Colbert and Sibert [140]. This pendulum, of mass m is animated by simultaneous planar oscillations, around the y axis, and stretching extensions, with respect to a fixed point O. Figure 6 demonstrates that the curvilinear coordinates r and Θ [see part (*a*)] follow these two natural motions, while the rectilinear x and y coordinates [see part (*b*)] mix them. In this figure, we have intentionally represented each time a pure bending motion from the equilibrium position $r = r_0$ and $\Theta = 0$. The stretching–bending mixing in the rectilinear representation is obvious, if one considers the extension Δr of the pendulum accompanying the pure

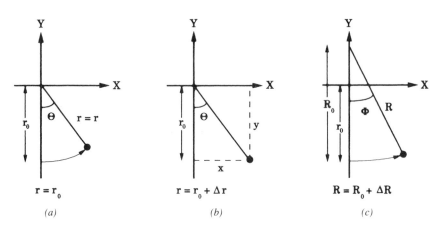

Figure 6. Comparison between (a) curvilinear coordinates (r, Θ), (b) rectilinear coordinates (x, y), and (c) variable-curvature coordinates (R, Φ) in the case of the 2D planar stretching pendulum (adapted from Colbert and Sibert [140]).

bending motion [see Fig. 6(b)]:

$$\Delta r = \frac{r_0}{\cos\Theta} - r_0 \tag{2.43}$$

Colbert et al. [140] expressed the vibrational Hamiltonians, as $\hat{H}_C$ and $\hat{H}_R$, corresponding to the curvilinear and rectilinear coordinate systems respectively, as

$$\hat{H}_C = -\frac{\hbar^2}{2m}\left[\frac{\partial^2}{\partial r^2} + \frac{1}{4(r+r_0)^2} + \frac{1}{(r+r_0)^2}\frac{\partial^2}{\partial\Theta^2}\right] + \frac{1}{2}[kr^2 + fr_0^2\Theta^2] \tag{2.44}$$

$$\hat{H}_R = \frac{1}{2m}(\hat{p}_x^2 + \hat{p}_y^2) + \frac{1}{2}(ky^2 + fx^2) + \frac{1}{2}k(r^2 - y^2) + \frac{1}{2}f[r_0^2\Theta^2 - x^2] \tag{2.45}$$

where the potential-energy operator has been chosen for simplicity as harmonic in the curvilinear representation. It thus appears that the transformation from curvilinear to rectilinear coordinates causes nondiagonal terms to appear in the potential-energy operator. Oppositely the kinetic energy operator is diagonal in the rectilinear representation but not in the curvilinear one. Similar migrations of coupling terms between the potential- and kinetic-energy operators, as occurring in this simple 2D model system, are commonplace in real systems. As pointed out, for instance, by Sibert et al. [54] in a study on CO_2, although there is no distinction between both representations through quadratic terms, in the small amplitude approximation, the difference is spectacular when anharmonicity is introduced. In their study of the prototype Fermi anharmonic resonance occurring in CO_2, these authors showed that, using the curvilinear representation, the anharmonic kinetic coupling dominates (77%) over the potential anharmonic coupling (23%), completely inverting the rectilinear situation.

Each rectilinear and curvilinear perspective thus brings its own advantages; the kinetic operator is diagonal and simpler in the rectilinear representation, but the potential coupling is usually smaller in the curvilinear one. A better separation between vibrational motions is therefore expected in curvilinear coordinates, leading to more rapidly convergent expansion in the description of highly excited vibrations and large-amplitude motions. This has motivated many authors to extensively investigate the use of different kinds of curvilinear coordinates, mostly but not exclusively in the framework of the forward trip. See, for instance the work of Sibert and co-workers [57,141,142], Mills, Carter, Handy, and co-workers [96,143], and Joyeux [144,145] on curvilinear backward trips.

c. Selected Curvilinear Coordinates

1. CURVILINEAR BOND-ANGLE COORDINATES. The curvilinear coordinate system, built from internal coordinates such as bond distances and bending angles, is an obvious choice for describing moderate amplitude vibrational motion within a single PES minimum. This is demonstrated, for instance, by the extensive variational work reported in the literature on triatomic and four-atom molecules (see, e.g., Refs. 49,92,93,95, and 146–149 and references cited therein). We refer to the acetylene Section (II.C.2) for detailed illustrations.

2. LOCAL-MODE COORDINATES. Local-mode coordinates are explicitly suited for problems related to overtone spectroscopy and are exploited by many authors (see, e.g., Refs. 150–157). Local modes are simple internal coordinates describing specific vibrations, ensuring that the diagonal anharmonicity parameter exceeds the nondiagonal coupling strength. Very simple Hamiltonians result from such a choice. The situation is thus opposite to the normal-mode approach, in which the nondiagonal coupling prevails, in the highly excited ranges. The local-mode approach concerns mainly the description of stretching modes overtones, as exemplified for instance by the basic work of Child [150] on H_2O, C_2H_2, and SO_2, and of Halonen [63,151] on AB_n molecules (with $n = 2,3,4$). Among many applications, one can also highlight the study of the bending overtone dynamics in some molecules [154,158,159]. Specific reviews are devoted to this topic, which is not further developed here [64,160–162].

3. HELIOCENTRIC-TYPE COORDINATES. Another class of curvilinear coordinates arises from modified heliocentric coordinate systems, like those constructed from Jacobi and Radau vectors [60]. A *heliocentric* system is a planetary system in which coordinates of all nuclei are defined with respect to a single central nucleus. A *Jacobi vector* describes the position of a given nucleus with respect to the centre of mass of a set of nuclei. *Radau vectors* define the positions of a set of light nuclei with respect to a heavier one. The origin of these vectors is not the position of the heavy atom, but rather a canonical center defined from the mass ratio. Such coordinate systems are obvious choices for molecules exhibiting large-amplitude vibrations involving atom orbiting motions. A typical example is the description of the acetylene-vinylidene isomerization process briefly discussed in Section II.C.2.*c.*

4. ADAPTED STRETCHING COORDINATES. Apart from the usual bond distance coordinate R_{AB}, other internal coordinates have been proposed for describing the stretching motion associated to a given bond between atoms A and B. They aim at improving the convergence of the power series expansions of the PES, such as the one in (2.67). We refer to the work of von Nagy-Felsobuki and coworkers, who carefully analysed the range of convergence and the efficiency of such expansions (see the reviews in Refs. 97 and 148 and references cited therein). One should point out the Dunham expansion [163], in which all coordinates are defined as variations $\Delta R_i = R_i - R_i^e$ of a bond length R_i from its equilibrium value R_i^e, the Simon–Parr–Finlan (SPF) [164] and the Ogilvie [165] expansions, in which the coordinates are respectively defined as

$$\rho_i^{SPF} = \frac{\Delta R_i}{R_i} \tag{2.46}$$

$$\rho_i^{Olg} = \frac{2(R_i - R_i^e)}{(R_i + R_i^e)} \tag{2.47}$$

Such PES expansions may be expected to present an horizontal asymptotic behaviour for an infinite value of the coordinate. They are thus expected to be more realistic at large internuclear distances and are widely used in the literature [85,87,97,166–174].

5. POTENTIAL-ADAPTED COORDINATES. The rectilinear normal coordinates are certainly the most popular potential-adapted coordinates, related to the harmonic oscillator potential:

$$\hat{V}(Q) = \tfrac{1}{2}kQ^2 \tag{2.48}$$

where Q is a normal coordinate and k the corresponding force constant. The eigenfunctions of the corresponding one-dimensional (1D) Schrödinger equation are the harmonic oscillator functions:

$$\chi_v(Q) = N_v e^{-(1/2)\gamma Q^2} H_v(\sqrt{\gamma}Q) \qquad (2.49)$$

with $\gamma = 2\Pi\sqrt{\mu k/h}$ and v the vibrational quantum number, N_v a normalization factor, H_v an Hermite polynomial of vth degree, and μ the oscillator's reduced mass. These functions form the usual basis set adopted in variational calculations using rectilinear normal coordinates. Dimensionless rectilinear coordinates are defined and used in Section III.

Morse *coordinates*, suggested by Meyer et al. [175] as an adapted alternative for describing stretching coordinates, are widely used in the literature, namely by Jensen and co-workers within the framework of the MORBID approach [49,176–178] (see Section II.B.2.f.6, and by Handy, Carter, and co-workers [e.g., 43,96,143,169]. The Morse potential describing the extension of a bond-stretching coordinate R is defined as [179]:

$$\hat{V}(R) = D(1 - \exp(-\beta \Delta R))^2 \qquad (2.50)$$

with ΔR the displacement from the equilibrium position $R = R_e$, D the bond dissociation energy, defined as the energy difference between $R = R_e$ and $R = \infty$; β is a parameter related to the anharmonicity of the oscillator. The corresponding Morse-adapted coordinate is

$$z = 1 - \exp(-\beta \Delta R) \qquad (2.51)$$

Such coordinates are also associated with corresponding basis sets in variational calculations [43,96,143,180–182]. Normalized Morse basis functions are given by [96,183]

$$\chi_v(z) = N_v e^{-(z/2)} z^{b/2} L_v^b(z) \qquad (2.52)$$

where N_v is a normalization factor, L_v^b is an associated Laguerre polynomial, and

$$\begin{aligned} b &= k - 2v - 1 \\ k &= \omega_e \omega_e x_e \end{aligned} \qquad (2.53)$$

where k defines, in this section, the anharmonic strength. Note that the parameters in (2.51) and (2.53) can be obtained from empirical data or can be variationally optimized [181,184].

An analytical form alternative to (2.52), obeying more correct boundary conditions, was proposed by Kauppi [185].

6.   ADAPTED BENDING COORDINATES.   Carter and Handy [169] proposed the following bending coordinate to build the potential of a nonlinear triatomic molecule:

$$\Delta\tilde{\Theta} = \Delta\Theta + \alpha\,\Delta\Theta^2 + \lambda\,\Delta\Theta^3 \qquad (2.54)$$

where $\Delta\Theta$ is the change in the bending angle Θ from its equilibrium value. The constants α and λ are defined by the value, at linear geometry, of the potential energy and of its first derivative with respect to Θ. The standard angular basis functions are given by [182,186]

$$\Theta_a^l(\theta) = (-1)^l \left[\frac{(2a+1)(a-l)!}{2(a+l)!}\right] P_a^l(\cos\theta) \qquad (2.55)$$

where P_a^l is an associated Legendre function. Quantum number a is related to the cutomary linear angle bend quantum numbers (v,l) by the relation

$$v = 2a - |l| \qquad (2.56)$$

7.   OPTIMISED COORDINATES.   Several authors have investigated the possibility to further improve the separability between nuclear degrees of freedom by defining two different kinds of optimized coordinates.

The first ensemble is based on optimization features in the SCF method (see Section II.B.2.g.2), used to determine optimal values of parameters included in the coordinate definitions [187–195]. Thompson and Truhlar [187] and Moiseyev [189] considered a rotation of the normal coordinates, and variationally optimized the rotation angle within the framework of the SCF approach. Bačić et al. [191] developed a similar treatment, using curvilinear coordinates. In the same spirit, hyperspherical coordinates were used by Gibson et al. [192]; Jacobi coordinates, by Zúñiga et al. [195]; and local mode coordinates, by Roth et al. [190].

The second class of coordinates, introduced by Sibert and co-workers [140,157] are the so-called *variable-curvature coordinates* (VCCs). Following Colbert and Sibert [140] once more, the basic idea of the variable curvature is exemplified with the 2D stretching pendulum model of Fig. 6. Part (c) of this figure shows a coordinate system defined by variables R and Φ, which presents a curvature intermediate to the standard bond-angle curvilinear system [part (a)] and the rectilinear system [part (b)]. The curvature κ is defined by a single parameter R_0. It appears that, for $R_0 \to r_0$, the VCCs tends to the bond-angle system and that, for $R_0 \rightarrow \infty$, the VCC tends to the rectilinear system. The VCC concept was demonstrated by Colbert and Sibert [140] to improve the separability of the C–H stretch–bend interaction in CHD_3, compared to a bond-angle system.

Mayrhofer and Sibert [157], in their study of H_2O and SO_2, have combined the VCC and SCF concepts to provide optimal VCC. An interesting analysis of the calculated eigenfunctions and of the coefficients of a spectroscopic Hamiltonian were developed at that occasion. They found for instance, that the curvature of the coordinate system was relevant when this Hamiltonian contained off-diagonal coupling terms.

8. CURVILINEAR NORMAL COORDINATES. Normal coordinates are not the privilege of the rectilinear systems. Curvilinear normal coordinates Q can also be defined [51,56,57,157]. They correspond [57] to a linear expansion of the bond-angle coordinates R:

$$R_t = \sum_{i=1}^{3n-6} L_{ti} Q_i \qquad (2.57)$$

in which the L transformation matrix is chosen to be the matrix relating the extension coordinates S to the rectilinear normal coordinates q, according to

$$S_t = \sum_{i=1}^{3n-6} L_{ti} q_i \qquad (2.58)$$

S is equivalent to R only in the small-amplitude approximation, as q is related to R by a nonlinear transformation.

This curvilinear normal approach was exploited by McCoy and Sibert [57] to apply canonical Van Vleck perturbation theory in a curvilinear perspective. These authors transformed in this way the curvilinear vibrational Hamiltonian into a block-diagonal effective Hamiltonian, comparable to the corresponding one written in rectilinear normal coordinates. Results obtained for acetylene are reported in Section II.C.2.*a*.3.

9. TRANSFORMATION OF COORDINATES. The impressive number of coordinate systems we just discussed makes algorithms transforming a given system into another one most welcome.

Allen et al. [196] recently derived general relations for transforming scalar and vector quantities between systems. These authors use the brace notation providing a simple and compact way to deal with derivatives of arbitrary nontensorial quantities.

Lukka [44] also discussed the transformations made easier through the use of infinitesimal rotation coordinates. Another approach is that by Schwenke [197], who wrote a general four-atom molecule program (see Section II.C.2.*a*.5), and investigated different systems by building the Hamiltonian matrix initially in Cartesian orthogonal coordinates and applying a sequence of transformations.

d. Approximate PES from Quantum Chemistry

1. *AB INITIO* LEVEL OF CALCULATION. Several general references deal with the *ab initio* calculation of the PES [72,73,76,198–201]. Benchmark calculations are published [e.g., 75,202–211] and a useful bibliography index is provided annually in the *Journal of Molecular Structure* (*Theochem*). This extensive literature demonstrates that very accurate solutions of (2.2) can be obtained nowadays from large-scale *ab initio* methods. However, it also shows that exact solutions are practically not accessible, due to numerical limitations in the resolution of the many-body problem. Variational calculations are limited in the size of the CI expansions, which increases very rapidly with the number of correlated electrons and the basis set size. The larger full-CI calculations reported in the literature, passing the several billions configurations limit [212,213], are not routinely applicable, even for small systems. They are moreover insufficient for nontrivial systems. Many-body perturbation theory calculations are similarly limited in the convergence of the perturbative expansions. It is, however, possible today to make relative energies converge to a sufficient accuracy for predictions of spectroscopic interest.

The relevant technical aspects of the *ab initio* calculations are very briefly summarized here. One-electron and *n*-electron basis sets are to be extended in order to obtain a convergence on spectroscopic quantities. At the *one-electron* level [72], molecular orbitals $\{\Phi\}$ are usually expanded as linear combinations of atomic orbitals (LCAO approximation), themselves expressed as linear combinations $\{\chi\}$ of Gaussian primitive functions $\{g\}$, according to the following expressions:

$$\Phi_i(\bar{r}_{ei}) = \sum_{A=1}^{n} \sum_{\substack{p=1 \\ CA}}^{D_A} a_{ip} \chi_p(\bar{r}_{ei} - \bar{r}_A) \tag{2.59}$$

$$\chi_p(\bar{r}_{ei} - \bar{r}_A) = \sum_{r=1}^{G_A} d_{pr} g_r(\alpha_{pr}, \bar{r}_{ei} - \bar{r}_A) \tag{2.60}$$

$$g_r(\alpha_{pr}, \bar{r}_{ei} - \bar{r}_A) = N_r x^u y^v z^w e^{\alpha_{pr}(\bar{r}_{ei} - \bar{r}_A)^2} \tag{2.61}$$

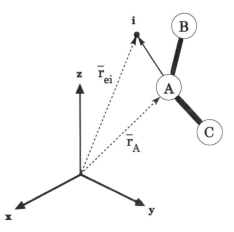

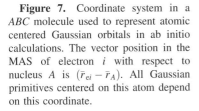

Figure 7. Coordinate system in a *ABC* molecule used to represent atomic centered Gaussian orbitals in ab initio calculations. The vector position in the MAS of electron *i* with respect to nucleus *A* is $(\bar{r}_{ei} - \bar{r}_A)$. All Gaussian primitives centered on this atom depend on this coordinate.

in which the molecular orbital $\Phi_i(\bar{r}_{ei})$ is a function of the coordinate $\bar{r}_{ei}$ of the *i*th electron, expressed in the MAS. $\chi_p(\bar{r}_{ei} - \bar{r}_A)$ is an atomic function centred on atom *A* and $g_r(\alpha_{ap}, \bar{r}_{ei} - \bar{r}_A)$ is a Gaussian primitive function, normalized by the factor *Nr*. The coordinate system is illustrated in Fig. 7 for an *ABC* molecule. The angular parts of the atomic functions χ are defined by the integral powers *u*, *v*, and *w* of *x*, *y*, and *z*, respectively. For instance, the angular parts of a p_0 and a $d_{\pm 1}$ orbitals are *z* and $(xz \pm i\, yz)$, respectively. D_A and G_A are the sizes of the atomic and Gaussian primitive basis sets for atom *A*, respectively. Values of the $\{d\}$ coefficients and of the Gaussians exponents $\{\alpha\}$ are optimized on model atomic or molecular systems and are gathered in specific basis set libraries [214–216], while $\{a\}$ coefficients are variationally optimized at a given geometry of the molecule studied.

The *radial flexibility* of such Gaussian basis sets is measured by the respective values of D_A and G_A describing the valence atomic orbitals of atom *A* (e.g., 2s and 2p orbitals for second-row atoms). In the usual *ab initio* nomenclature, an "*n*-zeta" basis set contains a number *n* of χ functions that are dedicated to each valence atomic orbital in the system. "Single-zeta" (SZ), "double-zeta" (DZ), "triple-zeta" (TZ), and so on basis sets thus mean that 1,2,3,... functions are actually used to describe each valence orbital. Polarization orbitals, characterized by higher values of the angular orbital quantum number (e.g., 3d,4f,... orbitals on second row atoms), are added to the standard *n*-zeta basis in order to introduce a further *angular flexibility*, required to properly represent pure molecular effects (bond and lone-pair formation). Diffuse Gaussian functions are also sometimes added to further improve the description of very diffuse electronic structures, namely, Rydberg orbitals.

Triple-zeta doubly polarized (TZ2P) basis sets are to be used, at the minimum, to calculate properties of spectroscopic interest. Up to quintuple zeta basis sets with d, f, and g polarization functions, augmented by s, p, and d diffuse functions, are required to match spectroscopic precision for vibrational properties, within less than $1\,\mathrm{cm}^{-1}$ [e.g., 210]. Proper description of anharmonic potentials also requires extended many-body expansions, far beyond the SCF level of approximation. Recall that at this level the n-electron basis set is limited to a single CSF (for configuration state function, i.e., a symmetry adapted linear combination of Slater determinants referring to a single electronic configuration). Applying the variational method to such a trial wavefunction leads to the well-known Hartree–Fock equations [72].

Methods beyond SCF introduce correlation energy contributions by taking excited-state interactions into account. Relevant methods are

1. Complete active-space self-consistent field (CASSCF) [217,218]. The trial wavefunction is a multiconfiguration expansion on a basis set of CSFs $\{\Psi_i\}$:

$$\Psi^{\mathrm{CASSCF}} = \sum_{i=1}^{b} c_i \Psi_i \qquad (2.62)$$

in which the $\{\Psi_i\}$ are built from the distribution of a given number of "active" electrons within an "active" set of molecular orbitals, usually chosen as the valence orbitals. Both orbitals $\{a\}$ coefficients and CSFs $\{c\}$ coefficients are optimized in this approach, leading to a fully optimized wavefunction (2.62). CASSCF calculations are often completed by MRCI calculations, for multireference configuration interaction [219–223]. The trial wavefunction is also a multi-configuration expansion such as (2.62), but extending on a larger CSF's basis set, obtained from the application of excitation operators on the multireference wavefunction resulting from a previous CASSCF calculation. One usually creates in this way single and double-electron replacements from the orbitals occupied in the reference wavefunction to the unoccupied ones. The trial wavefunction can thus be written as

$$\Psi^{\mathrm{MRCI}} = (\hat{T}_1 + \hat{T}_2)\Psi^{\mathrm{CASSCF}} \qquad (2.63)$$

where $\hat{T}_1$ and $\hat{T}_2$ are the single- and double-excitation creation operators, respectively. The molecular orbitals arising from the preliminary CASSCF calculation are not further varied and the linear variational method, developed in Section II.B.1.f.2, is strictly applied to (2.63). Note that the CSF basis set, generated by (2.63), involves electron excitations from the valence to nonvalence orbitals, introducing an important set of electron correlation contributions, ignored at the CASSCF level.

2. Single- and double-configuration interaction (SDCI). This is a particular case of the MRCI, in which the CASSCF step is replaced by an SCF calculation, leading to a single reference trial function in (2.63).

3. SD(Q)CI. This is a SDCI calculation in which the total energy has been corrected by adding Davidson's correction [224] taking the effect of four particles unlinked clusters into account.

4. Coupled cluster, including all single and double excitations (CCSD) [225–229]. The CCSD approach is another way to include single and double excitations. The wavefunction is in this case defined by

$$\Psi^{CCSD} = e^{(\hat{T}_1 + \hat{T}_2)} \Psi^{SCF} \tag{2.64}$$

where the exponential operator is used in order to ensure a multiplicative separability of the wavefunction, which guarantees the size consistency of the calculation. Quasi-perturbative treatment of the connected triples (T) and quadruples (Q) leads to the CCSD(T) and CCSD(TQ) approaches [230–232].

5. Many-body perturbation theory [79], usually based on the Møller–Plesset partition of the Hamiltonian [233] (MP2 [234] and MP4 [235,236] for second- and fourth-order treatments, respectively). The zero-order Hamiltonian is in this case the Hartree–Fock Hamiltonian.

6. Complete active-space perturbation theory at second order (CASPT2). This is an extension of the MP2 method to a multireference CASSCF zero-order wavefunction.

All of these methods take the electron correlation effects into account. The number of electrons explicitly correlated from these approaches is of major importance. Pure valence electron correlation, referred to as the *non-dynamical correlation*, is usually insufficient, and *dynamical correlation*, involving excitations to outer valence orbitals, is to be considered. The opening of core orbitals, leading to core–valence correlation effects, also appears to be of decisive importance in some systems and for some properties [e.g., 83,210,237].

Eventually, for completeness, *density function theory* (DFT) must be mentioned [238], because it gains more and more popularity within the quantum chemistry community, and starts to produce results of spectroscopic interest [239–242]. This theory, initially developed by solid-state physicists, obeys more favorable scaling laws with n_e, the number of electrons, and is thus applicable to larger molecules at reasonable computational cost. Note that the applicability of the CCSD and MRCI methods, scaling from n_e^5 to n_e^8, is limited by the molecule size. DFT is, however, restricted to the ground electronic state and does not provide the convergence guaranty of the variational techniques. Different density functionals (DFs) are used in the literature [238,243]. *Hybrid functionals* uses an SCF density in the calculation of the exchange correlation energy. The BLYP functional for instance includes the Becke long-range correction [244,245] together with the Lee–Yang–Parr correlation functional. Another form of DFT is based on the *Kohn–Sham* theory [246], in which one solves similar equations than in the Hartree–Fock theory, but with a local exchange potential as exchange contribution. The latter form of DFT allows the calculation analytical gradients required for geometry optimizations and harmonic frequency calculations [239,247].

2. ANALYTICAL EXPRESSION FOR THE PES. As discussed by Mezey [248], Murrel et al. [249], and von Nagy-Felsobuki [97], the overall topology of the PES and the relevant restricted domain $\mathcal{D}$ of this surface within a specific problem provides the guide lines for selecting the best-suited analytical expression of the PES. Many-body functions and polynomial expansions are commonly used.

The success of the many-body expansion, developed mainly by Murrel and co-workers [249–255], relies on the useful *global representation* it provides. The number of parameters is large enough for accurately reproducing the energy, geometry, and force constants at equilibrium, as well as the dissociation channels. To highlight this approach, we describe the form of the Sorbie and Murrell potential [250] for an *ABC*-type molecule:

$$V(R_1, R_2, R_3) = V_{AB}(R_1) + V_{BC}(R_2) + V_{AC}(R_3) + V_{ABC}(R_1, R_2, R_3)$$

$$(2.65)$$

with R_1, R_2 and R_3 representing the *AB*, *BC*, and *AC* internuclear distances, respectively, V_{XY} (with $X, Y = A, B, C$), the corresponding adiabatic diatomic potential curves, and V_{ABC} the three-body term, of the form

$$V_{ABC}(R_1, R_2, R_3) = \sum_{ijk} u_{ijk} s_1^i s_2^j s_3^k \prod_l \left(1 - \tanh\left(\frac{1}{2}\gamma_l s_l\right)\right) \qquad (2.66)$$

with s_i the displacement coordinates. The values for the coefficients u_{ijk} and γ_l are derived from experimental or *ab initio* data on the *ABC* molecule studied. A similar four-body expansion has been used to describe the ground-state PES of acetylene, as detailed in Section II.C.1.*f*.3.

In the absence of relevance of large-amplitude coordinates extending over more than a single minimum in the problem, one is usually satisfied with a *local representation* of the PES. Only small-amplitude vibrations are represented around a stationary point in the nuclear configuration, corresponding to a given minimum on the global PES. The analytical form usually adopted in this case is a Taylor series expansion of the potential energy in the displacements of the internal coordinates q, with respect to their equilibrium values at the stationary point considered:

$$E(q) = \tfrac{1}{2}\sum_{ij} f_{ij} q_i q_j + \tfrac{1}{6}\sum_{ijk} f_{ijk} q_i q_j q_k + \tfrac{1}{24}\sum_{ijkl} f_{ijkl} q_i q_j q_k q_l + \cdots \quad (2.67)$$

where the summations run over the $3n - 6(5)$ internal coordinates. In practice, the expansion is limited to a given order, assumed to be representative of the PES for the given problem. The quartic order approximation is most common. Higher orders are, however, sometimes required as detailed in Section III. The force constants f_{ij}, f_{ijk}, and f_{ijkl}, respectively refer to the second, third, and fourth derivatives of the energy, calculated at equilibrium reference configuration.

3. ADJUSTMENT OF AN ANALYTICAL FUNCTION TO *AB INITIO* CALCULATED POINTS. One can derive the parameters corresponding to a given analytical form of the PES by performing a least-squares adjustment to a set of *ab initio* energies calculated at different clamped nuclear geometries. As we pointed out in Section II.B.1.*b*, each configuration corresponds to a point of the Euclidian space E^{3n}, and we are concerned with a restricted domain $\mathcal{D}$ of this space, to which the analytical form is dedicated. A discretized representation of the PES is obtained on a grid of K points properly distributed in the domain

$$\{E(\bar{R}_n^{(1)}), E(\bar{R}_n^{(2)}), \ldots, E(\bar{R}_n^{(K)})\} \qquad \text{with} \qquad \{\bar{R}_n^{(1)}, \bar{R}_n^{(2)}, \ldots, \bar{R}_n^{(K)}\} \subset \mathcal{D}$$
$$(2.68)$$

The calculation of the potential energy aiming at producing precise parameters requires first-class *ab initio* methods, such as those mentioned in the previous section, and a large number of grid points. It is therefore a particularly demanding computational task, even for small systems. The number of points rapidly scales up with the number of nuclear degrees of freedom m_{int} ($3n - 5$ or $3n - 6$), and the computational effort at each point also scales with the number of electrons, that is, usually, with the number of atoms. Grids equally spaced in all directions lead to huge numbers of points, scaling as $w^{m_{int}}$, where w is the number of points in a given direction. Optimal selection of the most relevant points is, of course, of major importance. Various methods were suggested in the literature, first for *global* PES:

1. A moving interpolation technique was proposed by Ischtwan and Collins [256]. It uses *ab initio*–calculated energies and first and second derivatives to determine dynamically important configurations. The interpolant of the energy and its derivatives is shown to converge to the exact energy with increasing number of *ab initio* data. It was applied, for example, to the six-dimensional (6D) problem of a diatom-plus-diatom reaction [256].

2. Another general interpolation method is proposed by Ho and Rabitz [257], within the framework of the reproducing kernel Hilbert space and the inverse problem theory. It yields globally to smooth PES that are continuous, with derivatives up to the second or higher order. It generates correct symmetry properties and asymptotic behaviour. It can also be easily expanded, starting from low-dimensional problems. It has been applied to H_3^+ [257] for example.

For *local* PES, the problem is a slightly different. According to (2.67), one needs a grid able to provide accurate values of the successive derivatives of the potential energy at a reference geometry, through a least-squares procedure. Sana [258–261] demonstrated that *experiment planification methods*, derived from the field of operational research and applied to economy, in particular [262], are relevant to the determination of force constants from *ab initio* data points. Such methods lead to a critical analysis of the compatibility of a data set, which consists into the grid of ab initio points, to the model and the parameters used and fitted to reproduce them, here a power series expansion of the PES. The grid is built in such a way that the variance–covariance matrix V associated to the least-squares fit most closely fits the following criteria: (1) the variance on each parameter is minimal (*D*-optimality criterion), (2) V is diagonal (orthogonality criterion), (3) the variance is constant at a given distance from the origin of the grid in every direction of space (rotation invariance criterion), and (4) the ratio between the number of parameters and the number of calculated points is not too small (*R*-efficiency criterion). A bias analysis is simultaneously provided [260], accounting for the role of neglected higher-order terms in the truncated Taylor expansion, by comparing the coefficients with their unbiased estimates.

Composite designs, built from the superposition of equiradial geometric designs, were proposed by different authors [263–267] and applied to force constant calculations, at the second order (H_2O, CH_3, $CH_4 + OH$, and $HF + H$ transition states [258], excited state of C_2H_2 [268]), third order (H_2O, NH_2, HNO [260]), and fourth order (H_2O [173], C_2H_2 [269,270]).

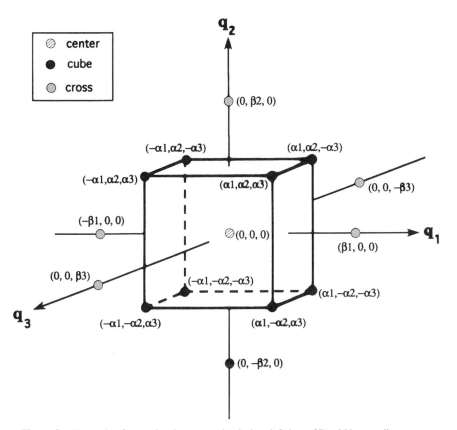

Figure 8. Example of second-order composite design defining a 3D grid in coordinates q_1, q_2, and q_3 (adapted from Liévin [268]).

They present the advantage to be extensible in dimension and in the model order. A second-order fit can, for instance, be achieved on a given grid, which can in a second run be extended to provide an adequate third- or fourth-order model. An illustrative example of second-order composite design [264] is given in Fig. 8 for a 3D problem, as used in Ref. 268. The 15 points grid is built from the superposition of a central point, a cross polytope and a cube, with an efficiency ratio of 40%. It obeys to the rotation invariance and quasi-orthogonality conditions, provided the central point is 9 times replicated, and that the size of the cross and of the cube are related by following scaling, in each coordinate q_i:

$$\beta_i = \alpha_i 2^{m_{\text{int}}/4} \tag{2.69}$$

42   THE FORWARD TRIP

where α_i and β_i are unit displacements along q_i, defining respectively the half-side of the cube and the arm length of the cross.

An additional example of a fourth-order 6D grid design, applied to acetylene, is provided in Section II.C.2.*b*.2.

4. ANALYTICAL VERSUS NUMERICAL DERIVATIVE CALCULATIONS. The force constants appearing in (2.67) are the successive derivatives of the energy with respect to the selected nuclear displacement coordinates, calculated at a reference geometry. These derivatives can be directly calculated at the *ab initio* level considered, without performing global fits, using *numerical finite differences* and *analytical* techniques.

The technique of numerical finite differences is more general. It can be achieved in principle with any kind of *ab initio* method and at any order of differentiation. The basic formulas of finite differences based on collocation polynomials are given in the literature [271,272]. They become complicated for higher-order derivatives and, as pointed out by Császár and Mills [93], are significantly facilitated by computer algebra programs. These authors derived a sextic force field for the water molecule. Single, double, and triple displacements of sizes $n \times 0.02$ (Å, rad) ($n = 0, 1, 2, 3$) along all internal bond-angle coordinates were adopted, giving a typical order of magnitude of the requirements.

The alternative technique is based on the analytical determination of the derivatives, from the resolution of the coupled Hartree–Fock equations (CPHF) [273–276]. It provides, in principle, more accurate derivatives than the numerical approach, although test calculations indicate similar precision [206]. First and second analytical derivatives are now available in most ab initio program packages (e.g., GAUSSIAN [277], HONDO [278], CADPAC [279]) at different orders and different levels of *ab initio* approximation. Higher-order derivatives are available at some levels of theory only, and are not routinely available. This holds for third and fourth derivatives calculated at the SCF level [280,281]. Combination of analytical and numerical finite differences is a good solution [206,282], when analytical derivatives are available only at a low order. Cubic and quartic derivatives can, for instance, be calculated from central differences of second analytical derivatives, by applying simple formulas such as

$$f_{ijk} = \frac{f_{ij}^+ - f_{ij}^+}{2\Delta x_k}$$

$$f_{ijkk} = \frac{f_{ij}^+ - f_{ij}^+ - 2f_{ij}^0}{(\Delta x_k)^2}$$

(2.70)

where Δx_k is the variation of coordinate x_k with respect to its equilibrium value and f_{ij}^0, f_{ij}^+, and f_{ij}^- are the second derivatives with respect to coordinates x_i and x_j, calculated at equilibrium geometry and at geometries displaced by $+\Delta x_k$ and $-\Delta x_k$, respectively.

For all techniques, a preliminary step is required: the determination of the reference geometry at which the derivatives are to be evaluated. One usually selects the equilibrium geometry calculated at the same level of theory as the force field to be estimated. Császár and Mills [93] demonstrated the strong limitations of that choice, as discussed later in this section. The determination of the equilibrium geometry at a given level of *ab initio* theory is performed from a so-called geometry optimization based on gradient techniques [283,284]. It uses steepest-descent algorithms [285–291], most of which are based on stepwise displacements on the PES, from an initial guess of the geometry to the actual equilibrium geometry. The geometric steps are determined from the calculated value of the gradient vector on all nuclear parameters and from an updated value of the Hessian. The stepwise process is stopped when the geometry satisfies simultaneous convergence tests based on thresholds defined by the user. The acceptance criteria include the absolute value of the total gradient vector, the value of the root mean squares of the individual gradients on all nuclear parameters, and the maximum value among these gradients. Such techniques thus also require first-derivatives calculations, and the same comments as before apply.

The selection of the reference geometry at which the force field is expanded is commented on by Császár and Mills [93], as we have pointed out. These authors argue that it is more efficient to use a geometry close to the experimental equilibrium geometry than to the one calculated at the same level of *ab initio* theory. The force constants they calculate are indeed in much better agreement with their experimental counterparts. The reason is that two terms of similar size but opposite sign contribute in the potential energy: the electronic V_{ee} and nuclear V_{nn} repulsion contributions. Only the latter contribution and its derivatives can be calculated exactly. The cancellation of errors therefore contributes in a rather unpredictable way. For the first and second derivatives, both contributions happen to nearly cancel each other, while for higher-order derivatives the V_{nn} terms are dominant. Therefore, the anharmonic part of the force field is less sensitive to the basis set and electron correlation effects than the quadratic part and than the equilibrium geometry. Several authors used this result [206,292,293] and calculated the harmonic contribution at a high level of *ab initio* theory in order to make these properties converge from the inclusion of large amounts of correlation energy, but calculated the cubic and quartic parts of the force field only at the SCF level. Another consequence is that the anharmonic constants are sensitive to the reference geometry through the contribution of

V_{nn}. This explains why a geometry more accurate than the one obtained from the optimization technique is required. A similar strategy was adopted by Martin [210], who derived an accurate force field for acetylene, further discussed in Section II.C.2.*b*.6.

5.   AB INITIO DVR APPROACH. We have highlighted that searching an analytical form for the PES necessarily introduces approximations in the potential-energy operator. Kauppi [294] showed that the discrete-variable representation (DVR) method (see Section II.B.2.*f*.4) allowed the vibrational problem to be solved, without introducing any explicit form of the PES. Only the knowledge of the potential energy at particular geometries corresponding to the DVR points is required. This approach was applied [294] to the calculation of the vibrational energy levels of the water molecule up to 14,000 cm^{-1} above the zero point energy (ZPE). In the largest calculation, the potential energy was calculated at 1172 points at the 6-311 + G**/ MP2 level of theory. The basis set associated with the corresponding DVR included 2057 functions and was found to provide vibrational energies converged to within 0.01 cm^{-1} for the fundamentals and, for the combination and overtone levels, to within 0.1 cm^{-1} up to 10,000 cm^{-1} and to within 1 cm^{-1} up to 14,000 cm^{-1}. Compared to the least-squares fitting technique discussed in Section II.B.2.*d*.3, the present method provides unique converged vibrational results, given a grid of *ab initio* points. It is, unfortunately, computationally very expensive and possibly not adapted for larger molecules.

e.   Ab Initio Electric Dipole Moment Surfaces. As described in Section II.B.1.*g*.3, the DMS and PES are usually calculated at the same level of *ab initio* theory and on the same grid of points. The DMS can also be fitted to an adapted analytical form. For instance, in the more general case of vibrations within a single stationary point, the three components in the MAS of the electric dipole moment function (2.40) $\mu_\xi^{(E)}(q)(\xi = x, y, z)$ are expanded as a Taylor series expansion in the displacements of the internal coordinates q from the stationary point equilibrium geometry:

$$\mu_\xi^{(E)}(q) = m_0^\xi + \sum_i m_i^\xi q_i + \frac{1}{2} \sum_{ij} m_{ij}^\xi q_i q_j + \frac{1}{6} \sum_{ijk} m_{ijk}^\xi q_i q_j q_k$$
$$+ \frac{1}{24} \sum_{ijkl} m_{ijkl}^\xi q_i q_j q_k q_l + \cdots \tag{2.71}$$

where m_0^ξ is the value of $\mu_\xi^{(E)}(q)$ at equilibrium geometry, and the constants $m_{ij}^\xi \ldots$ refer to the derivatives of $\mu_\xi^{(E)}$ with respect to $q_i, q_j, \ldots$, calculated at equilibrium geometry.

As pointed out by Le Sueur et al. [98], the vibrational band intensities calculated from a DMS depend on how the Cartesian axes of the dipole surface are defined. These authors suggest using the definition based on the rules proposed by Eckart for separating the vibration and rotation motions. To follow this prescription, the electric dipole moments, computed on each point of the grid, are rotated to the Eckart axis system and then expanded in a Taylor series like (2.71).

f. Variational Methods. The number of nuclear degrees of freedom to be handled causes the variational calculations to blow up for larger molecules. The general variational scheme depicted in Section II.B.1 thus cannot be applied in many cases. An extensive work has been devoted in the literature [e.g., see Refs. 87–95 and 295 and references cited therein], pioneered by Carney and Porter [86], in order to improve the convergence of the huge basis set expansions. Our review only highlights some interesting alternatives aiming at reducing the size of the vibration–rotation secular equation. Further details and examples are provided in Section II.C, concerning acetylene.

1. CONTRACTION OR DIAGONALIZATION–TRUNCATION. Different authors reduced the size of the product basis set (2.27) without significant loss of accuracy, using a *successive diagonalisation–truncation* procedure [92,94,143,146, 147,169,296–306]. All different contraction schemes progressively improve the basis set by spanning only the portion of the configuration space relevant for the set of eigensolutions of interest. Separate Hamiltonian matrices are built for limited sets of nuclear degrees of freedom. They are diagonalized, and the resulting set of eigenvectors is truncated, by elimination for instance from an energy criterion. The coupling between the different sets of degrees of freedom is then progressively introduced with, at each step, a diagonalization followed by a new basis truncation. The size of the final basis set is thus supposed to be dramatically smaller than the full initial product basis, and the most relevant information is supposed to be saved. A contraction scheme is to be defined, deciding the order in which the different coordinates couplings are introduced. Figure 9 illustrates schematically a six-mode problem. Two different contraction schemes (part (*b*) and (*c*) of the figure), corresponding to quite different coupling situations, are compared to the uncontracted full-CI approach [part (*a*)], in which a single diagonalization within the complete product basis set is performed. Cases (*a*), (*b*), and (*c*) involve one, three, and four contraction steps respectively. The contraction steps are numbered by Roman numerals in the figure. The first step, common to cases (*b*) and (*c*), is a prediagonalisation of the single-mode basis set, useful in many cases to introduce diagonal anharmonicity, such as when a standard

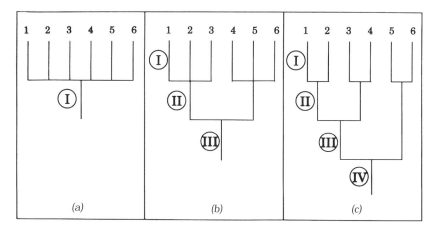

Figure 9. Schematic illustration of a successive diagonalization-truncation procedure applied to a six-mode problem. Two different contraction schemes, (*b*) and (*c*), are compared to the full-CI calculation (*a*). The successive contraction steps are numbered by capital roman numerals.

harmonic oscillator basis set is used (see the UAO approach, for uncoupled anharmonic oscillators, developed by Bowman and co-workers [85]). Step (II) then determines the more important couplings, which are ternary or binary for (*b*) and (*c*), respectively. The reduction of the basis set size can be roughly estimated. Assuming each individual mode to be initially represented by n primitive functions, leads to a basis size of n^6 for case (*a*). Assuming that typically truncations of 25% can be applied to step (I) and of 50% to the next steps, leads to basis set reductions of about a factor of 23, and 46 times for (*b*) and (*c*), respectively.

Different contraction schemes applied to the acetylene molecule are exemplified in Section II.C.2. Another interesting example is given by Yan et al. [303], who calculated vibration–rotation energy levels of the water molecule for high values of J (up to 20) from a two-step procedure [147,169,304,306]. The problem encountered with such high J values is the size of the final Hamiltonian matrix, even though it can be diagonalized into two parity different blocks, of size J and $(J+1)$. The first step of the contraction procedure used to overcome this problem is the solution of a series of secular problems depending on K, which is the projection of the total angular momentum onto the body-fixed z axis (MAS). In the second step, the N lowest eigenfunctions obtained in the first step are used to build the vibration–rotation Hamiltonian matrix. This procedure is actually quite similar to the process of solving in the first step the pure $J = 0$ vibrational problem, as done by Chen et al. [304].

2. GENERAL MEAN-FIELD OPTIMIZATIONS. Another way for reducing the size of the variational calculations is to optimize the basis sets. This can be done within the framework of the mean-field approach. The term *mean-field* refers to a class of variational methods for finding approximate solutions of a given many-body problem, which is reduced to a set of coupled one-body problems. Each body is considered as moving within a mean-field representing an effective interaction of this body with the other ones. The more popular mean-field approaches are the Hartree–Fock or SCF and the multiconfigurational self-consistent field (MCSCF) methods. They are widely used, in the framework of atomic [71] and molecular [72,73,75,76] electronic structures, and of nuclear structures [77].

The SCF method was first applied to the resolution of the vibrational problem by Carney [87] followed by others [84,85,127,173,194,303,307–317]. The *vibrational* SCF *method*, to be referred to as the *VSCF method*, uses, as trial function for the variational procedure, a single product of vibrational functions $\{\Phi_{v_i}(q_i)\}$:

$$\Psi^{\text{VSCF}}_{v_1,v_2,\ldots,v_{3n-6}}(q) = \prod_{i=1}^{3n-6} \Phi_{v_i}(q_i) \tag{2.72}$$

with the variational conditions

$$\frac{\partial E}{\partial \Phi_{v_i}} = 0 \tag{2.73}$$

One obtains in this way the system of VSCF-coupled equations

$$\left[\hat{h}_j + \left\langle \prod_{i \neq j} \Phi_{v_i}(q_i) \middle| \hat{H}_c \middle| \prod_{i \neq j} \Phi_{v_i}(q_j) \right\rangle \right] \Phi_{v_j}(q_j) = \varepsilon^{\text{VSCF}}_{v_j} \Phi_{v_j}(q_j) \tag{2.74}$$

$$j = 1, 2, \ldots, 3n - 6$$

in which the vibrational Hamiltonian has been written as a sum of (1) the $\hat{h}_i$ terms, depending on a single vibrational coordinate q_i, and (2) $\hat{H}_c$, regrouping all coupling terms between these coordinates:

$$\hat{H}_v = \sum_{i=1}^{3n-6} \hat{h}_i + \hat{H}_c \tag{2.75}$$

Each equation of the system (2.74) represents the vibrational motion along one given coordinate in an average field created by the others. This system is solved by iteration until convergence. The total energy of the vibrational state $(v_1, v_2, \ldots, v_{3n-6})$ is given by

$$E^{\mathrm{VSCF}}_{v_1, v_2, \ldots, v_{3n-6}} = \sum_{i=1}^{3n-6} \varepsilon^{\mathrm{VSCF}}_{v_i} - \left\langle \Psi^{\mathrm{VSCF}}_{v_1, v_2, \ldots, v_{3n-6}} \left| \hat{H}_c \right| \Psi^{\mathrm{VSCF}}_{v_1, v_2, \ldots, v_{3n-6}} \right\rangle \qquad (2.76)$$

Although the numerical resolution of (2.74) is possible in certain cases [312], analytical solutions are usually obtained from a projection on a basis set $\{\chi_d; d = 1, 2, \ldots, d_i\}$ in each vibrational coordinate q_i:

$$\Phi_{v_i} = \sum_{d=0}^{d_i} a^{v_i}_d \chi_d(q_i) \qquad \text{with } i = 1, 2, \ldots, 3n - 6 \qquad (2.77)$$

where d_i is the basis set size and $v_i = 0, 1, \ldots, d_i - 1$ refers to the quantum number associated with the considered vibrational mode. The so-called primitive basis set, $\{\chi\}$, is built from standard functions adapted to the coordinates system, including harmonic oscillators, Morse functions, and Legendre functions (see Section II.B.2.c). The variational conditions (2.73) reduce to

$$\frac{\partial E}{\partial a^{v_i}_d} = 0 \qquad (2.78)$$

Each equation in the VSCF system (2.74) takes a matrix form. These matrices are iteratively diagonalized until convergence, leading to d_i solutions $\{\Phi_{v_i}\}$ for each $3n - 6$ vibrational mode. As in any SCF calculation, one optimizes only the $\{\Phi_{v_i}\}$ "occupied" functions and not the other so-called virtual oscillator functions, within the single vibrational configuration trial function (2.72).

The VSCF method provides a way to prepare a basis set $\{\Phi\}$, which is an improved version of the primitive one, $\{\chi\}$, because it includes an effective information on the intermode coupling. The VSCF basis set is then used in a second step [84,85,173,303,313,316,317], to perform a *configuration interaction calculation*, in which the vibrational Hamiltonian matrix built on the product basis set of VSCF functions is diagonalized. The VSCF optimization allows in principle a severe truncation of the CI basis set. The two-step VSCF-CI procedure we just described is identical to the SDCI method widely used in electronic structure calculations (see Section II.B.2.d.1).

The VSCF method, however, introduces a correlation error due to the effective nature of the interactions between the oscillators. This error can be large when the interactions are strong [174,316]. It is mainly corrected at the VSCF-CI level if the basis set truncation is not too drastic. When strong mixing occur between vibration levels, which is often the case for highly excited levels, the convergence of the VSCF-CI expansion is very slow. More suitable mean-field optimization techniques are then required, as formally suggested many years ago by Tobin and Bowman [311].

3. THE VIBRATIONAL MULTICONFIGURATIONAL SCF METHOD. Learning from electronic calculations that face similar correlation effects, now between electrons, Culot and Liévin [174,316] developed a vibrational multi-configurational SCF method, referred to as the *VMCSCF method*. The trial function used in this method is now a multiconfigurational wavefunction

$$\Psi^{\text{VMCSCF}} = \sum_i c_i \Psi_i \qquad (2.79)$$

in which the Ψ_i represent vibrational configurations, specifically, single products of functions $\{\Phi_{v_i}(q_i)\}$. The mean field effective treatment can be applied to this particular trial function. Two different sets of variational conditions apply in this case. Indeed, in addition to the conditions (2.78) arising from the variation of the oscillators, as in the VSCF case, one has also to consider the variation of the $\{c\}$ coefficients, describing the configuration mixing:

$$\frac{\partial E}{\partial c_i} = 0 \qquad (2.80)$$

The VMCSCF solutions obeying the two variational conditions provide very compact and optimized wavefunctions. They describe the vibrational motions along each different coordinate in the average field created by the others, for a set of mixed states, defined by the reference function (2.79).

Solving a multiconfigurational mean-field problem is less straightforward than a single configurational one, because the two sets of coupled variational conditions must be handled. To our knowledge, only two solutions to this problem were proposed in the literature in the framework of the vibrational problem. The first one, by Schwenke [318], is a second order algorithm used in a single test calculation. Culot and Liévin proposed an alternative solution [174,269,316], based on the super-CI algorithm, directly translated from quantum chemistry algorithms [319,320]. It exploits the generalised Brillouin theorem [321], in its symmetry-adapted form [322], to obtain variational solutions. From this theorem, it is known that, at variational convergence, the matrix elements of the Hamiltonian cancel between the

VMCSCF function (2.79) and any single-excited function with respect to this reference function. The super-CI algorithm thus resolves two different secular problems at each iteration: the one corresponding to the linear variational problem related to the VMCSCF expansion (2.79), and the one in which the reference wavefunction is linearly combined to all the Brillouin-like single excited vibrational configurations. It was shown [316] that this algorithm is well adapted to the vibrational framework, with quadratic convergence, and that a *state-averaging technique*, also migrated from quantum chemistry, is efficient for obtaining optimized solutions averaged over a set of strongly mixed states. The method was extended [174] to the *complete active-space* SCF methodology (CASSCF). The so-called VCASSCF algorithm presents the advantage of a better convergence and of a nonambiguous definition of the active oscillators involved in the reference function (2.79). The active oscillators are defined in terms of all oscillators of a given mode covering a specified range of vibrational quantum numbers. All possible configurations built from these active oscillators are included in the reference function, supporting the "complete active-space" features of the method. The optimised $\{\Phi\}$ basis sets resulting from VMCSCF or VCASSCF calculations are expected to be very compact, and the rapid convergence of the VMCSCF-CI and CASSCF-CI configuration interaction expansions was demonstrated [174,270,316]. Such methods allow for a systematic analysis of the coupling effects between different sets of vibrational modes, such as the stretch–bend interactions in water [174,316] and acetylene [270]. The latter results are developed in Section II.C.2.b.5 of the present review.

As pointed out in Section II.B.2.c.7, the optimization features of the VSCF method were also exploited in order to define optimal coordinates (see, e.g., Refs. 187–195).

4.   DISCRETE-VARIABLE REPRESENTATION (DVR). Light and co-workers [92,323–325] introduced an elegant and powerful simplification of the variational calculation, through the so-called discrete-variable representation (DVR). It belongs to the class of *collocation* or *pseudospectral* methods [326]. A trial wavefunction Ψ is expanded as a linear combination of N basis functions $\{\Psi_1, \Psi_2, \ldots, \Psi_N\}$, and the coefficients are determined to satisfy the Schrödinger equation at N particular points $\{\bar{r}_1, \bar{r}_2, \ldots, \bar{r}_N\}$ in space. This yields to N algebraic equations that have the structure of a general secular problem and leads to the eigenvalues and coefficients. In the particular case of DVR, two representations of the wavefunctions are defined. The first one is the DVR itself: the wavefunction is defined by its value at the N points of the grid. Multiplicative operators such as the potential-energy operator are generally diagonal. The second representation is the *finite basis representa-*

tion (FBR): Ψ is expanded on a discrete, orthonormal, and complete basis set of N analytically known functions. The kinetic-energy matrix is usually simpler in this latter representation and can be determined by numerical quadrature over the DVR points. The two representations are thus related, and one defines a particular DVR using an appropriate basis set $\{\Psi_1, \Psi_2, \ldots, \Psi_N\}$ and quadrature conditions over the DVR set of points $\{\bar{r}_1, \bar{r}_2, \ldots, \bar{r}_N\}$. Exploiting the advantages of this double representation usually yields sparse Hamiltonian matrices that can be efficiently diagonalised. A wide class of mathematical functions is available to simplify the kinetic-energy matrix, including the Chebychev, Legendre, Laguerre, and Hermite polynomials with their associated weight functions [272,323]. The DVR method is widely used in the vibration–rotation literature [95,149,294,296,298,299,302,327–335].

Carrington and co-workers [95,149,299,327–329,334], applied the method, under various forms, to triatomic (H_2O, HCN, SO_2, H_3^+) and four-atom (CH_2O) systems. They address the following problems:

- The practicability of the DVR approach when the associated FBR is chosen on criteria other than the optimal simplification of the kinetic-energy matrices [327]. They show that accurate and efficient DVR may also be obtained from basis sets which are eigenfunctions of a one-dimensional Hamiltonian extracted from the vibration–rotation potential-energy operator $\hat{V}_{rv}$, defining so the so-called *potential optimized* DVR, or PODVR.

- The comparison between alternative diagonalization techniques adapted to sparse matrices [149,299,328,336–339], such as the Lancsoz algorithm [340,341], the recursive residue generation method (RRGM) [336], the Fourier–Lancsoz method [342], and the direct-operation Lancsoz approach [343]. Note that the Lancsoz algorithm allows the diagonalization of huge matrices (e.g., $1,000,000 \times 1,000,000$ for CH_2O [339]), as it does not store the whole matrix in core.

- The restriction of the calculation to energy levels in a specific range [99], using the filter-diagonalization method [100], the spectral transform method, and the guided Lancsoz method [99].

- The test of combined approaches: direct-product versus contracted DVR basis sets and Lancsoz versus conventional eigensolvers [299]. Their conclusion is that the contracted/Lancsoz method is consistently the most efficient, in agreement with earlier work by Friesner et al. [344]. A factor of 2 was gained in computer time with respect to the contracted/conventional approach of Bačić and Light [301] and of about 4 with respect to the direct-product/Lancsoz alternative.

- The analysis of the quadrature errors in DVR [335] yielding to unphysical "ghost" levels.

- The use of a *direct-operation Lancsoz algorithm*, in which the "direct" technique, widely used in quantum chemistry computer codes, is adopted. It calculates the action of the Hamiltonian on a vector, rather than computing matrix elements of the Hamiltonian and performing matrix-vector products [343].

To conclude about the DVR approach, let us cite some impressive results obtained on water, by Choi and Light [345]. These authors calculated converged energy levels up to $30,000\,\text{cm}^{-1}$ above ZPE, using different PES from the literature, and examined the vibrational energy spacing distribution. Bramley and Carrington [149] studied H_2O, H_3^+, and CH_2O, up to 22,000, 18,000, and $5700\,\text{cm}^{-1}$, respectively. Results concerning C_2H_2 are detailed in Section II.C.2.c.1.

5. ADIABATIC APPROACHES. The adiabatic approximation is the usual way to reduce the dimensionality of the secular problem, when the degrees of freedom of the problem can be separated into a class of fast and slow coordinates. One first solves the secular problem limited to the fast coordinates and then uses the averaged field defined in this way to solve the slow coordinates problem. The best-known adiabatic approximation is certainly the Born–Oppenheimer approximation, introduced in Section II.B.1.b. It is also applied successfully in some vibration–rotation problems [346,347]. Let us highlight three investigations, among many others. Harvey and Truhlar [348] used vibrational adiabatic basis sets in a scattering study of the system $He + H_2$. McCoy and Sibert [349] performed the adiabatic treatment of the bending dynamics in acetylene. They dealt with the three stretching vibrations as fast coordinates and averaged the total vibrational Hamiltonian over the ground state of these vibrations. This work is further detailed in section II.C.2.c.2. A last illustrative example concerns water, whose vibration–rotation energy levels were calculated using the adiabatisation of the rotational motion [350]. This means that the rotational energy has been added to the electronic energy to form an effective potential for describing the vibrational dynamics.

6. MORSE OSCILLATOR RIGID BENDER INTERNAL DYNAMICS (MORBID) APPROACH. Jensen [49,176–178] developed a variational method specifically adapted to "floppy" triatomic molecules *ABC*, in order to describe highly excited vibrational states involving large amplitude bending motion. This approach is based on a vibration–rotation Hamiltonian expressed in terms of two-bond-length displacement ΔR_{AB} and ΔR_{BC}, as defined by Hougen

et al. [59] and a bending coordinate ρ. The PES obeys the following analytical form:

$$
\hat{V}(\Delta R_{AB}, \Delta R_{BC}, \rho) = \hat{V}_0(\rho) + \sum_j F_j(\rho)z_j + \sum_{j\leq k} F_{jk}(\rho)z_j z_k
$$
$$
+ \sum_{j\leq k\leq l} F_{jkl}(\rho)z_j z_k z_l + \cdots \tag{2.81}
$$

in which all indices j,k,l refer to AB and BC and z_j, z_k, z_l are Morse coordinates, as defined in (2.51). The F expansion coefficients are defined as general cosine expansions of the bending variable ρ. The kinetic operator is approximated as in the nonrigid bender approach [351,352]. The variational calculation proceeds in three steps:

1. The pure stretching Hamiltonian spanned by a symmetrized basis of Morse oscillators functions is diagonalized.
2. The pure bending Hamiltonian matrix is calculated numerically through the Numerov–Cooley algorithm [351] and diagonalized.
3. The resulting stretching and bending basis functions are then combined with usual rigid rotor functions to form the complete basis, in which the MORBID Hamiltonian is diagonalized.

The MORBID approach was successfully used in the literature, to investigate many triatomic systems, including NH_2^+, HOC^+, C_3, CNC^+, CCN^+, CH_2, H_2F^+ [49], H_2O [353], NO_2 [353,354], BH_2^- [37], H_2P [355], H_2S [356], H_2Te [357], CH_2^+ [38,39], and references cited by Jensen [49].

g.   Perturbation Theory Methods

1.   BASIC FORMULAS. The variational method is certainly the most satisfactory approach for deriving energy levels from *ab initio* vibration–rotation Hamiltonians. Provided all necessary ingredients are input (adequate size and composition of the basis set, inclusion of all relevant coupling terms), the method generates accurate results. As discussed above, its performances are, however, limited by the rapid increase of the basis set size with the molecular size. The perturbation theory approach, widely used in the literature, provides an interesting alternative. It directly connects the various vibration–rotation constants fitted by the spectroscopist, discussed extensively in Section III, to the molecular geometry and the anharmonic force field. The necessary formulas have been derived from *second-order perturbation theory* or from the equivalent *contact transformation theory* by Wilson and Howard [358], Nielsen [359,360], Amat and co-workers [361–365], Oka and Morino [366,367], Kivelson and Wilson [368], and

54 THE FORWARD TRIP

Watson [369–372], in particular. They have been reviewed by Mills [373] and constitute the core of Section III. The formulas may also be found in Refs. 47 and 357 for asymmetric top molecules and for symmetric top ones including principal axes of order 3. Pliva has reported the formulas for principal axes of higher order [374]. Formulas for linear molecules are reviewed by Allen et al. [206]. Most relations are implemented in standard computer codes, as SPECTRO [375] developed by Handy and coworkers. They open up the study of large systems using second-order perturbation theory, without major computational limitations. An illustrative example is provided on benzene by Handy and Willetts [131].

2. TREATMENT OF RESONATING STATES. The perturbative approach fails when anharmonic resonances occur. Small denominators then appear in the expression of the anharmonic constants, leading to indefinitely large terms. One can overcome this problem, as originally suggested by Nielsen [359], by removing the undesired contributions from the diagonal matrix elements of the resonating states [47]. The remaining terms then describe unperturbed levels, which one couples using nondiagonal matrix elements. The small matrices regrouping states of a given, so-called polyad is then diagonalized. This method, further commented on in Section III.A.1.c is general. It is followed by theoreticians as well as by spectroscopists. Expressions to second order for the diagonal and nondiagonal matrix elements between resonating levels are developed in Section III. One characterizes a resonance by two numbers, which are those of created and annihilated quanta between the two interacting states. Anharmonic resonances such as those of Fermi and Darling–Dennison types, fully detailed in Section III, are classified as "1–2" and "2–2" resonances, respectively. Useful formulas for deriving resonance constants from *ab initio* force fields have been tabulated by Martin and co-workers for 2–2 [376] and 1–1, 1–2 and 1–3 resonance types [377]. They are included in an extended version of the program SPECTRO [377]. Higher order corrections to the 2–2 case are gathered by Law and Duncan (see Law and Duncan [378] and references cited therein).

3. HIGH-ORDER CANONICAL VAN VLECK PERTURBATION THEORY (CVPT). Standard applications of perturbation theory to vibration–rotation spectroscopy is based on the normal-mode harmonic oscillator approach taken as zeroth-order picture. It is therefore dedicated to the description of small-amplitude motion and to low vibration–rotation excitation. Sibert and co-workers [57,140,142,158,379–382] extensively investigated the use perturbation theory outside this general scope. They found that the *canonical Van Vleck perturbation theory* was a powerful tool in this different context. Applying this theory up to the nth order is referred to hereafter as CVPT(n). The

Hamiltonian is canonically transformed into a new representation using a reduced basis set. The information content of the Hamiltonian is assumed not to vary, and the form of the transformed Hamiltonian is selected to optimally suit the present needs. In departure from Nielsen's work [359,360], the zeroth-order part of the Hamiltonian describes a given number N of oscillators, under any form, local, curvilinear, or other. The pure $(J = 0)$ vibrational Hamiltonian is expanded in a perturbative series and rewritten algebraically in terms of the harmonic oscillator raising and lowering operators. Van Vleck transformations, consisting in a series of unitary transformations, are then applied up to the desired order. The goal of these transformations is to give a representation in which only quasi-degenerate states are coupled. The off-resonant anharmonic terms of the original Hamiltonian, which can lead to considerable energy shifts, are diagonal in the new representation. In one of his pioneering works, Sibert [380] applied CVPT(6) to the study of the highly excited overtones of CHD_3, with up to five quanta in the CH bond. He demonstrated the flexibility and efficiency of the method for systems facing the resonance problem and the divergence of the standard perturbative expansions. Sibert and co-workers successfully applied high-order CVPT to highly excited vibrations of H_2O [155,383,384], CH_2O [155,383,384], CO_2, HCN, and C_2H_2 [57].

CVPT was also implemented [57,142,158] in curvilinear normal coordinates (see Section II.B.2.*c*.8), after successively applying the following transformations:

1. The Hamiltonian is expanded in a Taylor series expansion in internal coordinates around the equilibrium configuration.
2. The expanded Hamiltonian is rewritten as a function of the curvilinear normal coordinates, using the L matrix [see equation (2.58)].
3. The Hamiltonian is then partitioned into $\hat{H}^{(0)} + \hat{H}^{(1)} + \hat{H}^{(2)} + \cdots$ up to the desired order and rewritten by means of harmonic oscillator raising and lowering operators.
4. Standard CVPT up to the nth order is applied and block-diagonal effective Hamiltonians are generated.

High-order CVPT thus appears as an efficient and flexible tool for investigating highly excited vibrations at low computational costs. Sibert and McCoy even claim that the overtone stretching energies of C_2H_2 and CH_2O, calculated from CVPT(6), are converged at $10,000\,cm^{-1}$ of excitation as well as are large variational calculations, but require a fraction of the computer time and memory [158]. It might, therefore, well be that CVPT will supersede the variational approach in large molecules or in highly excited energy regions where the latter method is computationally impracticable.

C. Acetylene: A Laboratory for Intramolecular Advances

1. The Acetylene Molecule

a. Both a Simple and Complex Molecule. Acetylene is undoubtedly one of the favorite molecules of both the theoretician and the spectroscopist, as attested by the huge list of references collected in Appendix B. Only papers concerned with the ground electronic state are mentioned, in agreement with the main goal of the present review. The list would have been dramatically enlarged if one had also considered the excited electronic states. Experimental and *ab initio* data were produced on more than 20 electronic states, providing detailed information on most known types of intramolecular couplings: anharmonic and Coriolis resonances (see references in Section III), Renner–Teller [385,386], Herzberg–Teller [387–389], electronic-vibration couplings [390,391] and conical intersections [392], axis switching [393–395], singlet–triplet intersystem crossings [396–398], and discrete-continuum interactions [390,392,399–401]. Excited electronic states of pure valence nature are found below 7eV above the ground state, of singlet [394,402–407] and triplet [408–412] multiplicity. Valence doubly excited [268,389,413] and Rydberg states coexist between 7 and 9eV, leading to strong Rydberg–valence mixing in the frontier region [268,392]. Rydberg states are observed above 9eV [388,391,414–418]. The dynamics associated to the different dissociation channels ($C_2H + H$, $C_2 + H_2$ and $CH + CH$), below or close to the first ionization limit, is not fully elucidated as yet [399,419–424]. Isomerization, predissociation, and autoionization thus occur, and in some regions of the PES compete with each other [425,426]. The picture on acetylene is made even more general, thanks to the existence of data concerning many different isotopomers, including the H, D, T, ^{12}C, and ^{13}C atomic isotopes. A literature review on the subject of the excited electronic states in acetylene is still to appear, and only a very restricted number of references were just mentioned.

The simultaneous development of sophisticated spectroscopic techniques and methodological and computational advances recently have today merged in a deep understanding of the energy pattern and of the intramolecular dynamics in this four-atom molecule. On the theoretical side, the literature indicates a rough increase of one atom per decade in the size of the molecules whose vibration–rotation energy-level structure can be precisely investigated; diatomic molecules were of concern in the 1970s; triatomics, in the 1980s; and four-atom ones, in the 1990s. Important advances were obviously also achieved in approximate methods dealing with molecules of larger sizes and for reduced dimensionality problems (pure vibration, low-energy region, frozen degrees of freedom, etc.). The scaling laws that govern the size of the Hamiltonian matrices to be constructed and diagonalized in vibration–rotation calculations can be estimated. Assuming that a basis set containing n_v functions is used to represent a single vibrational mode and n_r rotational functions, one needs a $n_r \cdot n_v^{3n-6}$ product basis set for a n-atom molecule. We disregard in this simplified estimation the reduction of basis set dimensionality from symmetry adaptation and contraction techniques (see Section II.B.2.f). Fixing the single-mode basis set size to a modest value, like $n_v = 10$, and limiting n_r to 5, leads to product basis sets of size $50, 5 \times 10^3, 5 \times 10^6$, and 5×10^9 for n varying from 2 to 5 by unit step. The initial setting for a accurate PES from *ab initio* calculations also rapidly scales with $3n - 6$ for the number of calculated points or the number of calculated derivatives and each pointwise calculation also rapidly scales up with the number of electrons. This rough picture helps in estimating the impressive amount of methodological work required to deal with four-atom molecules, compared to triatomics.

This is why acetylene, which contains only 14 electrons and 4 light nuclei, cannot be dealt with so easily. Thanks to those very same characteristics, however, acetylene is most appealing, actually for both the theoretician and the experimentalist, and complex features of the type expected to occur in larger systems can be investigated in detail.

The present section (II.C) focuses on the recent theoretical literature work on the vibration–rotation structure in the ground electronic state of acetylene. Its vibration–rotation levels today constitute literally a laboratory for experimenting microscopic-scale methodological developments.

b. Symmetry Properties. Acetylene is a four-atom molecule $ABCD$ with, considering only H, D, ^{12}C, and ^{13}C isotopes:

1. $A = D/B = C$ for C_2H_2 and C_2D_2 (with B and $C = {}^{12}C$ or ^{13}C)
2. $A \neq D/B = C$ for C_2HD (with B and $C = {}^{12}C$ or ^{13}C)
3. $A = D/B \neq C$ for $^{12}C^{13}CH_2$ and $^{12}C^{13}CD_2$
4. $A \neq B \neq C \neq D$ for $^{12}C^{13}CHD$

Building 3D geometric structures with these isotopes leads to a wide variety of symmetry species. Specifically, considering only stationary points encountered on the PES of acetylene, one finds the following structures: sequential linear ($D_{\infty h}$ and $C_{\infty v}$), sequential nonlinear (C_s), vinylidene (C_{2v}), bridged nonplanar (C_{2v}), bridged planar (D_{2h}), sequential *trans* bend (C_{2h}), and *cis* bend (C_{2v}). The Schoenflies notation has been used here to label the symmetry point groups [46,427]. We need to remember (see

TABLE II
Character Table of $D_{\infty h}$ Symmetry Point Group[a]

$D_{\infty h}$:	E	$2C_\infty^\varepsilon$	...	$\infty\sigma_v^{(\varepsilon/2)}$	i	$2S_\infty^{\pi+\varepsilon}$	...	$\infty C_2^{(\varepsilon/2)}$
$D_{\infty h}$(EM):	E_0	E_ε	...	∞E_ε^*	$(12)_\pi^*$	$(12)_{\pi+\varepsilon}^*$	...	$\infty(12)_\varepsilon$
Σ_g^+:	1	1	...	1	1	1	...	1
Σ_g^-:	1	1	...	-1	1	1	...	-1
Σ_u^+:	1	1	...	1	-1	-1	...	-1
Σ_u^-:	1	1	...	-1	-1	-1	...	1
Π_g:	2	$2\cos\varepsilon$	...	0	2	$2\cos\varepsilon$	...	0
Π_u:	2	$2\cos\varepsilon$	...	0	-2	$-2\cos\varepsilon$	...	0
Δ_g:	2	$2\cos 2\varepsilon$	...	0	2	$2\cos 2\varepsilon$	...	0
Δ_u:	2	$2\cos 2\varepsilon$	...	0	-2	$-2\cos 2\varepsilon$	...	0
⋮	⋮	⋮	...	⋮	⋮	⋮	...	⋮

[a] The elements of each class for both the point symmetry group and the corresponding EMS group are listed.

Source: Adapted from Bunker [46].

Appendix A of Ref. 46) that the *molecular symmetry point groups* are isomorphic to the *molecular symmetry group* (Bunker's MS groups) for nonlinear rigid molecules. For rigid linear molecules, isomorphism also applies, but in this case with the *extended molecular symmetry group* (Bunker's EMS groups).

As already pointed out in Section II.B.1, in quantum-mechanical applications, symmetry is generally synonymous of simplification, basis set reduction and computer time savings. This is obviously true in the framework of vibration–rotation calculations and in particular for a system having a high symmetry like linear acetylene. We refer the reader to Refs. 3 and 46 for reference character tables for the different groups cited in this section. Useful tables, including direct product tables and branching rules, are also provided in these reference books. We only report here (see Table II) the character table of the $D_{\infty h}$ symmetry group, corresponding to the linear symmetric geometry, of central interest in this review. Following Bunker [46], the elements of each class for both the point symmetry group and the corresponding EMS group are listed in Table II.

c. Topology of Ground-State PES. The equilibrium geometry of acetylene in its ground electronic state is linear symmetric ($D_{\infty h}$ symmetry group). The electronic state arises from the configuration $1\sigma_g^2 1\sigma_u^2 2\sigma_g^2 2\sigma_u^2 3\sigma_g^2 1\pi_u^4$, and is therefore $^1\Sigma_g^+$. The corresponding PES is well separated from any other one that could induce nonadiabatic interactions. There is thus no electronic vibration mixing to be taken into account. There is no Renner–Teller effect to consider, given the nondegeneracy of the total orbital angular momentum. There is furthermore, no crossing with any other PES. The reason is that the ground-state configuration is strongly stabilized by the presence of four electrons in the $1\pi_u$ bonding orbital, contributing to triple bonding nature of the C–C bond. The first excited valence configuration resulting from a single excitation from the $1\pi_u$ to the antibonding $1\pi_g$ molecular orbital is quite energetically unfavorable. Therefore, the states arising from this first excited configuration are high in energy above the ground state. Specifically, the first triplet $\tilde{a}^3B_2$ and singlet $\tilde{A}^1A_u$ states are respectively located at 35,800 [428] and 41,200 cm^{-1} [416] from the ground state. These states have both a *cis*-bend configuration (H–C–C = 128.7° and 122°, for the $\tilde{a}$ and $\tilde{A}$ states, respectively). As a global consequence, the adiabatic approximation is thus fully justified and the vibration–rotation treatment can be applied within a single Born–Oppenheimer PES, as described in the previous sections of the present section.

The general topology of the ground-state PES has been thoroughly explored up to 43,000 cm^{-1} by Halvick et al. [429], by means of *ab initio* calculations. Eight stationary points have been located and the minimum energy paths connecting them have been calculated. The geometries, energies, and vibrational frequencies at the stationary points have been calculated at the MP4/6-311G(*d*,*p*) level of *ab initio* theory and the connecting paths at the MP2/6-311G(*d*,*p*) level. In these calculations, all electrons were correlated at either the fourth (MP4) or second (MP2) order of Møller–Plesset perturbation theory and a medium-size polarized basis set was used. At this level of theory, a reliable qualitative picture of the topology is expected to be produced. The stationary, or critical points, are characterized by the annihilation of the gradient vector and by the number of imaginary eigenvalues of the corresponding Hessian matrix. This number is an index that defines the nature of each critical point. Minima on the hypersurface are characterized by an index equal to zero, saddle points by an index equal to 1. Higher values of the index (>1) refer to the so-called super–saddle points or transition points. Some of the characteristics of the eight calculated critical points are reproduced from [429] in Table 3, namely the geometrical structures and the symmetry point group to which they belong, the critical point indexes and their multiplicity. Accurate energies for the three lowest structures (labeled M$_1$, M$_2$ and S$_1$), calculated by Chang et al. [430] using CCSD(TQ) results extrapolated to infinite basis set limits, are also reported in Table III. Figure 10 illustrates the energy spacing between all critical points and the minimum energy pathways connecting them. To complete this energy picture, we have included in Fig. 10 some other

TABLE III
Characteristics of the eight Critical Points of the Ground PES of Acetylene, from Ab Initio Calculations [429,430]

Geometry[a]	Stationary Point Structure	Index	Label[b]	Symmetry	Relative Energy[c] (kcal/mol)	Multiplicity
	Linear acetylene	0	M_1	$D_{\infty h}$	0	2
	Vinylidene	0	M_2	C_{2v}	44.54 (42.95)	2
	Bridged symmetric	0	M_3	C_{2v}	74.61	2
	Planar bent	1	$S_1(M_1-M_2)$	C_s	45.09 (44.45)	4
	Bridged nonsymmetric	1	$S_2(M_2-M_3)$	C_s	74.47	2
	Planar bridged	2	$T_1(M_2-M_3)$	D_{2h}	108.9	1
	Bridged nonorthogonal	2	$T_2(M_1-M_3)$	C_2	75.53	4
	Planar bridged nonorthogonal	3	$T_3(M_1-T_1)$	C_{2h}	115.67	2

[a]Black and white balls refer to carbon and hydrogen atoms, respectively.
[b]M, S, and T refer to minima, saddle points and transition points, respectively; $S_1(M_1-M_2)$ indicates that S_1 is a saddle point separating minima M_1 and M_2.
[c]Relative energies are calculated with respect to M_1 and take ZPE into account. All values come from the *ab initio* calculation of [429], except values in parentheses from Ref. [430].

important reference energies below 150 kcal/mol: the first low-lying excited electronic states of singlet and triplet spin multiplicities, at 102.4 [428] and 117.8 [416] kcal/mol respectively, and the two more stable dissociation channels [211,399,421,431–434]:

$$C_2H\left(^2\Sigma^+\right) + H(^2 2S) \quad \text{and} \quad C_2\left(^2\Sigma_g^+\right) + H_2\left(^1\Sigma_g^+\right)$$

at 131.4 and 143 kcal/mol, respectively.

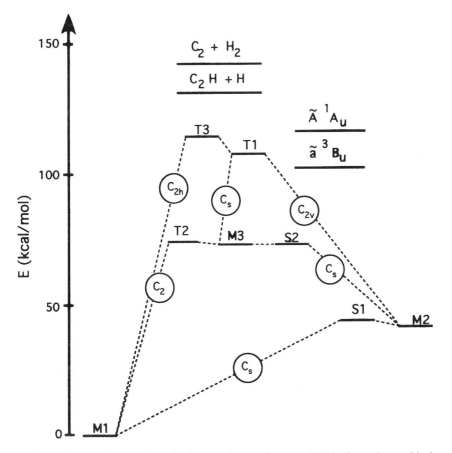

Figure 10. Relative energies of relevant points on the ground PES of acetylene: critical points (minima, saddle points, and transition points) and dissociation channels. Minimum energy pathways (dashed lines) and the symmetry point groups to which they belong are also shown (adapted from Ref. 429). Additional energetic data come from Refs. 211,399,416,428,430,433.

The dissociation limits

$$2CH\left(^2\Pi\right), \qquad C_2\left(^2\Sigma_g^+\right) + 2H(^2S), \qquad \text{and} \qquad 2C(^3P) + 2H(^2S)$$

are higher in energy [433,434]: 228.8, 243.7, and 388.7 kcal/mol, respectively. The first ionization limit is at 263 kcal/mol [435]. Note that all dissociation channels cited above involve product species in their ground electronic states and all correlate with the acetylene ground state, according to the Wigner–Witmer rules [3].

Halvick et al. [429] thus predict on the ground-state PES:

- Three minima:
    Linear acetylene
    Vinylidene
    Bridged symmetric acetylene
- Two saddle points:
    Planar bent
    Bridged nonsymmetric
- Two transition points of index 2:
    Planar bridged acetylene
    Bridged nonorthogonal
- One transition point of index 3:
    planar bridged and nonorthogonal

The multiplicity of each point, also reported in Table III, gives the number of times a given structure appears on the global PES. Each structure is distinguishable when identical atoms are labeled. This means, for instance, that the two linear acetylene forms are $H_1-C_1-C_2-H_2$ and $H_1-C_2-C_1-H_2$ and the two vinylidene forms are $H_{1,2}-C_1-C_2$ and $H_{1,2}-C_2-C_1$, connected by four identical saddle points.

The global topology of the acetylene ground state PES has been studied by Rerat et al. [436] by means of the Morse theory [248]. It is based on the PES features calculated by Halvick et al., completed by further semiempirical CASSCF-CI calculations. This work provides an interesting critical analysis of some analytical PES published in the literature aimed at characterizing mainly three low-lying critical points [151,437,438]. Corrections to these analytical functions, which are topologically consistent, in the mathematical sense, are proposed: the addition of a long-range penalty function converting repulsive valleys and ridges onto ordinary critical points.

d. Calculated Properties. Many properties of the ground-state PES have been calculated using *ab initio* means, as referred to in Appendix B. They include electronic, electric, magnetic, vibration, and rotation features of the major stationary points on this surface and of the different dissociation channels. Different levels of *ab initio* theory were used, making acetylene a favorite system for benchmark studies in which the convergence of calculated properties are investigated, [206,207,210,211,239,240,242,433, 439–441].

e. Coordinate Systems

1. 9D SYSTEMS. Heliocentric and modified heliocentric coordinate systems based on Jacobi and Radau vectors (see Section II.B.2.c.3) and set up for triatomic molecules [58,346] were extended to four-body systems [60,197,296,302,442]. Such systems are well suited for describing a wide variety of bonding situations, in particular when there is a mixture of heavy and light atoms. Figure 11 presents six different ways to combine Jacobi and Radau vectors for describing the vibration and rotation of the set of four nuclei A, B, C, and D.

These nine-dimensional (9D) coordinate systems are the

1. Standard heliocentric system, dedicated for one heavy atom A surrounded by three lighter ones.
2. Standard Jacobi system $A + (B + CD)$, in which vectors $\bar{r}_1$ and $\bar{r}_2$ are pointing toward X and Y, the center of masses of BCD and CD, respectively.
3. Diatom-diatom Jacobi system $(AB) + (CD)$, in which vector $\bar{r}_2$ links the center of masses of AB and CD.
4. Standard Radau system $A + B + C + D$, in which O is the so-called canonical center, lying on the same line as points A, X, and Y, where the distance OY is the geometric mean of AY and XY. Such a system is dedicated to the description of three light nuclei B, C and D and a heavier one, A. The center of the orbiting motion of B, C, and D is not A, as in the pure heliocentric system (1), but rather the canonical center O, taking into account the mass equilibrium within the system.
5. Jacobi/Radau system $(A + B + C) + D$, in which $\bar{r}_1$ and $\bar{r}_2$ are two Radau vectors for particles A, B, and C and $\bar{r}_3$ is a Jacobi vector linking the center of mass of ABC to D.
6. Another Jacobi/Radau system $A + B + CD$, in which $\bar{r}_3$ is a Jacobi vector for the couple of particles C and D, and $\bar{r}_1$ and $\bar{r}_2$ are Radau vectors for the triplet A, B, and CD.

2. 6D SYSTEMS. Eliminating rotation from the 9D coordinate systems discussed above leads to equivalent 6D systems. Two of them, the Jacobi systems 2 and 3, are better adapted to acetylene [see Fig. 12(a) and (b)]. They are particularly interesting for describing the large-amplitude bending motion spanning large regions of the PES, namely, those involving the isomerisation to vinylidene [298,349,442].

Other systems depicted in Fig. 12(c) and 12(d), are used for describing the low-energy region. The nuclear motion is then confined within the potential well corresponding to linear, sequential acetylene. The systems in

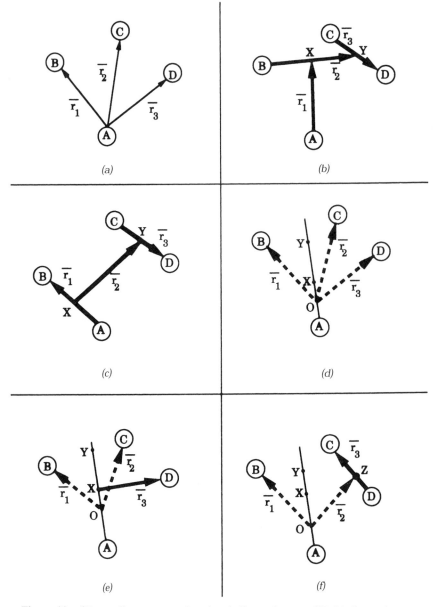

Figure 11. 9D coordinate systems based on heliocentric or modified heliocentric vectors (Jacobi and Radau) for the description of the vibration and the rotation of an $ABCD$ four-atom molecule. X, Y, and Z are the center of mass (CM) of sets of nuclei: (b) $X = \mathrm{CM}(BCD)$, $Y = \mathrm{CM}(CD)$; (c) $X = \mathrm{CM}(AB)$, $Y = \mathrm{CM}(CD)$; (d) $X = \mathrm{CM}(ABCD)$, $Y = \mathrm{CM}(BCD)$; (e) $X = \mathrm{CM}(ABC)$, $Y = \mathrm{CM}(BC)$; (f) $X = \mathrm{CM}(ABCD)$, $Y = \mathrm{CM}(BCD)$ and $Z = \mathrm{CM}(CD)$. (a) Heliocentric; (b) Jacobi $(A + (B + CD))$; (c) Jacobi diatom–diatom $(AB) + (CD)$; (d) Radau $(A + B + C + D)$; (e) Jacobi/Radau $(A + B + C) + D$; (f) Jacobi/Radau $(A + B + CD)$. O is the canonical center of Radau vectors. Jacobi and Radau vectors are drawn in plain and dashed lines, respectively.

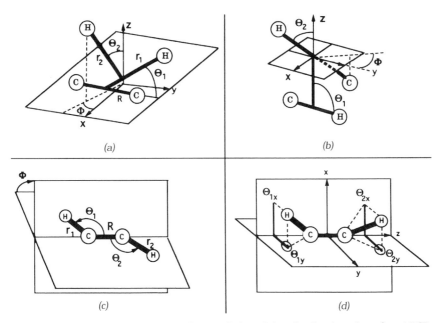

Figure 12. 6D coordinate systems for description of the vibrational motion of an *ABCD* four-atom molecule: Jacobi (*a,b*) [(*a*) Jacobi (A + (B + CD)); (*b*) Jacobi diatom–diatom (AB) + (CD)] and bond-angle valence (*c,d*) coordinate systems.

Fig. 12(*c*) and 12(*d*) are standard internal curvilinear coordinate systems, also referred to as bond-angle valence coordinate systems [51,52,443]. Both systems share the bond distance definition, but differing by the angular coordinates. The three-bond distance coordinates are R, r_1 and r_2, corresponding to the R_{CC} and the two R_{CH} bond distances, respectively. The bending motion is then defined by four coordinates, leading to a 7D system, which is further reduced, to 6D, by fixing one redundant angular coordinate.

System 12(*c*) is well adapted to sequential valence bonding [43,52,143,297], using two bonding angles, Θ_1 and Θ_2, defining the angles between the two adjacent bonds ($R_{CC}-R_{CH_1}$ and $R_{CC}-R_{CH_2}$, respectively), and one torsional angle ϕ, defined as the dihedral angle between planes CCH$_1$ and CCH$_2$.

System 12(*d*) [51,443] uses four linear angle bend coordinates θ_{1x}, θ_{1y}, θ_{2x}, and θ_{2y}. These are the projection of the two CH bonds on the x and y MAS axes, the CC bond lying on the z axis. One of the angular projection, say, θ_{2x}, is to be set to zero in order to fix the 6D system. It is related by trivial trigonometrical expressions [52] to system 12(c).

3. SYMMETRY-ADAPTED COORDINATES. Symmetry-adapted coordinates for linear acetylene are obtained by projecting the internal set of coordinates 12(d) [in Fig. 12(d)] on the irreducible representations of the $D_{\infty h}$ symmetry group (see Table II):

$$S_1 = \frac{1}{\sqrt{2}}(r_1 + r_2)$$

$$S_2 = R$$

$$S_3 = \frac{1}{\sqrt{2}}(r_1 - r_2)$$

$$S_{4x} = \frac{1}{\sqrt{2}}(\theta_{1x} - \theta_{2x}) \qquad (2.82)$$

$$S_{4y} = \frac{1}{\sqrt{2}}(\theta_{1y} - \theta_{2y})$$

$$S_{5x} = \frac{1}{\sqrt{2}}(\theta_{1x} + \theta_{2x})$$

$$S_{5y} = \frac{1}{\sqrt{2}}(\theta_{1y} + \theta_{2y})$$

with S_1 and S_2 transforming as σ_g^+, S_3 as σ_u^+, and S_4 and S_5 as the two-dimensional degenerate π_g and π_u species, respectively. This labeling follows the standard spectroscopic normal-mode notation in regular acetylene.

4. RECTILINEAR NORMAL COORDINATES. Rectilinear normal-mode coordinates are isotopomer-dependent. Arising from the diagonalization of the Wilson FG matrix [45], they are implicitly adapted to the actual symmetry of the corresponding isotopomers of acetylene ($D_{\infty h}$ or $C_{\infty v}$ for linear acetylene). A general form for the normal mode coordinates of the linear isotopomers $A-B-C-D$ is:

$$Q_i = \sum_{X=A,B,C,D} L_{i\xi_X}^{-1} \xi_X \qquad (2.83)$$

with $\xi = z$ for $i = 1$ to 3 and $\xi = x, y$ for $i = 4, 5$; ξ_X is the Cartesian displacement coordinate of atom X with respect to its equilibrium position (expressed in the MAS), and L is the Wilson matrix [45] relating the $3n$ ξ

coordinates to the $3n - 5$ normal coordinates Q. Following, as in (2.82), the standard spectroscopic labeling, the first three coordinates ($i = 1-3$) refer to the stretching modes, with ξ corresponding to the z axis, that is the linear molecular axis. The last two modes ($i = 4, 5$) refer to the bending coordinates, in which the two-dimensional degeneracy results in two sets of Cartesian nuclear displacements, orthogonal to the linear z axis ($\xi = x, y$). Expression (2.83) is valid for any linear isotopomer, but the following additional symmetry relations hold for centrosymmetric species such as C_2H_2 and C_2D_2 ($D_{\infty h}$):

$$L_{i\xi_A}^{-1} = \alpha_i L_{i\xi_D}^{-1}$$
$$L_{i\xi_B}^{-1} = \alpha_i L_{i\xi_C}^{-1}$$
(2.84)

in which $\alpha_i = +1$ and -1 for g and u symmetries, respectively.

Numerical values for the L matrix elements can be obtained from *ab initio* calculations, as shown in Fig. 13 illustrating the normal modes of C_2H_2, C_2D_2, and C_2HD, derived from SD(Q)CI calculations [270]. See Section II.C.2.*b*.2 for details on these calculations.

5. PLANAR 5D SYSTEMS. Coordinate systems of reduced dimensionality were also used in the literature, as the planar 5D Jacobi system represented in Fig. 14, which has been used to study the in-plane bending dynamics of acetylene [296,298,442]. This dynamics implies a large-amplitude motion for the hydrogen nuclei, with a possible isomerization from linear acetylene (M_1) to vinylidene (M_2) passing through the planar saddle-point structure (S_1). This nonrigid process is properly described [296] by the G_8 complete nuclear permutation inversion (CNPI) group [46], isomorphic to D_{2h}.

f. Vibration–Rotation Hamiltonian for Acetylene. Table IV gives an overview on the theoretical work that has been dedicated to the vibration–rotation spectrum of acetylene. The diversity of methods, coordinate systems, spectral ranges, and isotopomers is impressive. We therefore focus in the next sections on selected aspects from Table IV.

1. CHOICE OF A COORDINATE SYSTEM. A first analysis of Table IV concerns the use of different coordinate systems and the corresponding analytical form of the kinetic energy operator. Most of the coordinate systems presented in the previous section were applied to acetylene. As already pointed out, the choice of coordinates is usually dictated by the ability of the selected system to separate the relevant vibrational degrees of freedom. Optimal coordinates must minimize most of the Hamiltonian crossing terms, in either the kinetic-

68

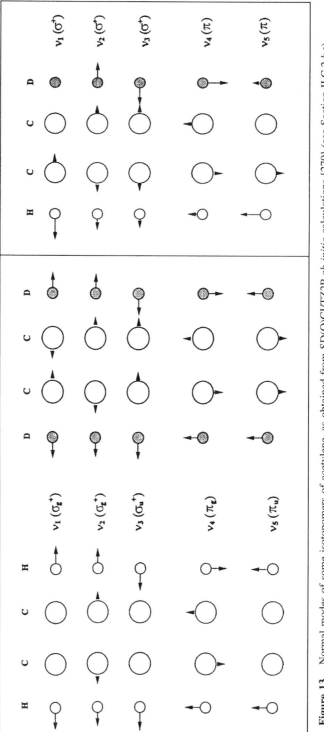

Figure 13. Normal modes of some isotopomers of acetylene, as obtained from SD(Q)CI/TZ2P *ab initio* calculations [270] (see Section II.C.2.*b*.2).

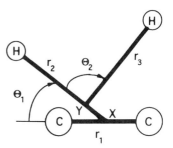

Figure 14. 5D Jacobi system describing the planar motion of the acetylene molecule. X is the CM of the CC bond.

or potential-energy operators. In the case of the stretching modes in acetylene, two major options are encountered, depending on the degree of vibrational excitation and on the nature of the motion in the problem: normal and local modes. One needs to introduce the stretching–bending coupling when large-amplitude bending motion is allowed. Curvilinear valence internal coordinates are an obvious choice for describing the vibrational motion confined within the potential well associated with the linear configuration stationary point. More flexible coordinates, such as those arising from Jacobi and Radau vectors, are, however, more adapted to the study of the pure bending dynamics, involving the acetylene–vinylidene isomerization process.

Other sets of coordinates prove to be simpler to use and are more general in application. This is the case for the rectilinear normal coordinates, which are often used in the spectroscopic Hamiltonians (see Section III). Let us also point out the original approach adopted by Schwenke [197], who developed a general variational computer code, based on a flexible description of the 9D vibration–rotation structure in any four-atom system. The strategy is to start from orthogonal Cartesian coordinates and transform the kinetic-energy operator to any kind of 9D heliocentric-like coordinate systems (see Fig. 11) by means of a mass factor matrix. This approach optionally includes rotational degrees of freedom as an extension of the vibrational ones. It is, moreover, adapted to a wide variety of bonding situations, simply handled by different choices of the mass factor matrix.

2.   KINETIC ENERGY OPERATOR. The exact form of the kinetic-energy operator associated with the different coordinate systems adopted for four-atom molecules can be found in the literature. We refer the reader to the references mentioned hereafter:

Normal coordinates: see the general form of the Watson Hamiltonian or Watsonian for nonlinear [371] and linear [372] molecules; see Section III for a detailed description.

TABLE IV

Overview of Theoretical Work in Literature Dedicated to Vibration–Rotation Spectrum of Acetylene [52,57,143,151,197,206,210,270, 296,298,349,438,442,444–448]

Method	Authors	Ref.	Coordinate System	PES[a] Refined (R) Ab Initio (A)	Vibration (V)[b] Rotation (R) Intensities (I)	Number of isotopomers
1. CVPT(2)	Strey and Mills	[52]	6D valence	R	V	5
2. Variational	Halonen et al.	[151]	3D stretching 6D 4-body	R	V	2
3. Classical trajectories/algebraic	Holme and Levine	[438]	5D planar	R	V	1
4. CVPT(6)	McCoy and Sibert	[57]	6D (valence) and normal	From Ref. [52]	V	1
5. DVR variational	Sibert and Mayrhofer	[298]	5D planar	From Ref. [151]	V	1
6. Adiabatic variational	McCoy and Sibert	[349]	3D bending	R	V	1
7. CVPT(8)	Sibert and McCoy	[158]	3D bending	From Ref. [349]	V	1
8. DVR variational	Bentley et al.	[296]	5D planar	From Ref. [151]	V	1
9. Ab initio quartic force field	Allen et al.	[206]	6D valence and normal	A	V,R	2
10. Ab initio quartic force field	Martin et al.	[210]	6D valence and normal	A	V,R	5
11. VMCSCF/CI variational	Culot and Liévin	[270]	6D valence and normal	A	V,I	3
12. Variational	Liévin et al.	[444]	3D stretching	A,R	V,I	3
13. Variational	Bramley et al.	[143]	6D valence	R	V,R,I	2
14. Variational	Schwenke	[197]	9D Jacobi and Radau	From Ref. [143]	V,R,I	1
15. Semiclassical	Promitis and Farantos	[442]	5D/6D/9D Jacobi	From Ref. [151]	V,R	1
16. Variational + PT	Lehmann et al.	[445]	3D streching	From Ref. [151]	V,R	5
17. Bifurcation theory	Rose and Kellman	[483]	Normal/local	From Ref. 473	V	1
18. Vibron	Iachello et al.	[446–448]	Vibron	R	V	3

[a]PES has been refined from spectroscopic data (R), calculated from *ab initio* methods (A), or is from another work, as listed in the first column of the table.

[b]Vibration (V), rotation (R) and intensity (I) data are provided

Valence internal coordinates (6D): $\hat{T}_{vr}$ and $\hat{T}_v$ have been derived by Handy [67] for the vibration–rotation and pure vibration kinetic-energy operator of linear acetylene. See Ref. [449] for a typographic correction to $\hat{T}_v$; see Ref. [57] for additional terms removing the singularity encountered when using a basis set of harmonic oscillators.

Jacobi coordinates (9D $J \neq 0$): see Prosmiti and Farantos [442].

Jacobi coordinates (6D $J = 0$): see Refs. 57 and 442 for the standard 6D Jacobi-like systems [Figure 12(*a*)] and 442 for the diatom–diatom 6D system [Fig. 12(*b*)].

Jacobi coordinates (planar 5D, $J = 0$): see Refs. 296 and 442 and in Ref. 298, in which the kinetic-energy cross-terms were eliminated by an appropriate affine transformation.

General heliocentric and modified heliocentric coordinates (9D $J \neq 0$) [(Figs. 11(*a*)–(*f*)]: general coordinate system for four-atom molecules, developed by Schwenke [197], obtained from orthogonal Cartesian coordinates transformed by a mass factor matrix.

Adiabatic bending (3D): see McCoy and Sibert [349].

3.   POTENTIAL ENERGY SURFACE. According to Table IV, the ground-state PES available in the literature arises from different sources: either from *ab initio* calculations [206,210,269,270], from the adjustment of a given analytical form to fit global topological features [151,438], or from spectroscopic data [52,143,158,444]. In the latter case, the least-squares fits are alternatively based on Van Vleck perturbation theory [52] and on variational calculations [143,158,444].

Most of the above PES [52,143,210,269,270,444] are *local ones*, which means that they aim at describing the vibrational motion in a very restricted part of the whole surface. This part is a single minimum on the PES, corresponding here to the more stable stationary point: the linear H–C–C–H structure (M_1 in Table III). The local character justifies the use of a polynomial analytical form such as (2.67), corresponding to a Taylor series expansion in an adapted set of coordinates, with respect to the stationary point geometry. This expansion is usually limited to the fourth order and is constructed in normal rectilinear or in valence internal coordinates.

The cylindrical symmetry relationships to which such quartic potential must obey determine the number of unique terms to be fitted from *ab initio* or spectroscopic data. If the molecule is considered to be rigid, the potential-energy operator must indeed be invariant to the elements of the point group of the most symmetric configuration in the internal coordinate space [143]. Found here is the symmetric linear configuration belonging to the $D_{\infty h}$ point group. The potential function thus belongs to the totally symmetric representation Σ_g^+. The number of linearly independent constants of order n in the potential is therefore given for all coordinate systems by the number n_1 of occurrences of the Σ_g^+ representation in the symmetric nth power of the total vibrational representation:

$$\left(2\Sigma_g^+ \oplus \Sigma_u^+ \oplus \Pi_g \oplus \Pi_u \right)^n = n_1\Sigma_g^+ \oplus n_2\Sigma_u^+ \oplus n_3\Pi_g \oplus \cdots \qquad (2.85)$$

It follows that the quartic force field of acetylene is entirely defined [52] by a total of 40 independent constants, namely, by 6, 11, and 23 quadratic ($n = 2$), cubic ($n = 3$), and quartic ($n = 4$) terms, respectively. The explicit form of the quartic potential in the valence internal coordinate system (r_1, r_2, R, θ_1, θ_2, ϕ) [see Fig. 12(c)] is given by Strey and Mills [52]. These authors also give the symmetry relations to translate this potential to the alternative internal coordinate system (r_1, r_2, R, θ_{1x}, θ_{1y}, θ_{2x}, θ_{2y}) [see Fig. 12(d)].

In rectilinear normal coordinates, the force field is mass-dependent, which means that the $D_{\infty h}$ counting given above only applies to the linear centrosymmetric isotopomers C_2H_2 and C_2D_2. For unsymmetric $C_{\infty v}$ species, such as $^{12}C^{13}CH_2$, $^{12}C^{13}CD_2$, or C_2HD, the number of independent constants grows to 9, 16, and 31 for orders 2, 3, and 4, respectively. According to the definition of the normal coordinates, the quadratic part of the potential is diagonal, restricting to 5 the number quadratic terms in all unsymmetric cases.

Some *global* PES of acetylene were developed [151,438]. Contrary to the quartic potentials just presented above, these PES are dedicated to the description of larger parts of the configuration space overlapping more than a single minimum. They are designed for describing the dissociation channels and the isomerisation pathway from linear acetylene to vinylidene. They cover therefore several stationary points: M_1, M_2, and S_1 (see Table III).

Halonen et al. [151] have developed a global PES (hereafter referred to as the HCC PES) defined by the following many-body expansion:

$$\begin{aligned}
V_{C_2H_2} &= V^{(2)} + V^{(3)} + V^{(4)} \\
V^{(2)} &= V_{H_2}^{(2)} + V_{C_2}^{(2)} + V_{CH}^{(2)} \\
V^{(3)} &= V_{CH_2}^{(3)} + V_{C_2H}^{(3)} \\
V^{(4)} &= V_{C_2H_2}^{(4)}
\end{aligned} \qquad (2.86)$$

We refer to [151] for the references of the two- and three-body terms and for the parameters of the four-body term, expanded in a very similar way than the three-body term of (2.65). This potential has been adjusted to include spectroscopic information on the dissociation limits as well as data from *ab initio* calculations. The latter data fix the energy difference between acetylene and vinylidene to $13,994\,\mathrm{cm}^{-1}$ and the isomerization barrier to $17,002\,\mathrm{cm}^{-1}$ (all vibrationless energies). The function (2.86) is assumed to be valid at all dissociation limits as well as in regions close to the equilibrium geometry. The global topology of this PES is criticized (see Section II.C.1.*c*) by Rerat et al. [436], in the sense of the Morse theory, some critical points others than M_1, M_2, and S_1 missing from the adjustment. Nevertheless, the resulting description of the isomerization process was used satisfactorily by most authors who investigated the large-amplitude bending motion in acetylene [296,298,442].

Another *global* PES was developed by Holme and Levine [438], restricted to 5D planar motion. Stretching CH coordinates are described by Morse oscillators with their parameters as defined by Halonen et al. [151]. The CC bond is represented using variable parameters, by another Morse oscillator defined by an hyperbolic tangent switching function, allowing for a smooth change of the parameters, from M_1 to M_2. The bending potential is expressed as a double Fourier series in the two bending angles. All parameters have been adjusted to give a reasonable fit to the known geometries and energies of the PES.

2. Forward and Backward Trips Applied to Acetylene

Most of the theoretical concepts developed from the beginning of this section have been applied to the study of the vibration–rotation structure of acetylene. The most illustrative applications are summarized in the present section.

a. Refined Quartic Force Fields in Valence Internal Coordinates

1. REFINEMENT FROM SECOND ORDER PERTURBATION THEORY: THE STREY–MILLS POTENTIAL. The first accurate quartic potential of acetylene was derived by Strey and Mills [52] (hereafter denoted as the SM potential). These authors performed least-squares calculations from spectroscopic constants of five different isotopomers. The potential was refined using standard second-order vibrational perturbation theory, already referred to in Section II.B.2.*g*. The quartic force field was derived in valence internal coordinates [Fig. 12.(c)], thanks to the L tensor approach [51] used to determine the nonlinear coordinate transformation from the curvilinear internal to the rectilinear normal coordinates. The general expression of this potential is

$$V = \sum_{ijklmn} u_{ijklmn} r_1^i r_2^j R^k \alpha_1^l \alpha_2^m \cos(n\phi) \tag{2.87}$$

where $\alpha_i = \sin(\theta_i)$, for $i = 1, 2$. This general expansion has been limited to fourth order by $i + j + k + l + m \leq 4$ and $l, m \geq n$. The total number of unique terms in this force field can, moreover, be compacted, as mentioned above, to 40 symmetry-unique force constants u, by making use of symmetry relations [52]. Only 25 of them were refined; 15 quartic constants were found unimportant or were poorly determined.

2. VARIATIONAL CALCULATIONS WITH A MODIFIED SM POTENTIAL. Bramley and Handy [297] have used the SM potential and a modified version of it, denoted SM(M), to perform variational calculations on the Σ states of C_2H_2, C_2D_2, and C_2HD, up to two quanta in the CH stretches. The variational approach, especially designed for four-atom molecules and involving the 9D vibration–rotation degrees of freedom, was implemented in the VISTA computer code. It uses the exact kinetic-energy operator, expressed in valence internal coordinates and Euler rotational coordinates, as derived in Ref. 67 and rearranged in a computationally more convenient form in Ref. 297. The variational basis sets, which are fully symmetry-adapted, are the standard Morse oscillator, Legendre, and Wigner functions for describing the stretching, bending, and rotational degrees of freedom respectively (see Section II.B.2.c). The rotational quantum number J was limited to $J \leq 2$ in the larger basis set calculations. Owing to a very efficient basis set contraction technique, the size of the Hamiltonian matrices could be maintained to a reasonable order of magnitude (lower than 4500×4500 for the less symmetric isotopomer C_2HD), allowing standard full-matrix diagonalisation techniques to be applied, and ensuring the convergence of hundreds of vibration–rotation levels to better than $1.5\,\mathrm{cm}^{-1}$. The contraction proceeds into six successive variational steps, introducing progressively the coupling between vibrational coordinates, with at each step elimination of the less important eigenvectors. The steps are as follows:

1. Variational resolution of an independent 1D problem for each internal coordinate, in an adapted basis set of primitive functions (Morse and Legendre functions for stretchings and bendings, respectively).
2. Bending (θ_1, θ_2) 2D-variational calculation in which the eigenfunctions of step (1) are coupled together.
3. Bending–torsion $\{(\theta_1, \theta_2), \phi\}$ 2D-variational calculation in which the eigenfunctions of steps 1 and 2 are coupled together.
4. Stretching CH(r_1, r_2) 2D-variational calculation.
5. Stretching CH–CC $\{(r_1, r_2), R\}$ 3D-variational calculations.
6. Full 9D vibration–rotation variational calculations, coupling eigenfunctions of steps 3 and 5 with rotational primitive functions.

The final eigenfunction constructed in this way is expected to be very compact. Tests on the convergence of the desired eigenstates are required to validate the truncation procedure.

The SM(M) force field differs from the SM one (2.87) in two ways: (1), by the use of $\delta\theta$ rather than $\alpha = \sin\theta$ for the bends and (2), by the use of Morse coordinates z [see (2.51)] instead of bond extension coordinates for the stretches, leading to the following general expression:

$$V = \sum_{ijklmn} u_{ijklmn} z_{r_1}^i z_{r_2}^j z_R^k (\delta\theta_1 - \pi)^l (\delta\theta_2 - \pi)^m \cos(n\phi) \qquad (2.88)$$

with a restriction of this expression to $i + j + k + l + m \leq 4$ and $l, m \leq n$.

The former modification to SM avoids any tunneling effect in artificially low-energy regions at $\theta = 0$, while the latter aims at improving the description of the stretching levels. However, it appears from the numerical tests performed in [297] on the main isotopomer that the pure bending levels are also significantly improved, as a result of the importance of the stretch–bend coupling in C_2H_2. These tests show that the low-lying SM(M) levels agree with experiment to about $10\,cm^{-1}$, while the SM CH stretch fundamentals were found in error by $35\,cm^{-1}$ and the two-quanta bends by up to $18\,cm^{-1}$. The overall agreement of the SM(M) potential on the Σ states up to $2\nu_1$ in the three isotopomers correspond to a root mean square of $10\,cm^{-1}$.

3.   SIXTH-ORDER CVPT WITH THE SM POTENTIAL. McCoy and Sibert [57] have applied sixth-order canonical Van Vleck perturbation theory [CVPT(6)] to calculate the vibrational energy levels of acetylene up to $17{,}000\,cm^{-1}$ above the zero point energy. They used an Hamiltonian based on Handy's pure vibration kinetic-energy operator [67] and another modified version of the SM potential, using SPF coordinates instead of bond extension coordinates for stretching modes. One major advance of this work was to transform, by means of CVPT, the *curvilinear normal* and *rectilinear normal* coordinate vibrational Hamiltonians to block-diagonal effective Hamiltonians, including the stretch–bend resonance interactions. A comparison of the perturbative vibrational energies obtained from both formulations is thus provided. We refer to Section II.B.2.c.8 for the definition of the curvilinear normal coordinates and to Section II.B.2.g.3 for the implementation of nth order CVPT(n). The results show that [57]:

1. Up to $5000\,cm^{-1}$, the perturbative energies obtained from both representations differ only by at most $0.3\,cm^{-1}$. An algebraic equivalence between the curvilinear and rectilinear normal coordinates is thus

numerically demonstrated. A similar equivalence was also found for HCN, but not for CO_2. In the latter case, more pronounced interactions between nearly degenerate states occur, and the convergence of the perturbative energies is significantly faster in the curvilinear representation. In any case, the present approach presents the advantage to include the stretch–bend resonance interactions in the effective Hamiltonian.

2. High order CVPT(n) ($n = 6$ here) properly describes highly excited states (with as many as five quanta in the stretching modes and four quanta in the bendings). Convergence problems reported for states with more than four quanta in the bends are probably coming more from deficiencies in the SM force field that was used, than from the CVPT approach.

4. REFINEMENT FROM VARIATIONAL CALCULATIONS: THE BRAMLEY–CARTER–HANDY–MILLS POTENTIAL. Bramley, et al. [143] used the variational technique developed in [297] (see Section II.C.2.*a*.2) to refine the SM(M) potential against experimental term energy values up to three and two quanta in the C–H stretch vibrations for C_2H_2 and C_2D_2, respectively. The resulting force field is referred to hereafter as the BCHM potential. Two different refinements, denoted R_1 and R_2, have been performed, characterized by different strategies adopted in the least-squares fitting procedure. The refined potentials have essentially the same form that the SM(M) one has, with the use of Morse and $\delta\theta$ coordinates as expansion variables. The same set of quartic constant as in SM or SM(M) is constrained to zero, except for the u_{000222}, which was relaxed in order to reproduce the vibrational l-doubling separation. The parameters to be refined are thus the u coefficients of (2.88), but those constrained to zero, as explained above.

Both refined R_1 and R_2 potentials have about the same global accuracy. They reproduce the energies of all known (at the time) Σ, Π, and Δ vibrational states, up to $2v_1$, with a root mean square error of $3\,cm^{-1}$. R_1 is more accurate than R_2 in the energy range $7000-10,000\,cm^{-1}$, due to an improved cubic shape. R_2 is probably slightly better than R_1 at low energies. The refined force field was used to study the resonances associated with the degeneracies $(v_2 + v_4 + v_5, v_3)$ and $(v_2 + 2v_5, v_1)$ in $^{12}C_2H_2$, helping at the time being to assign the states observed around $2v_3$ and $3v_3$. It was also shown that the corresponding resonances were due mainly to vibrational l-type interactions (see Section III.B) rather than to quartic anharmonicities in the potential.

The BCHM potential is undoubtedly the best empirical potential available for acetylene at the time of the present review. It is limited by its analytical form, restricting its range of application to three C–H quanta in the stretching modes and four quanta in the bends. The latter restriction excludes the description of large amplitude bending motion, involving orbiting of the H

TABLE V
Comparison of BCHM Predictions [143] with New Experimental Observations [451–453] and with Predictions of El Idrissi et al. from a Global fit using the Cluster Model Described in Section III.C [450]

v_1	v_2	v_3	v_4	l_4	v_5	l_5	R_1^a	R_2^a	Exp.	From Global Fit[b]
0	1	0	1	−1	1	1	3299.9	3302.9	3300.637[c]	3300.0
1	0	0	2	2	1	−1	5270.7	5272.9	5269.729[d]	5270.2
0	2	0	3	1	1	−1	6406.9	6411.9	6413.898[e]	6413.6
0	1	1	2	0	0	0	6452.4	6455.3	6449.105[e]	6449.2
0	2	0	1	1	3	−1	6656.1	6657.0	6654.253[e]	6654.8

[a] Predicted by Bramley et al. [143], from variational calculations on R_1 and R_2 refined PES.
[b] Predicted by El Idrissi et al. [450], from global fits (see Section III).
[c] From Vander Auwera et al. [451].
[d] From Palmer et al. [452].
[e] From Kou et al. [453]

atoms and formation of the vinylidene-like structure. It presents other limitations: (1) the ability of the force field to accurately reproduce the experimental data included in the refinement procedure does not prove that it can account for other spectral information with the same precision, (2) the model potential used in the initial step of the refinement (the SM field in this case) may carry major defaults that could have influenced the refinement procedure in some unpredictable way, and (3) the parameters that have been constrained to zero certainly bear some arbitrariness. A posteriori checks of the BCHM potential can be today attempted by comparing at the time predicted transition energies below $6700\,cm^{-1}$ for $^{12}C_2H_2$ (see Table 14 of Ref. 143) with new observations. The five possible comparisons are most satisfactory, as demonstrated in Table V, with a mean absolute deviation of 2.76 and $2.95\,cm^{-1}$ for R_1 and R_2, respectively. These predictions can also be compared to those resulting from very recent global fits [450] performed on 219 spectroscopic data, by means of the cluster model discussed in Section III.C. See Appendix C for a complete list of energy-level predictions up to $10,000\,cm^{-1}$. The absolute mean deviation of the latter results, for the set of data of Table 5, [450–453], is $0.4\,cm^{-1}$. These predictions agree to those of Bramley et al. for the 24 band origins of Table 14 in Ref. 143 within 2.2 and $1.5\,cm^{-1}$, for R_1 and R_2, respectively. The latter figures confirm the agreement between both totally independent adjustments (spectroscopic normal-mode Hamiltonian as detailed in Section III.C vs. variational valence coordinate approach), based on similar sets of input data, in the same energy range. Also given in Appendix C is the level assignment obtained by El

Idrissi et al. [450], which confirms and extends the BCHM predictions. Let us note a misprint in Table 14 of Ref. 143 concerning the level at 4589.7 cm^{-1}, which is to be assigned as $(v_1 v_2 v_3 v_4^{l_4} v_5^{l_5})$ $(0102^2 2^{-2})$ rather than $(1002^0 0^0)$.

Electric dipole transition intensities were also calculated by Bramley et al. [143], from an *ab initio* DMS expanded as a cubic Taylor series in valence internal coordinates. Eckart conditions were applied [98], defining the three components of dipole field. The cubic dipole field originates from different levels of *ab initio* calculations used to calculate the analytical successive derivatives, according to the method of Jayatilaka et al. [454]. The Cartesian first derivatives obtained at the MP2 level were added to second and third SCF derivatives. All calculations were performed with a TZ2P basis set augmented by a single polarization *f*-type Gaussian primitive centered on each carbon atom. The transition dipole moment matrix elements were calculated from the variational wavefunctions, according to the general expression (2.36), and the absorption coefficients were calculated as detailed in Section III.E. The calculated values of these absorption coefficients helped predicting the vibrational bands of interest to new experimental investigation.

5.   VARIATIONAL CALCULATIONS WITH THE BCHM POTENTIAL. The BCHM force field was used by Schwenke as a test case for a new general program for four-atom molecules [197]. The R_2 potential is reexpanded in two different ways, denoted by the author as R_2^T and R_2^j. The first version was obtained by projecting out the cos $n\phi$ components of (2.88), and the second one uses the CH + CH diatom–diatom Jacobi coordinates, represented in Fig. 11(c). The R_2^T potential contains many more angular terms than the original R_2 one and is thus computationally more expensive, while R_2^j is better adapted to the general strategy of the program for dealing with angular coordinates. Two choices of rotation–bending functions were investigated: the *llk* and *ljk* basis, respectively corresponding to *local-* and *normal-mode* motions and both using contracted basis functions. These functions were generated in two steps: (1) a SCF calculation using a truncated potential and (2) a calculation on the full potential introducing the coupling between the functions from the first step. The efficiency of the basis set contraction was verified, for both *llk* and *ljk* basis sets, by measuring the amount of coupling in the second step just mentioned, using the magnitude of the largest components of the eigenvectors corresponding to the 10 lowest energy states. This test demonstrated the superiority of the *ljk* basis set.

The comparison [197] of converged vibrational energies for vibrational levels of $^{12}C_2H_2$ up to $v = 2$ in all modes obtained from the original R_2 version and the bend-transformed R_2^T and R_2^j versions, is worthwhile. The

R_2^T results agree with those of R_2 [143] with a maximum difference of $5\,\mathrm{cm}^{-1}$. Larger differences are, however, observed between R_2^T and R_2^j, mainly for the bending modes (maximum difference of $17\,\mathrm{cm}^{-1}$). This discrepancy might indicate [197] some weakness in the quartic part of the BCHM force field.

b. Ab Initio Quartic Force Fields

1. SCF QUARTIC FORCE FIELD FROM ANALYTICAL DERIVATIVES. Schaefer and co-workers [206,455] have carefully investigated the use of *ab initio* calculations for predicting a wide variety of vibration–rotation constants of semirigid molecules. In particular, Allen et al. [206] systematically investigate how SCF analytical high-derivative methods help in the study of vibration anharmonicity and vibration–rotation interaction in linear molecules. That paper is very complete and detailed, and therefore most helpful. Two of the main isotopomers of acetylene ($^{12}C_2H_2$ and $^{12}C_2D_2$) were selected, among other species, by these authors. They calculated *ab initio* the successive derivatives of the energy with respect to nuclear displacement coordinates, defining the different terms of the Taylor series expansion of the potential. Low-order perturbation theory [456] was then used to relate these potential constants to vibration–rotation spectroscopic constants. The computation proceeds in four successive steps: (1) determination of the equilibrium geometry of the studied molecule by optimization using SCF analytical gradient methods [see references in [206], (2) analytical calculation of the quadratic and cubic force constants at the SCF or SDCI levels, (3) determination of the quartic force constants by finite differences of SCF third-order derivatives, and (4) calculation of the spectroscopic constants from the previously derived quantities.

The quartic force field of acetylene was calculated at the SCF level with a TZ2P basis set. The harmonic force field was also calculated at the SDCI level with the same basis set. Excluding the *trans*-bending frequency ω_4, the other harmonic frequencies of both isotopomers are, as expected, overestimated by 8–9% and 2–3% for the SCF and SDCI methods, respectively. The case of ω_4 is pathological, with an overestimation of 23–25% at the SCF level and an underestimation of 3–4% at the SDCI level. The curvature of the PES along this mode is, in addition, very sensitive to the basis set effect. Simandiras et al. [457] demonstrated that the inclusion of a single f polarization function was decisive for fixing ω_4 to a reasonable value (1% too large at *CCSD* level). The quartic force field is given in valence internal coordinates in the form (2.87) and in normal coordinates, for the principal isotopomer. It is the first published *ab initio* quartic force field for acetylene. Most of the SCF calculated spectroscopic constants are found to be lower but in reasonable agreement with the corresponding experimental values, despite the low-order level of theory that is used. An order of magnitude of the deviation from experiment is 8, 11, and 11.5% for the vibration–rotation interaction, quartic centrifugal distortion and rotational l-doubling constants, respectively. The fundamental frequencies are predicted some 2–4% below the observations.

2.   SD(Q)CI QUARTIC FORCE FIELD AND SDCI CUBIC ELECTRIC FIELD FROM GRID CALCULATIONS. A full quartic force field in valence internal and normal coordinates was determined by Culot and Liévin [269,270] at the SDCI and SD(Q)CI levels of theory, using the same TZ2P basis set as in Allen's work [206]. The constants of the potential in the two coordinate systems were obtained from least-squares adjustments of the corresponding quartic polynomials on a grid of *ab initio* calculated points. The grid is constructed [270], as explained in section II.B.2.*d*.3, by means of a statistical approach based on experiment planification methods [258,260]. It is a fourth-order rotatable design optimized to give a quasi-diagonal variance–covariance matrix and a small standard deviation of the fitted constants. The grid contains a total of 545 points which by symmetry reduce to 141 unique points, leading to a D efficiency of the grid of 30%. The normal-mode coordinates, obeying to (2.83), calculated from the SD(Q)CI PES, have been presented in Fig. 13 for $^{12}C_2H_2$, $^{12}C_2D_2$, and $^{12}C_2HD$. Note the isotope mass effect on the stretching nuclear displacement, in particular for $^{12}C_2HD$, exhibiting quasi-local stretching vibrations.

A DMS was calculated at the TZ2P/SDCI level of theory on the same grid and adjusted to a cubic Taylor series expansion in the normal coordinates derived from the potential.

3.   TESTING THE QUALITY OF THE SDCI AND SD(Q)CI FORCE FIELDS. The PES and DMS described in the previous section were calculated at a medium level of *ab initio* theory, not requiring too much computational effort. Their major deficiencies arise from the single reference used in the CI treatment and the lack of f polarization orbital to improve the description of the bending motion [457]. These surfaces are thus not expected to be of spectroscopic accuracy. Nevertheless, Culot and Liévin [269,270] showed that such an economic full *ab initio* approach gives interesting insight into the complex anharmonic behaviour of a molecule such as acetylene and, owing to the low computational costs, it is applicable to larger systems.

The superiority of the SD(Q)CI PES over the SDCI one is clearly demonstrated from a comparison of the equilibrium geometry and of the harmonic part of the force field:

- The SDCI and SD(Q)CI bond equilibrium distances agree with the experimental values within 0.7 and 0.07% (7×10^{-4} Å), respectively.
- The stretching harmonic frequencies agree with experiment within 4 and 1.5%, respectively, for both methods of calculation. The SD(Q)CI ω_4 bending frequency is however underestimated by 94 cm^{-1}, due the basis set deficiency already pointed out above.

The importance of the Davidson's correction [224], accounted for at the SD(Q)CI level, on the shape of the PES has been reported in other cases [173,174,458].

The quality of the anharmonic part of the quartic force field has been investigated by comparing calculated transition energies, up to two quanta of vibrational excitation, to the corresponding experimental values. The calculated energies depend, however, on other factors than the accuracy of the potential. As already pointed out at several occasions previously, the coordinates in which the potential is expanded and the method used for solving the vibrational problem are also determinant. The potential has thus been tested first, with respect to the best empirical force field, specifically, the R$_2$ potential of BCHM, using the same coordinates and variational method as adopted by Bramley et al. [143] (see Section II.C.2.a.4). The mean absolute deviation is 25 and 37 cm^{-1} for the fundamental bands and for all the transitions up to $v = 2$ respectively, demonstrating the reasonable accuracy of the SD(Q)CI quartic potential.

Calculations were also performed, in a second time, within the rectilinear normal coordinates approach, using the vibrational VCASSCF-CI variational method [269,270], described in Section II.B.2.f.2. The convergence of this two-step procedure, with respect to the primitive, active oscillators, and CI basis sets have been carefully verified, which means that the compact wavefunctions obtained in this way can be considered as converged for the levels of interest. Comparison with the BCHM internal coordinate results is thus essentially a test of the curvilinear versus normal approaches. The superiority of the former is evident; the mean deviation to the experimental values for the fundamental bands is 25 cm^{-1} in internal coordinates and of 64 cm^{-1} in normal coordinates. The deviations also affect the stretching as the bending modes. The former are better described by the Morse variables and the latter by the curvilinear bending coordinates than by the corresponding rectilinear displacements along the Cartesian axes.

4. OVERTONE SPECTRUM OF STRETCHING MODES. The SD(Q)CI PES and the SDCI DMS were used [270] to analyze the overtone spectrum of the stretching modes of three isotopomers $^{12}C_2H_2$, $^{12}C_2D_2$, and $^{12}C_2HD$. The

rough approximation followed long ago by Botschwina [459], consisting in completely neglecting the bending motion, was initially adopted. The full *ab initio* treatment of Fig. 5 was applied, restricting the SD(Q)CI PES and SDCI DMS to the 3D coordinate space defining the pure stretching motion. A normal coordinate full-CI variational calculation, as implemented in the POLYMODE program [460], was performed, and the vibrational transition moments were calculated [173,458]. Figure 15 shows the correlation between *ab initio* and observed transition energies on the three isotopomers considered. The observed values are all today's available experimental values, completed by the values recalculated from the parameters of a recent global fit [450] (see Section III and Appendix C), when no experimental value was available. The number of unobserved transitions up to four quanta in the stretching modes is of 15, 17, and 27 among 34 for $^{12}C_2H_2$, $^{12}C_2D_2$, and $^{12}C_2HD$, respectively.

The correlation shown in Fig. 15 is satisfactory for $^{12}C_2D_2$ and $^{12}C_2HD$, owing to the rough approximation used. The absolute mean deviation is of 15

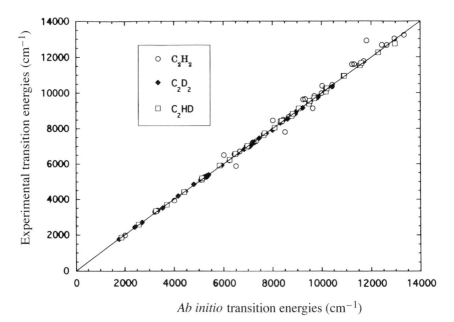

Ab initio transition energies (cm^{-1})

Figure 15. Correlation between *ab initio* and observed stretching modes transition energies in $^{12}C_2H_2$, $^{12}C_2D_2$, and $^{12}C_2HD$, up to four quanta in the stretching modes. Calculated values come from a full *ab initio* forward trip [pure stretching modes model, SD(Q)CI PES and full-CI variational calculation] [270]. Experimental values come from the literature; unobserved transitions are recalculated from the parameters of a recent global fit [450] (adapted from Ref. 270).

(30), 23 (42), 31 (40), and 63 (61) cm^{-1}, for the $v = 1, 2, 3$, and 4 vibrational excitations on $^{12}C_2D_2$ ($^{12}C_2HD$), respectively. The slow increase of the deviation with v tends to prove the consistency of the SD(Q)CI quartic pure stretching PES. The same conclusion does not hold, however, for $^{12}C_2H_2$, for which larger deviations are observed (45,125, 209, and 320 cm^{-1} for $v = 1, 2, 3$ and 4, respectively), some transition energies being completely mismatched, as shown on the figure. This illustrates the bigger importance of the stretch–bend interactions in the case of the $^{12}C_2H_2$ isotopomer.

   The consistency of the 3D pure stretching model has been further investigated [270] in the case of $^{12}C_2HD$, for which transition intensities were calculated by means of the SDCI cubic stretching DMS. The calculated intensities are compared in Fig. 16 with those obtained [444] from the measured absolute induced dipole transition moments. That figure illustrates the evolution of the intensity as a function of the vibrational energy for $v = 1, 2, 3$ quanta of excitation in the stretching modes. The correlation is

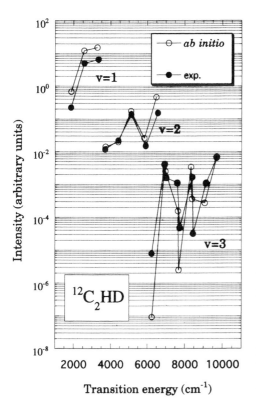

Figure 16. Correlation between *ab initio* and observed stretching modes transition intensities in $^{12}C_2HD$, up to three quanta of excitation in the stretching modes. Calculated values come from a full *ab initio* forward trip [pure stretching modes model, SD(Q)CI PES, SDCI DMS, and full-CI variational calculation] [270]. Experimental values come from Ref. 444 (adapted from Ref. 270).

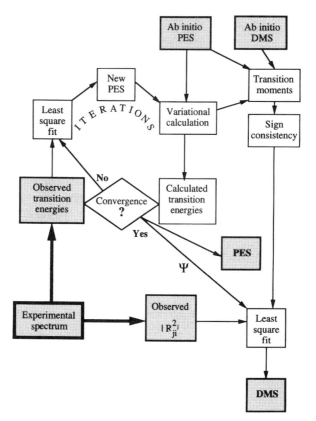

Figure 17. Iterative procedure for refining both the PES and DMS from variational calculations. *Ab initio* information are used as input data, providing an initial guess for the PES, and relative signs consistency for the DMS refinement (adapted from Ref. 444).

again remarkably good. One sees that the overall evolution of the intensity is qualitatively well reproduced by the *ab initio* calculation.

The success of the 3D stretching model in $^{12}C_2HD$, and the availability of a large set of observed transition energies and intensities, motivated Liévin et al. [444] to achieve a systematic analysis of the overtone stretching intensities in terms of the mechanical and electrical anharmonicities. This analysis was based on the refinement of the 3D SD(Q)CI PES and SDCI DMS, leading to effective PES and DMS. The refinement procedure, detailed in Ref. 444, is depicted in Fig. 17. The originality of this approach lies in the way effective DMS have been refined from experimental intensity values. The major problem encountered in any attempt to derive parameters of the DMS from experiment is the ambiguity arising from the relative signs of the transition dipole moments. Only the square of these quantities can indeed be determined from the experimental intensities. To overcome this bottleneck, one can directly use, as input of the refinement procedure (see Fig. 17), the *ab initio* information on the transition moment relative signs. Let us cite another approach adopted by Teffo and co-workers on N_2O and CO_2 [461– 463], in which an effective dipole moment is adjusted from experimental $|R^2|$ values. The sign problem is overcome by adjusting the DMS parameters for different relative signs combinations and choosing the one corresponding to an absolute minimum of the norm of the deviations.

Different effective models were investigated in this way, allowing the evolution of the intensity along different overtone series to be interpreted. As a relevant example, let us focus on the differences observed in the intensity drop along the CH and CD stretches overtone series. As seen in Fig. 18, a

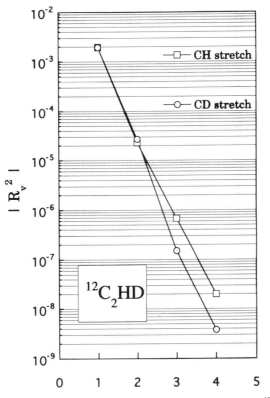

Figure 18. Evolution of the CH (v_1) and CD (v_3) stretch intensities in $^{12}C_2HD$ (adapted from Ref. 444).

more severe drop of the square modulus of the transition moment is observed in the CD series for $v \geq 3$. This feature was attributed to the shape of the DMS. It is indeed quite different along both stretching coordinates, with a sign change in the cubic dipole moment terms. The effect of this change on the overtone intensity is illustrated from the simplest 1D effective models in Fig. 19, in which the relative contributions of the different terms of electrical

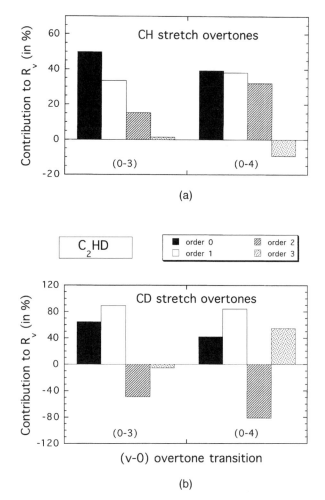

Figure 19. Contribution in percent of the different orders of electric anharmonicity to the second and third overtones of the CH (v_1) and CD (v_3) stretching modes of $^{12}C_2HD$. (a) CH stretch overtones; (b) CD stretch overtones. Calculated from a refined effective PES and DMS (adapted from Ref. 444).

anharmonicities to the total transition dipole moment values are visualized for the CH and CD overtone transitions up to $v = 4$. One clearly sees that the negative sign of the second-order anharmonicity contribution (cubic term in the dipole moment expansion) in the CD series is responsible for the intensity drop at $v = 3, 4$.

5.   ANALYSIS OF STRETCH–BEND INTERACTIONS FROM THE VCASSCF APPROACH. The importance of the stretch–bend interactions in the principal isotopomer $^{12}C_2H_2$ was further studied using the VCASSCF method applied to the full 6D problem [270]. This study had essentially a methodological motivation and only considered the fundamental bands. As demonstrated in a previous study on the water molecule [174], mean-field approaches are well suited for analyzing the effects of mode couplings. Three levels of calculation were actually defined:

- *Effective treatment of all mode couplings*, by a single-configuration VSCF calculation, in which states $(v_1, v_2, v_3) = (0\ 0\ 0), (1\ 0\ 0), (0\ 1\ 0)$ and $(0\ 0\ 1)$ were optimised separately (with $v_4 = v_5 = 0$).

- *Direct treatment of the stretch–stretch interactions* and *effective treatment of the stretch–bend interactions*, by a VCASSCF calculation. Stretch–stretch and bend–bend excitations are explicitly included in the optimized multiconfiguration functions. The remaining interactions (stretch–bend) are involved through the effective mean-field treatment.

- *Direct treatment of the bend–bend and stretch–bend interactions*, by a VCASSCF-CI calculation based on the optimized wavefunction of the previous point, but enlarging the configuration space to stretch–bend interactions at the CI level.

The general trends are illustrated in Fig. 20, where one sees that the kind of relaxation considered at the different levels of mean-field approximation produces significant energy effects on the low-lying levels. While not investigated here, one expects larger effects at higher stretching excitations, explaining the large discrepancies observed in the pure stretching calculations (see Fig. 15).

6.   CCSD(T) QUARTIC FORCE FIELD FROM NUMERICAL DIFFERENCES. Large-scale *ab initio* calculations were performed by Martin et al. [210]. They use the coupled cluster method including all single and double excitations and treating the connected triples in a quasi-perturbative way (CCSD(T)). The force constants were determined by repeated central differences in the system of symmetry-adapted internal coordinates. This numerical procedure required 101 calculations corresponding to the symmetry unique points

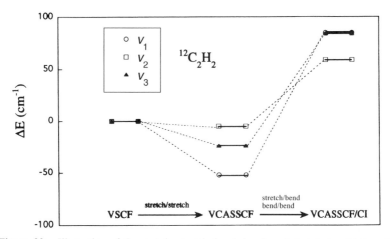

Figure 20. Illustration of the stretch–stretch, bend–bend, and stretch–bend interactions on the fundamental bands of the stretching modes v_1, v_2, and v_3 in $^{12}C_2H_2$, from mean-field variational calculations; ΔE is the energy shift with respect to the VSCF energies (adapted from Ref. 270).

needed for calculating the derivatives up to the fourth order. The force field initially derived in the internal system of coordinates was transformed in dimensionless normal coordinates. The spectroscopic analysis was carried out by second-order vibration–rotation perturbation theory with the SPECTRO program [375]; the resonance polyads were treated by diagonalization of small matrices in appropriate zeroth-order basis sets.

Several extended one-particle basis sets of the atomic natural orbital (ANO) [464], correlation consistent [465], and augmented correlation consistent [466,467] types were used to perform the CCSD(T) calculations. The effect of inner-shell correlation was investigated. The final *ab initio* quartic field computed by Martin et al. (M) [210] therefore combines the cubic and quartic force fields calculated at the CCSD(T) level with a quadratic force field and a geometry that additionally includes inner-shell correlation effects arising from the carbon 1s atomic orbitals. This force field is the most accurate one presently available for acetylene. It reproduces the fundamentals of five isotopomers with a mean absolute error of $1.3\,cm^{-1}$, and agrees with the equilibrium rotational constants to four decimal places. Particularly interesting is the comparison of the calculated interaction constants for the quartic resonances with those arising from the experimental study of Abbouti Temsamani and Herman (AH), further discussed in Section III.C [468,469] (see also the recent values by El Idrissi et al. [450]). Martin's values for the constants $K_{3/245}$, $K_{1/255}$, $K_{1/244}$, $K_{11/33}$, and $K_{44/55}$ agree within 15% with those of AH. Some of these constants computed by Borro et al. [470], from the SM force field, were in complete disagreement (e.g., $K_{3/245} = -48.13$ and $-16.7\,\mathrm{cm}^{-1}$ from Borro et al. and AH, respectively). Martin et al. show from a decomposition of the constants into cubic, quartic, and Coriolis contributions that the quartic component in the SM potential is responsible for the difference. This probably arises because of some distortion in the quartic part of the SM field, due to the severe restrictions imposed to the fit. As we pointed out in Section II.C.2.a.1, indeed, 15 among the 23 quartic constants were constrained to zero. The BCHM force field, refined using the same restrictions as the SM one, could be affected by similar errors. Comparing individual values of cubic and quartic constants from the R_1 and R_2 versions of BCHM with those of SM and of M. clearly indicates that some force fields constants are not yet well defined. As a single example, the f_{2211} force constant takes covers a wide range of values: -6.63, -28.35, 3.4, and $-1.29\,\mathrm{aJ/\AA^4}$, as calculated in R_1, R_2, SM, and M force fields, respectively, and therefore remains quite undetermined. This example clearly illustrates the difficulty for making the forward/backward approaches converge. It also illustrates how two highly refined *ab initio* force fields respectively, both carefully optimized to get converged results, the BCHM and M force fields, may diverge in their results. It is even probable that applying the same state-of-the-art variational method to calculate the spectrum from both force fields would lead to different predictions for the unobserved energy levels. Such calculations would be very interesting to perform, with comparisons to the observed eigenstates, to better characterize the *ab initio* PES. This merging of experimental and theoretical investigations remains an essential step, in any case.

c. Bending Dynamics. In this section we summarize another series of theoretical investigations, aiming at describing highly excited bending vibrational states. They are based on methods adapted to so-called floppy molecules. Along the excitation of the bending states, one has indeed to face H-orbiting states, involving the isomerization process from linear acetylene to vinylidene. The methods presented in the previous sections, based on a Taylor series expansion around the linear stationary point, are clearly not adapted. Specific coordinates and basis set are required, as well as a more global form of the PES covering at least the regions of the linear and vinilydene minima (M_1 and M_2) and the saddle point S_1 connecting them.

1. DVR VARIATIONAL CALCULATIONS. Two close variational calculations were published almost simultaneously, by Bentley et al. [296] and by Sibert and Mayrhofer [298]. They are both limited to the 5D planar motion and use the Jacobi 5D system of coordinates. These are well suited for describing highly excited bent states with one or two mobile H atoms. The two studies also

share the global HCC PES presented in Section II.C.1.f.3. The variational treatment is not identical but has common features. For instance, they both use DVR instead of the more common *finite basis* representation (FBR) for some of the nuclear degrees of freedom.

The calculations of Sibert and Mayrhofer are more successful, with converged eigenstates to within $2\,cm^{-1}$ up to $8770\,cm^{-1}$ above ZPE, and to within $5\,cm^{-1}$ up to $11740\,cm^{-1}$ above ZPE, respectively, corresponding to 12 and 16 quanta of bending excitation. The agreement with the results from El Idrissi et al. [450], who consider up to 12 quanta in the bends, is moreover very satisfactory (within better than $1\,cm^{-1}$).

The main interest in Bentley et al. lies in some methodological aspects, two of them briefly reported hereafter. In the first place, the kinetic-energy coupling term is neglected in the calculation. This approximation is justified by the mass factor multiplying this term, which is 25 times smaller than those multiplying the diagonal kinetic-energy terms. The resulting error on the lower energy states is estimated to be not larger than a few wavenumbers. In the second place, they incorporated symmetry adaptation using the G_8 CNPI group (see Section II.C.1.e.5) in their basis set and using a diagonalization–truncation scheme, leading to reduce the initial basis set size, of 2.7 millions, to only 647 5D basis functions. Their procedure nicely exemplifies one of the alternative strategies for avoiding the diagonalization of huge matrices, without significant loss of precision on the energy levels of interest. It is detailed in Section II.B.2.f.1.

The central difference between the two investigations lies in the representation of the bending motion, for which Sibert and Mayrhofer also use DVR. These authors demonstrate the drawbacks of the symmetry-adapted FBR used by Bentley et al. and show that DVR is nicely adapted to an adiabatic separation of the high-frequency stretching modes and the low-frequency bending modes. Sibert and Mayrhofer do not neglect the kinetic energy cross-term. They rather make it disappear by means of an appropriate affine transformation, leading to enhance the stretch–bend coupling. Their contraction scheme proceeds in two steps. In the first step, the 3D stretching Hamiltonian is solved at each bend DVR point, with a prediagonalization in the three individual 1D stretch DVR, defined by equally spaced points for all coordinates. A cutoff in the number of functions in the 3D stretching secular problem is applied to reduce the dimension N_c of the stretching basis set $\{\chi_1^n; l = 1, N_c\}$, where the n superscript refers to a given bent DVR point. In the second step, the full 5D vibrational H matrix is diagonalized, in the product basis set constructed by coupling the 3D $\{\chi\}$ stretching basis set with the 2D symmetrized bend DVR basis set of size $N_a \times N_s$. The convergence of the bend DVR basis was examined as a function of N_a and N_s, with a further restriction of the basis set by an energy cutoff. Interesting tests of adiabaticity were performed. The adiabatic approximation is obtained by constraining N_c to 1, with the following result: the maximum error induced by this approximation, for the pure bending states up to $v_4 + v_5 = 16$, is $400\,\mathrm{cm}^{-1}$. This error drops to $5\,\mathrm{cm}^{-1}$ when N_c is increased to 10, which means that a relatively large number of stretching channels is required for a proper description of the bending states. The mixing of the $N_c = 1$ channel is, however, found to be very weak, allowing a further contraction of the bend wavefunction to be performed.

Another interesting result in [298] is a test of the 5D planar approximation from a comparison of 5D results, using the BCHM potential with the 6D results of Bramley et al. [143]. Deviations do not exceed $10\,\mathrm{cm}^{-1}$.

2. ADIABATIC VARIATIONAL REFINEMENT. McCoy and Sibert [158,349] further investigated the bending dynamics, going beyond the planar geometry approximation. Such nonplanar calculations properly describe the important l-doubling resonances. A 6D Jacobi system is adopted for these calculations. It is somewhat different from the one reported in Fig. 12, by the fact that both Jacobi vectors pointing to the H atoms have a common origin, at the middle of the C–C bond. This system is identical to the one used by Kolos and Wolniewicz [471] in their nonadiabatic calculations on the H_2 molecule. The two heavy particles are in this case the hydrogen atoms and the two light, orbiting ones, are the electrons. The kinetic-energy operator of acetylene is thus identical to the one used for H_2, except that an affine transformation is applied to eliminate the kinetic-crossing term, as in previous work [298]. The goal of this work was not to solve the full 6D problem in this coordinate system, but rather to derive an adiabatic Hamiltonian for the bending motion and to demonstrate the adequacy of this approximation to represent the bending overtone structure of acetylene, up to sufficiently excited bending vibrations for exhibiting vinylidene-like motions. The adiabatic bending Hamiltonian is obtained by averaging the total vibration Hamiltonian over the ground state in the three stretch coordinates. The latter coordinates are treated as the fast coordinates, and the lowest eigenvalue of the stretch contribution to the Hamiltonian is solved as a function of the bending coordinate. The same DVR approach as in Ref. 298 is adopted. The adiabatic bending Hamiltonian is rewritten as an expansion in normal curvilinear normal coordinates, and expanded, up to the sixth order, found sufficient for obtaining converged energies to within $1\,\mathrm{cm}^{-1}$. Variational calculations are used to solve the secular equation. The bending basis set includes all the zero order states with $l = 0$ and a total number of quanta in the bending excitations $n_{\mathrm{bend}} \leq 24$, corresponding to about $18{,}000\,\mathrm{cm}^{-1}$ of vibrational excitation above ZPE. Note the small size of this basis set (252 states in the Σ_g^+ block of the Hamiltonian matrix). Enlarging the basis set to $n_{\mathrm{bend}} \leq 30$

lowers the vibrational energies by less than $0.5\,cm^{-1}$, up to $9000\,cm^{-1}$, which demonstrates the convergence of the variational calculations. The method was used to adjust an effective vibrational bend force field to improve the agreement between experimental and calculated eigenvalues. Only five experimental vibration transition energies for the bends in acetylene were used to adjust four quartic and a single sextic force constants; the quadratic terms were constrained to the values of Pliva [472]. The quality of the resulting bend force field is surprisingly good despite its simplicity. It qualitatively describes the vinylidene vibrations and quantitatively reproduces the acetylene vibrations. The results were successfully applied to assign experimental data involving highly excited bent states, recorded using simulated emission pumping (SEP) and dispersed fluorescence (DF) [473,474]. The observed clumps of lines are properly reproduced by the effective adiabatic model and assigned to states with 8–20 quanta of excitation in the bend. The deviations with the DF experimental transition energies [474] are smaller than $10\,cm^{-1}$, up to $10{,}600\,cm^{-1}$. The same data are considered in Section III.C, including rotation features.

Compared to the 6D BCHM refined force field, the present one applies to much higher energies. Within the limit of applicability of BCHM, the differences are smaller than $5\,cm^{-1}$. Compared to the HCC force field, operational up to at least $11{,}000\,cm^{-1}$ [151], the agreement is excellent (2% of accuracy up to $9000\,cm^{-1}$), but a slow degradation is observed at high excitation energies.

The simple 3D adiabatic Hamiltonian was used by Sibert and McCoy [158] to develop analytical models able to describe the bending dynamics. They reduced, by means of perturbation theory, the bending Hamiltonian to a local-mode Hamiltonian corresponding to two coupled hindered rotor Hamiltonians. They predicted that the local-mode dynamics of the bending motion first occurred at about $6000\,cm^{-1}$ of excitation.

d. Classical and Semiclassical Approaches of Intramolecular Dynamics. Some classical and semiclassical methods were used to relate the spectroscopy of highly vibrationally excited acetylene to some intramolecular dynamics features. This is only possible with this class of methods that circumvent the problems of dimensionality and the range of excitation required in the quantum-mechanical approaches.

1. CLASSICAL TRAJECTORIES. Holme and Levine [438,475] carried out classical trajectory calculations in the time domain together with quantum-algebraic computations on the 5D planar PES described in Section II.C.1.*f*.3, in order to simulate the SEP spectrum of acetylene. The existence of different timescales is suggested by the algebraic Hamiltonian and verified

by the statistical Fourier transform analysis of the classical trajectories. Four nested relaxation stages following the initial state preparation are differentiated and used to interpret some of the features of the experimental SEP spectrum [476–480]. At first, energy is predicted to flow, within less than 1 ps, between the two CH stretches only, then followed by a migration of the energy from the CH stretches to the CC stretch. Next, at some 10 ps, the energy reaches the *trans*-bend levels. Finally, at about 15 ps, total relaxation takes place with a concerted orbiting motion of both H atoms about the CC bond.

2. SEMICLASSICAL APPROACH. Prosmiti and Farantos [442] characterized the principal families of periodic orbits that emerge from the stationary points of the full 6D PES of acetylene, as well as periodic orbits from saddle-point bifurcation. The acetylene to vinylidene isomerization is of main concern in this analysis. The global PES of Murell et al. [249] and Jacobi 9D, 6D, and 5D coordinates are used in the calculations. The bifurcation diagrams of the periodic orbits reveal the regions of phase space where the dynamics are regular or chaotic, for linear acetylene, vinylidene and the region over these two isomers. The Gutzwiller's trace formula and the classical survival probability functions were also calculated and related with the spectroscopic findings. Further studies on this subject are being accomplished by Gaspard and co-workers [444,481,482].

3. NORMAL, LOCAL, AND PRECESSIONAL BENDING MODES. Another approach of the dynamics at high excitation level was carried out by Rose and Kellman [483], who analyzed the bending dynamics starting from the so-called spectroscopic fitting Hamiltonians. A standard single-resonance integrable quantum Hamiltonian is used, as for fitting experimental spectra. The primary pathway of energy transfer within the bends is assumed to be controlled by a single anharmonic resonance, of Darling–Dennison type as detailed in Section III. The related Hamiltonian is transformed into a classical Hamiltonian [484,485]. The resulting phase-space representation of the dynamics is plotted on the polyad phase sphere [486]. This classical Hamiltonian is subjected to bifurcation analysis and the results are represented with the help of molecular catastrophe maps [487]. It is shown that, in addition to normal and local bending modes, so-called precessional modes are encountered in acetylene, at high vibrational excitation. The difference between the three kinds of bending motion is illustrated in Fig. 21. The precessional motion, first characterized by Gray and Child [488] from classical trajectories on a model Hamiltonian, got its name from the appearance of the trajectories plotted in the vibrational coordinates (see Fig. 1 of Ref. 483). Polyads with the same number of bend quanta n_{bend} are

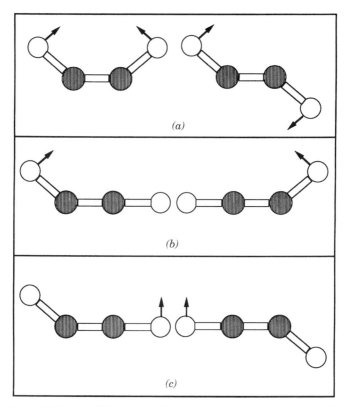

Figure 21. (*a*) Normal, (*b*) local, and (*c*) precessional modes in acetylene (adapted from Ref. 483).

plotted in phase space on the polyad phase sphere, allowing a determination of the normal, local and precessional character of the motion for any given quantum number. Local modes are found in the experimentally observed bend polyads when $n_{bend} \geq 14$ and precessional modes when $n_{bend} \geq 20$.

e.   Algebraic Approaches. Algebraic approaches, like the *vibron model*, were also used to analyze and interpret the vibration–rotation spectrum of acetylene [446–448,489]. Such methods are not detailed here, and we refer the reader to the books of Iachello and Levine [490] and Frank and Van Isacker [491] and also to the review paper of Oss [492]. Let us also cite the review by Kellman [493], who discusses and compares the usual spectroscopic fitting Hamiltonian models with other kinds of algebraic approaches, namely, the vibron. The vibron model is based on the concept of dynamical symmetries and uses the power of unitary Lie algebra, such as U(4), to provide compact effective Hamiltonians, which are well suited for applying fitting to spectroscopic data. The number of parameters defining such algebraic Hamiltonians is relatively small. For instance, Iachello et al. [447] have developed 8- and 16-parameter Hamiltonians able to describe 96 levels acetylene with root mean square deviations of 17 and $11\,\text{cm}^{-1}$, respectively. Iachello and Oss [448] also investigated the bending structure of acetylene. They used a two-dimensional approach based on the U(3) Lie algebra for representing the two coupled local benders and calculated the pure bending spectrum up to four quanta in the bending modes.

III. THE BACKWARD TRIP: FROM THE VIBRATION–ROTATION DATA TO THE HAMILTONIAN

A. Introduction

1. Strategy

a. Aim. The present section intends to provide the reader with a connection between the vibration–rotation spectrum and a full, effective Hamiltonian. For the first time in the review, the rotation degrees of freedom are introduced, both to account for additional effects and to get closer to the experimental spectroscopic accuracy and reality.

Concerning vibration, as discussed in Section II.B.2.*b*, a wide selection of models and coordinates exist, actually identical for the forward and backward trips. One might argue, as already pointed out in Section II, that one ought to consider problems of reduced size and each time extract related spectroscopic parameters, using the most adequate model. This procedure, however, produces constants that are more effective and hardly contribute to the global picture. A typical example is the production of the energy of a vibrational level from the rotation analysis of a single vibration–rotation band, only. Such a vibrational energy is most effective and do not truly help in providing useful predictions. Dealing with several overtones of the vibration helps a lot more in that respect, and possibly reveals internal couplings. Including other modes of vibration further allows the anharmonic couplings to be worked out. Eventually, a global treatment of all existing experimental data involves the full Hamiltonian and leads to reliable predictions. This is precisely our aim. We wish the selected model to allow for a description of the vibration–rotation energy levels as universal as possible, addressing all vibration degrees of freedom, in the "well behaved" as well as the "non-well-behaved" ranges of the PES. We furthermore ask the model to account as closely as possible for full experimental accuracy.

b. Selection of Coordinates. Any global perspective implies a zeroth-order model and coupling within the zeroth-order basis set. The zeroth-order set we selected in this section is the rectilinear normal mode basis set, for four reasons:

1. This set was demonstrated to handle full spectroscopic accuracy, in the lower infrared spectral range.
2. It merges all vibration degrees of freedom and can therefore account for the full density of vibration levels at higher excitation energy.
3. The problems encountered with anharmonicity and highly excited vibrations can be efficiently overcome using well-formulated methods.
4. The global perspective using rectilinear normal modes was recently reported in the literature with acetylene, opening interesting questions about new constants of the motion.

The rectilinear normal mode basis set is therefore adopted here to build the global model of the vibration–rotation energy levels. As pointed out before, we assume an isolated ground electronic state and the absence of large-amplitude motion, as occurring with acetylene if one excludes the isomerization process toward vinylidene.

The vibration–rotation Hamiltonian in rectilinear coordinates was actually built already almost 50 years ago, around the work of Nielsen and Amat, Howard, and Wilson. It was since significantly reshaped by Watson, as recalled in the next section in this introduction. This Hamiltonian is the one developed in the present chapter, following many published research books and review papers, such as those to which contributed, in alphabetic order, Aliev, Bunker, Kroto, Mills, Papousek, and Watson, all of which are fully referenced to, in the core of this Section. Such currently well-known developments are presented in this review in the less usual framework of a global analysis.

c.   Matrix Image of the Molecule.   The strategy is as follows. We model the vibration–rotation levels using a squared, diagonal matrix. We refer to this matrix as to the *matrix image of the molecule* (MIME). There is one such matrix for each J-value, with J taken, not considering nuclear and electron spins, as the full quantum number. Diagonal elements in the matrix correspond to the unperturbed energies expressed with the help of zeroth-order basis states. Off-diagonal elements represent interactions between these basis states. The full MIME thus contains as many rows and columns as there are energy levels in the molecule. Practically, the size of the MIME has to be reduced, depending on the available experimental information. The standard vibration and vibration–rotation MIME are built in Sections III.B and III.D, respectively. In section III.E, vibration–rotation transition intensities are discussed. All three sections devote a large place to the step by step, building of the Hamiltonian from the early stages. The emphasis is set in each section on internal couplings. Section III.C focuses on how to build the full MIME successfully. As in Section II, most features are illustrated with acetylene, among other species.

d. Matrix Treatment. As already discussed in Section II, it is today commonplace to use Hamiltonian matrices in high-resolution spectroscopy. This technique brings at least three decisive advantages over perturbation treatments:

1. The availability of high-speed computers with adequate subroutines is a strong support for the use of matrix Hamiltonians. It allows, in principle, problems of any dimension to be treated. Matrix diagonalization is to be performed once to calculate eigenvalues, and several times during fitting procedures. This step may, however, not always be trivial.

2. Matrices provide the required adaptability to deal with the complexity of the vibration–rotation pattern of the energy levels. Adequate off-diagonal terms can be plugged into the matrix to account for the relevant couplings within the experimental precision. The resulting eigenvalues integrate all such terms whose weight is accessible through the related eigenvectors produced during the diagonalization process of the matrix Hamiltonian. One should, however, remain careful not to immediately set up very large matrices that provide so many degrees of freedom in the fitting procedure that the final constants simply lose any physical context, even though they reproduce all data within the selected model. One should rather end up with such a global treatment only, once the need for the various off-diagonal matrix elements has been carefully tested on the basis of small-scale investigations performed in various energy ranges of the spectrum.

3. The third, decisive advantage of matrices, is that they include *all* orders of perturbation and account for coupling cases in which the difference between the zeroth-order energies E_i^0 of the interacting levels approaches zero: $E_m^0 - E_n^0 = \delta \to 0$. The perturbation treatment is then simply inadequate. When $\delta \to 0$ and $\alpha \gg \delta$, with α the coupling element, the energy of the eigenstates is approximated by $E \approx \frac{1}{2}(E_m^0 + E_n^0) \pm \alpha$. This formula is useful for estimation purposes. Second-order perturbation theory also remains an useful predictive tool when $\alpha \leq \delta$. For instance, it predicts that the effect of the coupling is proportional to δ^{-1}; that is, the smaller is the energy separation between zeroth-order states, the larger is the energy shift, for a given perturbation mechanism. This simple picture provides an important insight in, for instance, the anharmonic resonance problems dealt with later in this section. The diagonal anharmonicities may indeed be such that the coupled levels are closer in one range of the potential well than in another. The same coupling may thus show up efficiently in some regions of

the spectrum, but not in others. We also adopt the perturbation approach in this section to provide some insight into the potential constants.

Matrix elements are defined and matrix Hamiltonians are developed in this section, in agreement with both early and recent developments in the literature. The inclusion of off-diagonal matrix elements to account for systematic couplings often requires linear combinations of the initial basis states to be performed, to ensure the linear independence of the final basis states. As a result, the full matrix is block-diagonalized. This important simplification step may also be obtained, in some cases, by simply rearranging the ordering of the columns and lines. This technique is used when presenting the so-called vibrational level clustering, in Section III.C. The most faithful and physical MIME is defined to be the one containing the coupling terms of smallest value, in the optimal block-diagonalized form.

The developments detailed in the present section assume that, as stated in Section II, starting from the right-handed X, Y, Z laboratory axis system (LAS), one has defined the molecular axis system (MAS), both of which are set at the nuclear center of mass. The MAS contains the x, y, z axes related to those of the LAS using three (two for linear tops) Euler angles and cosine directions, as defined in Figs. 1 and 2 and in Table I. As detailed in Section II again, the most conventional translation and rotation Eckart conditions [48,456] are adopted. This choice allows the translation to be separated from the vibration–rotation degrees of freedom, and the inertia tensor to be diagonalized. Terms remain, however, that couple vibration and rotation, as we detail at the end of this section (III).

The reader is reminded that a list of symbols used in this review is provided in Appendix A. It is supposed to agree with the most recent recommendations on spectroscopic conventions and notations [494–496]. Also some units in this section are different from those in Section II, in particular energies are given in cm^{-1}.

2. Watsonian

a. *General Form.* Following the work of various authors including Wilson and Howard [358], Nielsen [359,360], Amat and co-workers [361–365], Oka and Morino [366,367], Kivelson and Wilson [368], and Watson [369–372], the vibration–rotation Hamiltonian for an asymmetric top molecule is, according to Watson [371] (discarding any type of tunnelling effects between like equilibrium configurations):

$$\hat{H} = \sum_{\alpha,\beta} \tfrac{1}{2}\hbar^2 (\hat{J}_\alpha - \hat{\pi}_\alpha)\mu_{\alpha\beta}(\hat{J}_\beta - \hat{\pi}_\beta) + \tfrac{1}{2}\sum_i \hat{P}_i^2 + V^{(Q)} + U \qquad (3.1)$$

with the following ensemble of definitions (with comments specific to linear tops added between parentheses):

$\mu_{\alpha\beta}$ — $(\alpha\beta)$ component of a modified reciprocal inertia tensor, function of Q

$\hat{J}_\alpha$ — αth component of the total angular momentum, in units of $\hbar$, excluding electronic and nuclear spins

$\hat{\pi}_\alpha$ — component of the vibrational angular momentum operator, in units of $\hbar$, along the rotating direction α, with $\hat{\pi}_\alpha = \sum_{i,j} \zeta_{ij}^\alpha \hat{Q}_i \hat{P}_j / \hbar = \sum_{i,j} \zeta_{ij}^\alpha q_i \hat{p}_j (\tilde{\omega}_j / \tilde{\omega}_i)^{1/2}$, with ζ_{ij}^α the Coriolis zeta constant, and $\zeta_{ij}^\alpha = -\zeta_{ji}^\alpha$

α, β — summed over the principal axis of rotation, a, b, and c, (b and c for a linear top), defined as those diagonalizing the regular inertia tensor (and perpendicular to the molecular axis for a linear top)

i — runs over the $3n - 6$ ($3n - 5$ for a linear top) normal modes of vibration, with n the number of nuclei

Q_i — ith normal coordinate

q_i — ith dimensionless normal coordinate $(q_i = \lambda^{1/4} \hbar^{-1/2} Q_i = \gamma_i^{1/2} Q_i)$

$\hat{P}_i$ — conjugate momentum of the ith normal coordinate $(= -i\hbar \partial/\partial Q_i)$

$\hat{p}_i$ — dimensionless conjugate momentum of the ith normal coordinate $(\hat{p}_i = \lambda_i^{-1/4} \hbar^{1/2} \hat{P}_i = \gamma_i^{-1/2} \hbar^{-1} \hat{P}_i)$, with $\gamma_i = \lambda_i^{1/2} / \hbar = \tilde{\omega}_i / \hbar = 2\pi\omega_i / \hbar = c\tilde{\omega}_i / \hbar = 2\pi c\tilde{\omega}_i / \hbar$ with the commutation relations $[q_i, \hat{p}_j] = i\delta_{ij}, [q_i, q_j] = 0, [\hat{p}_i, \hat{p}_j] = 0$

$V^{(Q)}, V^{(q)}$ — nuclear potential energy, function of $3n - 6$ vibrational coordinates (even for linear tops because of their cylindrical symmetry about the molecular axis)

$[V^{(en)}(q_r)] = V^{(en)}(Q_r) V^{(Q)}[V^{(q)}] + E_n^e(Q)[E_n^e(q)]$ — total potential energy, including the electronic contribution (E_n^e) within the Born Oppenheimer approximation

U — small, mass-dependent correction to the vibrational potential energy, with $U = -(\hbar^2/8) \sum_\alpha \mu_{\alpha\alpha}$ ($= 0$ for a linear top)

$\bar{\omega}_i$	harmonic angular vibration frequency (Hz), associated with ith normal mode of vibration
$\tilde{\omega}_i$	harmonic angular vibration frequency (cm^{-1}), associated with ith normal mode of vibration
ω_i	harmonic vibration frequency (Hz), associated with ith normal mode of vibration
$\tilde{\omega}_i$	harmonic vibration frequency (cm^{-1}), associated with ith normal mode of vibration

[The term *vibration frequency* will be used throughout the review to characterize the vibrational energy, independently of the unit (Hz or cm^{-1}).]

Several relationships between the parameters in the Hamiltonian exist, called the sum rules [371,497]. The vibration–rotation interaction terms appear in the centrifugal coupling, through the Q dependence of the $\mu_{\alpha\beta}$ tensor, and in the Coriolis coupling, through the cross-terms $-\hbar^2 \sum_{\alpha\beta} \mu_{\alpha\beta} \hat{J}_\alpha \hat{\pi}_\beta$. The present choice of coordinates allows the $\mu_{\alpha\beta}$ terms to be evaluated as a product of tensors $\mu = (I'')^{-1} I^e (I'')^{-1}$, with the following components:

$$I^e_{\alpha\beta} = \sum_i m_i [\delta_{\alpha\beta} \sum_\gamma (r^e_{i\gamma})^2 - r^e_{i\alpha} r^e_{i\beta}] \tag{3.2}$$

where the superscript "e" refers to the equilibrium nuclear configuration, and

$$I''_{\alpha\beta} = I''_{\beta\alpha} = I^e_{\alpha\beta} + \frac{1}{2}\sum_i \left(\frac{\partial I_{\alpha\beta}}{\partial Q_i}\right)_e Q_i = I^e_{\alpha\beta} + \frac{1}{2}\sum_i a^{\alpha\beta}_i Q_i \tag{3.3}$$

The last term in (3.2) vanishes if the principal inertial axis system is selected. In symmetric tops, the problem is simpler than in asymmetric tops because of the equality of two of the principal moments of inertia. It is at the same time more complex because of the existence of degenerate modes of vibration. In the special case of linear species, the situation is extreme. One of the principal moments of inertia is zero, in the fixed equilibrium configuration, and one rotational degree of freedom is replaced by one vibrational degree of freedom. The detailed structure of the Watsonian is, in the end, more complicated for linear and symmetric tops than for asymmetric tops.

It is conventional to write the Watsonian for a linear molecule using an isomorphic Hamiltonian [372,498] to describe the kinetic energy, with the undetermined Euler angle set to zero:

$$\hat{H}^{\text{lin}} = \frac{1}{2}\frac{\hbar^2}{I'} [(\hat{J}_b - \hat{\pi}_b)^2 + (\hat{J}_c - \hat{\pi}_c)^2] + \frac{1}{2}\sum_{i=1}^{3n-5} \hat{P}_i^2 + V^{(Q)} \tag{3.4}$$

TABLE VI
Lower-order Terms of the Vibration–Rotation Hamiltonian[a]

$$H_{20}/hc = \frac{1}{2}\sum_i \tilde{\omega}_i(q_i^2 + \hat{p}_i^2)$$

$$H_{30}/hc = \frac{1}{6}\sum_{ijk} \phi_{ijk} q_i q_j q_k$$

$$H_{40}/hc = \frac{1}{24}\sum_{ijkl} \phi_{ijkl} q_i q_j q_k q_l + \sum_{ijkl\alpha} B_\alpha^e \zeta_{ij}^\alpha \zeta_{kl}^\alpha \left(\frac{\tilde{\omega}_j \tilde{\omega}_l}{\tilde{\omega}_i \tilde{\omega}_k}\right)^{1/2} q_i \hat{p}_j q_k \hat{p}_l$$

$$H_{21}/hc = \sum_{ij} R_i^j q_i \hat{p}_j$$

$$H_{31}/hc = \sum_{ijk} R_{ij}^k q_i \hat{p}_k q_j$$

$$H_{02}/hc = \sum_\alpha B_\alpha^e \hat{J}_\alpha^2$$

$$H_{12}/hc = \sum_i R_i q_i$$

$$H_{22}/hc = \sum_{ij} R_{ij} q_i q_j$$

$$H_{32}/hc = \sum_{ijk} R_{ijk} q_i q_j q_k$$

R Operators

$$R_i^j = -2\left(\frac{\tilde{\omega}_j}{\tilde{\omega}_i}\right)^{1/2}\sum_\alpha B_\alpha^e \zeta_{ij}^\alpha \hat{J}_\alpha = -\left(\frac{\tilde{\omega}_j}{\tilde{\omega}_i}\right) R_j^i$$

$$R_{ij}^k = R_{ji}^k = \left(\frac{\tilde{\omega}_k}{\tilde{\omega}_i \tilde{\omega}_j}\right)^{1/2}\sum_{\alpha\beta} (\tilde{\omega}_i^{3/2} C_i^{\alpha\beta}\zeta_{jk}^\beta + \tilde{\omega}_j^{3/2} C_j^{\alpha\beta}\zeta_{ik}^\beta)\hat{J}_\alpha$$

$$R_i = -\tilde{\omega}_i \sum_{\alpha\beta} C_i^{\alpha\beta}\hat{J}_\alpha \hat{J}_\beta$$

$$R_{ij} = R_{ji} = \frac{3}{8}\tilde{\omega}_i \tilde{\omega}_j \sum_{\alpha\beta\gamma} \frac{(C_i^{\alpha\gamma} C_j^{\gamma\beta} + C_j^{\alpha\gamma} C_i^{\gamma\beta})\hat{J}_\alpha \hat{J}_\beta}{B_\gamma^e}$$

$$R_{ijk} = R_{kji} = -\frac{1}{4}\tilde{\omega}_i \tilde{\omega}_j \tilde{\omega}_k \sum_{\alpha\beta\gamma\delta} \frac{(C_i^{\alpha\gamma} C_k^{\delta\beta} + C_k^{\alpha\gamma} C_i^{\delta\beta})C_j^{\gamma\delta}\hat{J}_\alpha \hat{J}_\beta}{B_\gamma^e B_\delta^e}$$

Coefficients

$$B_\alpha^e = \hbar^2 \mu_{\alpha\alpha}^e / 2hc = \hbar^2/2hcI_\alpha$$

$$C_i^{\alpha\beta} = a_i^{\alpha\beta}/2\gamma_i^{3/2}I_\alpha I_\beta = -B_i^{\alpha\beta}/\tilde{\omega}_i$$

$$\phi_{ijk} = (\delta^3 V/\delta q_i \delta q_j \delta q_k)_e/hc$$

$$\phi_{ijkl} = (\delta^4 V/\delta q_i \delta q_j \delta q_k \delta q_l)_e/hc$$

[a]All constants in reciprocal centimeters (cm^{-1}).

Source: Adapted from Aliev and Watson [456].

with all terms defined as previously ($z = a$ is selected here as the internuclear axis), but $I' = (I'')^2/I^e$ and $I^e = I_{aa} = I_{bb}$. For this isomorphic Hamiltonian, only those eigenvalues belonging to the zero eigenvalues of the operator $(\hat{J}_a - \hat{\pi}_a)$ are physically significant. The condition actually concerns the full operator, which is $(\hat{J}_a - \hat{\pi}_a - \hat{L}_a)$, taking into account the electronic angular momentum operator $(\hat{L}_a)$. This condition, called the *Sayvetz condition* [499] implies that the quantum numbers generated by these operators for a symmetric or a linear top; thus respectively k, l, and Λ (in units of $\hbar$), satisfy the rule $k = l + \Lambda$, or $k = l$, if $\Lambda = 0$. If various degenerate modes of vibration generate angular momenta, then $k = \sum_i l_i$, if $\Lambda = 0$. Both k and l are signed numbers.

b. Fractionation. One can split the Watsonian for both asymmetric and linear tops in various terms:

$$\hat{H} = \hat{H}_v^{\text{har}} + \hat{H}_v^{\text{anh}} + \hat{H}_r^{\text{rig}} + \hat{H}_r^{\text{nr}} + \hat{H}_{\text{rovib}}^{\text{Cor}} \tag{3.5}$$

Using $(q, \hat{p})^n \hat{J}_\alpha^m$, with n and m thus referring to the power of the vibrational operators q and $\hat{p}$ and to the power of the total angular momentum $\hat{J}$, respectively:

$$\hat{H} = \sum_{n,m} H_{nm}, \quad \text{with} \quad \hat{H}_v^{\text{har}} = H_{20}, \quad \hat{H}_v^{\text{anh}} = H_{30} + H_{40},$$
$$\hat{H}_r^{\text{rig}} = H_{02}, \quad \hat{H}_r^{\text{nr}} = H_{12} + H_{22}, \quad \hat{H}_{\text{rovib}}^{\text{Cor}} = H_{21} \tag{3.6}$$

considering the first terms in the development only. The correspondence was explicitly given by Aliev and Watson [456] and is adapted to the present notation in Table VI (see also Ref. 500).

We have now basic terms to plug in the MIME, which we need to develop and classify in order to provide a tractable picture of the energy levels.

B.   Vibrational Terms

1.   Basic Features

In a classical picture, applying Newton's second law to a spring with restoring force $-kx$ and no external force (k is the label for a force constant in this Section III.B.1.) leads to

$$F = \dot{p} = \frac{d}{dt}mu = m\frac{d^2}{d^2t}x = a_0 + a_1x + a_2x^2 + \cdots = -kx = m\frac{d^2x}{dt^2} \tag{3.7}$$

which is Hooke's law if keeping only the linear term in the development, after setting a_0 to zero by an appropriate choice of coordinates, with m the mass and u the speed. Therefore:

$$V^{(x)} = -\int F\,dx = \frac{1}{2}kx^2 + c \quad \text{and} \quad k = \frac{1}{x}\frac{dV^{(x)}}{dx} \tag{3.8}$$

thus leading to the well-known parabola dependence of the motion.
Using the solution $x = x_0\cos(\bar{\omega}t + \varphi)$, then

$$F = -kx = m\frac{d^2x}{dt^2} = -mx_0\bar{\omega}^2\cos(\bar{\omega}t + \varphi)$$

$$\text{and} \quad k = m\bar{\omega}^2 \text{ or } \bar{\omega} = 2\pi\omega = \sqrt{\frac{k}{m}} \tag{3.9}$$

These elementary results can be exploited in three different ways:

1. Introducing the potential- and kinetic-energy contributions and expressing the work W_{ab} performed when moving along the parabola from point a to point b:

$$W_{ab} = \int_a^b F\,dx$$

$$\underline{V^{(x)}(x)} \qquad\qquad\qquad \underline{T(\dot{x})}$$

$$F(x) = \frac{-dV^{(x)}(x)}{dx} \qquad\qquad F = \frac{dp}{dt} = \frac{d(mu)}{dt}$$

$$W_{ab} = -\int_a^b \frac{dV^{(x)}dx}{dx} \qquad\qquad W_{ab} = \int_a^b \frac{d(1/2)mu^2}{dt}$$

$$V^{(a)} - V^{(b)} \qquad = W_{ab} = \qquad T^{(b)} - T^{(a)}$$

$$E_{\text{tot}} = T + V^{(x)} = T_a + V^{(a)} = T_b + V^{(b)} = \frac{1}{2}mu^2 + \frac{1}{2}kx^2 = \frac{p^2}{2m} + \frac{1}{2}m\bar{\omega}^2x^2 \tag{3.10}$$

The classical energy of the harmonic oscillator is thus a constant term, as expected for a conservative system, expressed as the sum of the potential and kinetic terms. One has

At the turning points $(x = \pm A)$: $\quad V^{x_{\max}} = E_{\text{tot}} \quad\quad T = 0$
At the bottom of the parabola $(x = 0)$: $V^{(x)} = 0 \quad\quad T_{\max} = E_{\text{tot}}$
In between $(|A| > |x| > 0)$: $\quad\quad T + V^{(x)} = E_{\text{tot}}$ with $T > 0, V^{(x)} > 0$

Thus, classically, the relative contribution of these terms to the energy depends on the value of the stretching coordinate. The same comment applies to the quantum-mechanical energy of the harmonic and anharmonic oscillators. As discussed in Section II, more generally, the expression and relative contribution of the kinetic and potential energies depend on the type of coordinate selected. This applies, in particular, to the anharmonic resonant terms [54]. Using normal mode coordinates built on Cartesian coordinates, as in the present chapter, such terms appear exclusively in the potential energy contribution [57].

2. We derived

$$\bar{\omega} = 2\pi\omega = \sqrt{\frac{k}{m}} = \sqrt{\frac{1}{mx}\frac{dV^{(x)}}{dx}} \qquad (3.11)$$

The classical parameters $\bar{\omega}$ and k are thus related to the first derivative of the shape of the potential. They therefore both effectively *decrease* when the shape of $V^{(x)}$ opens up toward dissociation at larger internuclear distances, as described, for example, by the familiar Morse-type potential curve. This picture highlights the role of anharmonicity. It also tells that the relative vibrational energies, say, at the bottom of the well, might change when climbing up the vibrational ladder, because of anharmonicity, which differs from one vibration to another. The efficiency of the resonance terms coupling vibrations, which is related to their relative energies in a simple perturbative approach, may therefore vary within the same electronic state.

3. The classical and semiclassical pictures are most relevant in the present context. They favor, in particular, the correspondence between the time- and energy-dependent status of the vibrational problem. A wide literature becomes available on this subject [111,493,501], which deals, among other topics, with periodic orbits, vibrograms, bifurcation toward local modes and chaos, and new constants of the motion. The latter developments have provided relevant connections toward effective molecular spectroscopic Hamiltonians, and we refer to some of them when dealing with vibrational clusters or polyads, in Section III.C.

There are four different types of contribution in the vibrational Watsonian:

1. Some of the operators in the vibrational Watsonian generate diagonal contributions in the MIME. This is the case of the terms leading to the harmonic oscillator expression which is considered hereafter in Section III.B.2.*a*.

2. Other terms lead to off-diagonal contributions that are merged into the diagonal terms in the MIME, after applying perturbation or contact

transformation. This is illustrated in section III.B.2.*b* with the derivation of the Dunham expression.

3. A third category involves all those coupling terms that are systematically present in the Hamiltonian matrix, which we refer to in the present work as *structural resonances*. They include the interaction terms between the levels with the same energy such as the symmetry degenerate levels. Contributions arising from degenerate vibrations, such as the *l*-resonance interaction are of this kind. They are considered in Section III.B.4. We also like to call structural resonances the systematic CH interbond anharmonic couplings, such as those of the Darling–Dennison type (see Section III.B.3), occurring in molecules as C_2H_2 and C_2H_4, which we further detail in Section III.C. Those were previously included in the so-called accidental resonances. It is, however, one objective of the present review to demonstrate that this label has to be revised and that the initial Hamiltonian matrix has to be built to include such coupling terms in the initial blocks.

4. Whenever vibrational levels that do not belong to the same block of the initial Hamiltonian are degenerate or nearly degenerate in energy, one usually refers to *accidental resonances*. In those cases, off-diagonal matrix elements have to be inserted in the MIME, à la carte. Their role is accounted for when performing numerical diagonalization, during fitting procedures to experimental data. Most anharmonic resonances belong to this latter type of contribution. They are considered in Section III.B.3. A full Section, III.C, is devoted to their role in shaping the MIME into blocks, as to the one of the structural resonances.

2. Diagonal Terms

We first consider those diagonal terms of vibrational nature in the Watsonian, arising from the harmonic oscillator (H_{20}). We then include the diagonal contribution arising from anharmonic terms. The present section and the next one deal only with nondegenerate modes of vibration and are therefore appropriate for asymmetric tops. Two-dimensional degenerate modes are considered in Section, III.B.4. As a general introductory picture to the present section, Fig. 22 presents the transmittance spectrum of acetylene $(^{12}C_2H_2)$ between 600 and 3400 cm^{-1}.

a. The Harmonic Oscillator $(\hat{H}_v^{har})$. To obtain the quantum model, we need to insert the quantum-mechanical correspondents of x and p in the classical expressions. Therefore, with adequate elementary transformations, for a diatomic species AB, with s the unique internal coordinate, μ the reduced

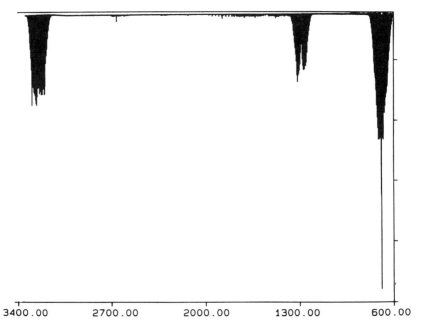

3400.00   2700.00   2000.00   1300.00   600.00

Figure 22. Transmittance spectrum of $^{12}C_2H_2$ (Herman, unpublished data) showing the ν_5 (*cis*-bend mode) and ν_3 (antisymmetric CH stretch mode) fundamental bands as well as the $\nu_4 + \nu_5$ (*trans* + *cis*-bend modes) combination band, around 720, 3300, and 1300 cm^{-1}, respectively (energy scale in cm^{-1}).

mass, defined as $\mu^{-1} = m_A^{-1} + m_B^{-1}$, $Q = \mu^{1/2}s$, and $\lambda = 2\Pi c\tilde{\omega}$:

$$\hat{H}_v^{\mathrm{har}} = -\frac{\hbar^2}{2\mu}\frac{d^2}{ds^2} + \frac{1}{2}ks^2 = \frac{1}{2}\mu^{-1}(-i\hbar\partial_\partial)^2 + \frac{1}{2}ks^2 \qquad (3.12)$$

$$\hat{H}_v^{\mathrm{har}} = \frac{1}{2}(-i\hbar\partial_Q)^2 + \frac{1}{2}\lambda Q^2$$

$$\hat{H}_v^{\mathrm{har}} = \frac{1}{2}\hat{P}^2 + \frac{1}{2}\lambda Q^2$$

$$\hat{H}_v^{\mathrm{har}} = \frac{1}{2}\lambda^{1/2}\hbar(\hat{p}^2 + q^2)$$

$$\hat{H}_v^{\mathrm{har}} = \frac{1}{2}hc\tilde{\omega}(\hat{p}^2 + q^2) \qquad (3.13)$$

Defining the shift-up and shift-down operators a^+ and a^-:

$$a_i^+ = \frac{1}{\sqrt{2}}(q_i - i\hat{p}_i) \tag{3.14}$$

$$a_i^- = \frac{1}{\sqrt{2}}(q_i + i\hat{p}_i) \tag{3.15}$$

with

$$\hat{N} = a^+a^-, \qquad \hat{N}+1 = a^-a^+, \qquad [a^-a^+] = 1,$$
$$[\hat{N}, a^+] = a^+, \qquad [\hat{N}, a^-] = -a^- \tag{3.16}$$

the harmonic oscillator Hamiltonian is

$$\hat{H}_v^{\text{har}} = a^+a^- + \frac{1}{2} = a^-a^+ - \frac{1}{2} = \hat{N} + \frac{1}{2} \tag{3.17}$$

The action of the operators, with $|v\rangle$ the normalized vibrational wavefunction and v the vibrational quantum number, is [18]

$$a^+|v\rangle = \sqrt{v+1}|v+1\rangle \qquad a^-|v\rangle = \sqrt{v}|v-1\rangle \tag{3.18}$$

and the matrix elements are

$$\langle v'|a^-|v\rangle = \sqrt{v}\,\delta_{v',v-1}$$
$$\langle v'|a^+|v\rangle = \sqrt{v+1}\,\delta_{v',v+1} \tag{3.19}$$

Thus, the product a^+a^- will successively lower and raise v by 1 and the related matrix element will be nonzero for $v' = v$, only. Similarly

$$\langle v'|q|v\rangle = \sqrt{\frac{\hbar}{2m\tilde{\omega}}}[\sqrt{v+1}\,\delta_{v',v+1} + \sqrt{v}\,\delta_{v',v-1}]$$
$$\langle v'|\hat{p}|v\rangle = i\sqrt{\frac{\hbar}{2m\tilde{\omega}}}[\sqrt{v+1}\,\delta_{v',v+1} + \sqrt{v}\,\delta_{v',v-1}] \tag{3.20}$$

Those various expressions allow all required Hamiltonian matrix elements to be calculated. The most useful ones are listed in Table VII, with only one equation for all operators each time, separated by a comma, giving the same eigenvalues (see also Refs. 502 and 503).

TABLE VII
Matrix Elements of Vibrational Operators

$\langle v|q, -i\hat{p}|v-1\rangle = (v/2)^{1/2}$

$\langle v|q, i\hat{p}|v+1\rangle = [(v+1)/2]^{1/2}$

$\langle v|q^2, -\hat{p}^2|v-2\rangle = \frac{1}{2}[v(v-1)]^{1/2}$

$\langle v|q^2, \hat{p}^2|v\rangle = v + \frac{1}{2}$

$\langle v|q^2, -\hat{p}^2|v+2\rangle = 1/2[(v+1)(v+2)]^{1/2}$

$\langle v|q\hat{p}, \hat{p}q|v-2\rangle = i/2[v(v-1)]^{1/2}$

$\langle v|q\hat{p}, -\hat{p}q|v\rangle = i/2$

$\langle v|q\hat{p}, \hat{p}q|v+2\rangle = -i/2[v(v+1)(v+2)]^{1/2}$

$\langle v|q^3, i\hat{p}^3|v-3\rangle = [v(v-1)(v-2)/8]^{1/2}$

$\langle v|q^3, -i\hat{p}^3|v-1\rangle = 3[v/2]^{3/2}$

$\langle v|q^3, i\hat{p}^3|v+1\rangle = 3[(v+1)/2]^{3/2}$

$\langle v|q^3, -i\hat{p}^3|v+3\rangle = [(v+1)(v+2)(v+3)/8]^{1/2}$

$\langle v|q^4, \hat{p}^4|v-4\rangle = \frac{1}{4}[v(v-1)(v-2)(v-3)]^{1/2}$

$\langle v|q^4, -\hat{p}^4|v-2\rangle = \frac{1}{2}(2v-1)[v(v-1)]^{1/2}$

$\langle v|q^4, \hat{p}^4|v\rangle = \frac{3}{4}(2v^2+2v+1)$

$\langle v|q^4, -\hat{p}^4|v+2\rangle = \frac{1}{2}(2v+3)[(v+1)(v+2)]^{1/2}$

$\langle v|q^4, \hat{p}^4|v+4\rangle = \frac{1}{4}[(v+1)(v+2)(v+3)(v+4)]^{1/2}$

$\langle v|q^2\hat{p}^2, \hat{p}^2q^2|v\rangle = \left[\frac{1}{2}\left(v+\frac{1}{2}\right)^2 + \frac{1}{8}\right]$

The eigenvalues of the harmonic oscillator are

$$\langle v|\hat{H}_v^{\text{har}}|v\rangle = \frac{1}{2}hc\tilde{\omega}\langle v|\hat{p}^2 + q^2|v\rangle = E_v^{\text{har}} = hc\tilde{\omega}\left(v + \frac{1}{2}\right) \qquad (3.21)$$

identical to

$$\langle v'|\hat{H}_v^{\text{har}}|v\rangle = hc\tilde{\omega}\left\langle v\left|a^+a^- + \frac{1}{2}\right|v\right\rangle = hc\tilde{\omega}\left\langle v\left|\hat{N} + \frac{1}{2}\right|v\right\rangle = E_v^{\text{har}}$$

$$\frac{E_v^{\text{har}}}{hc} = \tilde{\omega}[\sqrt{v}\langle v|a^+|v-1\rangle] + \frac{1}{2} \qquad (3.22)$$

$$\frac{E_v^{\text{har}}}{hc} = \tilde{\omega}\left(v + \frac{1}{2}\right)$$

b. *The Anharmonic Oscillator* ($\hat{H}_v^{\text{anh diag}}$). The vibrational Watsonian includes higher-order terms, H_{30} and H_{40}. Their contribution can be calculated by perturbation theory or by applying contact transformation [47], with both methods generating identical results. In either case, higher-order, sometimes off-diagonal contributions in the Watsonian are merged into diagonal terms in a transformed Hamiltonian. This is a crucial step toward a more compact global model. The parameters become more effective, however, and their physical content becomes less easy to evaluate.

As highlighted by Mills [373], if two off-diagonal terms of the Watsonian, H_{kl} and H_{mn}, are treated by second-order perturbation theory; then

$$\frac{1}{hc} \frac{\langle v|H_{kl}|v'\rangle \langle v'|H_{nm}|v\rangle}{(E_v^0 - E_{v'}^0)} \tag{3.23}$$

they generate terms in the diagonal matrix elements of the MIME:

$$\langle v|\tilde{h}_{k+n-2,l+m}|v\rangle \quad \text{and} \quad \langle v|\tilde{h}_{k+n,l+m-1}|v\rangle \tag{3.24}$$

The superscript $\sim$ is inserted, as in the literature, to identify the transformed Hamiltonian. In the present review, the factor $1/hc$ is included in $\tilde{h}_{nm}$. We also switch from capital letters (H) to lowercase (h), when going from the initial terms in the Watsonian to the effective terms including the factor $1/hc$.

We illustrate this transformation hereafter by generating the first anharmonic term in the diagonal expression of the vibrational energy for a diatomic molecule. We use the H_{30} and H_{40} contributions, which are merged into one single $\tilde{h}_{40}$ term using first- and second-order perturbation theory, respectively.

1. ONE-DIMENSIONAL DUNHAM EXPANSION. Developing the potential $V^{(q)}$ in series, and considering a system with one single vibrational degree of freedom, we obtain

$$\frac{V^{(s)}}{hc} = \frac{1}{2} f_{ss} s^2 + \frac{1}{6} f_{sss} s^3 + \frac{1}{24} f_{ssss} s^4 + \cdots \tag{3.25}$$

$$\frac{V^{(s)}}{hc} = \hat{H}_v^{\text{har}} + \hat{H}_v^{\text{anh}} \tag{3.26}$$

$$\frac{V^{(s)}}{hc} = \hat{H}_v^{\text{quadratic}} + \hat{H}_v^{\text{cubic}} + \hat{H}_v^{\text{quartic}} \tag{3.27}$$

where $f_{s^n} \equiv (\partial^n V^{(s)}/\partial s^n)_{s=s_0}$ and is therefore zero for $n=1$ because corresponding to the first derivative of the parabola at the bottom of the well. (V_0^E) which is arbitrarily set to zero.

$$\hat{H}_v^{\text{anh}} = \frac{1}{6} f_{sss} s^3 + \frac{1}{24} f_{ssss} s^4 + \cdots \tag{3.28}$$

$$\hat{H}_v^{\text{anh}} = hc \frac{1}{6} \phi_3 q^3 + \frac{1}{24} \phi_4 q^4 + \cdots \tag{3.29}$$

with

$$q = \gamma^{1/2}Q = \gamma^{1/2}\mu^{1/2}s = \alpha^{1/2}s, \phi_3 = \frac{f_{sss}}{hc\alpha^{3/2}}(cm^{-1})$$

$$\text{and}\quad \phi_4 = \frac{f_{sss}}{hc\alpha^2}(cm^{-1}) \tag{3.30}$$

One defines the transformed, effective operator $\tilde{h}_{40}$ by developing $\hat{H}_v^{\text{anh}}$, namely, H_{30} and H_{40}, in perturbation theory:

$$(hc)\tilde{h}_{40} = H_{40} + \frac{H_{30} \times H_{30}}{\Delta E_v} \tag{3.31}$$

that is, using the first order of perturbation for the term in q^4 (H_{40}) and the second order for the term in q^3 (H_{30}). Therefore

$$\frac{1}{24}\phi_4\langle v|q^4|v\rangle = \frac{1}{16}\phi_4\left(v+\frac{1}{2}\right)^2 + \frac{1}{64}\phi_4 \tag{3.32}$$

$$\left\langle v\left|\frac{1}{6}\phi_3 q^3\right|v\right\rangle = \left(\frac{1}{6}\phi_3\right)^2 \times \left[\frac{\langle v|q^3|v+1\rangle\langle v+1|q^3|v\rangle}{(E_v^0 - E_{v+1}^0)/hc}\right.$$

$$+ \frac{\langle v|q^3|v-1\rangle\langle v-1|q^3|v\rangle}{(E_v^0 - E_{v-1}^0)/hc} + \frac{\langle v|q^3|v+3\rangle\langle v+3|q^3|v\rangle}{(E_v^0 - E_{v+3}^0)/hc}$$

$$\left. + \frac{\langle v|q^3|v-3\rangle\langle v-3|q^3|v\rangle}{(E_v^0 - E_{v-3}^0)/hc}\right] \tag{3.33}$$

$$\left\langle v\left|\frac{1}{6}\phi_3 q^3\right|v\right\rangle = -\frac{5}{48}\frac{\phi_3^2}{\tilde{\omega}}\left(v+\frac{1}{2}\right)^2 - \frac{7}{288}\phi_3^2 \tag{3.34}$$

Thus

$$\frac{E_v^{\text{har}} + E_v^{\text{anh}}}{hc} = G(v) = \tilde{\omega}\left(v+\frac{1}{2}\right) + x\left(v+\frac{1}{2}\right)^2 \tag{3.35}$$

with

$$x = -\frac{5}{48}\frac{\phi_3^2}{\tilde{\omega}} + \frac{1}{16}\phi_4. \tag{3.36}$$

The development of further terms in $\hat{H}_v^{\text{anh}}$ adds contributions in the energy in successive powers of $(v+\tfrac{1}{2})$:

$$G(v) = \tilde{\omega}(v+\tfrac{1}{2}) + x(v+\tfrac{1}{2})^2 + y(v+\tfrac{1}{2})^3 + \cdots \qquad (3.37)$$

which is the well-known Dunham expansion [163].

The vibrational energy $[G(v)]$ and the rotational energy, to be introduced later $[F(J)]$, are given (in cm^{-1}) in the present review, except when stated otherwise.

In terms of the second quantization, the Dunham expression can be set using the following Hamiltonian operators, with the summation over $3n-6$ normal modes of vibration for a nonlinear system with n nuclei:

$$\begin{aligned}
\frac{\hat{H}}{hc} &= \sum_i \tilde{\omega}_i\left(a_i^+ a_i^- + \frac{1}{2}\right) + \sum_{i,j} x_{ij}\left(a_i^+ a_i^- + \frac{1}{2}\right)\left(a_j^+ a_j^- + \frac{1}{2}\right) \\
&+ \sum_{i,j,k} y_{ijk}\left(a_i^+ a_i^- + \frac{1}{2}\right)\left(a_j^+ a_j^- + \frac{1}{2}\right)\left(a_k^+ a_k^- + \frac{1}{2}\right)
\end{aligned} \qquad (3.38)$$

Using (3.37); we obtain

$$\begin{aligned}
G(0) &= \tfrac{1}{2}\tilde{\omega} + \tfrac{1}{4}x + \tfrac{1}{8}y + \cdots \\
G(1) &= \tfrac{3}{2}\tilde{\omega} + \tfrac{9}{4}x + \tfrac{27}{8}y + \cdots \\
G(2) &= \tfrac{5}{2}\tilde{\omega} + \tfrac{25}{4}x + \tfrac{125}{8}y + \cdots
\end{aligned} \qquad (3.39)$$

Therefore, ΔG is

$$\begin{aligned}
G(1) - G(0) &= \tilde{\omega} + 2x + \tfrac{26}{8}y + \cdots \\
G(2) - G(1) &= \tilde{\omega} + 4x + \tfrac{98}{8}y + \cdots
\end{aligned} \qquad (3.40)$$

The values of $\tilde{\omega}$, x, and y usually decrease by two orders of magnitude each time [455]. The classical treatment of Section III.B.1 predicts that the spacing ΔG, namely, the *effective* vibrational frequency, decreases when v increases. This implies that the contribution of the first diagonal anharmonicity, $+x(v+\tfrac{1}{2})^2$ as written here and $-x(v+\tfrac{1}{2})^2$ as sometimes written in the literature, is negative. The contribution of the next term, in y, somewhat compensates this decrease in energy, in most cases. Sometimes, however, the overall contribution of the anharmonicity on the energy is positive and the *effective* vibrational frequency increases with v. This

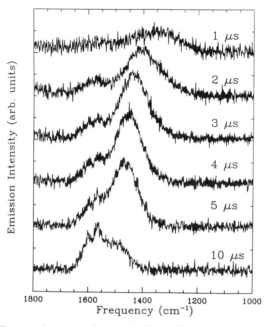

Figure 23. Time- and wavenumber-resolved laser-induced infrared fluorescence from highly vibrationally excited NO_2 (reproduced from Dai and co-workers [506], with permission from the authors and from the AIP). As time increases, collision deactivation progresses and, simultaneously, $\Delta v_3 = -1$ emission occurs between pairs of decreasingly excited vibrational levels, with v_3 the antisymmetric NO stretching mode. The transition is progressively blue-shifted toward its fundamental value, illustrating the role of mechanical anharmonicity.

situation is encountered, for instance, for the *trans*-bending mode in C_2H_2 [504], $^{12}C_2D_2$ [505], and $^{12}C_2HD$ [444]. The sign of this effect is determined by the relative size of the third- and fourth-order potential constants appearing in the development of x, see (3.36).

The influence of the anharmonicity we have just detailed, which is the mechanical anharmonicity, is illustrated in Fig. 23 [506].

It is common feature to use vibrational energies after removing the zero point energy, thus referred to the $v = 0$, ground level rather than to the bottom of the potential well. They are defined as

$$G_0(v) = G(v) - G(0)$$
$$G_0(v) = \tilde{\omega}v + xv + xv^2 + \tfrac{3}{4}yv + \tfrac{3}{2}yv^2 + yv^3 \qquad (3.41)$$
$$G_0(v) = \tilde{\omega}^0 v + x^0 v^2 + yv^3$$

with

$$\begin{aligned}
\tilde{\omega}^0 &= \tilde{\omega} + x + \tfrac{3}{4}y \\
x^0 &= x + \tfrac{3}{2}y
\end{aligned} \tag{3.42}$$

One seldom writes the superscript "0" for the last term in the development, which is y in this case, as $y^0 = y$.

The energy of the fundamental band, in cm^{-1}, that is, the energy of the transition between the $v = 0$ and $v = 1$ levels, is

$$\tilde{\nu}^0 = \tilde{\omega}^0 + x^0 + y = \tilde{\omega} + 2x + \tfrac{13}{4}y \tag{3.43}$$

The different label $\tilde{\nu}_0$ is used to refer to a vibration–rotation band origin, in general.

It is worth pointing out at this stage already that, for symmetric and linear tops, the relationship between the band center resulting from the rotational analysis of the band, $\tilde{\nu}_0$ or G_c, and the vibrational origin, G_0, must include the contributions of vibration–rotation nature, whenever $k \neq 0$ in such a way that

$$G_c = G_0 - B_v' k^2 + D_v' k^4 + B_v'' k^2 - D_v'' k^4 \tag{3.44}$$

as detailed in Section III.D.

2. MULTIDIMENSIONAL DUNHAM EXPANSION. The extension is made to a $3n - 6$ multimode system for a nonlinear species, with n the number of nuclei constituting the molecule, and with

$$\hat{H}_v^{\text{har}} = \sum_{i=1}^{3n-6} \tfrac{1}{2} hc\tilde{\omega}_i(\hat{p}_i^2 + q_i^2) \tag{3.45}$$

using:

$$V^{(q)} = hc\tfrac{1}{2}\sum_i \tilde{\omega}_i q_i^2 + \tfrac{1}{6}\sum_{ijk} \phi_{ijk} q_i q_j q_k + \tfrac{1}{24}\sum_{ijkl} \phi_{ijkl} q_i q_j q_k q_l + \cdots \tag{3.46}$$

to give

$$\begin{aligned}
G(v) = \sum_{i=1}^{3n-6} \tilde{\omega}(v_i + \tfrac{1}{2}) &+ \sum_{i\leq j=1}^{3n-6} x_{ij}(v_i + \tfrac{1}{2})(v_j + \tfrac{1}{2}) \\
&+ \sum_{i\leq j\leq k=1}^{3n-6} y_{ijk}(v_i + \tfrac{1}{2})(v_j + \tfrac{1}{2})(v_k + \tfrac{1}{2})
\end{aligned} \tag{3.47}$$

Further terms in the expansion, in z_{ijkl} and even of higher order, may be required depending on the experimental data considered.

Referring to the ground level $\sum_i v_{i=0}$; we have

$$G_0(v_i) = \sum_i \tilde{\omega}_i^0 v_i + \sum_{i \leq j} x_{ij}^0 v_i v_j + \cdots \qquad (3.48)$$

with

$$\tilde{\omega}_i^0 = \tilde{\omega}_i + x_{ii} + \sum_{i<j} \tfrac{1}{2} x_{ij} + \cdots \qquad (3.49)$$

and the origin of the ith fundamental band (thus of the transition between the levels $v_i = 1$ and $\sum_i v_i = 0$) is (in cm^{-1})

$$\tilde{v}_i^0 = \tilde{\omega}_i^0 + x_i^0 + \cdots = \tilde{\omega}_i + 2x_{ii} + \sum_{i<j} \tfrac{1}{2} x_{ij} + \cdots \qquad (3.50)$$

A perturbation treatment similar to the one described earlier for the single oscillator, now applied to the full Watsonian, gives [373]

$$x_{ij} = \frac{1}{16} \phi_{iiii} - \frac{1}{16} \sum_j \frac{\phi_{iij}^2 (8\tilde{\omega}_i^2 - 3\tilde{\omega}_j^2)}{\tilde{\omega}_j(4\tilde{\omega}_i^2 - \tilde{\omega}_j^2)} \qquad (3.51)$$

and

$$x_{ij} = \frac{1}{4} \phi_{iijj} - \frac{1}{4} \sum_k \left(\frac{\phi_{iik}\phi_{kjj}}{\tilde{\omega}_k} \right)$$

$$- \frac{1}{2} \sum_k \left[\frac{\phi_{ijk}^2 \tilde{\omega}_k(\tilde{\omega}_k^2 - \tilde{\omega}_i^2 - \tilde{\omega}_j^2)}{\Delta_{ijk}} \right] \qquad (3.52)$$

$$+ [A(\zeta_{i,j}^{(a)})^2 + B(\zeta_{i,j}^{(b)})^2 + C(\zeta_{i,j}^{(c)})^2] \left[\left(\frac{\tilde{\omega}_i}{\tilde{\omega}_j} \right) + \left(\frac{\tilde{\omega}_j}{\tilde{\omega}_i} \right) \right]$$

with

$$\Delta_{ijk} = (\tilde{\omega}_i + \tilde{\omega}_j + \tilde{\omega}_k)(\tilde{\omega}_i - \tilde{\omega}_j - \tilde{\omega}_k)(-\tilde{\omega}_i + \tilde{\omega}_j - \tilde{\omega}_k)(-\tilde{\omega}_i - \tilde{\omega}_j + \tilde{\omega}_k) \qquad (3.53)$$

in which the rotation and vibration–rotation parameters arise from contributions of the form $q_i \hat{p}_j q_k \hat{p}_l$ in the product $\hat{\pi}_\alpha \hat{\pi}_\beta$ appearing in H_{40} [456]. The constants A, B, and C, which we generically label B_α, are the principal rotation constants $[B_\alpha = (8\pi^2 I_\alpha c/h)^{-1}$ (in cm^{-1})]. It is interesting to note that terms arising from vibration–rotation operators contribute in the Dunham expansion providing the diagonal *vibrational* energies in the MIME.

At the present stage, the MIME appears as a squared matrix containing only diagonal elements $G_0(v)$ of the form of (3.37) for an asymmetric top, with as many elements as there are observed vibrational energy levels in the isolated ground electronic state. We label each such zeroth-order vibrational energy level using several possible notations. We provide hereafter examples referring to the level with five quanta in mode 3, three quanta in mode 4, and zero quanta in all other modes, with m_i the number of quanta in the ith mode:

$$v_3 = 5, v_4 = 3$$

$$i_1^{m_{i1}}, \ldots, i_{3n-6}^{m_{i(3n-6)}} \qquad (3^5, 4^3)$$

$$\sum_i m_i v_i \qquad (5v_3 + 3v_4) \qquad (3.54)$$

$$(m_1, m_2, \ldots, m_{3n-6}) \qquad (0, 0, 5, 3, \ldots, 0)$$

Bands are labeled using one of the following notations [507], with the example hereafter corresponding to the hot band connecting the upper ($'$) $5v_3 + 3v_4$ and lower ($''$) v_2 vibrational levels:

$$\sum_i (m_i v_i)' - \sum_i (m_i v_i)'' \qquad 5v_3 + 3v_4 - v_2$$

$$i_{1m_{i1}''}^{m_{i1}'}, \ldots, i_{3n-6_{m_{i(3n-6)}''}}^{m_{i(3n-6)}'} \qquad 2_1^0 3_0^5 4_0^3 \qquad (3.55)$$

3. Off-Diagonal Terms

a. Anharmonic Resonances ($\hat{H}_v^{\text{anh off-diag}}$)

1. TERMINOLOGY. We now consider those coupling terms off-diagonal in v, called *anharmonic resonances*. According to the present choice of normal coordinates, they occur from the terms in the development of the potential energy. We discuss only those terms that are of cubic and quartic order, because they are of dominant numerical importance. Two of those carry special names:

1. When the coupling operator is of cubic form and causes vibrational levels with $\Delta v_i = \pm 2$, $\Delta v_j = \mp 1$ to interact, it is called a *Fermi* [508] resonance, which we abbreviate "*F* resonance."

2. When the transformed coupling operator is of quartic form and makes vibrational levels with $\Delta v_i = \pm 2$, $\Delta v_j = \mp 2$ interact, it is called a Darling–Dennison resonance [509], which we abbreviate "*DD* resonance."

All vibrational interactions of similar nature but fulfilling other selection rules should be referred to under the generic term of anharmonic resonance. Among all possible anharmonic resonances, some can expected to always

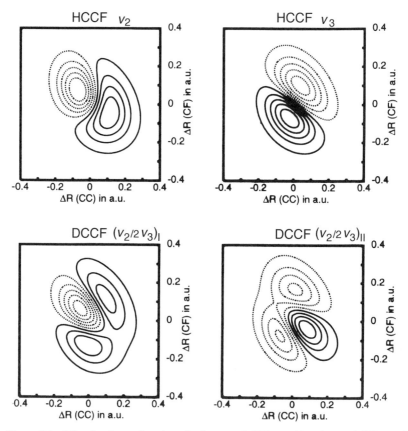

Figure 24. Vibrational wavefunctions for the $v_2 = 1$ (CC stretch) and $v_3 = 1$ (CF stretch) vibrational states in $^{12}C_2HF$ (upper part), and for the Fermi-coupled $v_2 = 1$ and $v_3 = 2$ vibrational states in $^{12}C_2DF$ [Ref. 510, reprinted from *Chemical Physics*, Vol. 190, Botschwina et al., "*Ab initio* calculations of stretching vibrational transitions for the linear molecules HCN, HNC, HCCF and HC_3N up to high overtones," pp. 345–362, copyright (1995), with permission from Elsevier Science and from the authors]. The fairly pure wavefunctions in one species appear intimately mixed in the other one, illustrating the role of anharmonic coupling.

occur with significant impact on the spectrum, given the molecular structure. This is the case for the DD interaction coupling structurally identical CH-type stretchings, which we refer to when detailing acetylene, later in Section III.C. We already suggested labeling them *structural* anharmonic resonances in section III.B.1.

All anharmonic couplings generate non-zero off-diagonal matrix elements in the model. Therefore, the general perturbation formulas for the anharmonic constants x_{ij} and y_{ijk} arising in the Dunham development must be modified to exclude terms whose effect is now also included in the off-diagonal elements. Formally, such terms ought to be labeled using an asterisk such as x_{ij}^*. This distinction is most relevant if one attempts connecting the effective constants resulting from the fit and the parameters in the potential-energy development. This relationship is, however, very difficult to reliably establish in larger molecules. We shall omit this distinction in the present review.

Anharmonic resonances are central in the remaining sections and their role is exemplified at several occasions, starting with Fig. 24.

2. MATRIX ELEMENTS. Given the general formula of the action of the vibrational operators (Table VII), one can generate the matrix element related to F, other cubic, DD and other quartic or higher order anharmonic resonances, as exemplified in Table VIII.

In Table VIII, we use, v_n for the unaffected vibration modes and $K_{ij/kl}$ for the coefficient for the coupling term between the levels $nv_i + nv_j$ and $nv_k + nv_l$. Both terms are hermitian conjugates of each other in the second quantization form. The labeling $K_{ij/kl}$ thus implies that the levels coupled by the interaction are such that v_i and v_j are decreased by one while v_k and v_l are increased by one, or inversely.

The anharmonic resonances push the interacting levels apart. The effect of each interaction depends on the coupling term, K_i, which is a constant factor weighted by some function of the vibrational quantum numbers, in v^3 or v^4, for cubic or quartic resonances, respectively. The coupling also depends on the energy difference between the zeroth-order states, the closer the levels, the stronger the perturbation, as we have already stressed in the introduction to the present section.

Development of the $\tilde{h}_v^{\text{anh}}$ terms in functions of the original contributions in the potential can be obtained using perturbation theory, as we have proceeded for the x anharmonicity coefficient in the Dunham series:

$$\langle v_i | \tilde{h}_{40} | v_j \rangle = \langle v_i | H_{40}/hc | v_j \rangle + \sum_k \frac{\langle v_i | H_{30}/hc | v_k \rangle \langle v_k | H_{30}/hc | v_j \rangle}{(E_{ij}^0 - E_k^0)/hc} \quad (3.56)$$

TABLE VIII
Anharmonic Resonance Matrix Elements

$$\langle v_i, v_j, v_n | \tilde{h}_v^{\text{Fermi}} | v_i + 2, v_j - 1, v_n \rangle = 1/(2\sqrt{2}) K_{ii/j}[(v_i + 1)(v_i + 2)v_j]^{1/2}$$

$$\langle v_i, v_j, v_k, v_n | \tilde{h}_v^{\text{cubic}} | v_i + 1, v_j - 1, v_k - 1, v_n \rangle = 1/(2\sqrt{2}) K_{i/jk}[(v_i + 1)v_j v_k]^{1/2}$$

$$\langle v_i, v_j, v_n | \tilde{h}_v^{\text{DD}} | v_i + 2, v_j - 2, v_n \rangle = \tfrac{1}{4} K_{ii/jj}[(v_i + 1)(v_i + 2)v_j(v_j - 1)]^{1/2}$$

$$\langle v_i, v_j, v_k, v_l, v_n | \tilde{h}_v^{\text{quartic 1}} | v_i + 1, v_j - 1, v_k - 1, v_l - 1, v_n \rangle = \tfrac{1}{4} K_{i/jkl}[(v_i + 1)v_j v_k v_l]^{1/2}$$

$$\langle v_i, v_j, v_k, v_l, v_n | \tilde{h}_v^{\text{quartic 2}} | v_i + 1, v_j + 1, v_k - 1, v_l - 1, v_n \rangle = \tfrac{1}{4} K_{ij/kl}[(v_i + 1)(v_j + 1)v_k v_l]^{1/2}$$

Or, using the second quantization form:

$$\tilde{h}_v^{\text{F}} = K_{\text{F}}(a_i^+ a_i^+ a_j^- + a_j^- a_i^- a_i^+)$$

$$\tilde{h}_v^{\text{DD}} = K_{\text{DD}}(a_i^+ a_i^+ a_j^- a_j^- + a_i^- a_i^- a_j^+ a_j^+)$$

with v_i, v_j, and v_k in (3.56) set for an appropriate combination of vibrational quantum numbers. It must be emphasized that the coefficients of the anharmonic resonance terms, just as the full Hamiltonian, must be totally symmetric, therefore restricting the number of terms in the development and, hence, the number of resonances to be considered. Symmetric behavior must actually also hold with respect to the time reversal symmetry operation, leading to define only real coefficients in the development. The relative contribution of all possible terms in the sum over k depends on the relative vibrational energies and the final expression for K_{v_i} therefore strongly depends on the case study [e.g., 377]. In any case, cubic resonance coefficients correspond to a single cubic term in the anharmonic development of the potential while the content of quartic resonance coefficients is made of a quartic term and a number of cubic terms in second order perturbation. Both types of contributions are usually of the same order of magnitude and may cancel each other [472,473].

3. BRIGHT AND DARK STATES. Let us assume two levels such that transition to level 1 has observable intensity while transition to level 2 has unobservable intensity, in the zeroth-order model:

$$|\langle \Psi_1^0 | \hat{\mu} | \Psi_0 \rangle|^2 \neq 0$$
$$|\langle \Psi_2^0 | \hat{\mu} | \Psi_0 \rangle|^2 = 0 \tag{3.57}$$

with $\hat{\mu}$ the electric dipole moment operator, as introduced in Section II and further detailed in Section III.E.8.c.2.

Levels 1 and 2 are called, respectively, "bright" and "dark" states (see, e.g. Refs. 511 and 512). Thus, although it is really the transition highlighting the level that is bright or dark, it is the state that one calls bright or dark. This concept is therefore not absolute. Indeed, the same two levels can also be monitored using transitions obeying different selection rules, say, of Raman or two-photon type, which might highlight the one-photon dark state, level 2 in this case. Level 2 is under such conditions the bright state. This change is typical when comparing the data recorded for the same centrosymmetric molecule from infrared absorption spectra and from laser-induced dispersed fluorescence experiments involving another, intermediate excited electronic state, reaching vibrational levels with u and g symmetry, respectively. Interesting observations in acetylene are detailed by Crim [513], Orr [514], and their co-workers, and exemplified in Fig. 25.

The anharmonic resonances, if any, mix the zeroth-order wavefunctions. The degree of mixing is given by the squared value of the coefficients in the eigenvector from the diagonalization of the MIME. In a 2×2 coupling between the zeroth-order functions Ψ_1^0 and Ψ_2^0, assuming the following matrix Hamiltonian:

$$\begin{bmatrix} E_1^0 & \alpha \\ \alpha & E_2^0 \end{bmatrix}, \quad \text{with} \quad \alpha = \langle \Psi_1^0 | \tilde{h} | \Psi_2^0 \rangle \tag{3.58}$$

the resulting wavefunctions representing the eigenstates are given by the standard formulation [515,516]:

$$\begin{aligned} \Psi_+ &= a\Psi_1^0 + b\Psi_2^0 \\ \Psi_- &= b\Psi_1^0 - a\Psi_2^0 \end{aligned} \tag{3.59}$$

and $E_\pm = \frac{1}{2}(E_1^0 + E_2^0) \pm \frac{\Gamma}{2}$
with

$$\Gamma = (\delta^2 + 4\alpha^2)^{1/2} \quad \delta = E_1^0 - E_2^0$$

$$a = \left(\frac{\Gamma + \delta}{2\Gamma}\right)^{1/2} \quad b = \left(\frac{\Gamma - \delta}{2\Gamma}\right)^{1/2} \quad a^2 + b^2 = 1$$

As a consequence, all properties of the wavefunctions are mixed, leading to their averaging. The intensity of the bands reaching the two interacting levels is, in particular, affected. Indeed, the intensity of the absorption from the ground level to both interacting levels is now given by

$$\begin{aligned} |\langle a\Psi_1^0 + b\Psi_2^0 | \hat{\mu} | \Psi_0 \rangle|^2 \\ |\langle b\Psi_1^0 - a\Psi_2^0 | \hat{\mu} | \Psi_0 \rangle|^2 \end{aligned} \tag{3.60}$$

with μ or D the dipole moment operator, introduced in Section II and further used in Section III.E.8.c.2. The result of the mixing is to share the initial

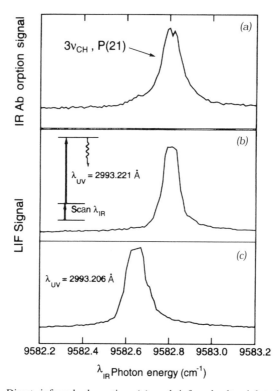

Figure 25. Direct infrared absorption (a) and infrared–ultraviolet double-resonance signals (b, c), recorded while scanning the same infrared laser (Scan λ_{IR}) around the $P(21)$ $3\nu_3$ vibration–rotation transition, in the ground electronic state of $^{12}C_2H_2$, with ν_3 the CH antisymmetric stretching mode [Ref. 513, reprinted from *Chemical Physics*, Vol. 190, Utz et al., "Direct observation of weak state mixing in highly vibrationally excited acetylene," pp. 311–326, copyright (1995), with permission from Elsevier Science and from the authors]. When the UV laser is tuned to coincide with the $P(21)$ $3\nu_3$ transition, the infrared absorption occurs on that infrared line. The $J = 20$, $3\nu_3$ state is then a bright state. When the UV laser is slightly detuned, by $< 0.2\,\mathrm{cm}^{-1}$, the infrared absorption occurs at a wavenumber where there is very little absorption in the one-photon spectrum, and the same $J = 20$, $3\nu_3$ state is now a dark state. Another close state has thus the bright/dark characteristics opposite to $J = 20$, $3\nu_3$. This figure illustrates the dependence of the bright/dark labeling in the experimental technique used to highlight the states.

intensity between both transitions in such a way that it is now given by

$$
\begin{aligned}
& a^2 |\langle \Psi_1^0 | \hat{\mu} | \Psi_0 \rangle|^2 \\
& b^2 |\langle \Psi_1^0 | \hat{\mu} | \Psi_0 \rangle|^2
\end{aligned}
\tag{3.61}
$$

and $I_{rel} = a^2/b^2$

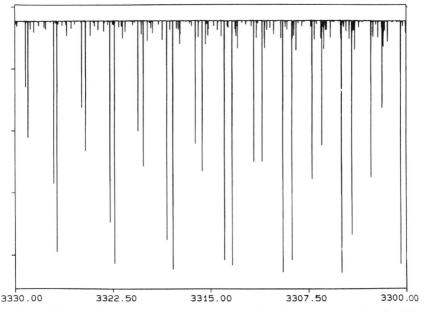

3330.00   3322.50   3315.00   3307.50   3300.00

Figure 26. Doubling of the fine structure in the region of the fundamental ν_3 band in $^{12}C_2H_2$ demonstrating the presence of a zeroth-order vibrational dark state ($\nu_2 + \nu_4 + \nu_5$) in anharmonic resonance with the zeroth order vibrational bright state (ν_3), leading to intensity borrowing. The interaction is strong enough that the mixing is 50:50 (see Ref. 451) with modes 2–5, the CC stretch, the CH antisymmetric stretch, and the *trans* and *cis* bends, respectively (energy scale in cm^{-1}).

It becomes therefore possible, under such specific conditions, to determine the zeroth-order parameters:

$$\alpha = ab(E_+ - E_-)$$
$$E_1^0 = E_+ - b^2(E_+ - E_-) \qquad (3.62)$$
$$E_2^0 = E_- + b^2(E_+ - E_-)$$

This intensity borrowing from the initially bright to the zeroth-order dark transitions is most often the criterion for identifying an anharmonic resonance from spectral data. Examples are provided in Fig. 26 and in the next section. In more complex cases, that is, if none of the zeroth-order transitions were dark, or when there are more than three interacting levels, cancellation effects may occur. Bright and dark states are further discussed in Section III.E.8.*c*.2.

As another consequence of the coupling induced by the anharmonic resonances, the rotational parameters of the eigenstates also change. The change is described as follows, for a linear polyatomic species [515]:

$$B_1 = a^2 B_1^0 + b^2 B_2^0$$
$$B_2 = b^2 B_1^0 + a^2 B_2^0$$

(3.63)

This method is extended to multilevel coupling [517] and to centrifugal distortion constants in Section III.D.4.*d*.

b. Quartic Resonances

1. DETAILED CONTENT. The content of quartic resonance coefficients can be determined using relation (3.56). Its application to acetylene and haloacetylenes is detailed by Mills and co-workers [470,518]. As an example, for $K_{11/33}$ [518], with modes 1, 3, and 2 the symmetric and antisymmetric CH stretches and the CC stretch, respectively, and modes b the bends, the relevant terms in the potential development are

$$\frac{H_{40}}{hc} = \frac{1}{4}\phi_{1133}q_1^2 q_3^2$$
$$\frac{H_{30}}{hc} = \frac{1}{6}\phi_{111}q_1^3 + \frac{1}{2}\phi_{112}q_1^2 q_2 + \frac{1}{2}\phi_{133}q_1 q_3^2 + \frac{1}{2}\phi_{233}q_2 q_3^2$$

(3.64)

which must be dealt with using first- and second-order perturbation theory

$$\langle v_i|\tilde{h}_{40}|v_j\rangle = \left\langle v_1, v_2, v_3, v_b \left|\frac{H_{40}}{hc}\right| v_1 + 2, v_2, v_3 - 2, v_b \right\rangle$$
$$+ \sum_k \frac{\langle v_1, v_2, v_3, v_b|H_{30}/hc|v_k\rangle\langle v_k|H_{30}/hc|v_1 + 2, v_2, v_3 - 2, v_b\rangle}{(E_{ij}^0 - E_k^0)/hc}$$
$$= \langle v_i|\tilde{h}_{vib}^{DD}|v_j\rangle = \frac{1}{4}K_{11/33}[(v_1 + 1)(v_1 + 2)v_3(v_3 - 1)]$$

(3.65)

Developing the term in the first order of perturbation leads to a contribution of $\frac{1}{4}\phi_{11/33}$ to $K_{11/33}$. As an example, one of the terms in the second order of perturbation is

$$[\langle v_1, v_2, v_3|\tfrac{1}{2}\phi_{133}q_1 q_3^2|v_1 - 1, v_2, v_3\rangle$$
$$\langle v_1 - 1, v_2, v_3|\tfrac{1}{2}\phi_{133}q_1 q_3^2|v_1 - 2, v_2, v_3 + 2\rangle]/\tilde{\omega}_{13}$$
$$+[\langle v_1, v_2, v_3|\tfrac{1}{2}\phi_{133}q_1 q_3^2|v_1 - 1, v_2, v_3 + 2\rangle$$
$$\langle v_1 - 1, v_2, v_3 + 2|\tfrac{1}{2}\phi_{133}q_1 q_3^2|v_1 - 2, v_2, v_3 + 2\rangle]/(-\tilde{\omega}_{13})$$

(3.66)

which contributes for a coefficient $\frac{1}{2}(\phi^2_{133}/\tilde{\omega}_{13})$. Developing all such terms and adding all coefficients leads to

$$K_{11/33} = \frac{1}{4}\phi_{133} + \frac{1}{12}\frac{\phi_{111}\phi_{133}}{\tilde{\omega}_{13}} + \frac{1}{4}\frac{\phi_{112}\phi_{233}\tilde{\omega}_2}{(4\tilde{\omega}^2_{13} - \tilde{\omega}^2_2)} - \frac{1}{2}\frac{\phi^2_{133}}{2\tilde{\omega}_{13}} \qquad (3.67)$$

2. x–K RELATIONS. The x-anharmonicity coefficients in the Dunham expansion and the $K_{ij/kl}$ quartic anharmonic resonance constants are related to the same cubic and quartic potential parameters; see (3.51), (3.52), and (3.67). One can therefore expect specific relationship to occur between them. One could work them out, which would actually be most interesting but highly specific to each species considered. One can also relate all parameters more generally, using the x–K relations [152,518–523].

We will reproduce almost exactly the elegant presentation of Lehmann [519], which is based on the dual description of two identical, coupled anharmonic oscillators, using normal- and local-mode approaches. In the first one, which we have described in this section, six parameters are required: $\tilde{\omega}_1$, $\tilde{\omega}_2$, x_{11}, x_{22}, x_{12}, and $K_{11/22}$. In the second set of coordinates, Child and Lawton [55,150] (see also Ref. 154) need only three parameters: $\tilde{\omega}_{local}$, χ_{local}, and λ, the effective bond mode coupling. In the local-mode limit defined by these authors, both oscillators are strictly identical, having the same frequency ($\tilde{\omega}_{local}$) and anharmonicity (x_{local}). Obviously, there must be some redundancy leading to constraints between the x–K normal-mode parameters. The matrix elements in the local-mode limit are

$$\langle n, m|\tilde{h}_{local}|n, m\rangle = \tilde{\omega}_{local}(n + \tfrac{1}{2}) + \tilde{\omega}_{local}(m + \tfrac{1}{2}) + x_{local}(n + \tfrac{1}{2})^2$$
$$+ x_{local}(m + \tfrac{1}{2})^2 \qquad (3.68)$$
$$\langle n + 1, m - 1|\tilde{h}_{local}|n, m\rangle = \lambda\sqrt{(n + 1)m}$$

with n and m the number of quanta in each bond, and with λ giving the required degree of freedom to match the observed energy levels. Introducing the pseudo–angular momentum eigenvalue labels $2J = m + n$ and $2K = n - m$, then the matrix elements are

$$\langle J, K|\tilde{h}_{local}|J, K\rangle = \tilde{\omega}_{local}(2J + 1) + \tfrac{1}{2}x_{local} + 2x_{local}[J(J + 1) + K^2]$$
$$\langle J, K \pm 1|\tilde{h}_{local}|J, K\rangle = \lambda\sqrt{J(J + 1) - K(K \pm 1)} \qquad (3.69)$$

Introducing rotation operators $\hat{J}_z$ and $\hat{J}_x$, we obtain

$$\tilde{h}_{local} = C(J) + 2x_{local}\hat{J}_z^2 + 2\lambda\hat{J}_x$$
$$C(J) = \tilde{\omega}_{local}(2J + 1) + \tfrac{1}{2}x_{local} + 2x_{local}J(J + 1) \qquad (3.70)$$

with $C(J)$ a constant. Performing a rotation of $\pi/2$ around the y axis, one changes from local to normal coordinates [524]. Then

$$
\begin{aligned}
\langle J, K | \tilde{h}_{\text{normal}} | J, K \rangle = \tilde{\omega}_{\text{local}}(2J+1) + \tfrac{1}{2}x_{\text{local}} + 2\lambda K \\
+ 3x_{\text{local}}J(J+1) - x_{\text{local}}K^2
\end{aligned} \tag{3.71}
$$

with, from now on, all wavefunctions adapted to the normal-mode coordinates.

Using the normal mode labels v_1 and v_2, we have

$$
\begin{aligned}
\langle v_1, v_2 | \tilde{h}_{\text{normal}} | v_1, v_2 \rangle = (\tilde{\omega}_{\text{local}} + \lambda)(v_1 + \tfrac{1}{2}) + (\tilde{\omega}_{\text{local}} - \lambda)(v_2 + \tfrac{1}{2}) \\
+ \tfrac{1}{2}x_{\text{local}}[(v_1 + \tfrac{1}{2})^2 + (v_2 + \tfrac{1}{2})^2] \\
+ 2x_{\text{local}}[(v_1 + \tfrac{1}{2})(v_2 + \tfrac{1}{2})] + \tfrac{1}{4}x_{\text{local}}
\end{aligned} \tag{3.72}
$$

which is the conventional local-mode expression, with an additional term $(\tfrac{1}{4}x_{\text{local}})$, provided

$$
\begin{aligned}
\tilde{\omega}_{1,2} &= (\tilde{\omega}_{\text{local}} \pm \lambda) \\
4x_{11} &= 4x_{22} = x_{12} = 2x_{\text{local}}
\end{aligned} \tag{3.73}
$$

The off-diagonal matrix elements become

$$
\begin{aligned}
\langle J, K+2 | 2x_{\text{local}}\hat{J}_x^2 | J, K \rangle &= \tfrac{1}{2}x_{\text{local}}[J(J+1) - K(K+1)]^{1/2} \\
&\times [J(J+1) - (K+1)(K+2)]^{1/2} \\
\langle J, K-2 | 2x_{\text{local}}\hat{J}_x^2 | J, K \rangle &= \tfrac{1}{2}x_{\text{local}}[J(J+1) - K(K-1)]^{1/2} \\
&\times [J(J+1) - (K-1)(K-2)]^{1/2}
\end{aligned} \tag{3.74}
$$

or

$$
\begin{aligned}
\langle v_1+2, v_2-2 | \tilde{h}_{\text{normal}} | v_1, v_2 \rangle &= \tfrac{1}{2}x_{\text{local}}[(v_1+1)(v_1+2)v_2(v_2-1)]^{1/2} \\
\langle v_1-2, v_2-2 | \tilde{h}_{\text{normal}} | v_1, v_2 \rangle &= \tfrac{1}{2}x_{\text{local}}[(v_1-1)v_1(v_2+1)(v_2+2)]^{1/2}
\end{aligned} \tag{3.75}
$$

specifically, the DD matrix element, provided

$$
K_{11/22} = x_{\text{local}} \tag{3.76}
$$

The important role of the DD coupling, which connects the normal- and local-mode pictures, is highlighted by this comparison. It was actually

TABLE IX
Selected Vibrational Constants in Various Isotopomers of Acetylene[a]

	$^{12}C_2H_2$	$^{12}C_2HD$	$^{12}C_2D_2$
x_{11}	− 24.87	− 48.48	− 12.30
x_{33}	− 27.70	− 23.30	− 15.52
x_{13}	− 107.66	− 4.71	− 47.52
$K_{11/33}$	− 105.97	0	− 47.2
x_{44}	3.46	0.25	1.84
x_{55}	− 2.35	− 2.92	− 1.62
x_{45}	− 2.20	1.29	− 1.64
$K_{44/55}$	− 8.37	0	− 7.96

[a] Modes 1 and 3 are the symmetric and antisymmetric CH stretchings and modes 4 and 5, the *trans* and *cis* bendings, respectively.
Source: From Herman and co-workers [444,450,468,469,525].

already raised when defining structural resonances, in Section III.B.3.*a*.1. The $x-K$ relations are most helpful in guiding and checking the results of the fitting procedures in larger molecules. An example of how well the relations are fulfilled is provided by acetylene. The related values are presented in Table IX [525]. We have also listed therein, for information, the up-to-date constants for the two bending vibrations, in *trans*, mode 4, and in *cis*, mode 5, of acetylene.

As could be expected, $^{12}C_2HD$, with two different carbon–hydrogen bonds, does not fulfill the requirements for the $x-K$ relations. They are respected for the C–H stretch parameters (modes 1 and 3) in all other cases, within 10%. Other successful applications of the $x-K$ relations for C–H bond parameters are presented in the literature [518], concerning various case studies, such as ethylene (C_2H_4) [526,527]. Other $x-K$ relations were adapted by Lehmann to deal with the bending modes of acetylene [528]. They failed to reproduce the experimental constants, due to large cancellation effects between cubic and quartic contribution to the constants.

We provide for completeness in Fig. 27 the correlation between local and normal-mode labeling, applied to a molecule with two identical CH bonds, such as symmetric acetylene. The local-mode labels in Fig. 27 are such that the local-mode wavefunctions are defined according to, for instance

$$\Psi[3,0]_- = \frac{1}{\sqrt{2}}[|\Psi_{m=3}\rangle|\Psi_{n=0}\rangle - |\Psi_{m=0}\rangle|\Psi_{n=3}\rangle$$

$$\Psi[2,1]_+ = \frac{1}{\sqrt{2}}[|\Psi_{m=2}\rangle|\Psi_{n=1}\rangle + |\Psi_{m=1}\rangle|\Psi_{n=2}\rangle \tag{3.77}$$

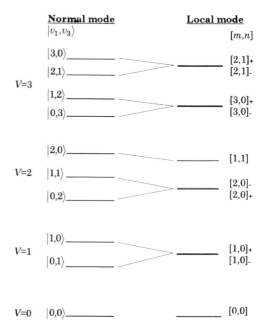

Figure 27. Correlation between local and normal mode labeling in a molecule with two identical CH bonds, such as symmetric acetylene. Modes 1 and 3 represent the symmetric and antisymmetric CH stretching normal vibrations, respectively. Numbers m and n indicate the number of quanta in each of the two identical CH local modes.

thus with symmetric (+) and antisymmetric (−) local-mode combinations corresponding to Σ_g^+ and Σ_u^+ states, respectively.

3.   UNUSUAL QUARTIC RESONANCES. Some unusual resonances were reported in the literature for molecules including methanol [529] and monodeuteroacetylene [444], which were considered by Lehmann [530,531] and Mills [532]. They correspond to a matrix element of the form

$$\langle v_1, v_2 | q_1^3 q_2 | v_1 - 1, v_2 + 1 \rangle = \tfrac{3}{4} K_{11/12} (v_1)^{3/2} (v_2 + 1)^{1/2} \qquad (3.78)$$

The peculiarity of such interactions is that they couple levels with $\Delta v_1 = \pm 1, \Delta v_2 = \mp 1$. Such couplings may not arise from off-diagonal quadratic terms, which are eliminated when selecting the normal mode basis set. Their strength grows rapidly with v_1, along a series of overtones of v_1:

$$\frac{1}{hc} \langle v_1, 0 | \hat{H} | v_1 - 1, 1 \rangle \propto (v_1)^{3/2} K_{11/12} \qquad (3.79)$$

Therefore such a coupling presents a different v dependence than other ones and is likely to play a more important role with the vibrational excitation,

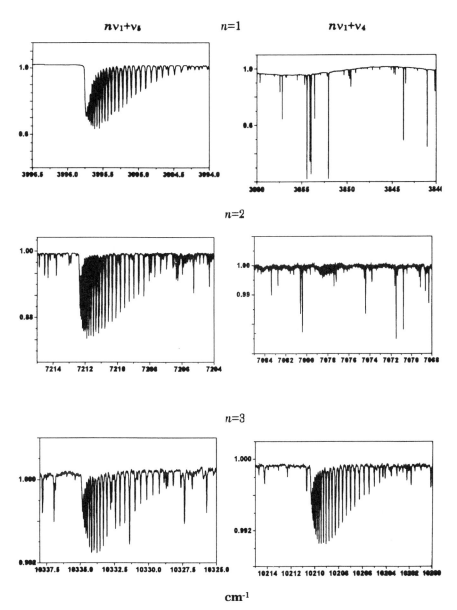

Figure 28. Evolution of the absorption band intensity for the series $n\nu_1 + \nu_4$ and $n\nu_1 + \nu_5$ in $^{12}C_2HD$ with modes 1, 4, and 5 close to the pure CH stretch, the CCD and the CCH bends, respectively. The wavenumber scale is different in each range. The intensity scale is directly comparable in the spectra of the first line, on one hand, and in the four last cases, on the other hand. In some spectra, extra lines arise from impurities. The $\nu_1 + \nu_5$ band is observed but not $\nu_1 + \nu_4$. With increasing values of n, the $n\nu_1 + \nu_{4/5}$ band intensity becomes increasingly similar, demonstrating the increasing effect of the anharmonic resonance between the $n\nu_1 + \nu_4$ and $n\nu_1 + \nu_5$ states (adapted from Liévin et al. [444] and Abbouti Temsamani [533]).

127

provided the interacting levels are not getting further apart in the overtone range, because of the role of mechanical anharmonicity.

As an example of these unusual quartic resonances, Fig. 28 [533], from Liévin et al. [444], shows portions of the absorption spectrum in $^{12}C_2HD$. It concerns the series of bands reaching the levels $nv_1 + v_4$ and $nv_1 + v_5$ from the ground state, with mode 1 the CH stretch and modes 4 and 5 the CCD and CCH bends, respectively. The resonance with $\Delta v_4 = \pm 1, \Delta v_5 = \mp 1$ is observed through the increasing intensity borrowing from the zeroth-order bright $nv_1 + v_5$ series of transitions by the zeroth-order dark $nv_1 + v_4$ series of transitions. The notion of bright and dark states was just introduced in Section III.B.a.3. In the present case, the coupling is greatly enhanced with increasing excitation, due to the related decrease in the energy difference between the interacting levels, induced by the ordering of the values of $\tilde{\omega}_4, \tilde{\omega}_5$ on one hand and x_{14}, x_{15} on the other hand [444].

4. Linear Tops

a. General Features. The developments in the previous sections were all adequate for nondegenerate vibration modes, that is, for asymmetric top-type molecules. We consider linear tops in the present section, and follow the same strategy in the presentation as previously. The intermediate developments are very similar in nature for both case studies but, because of the presence of degenerate vibrations, they are more complicated for linear tops and, in some cases, not detailed in the literature, yet. We therefore converge hereafter more directly to energy expressions including higher-order parameters, which are conveniently obtained by empirical extension of the lower-order terms. The same procedure was actually implicitly followed, for example, when stating that higher-order terms in y_{ijk} and z_{ijkl} could be introduced in the Dunham expression in (3.47). Effective Hamiltonians are often built empirically and adapted to the problem studied, with as many terms as required by the quality of the experimental data and by the couplings they highlight. Ambiguity problems, such as the generation of parameters whose physical content is either inappropriate or already included in other terms, can result from this procedure. Such problems can hardly be spotted without a detailed theoretical investigation, as illustrated in Section III.D.4.b, or, sometimes, without complementary data. This problem applies in general to all effective, transformed Hamiltonians and directly affects the reliability of procedures connecting potential parameters or geometric structures to the effective constants resulting from the fit.

In a linear polyatomic molecules with n nuclei, there are $(n - 1)$ nondegenerate stretching vibrations, with σ^+ symmetry and that we label s, and $(n - 2)$ two-dimensional degenerate bending vibrations, with π symmetry, which we label b. Each b two-dimensional degenerate mode has coordinates b_1 and b_2, with $1, 2 \equiv x, y, z$ as the molecular axes.

The vibrational angular momentum for a doubly degenerate vibration

$$\hat{L} = q_{b_1}\hat{p}_{b_2} - q_{b_2}\hat{p}_{b_1} \tag{3.80}$$

can be understood as arising from the superposition of the movements corresponding to the normal vibrations q_{b_1} and q_{b_2}, with a $90°$ phase shift. As a result the molecule starts rotating, thus generating an angular momentum. This picture illustrates the severe mixing between some vibrational and rotational degrees of freedom in a linear top. The vibrational angular momentum quantum number is l, the eigenvalue of $\hat{L}$. The relationship between $\hat{L}$ and $\hat{\pi}$ is rather direct. Indeed, for a degenerate vibration with coordinates b_1, b_2 and associated vibration frequencies $(\tilde{\omega}_{b_1} \equiv \tilde{\omega}_{b_2})$:

$$\hat{\pi}_\alpha = \sum_{b_1,b_2} \zeta^\alpha_{b_1 b_2} \left[\left(\frac{\tilde{\omega}_{b_1}}{\tilde{\omega}_{b_2}} \right)^{1/2} q_{b_1}\hat{p}_{b_2} - \left(\frac{\tilde{\omega}_{b_2}}{\tilde{\omega}_{b_1}} \right)^{1/2} q_{b_2}\hat{p}_{b_1} \right] = \sum_b \zeta^\alpha_b \hat{L}_b \tag{3.81}$$

We use phase conventions for the degenerate harmonic oscillator such that

$$\sigma_{xz}|v_b^{l_b}\rangle = (-1)^{l_b}|v_b^{-l_b}\rangle \tag{3.82}$$

or, more generally

$$E^*|v_b^{l_b}, J\rangle = (-1)^{J-l_b}|v_b^{-l_b}, J\rangle \tag{3.83}$$

with σ_{xz} the point group symmetry operation corresponding to the reflection of the nuclear coordinates through the (xz) plane of the molecule, and E^* the molecular symmetry operation corresponding to the inversion of all coordinates through a reference, space-fixed point [46].

b.   Diagonal Terms

1.   HARMONIC CONTRIBUTIONS ($\hat{H}_v^{\mathrm{har(lin)}}$). The harmonic contribution to the vibrational terms (H_{20}) arises from the following Hamiltonian:

$$\hat{H}_v^{\mathrm{harm(lin)}} = \frac{1}{2}hc \left[\sum_{s=n-1}^{2n-3} \tilde{\omega}_s(\hat{p}_s^2 + q_s^2) + \sum_{b=1}^{n-2} \tilde{\omega}_b(\hat{p}_{b_1}^2 + \hat{p}_{b_2}^2 + q_{b_1}^2 + q_{b_2}^2) \right] \tag{3.84}$$

and

$$\hat{H}_v^{\text{harm(lin)}} = \frac{1}{2}hc\left[\sum_{s=n-1}^{2n-3}\tilde{\omega}_s(\hat{p}_s^2 + q_s^2) + \sum_{b=1}^{n-2}\tilde{\omega}_b(\hat{p}_+\hat{p}_- + q_+q_-)\right] \quad (3.85)$$

with

$$q_\pm = q_{b_1} \pm iq_{b_2} \quad \text{and} \quad \hat{p}_\pm = \hat{p}_{b_1} \pm i\hat{p}_{b_2} \quad (3.86)$$

and, as already stated, with s running over all $(n-1)$ nondegenerate normal modes of vibration, and b running over all other $(n-2)$ two-dimensional degenerate ones. The resulting energies are

$$G(v) = \sum_{i=1}^{2n-3}\tilde{\omega}_i\left(v_i + \frac{d_i}{2}\right) \quad (3.87)$$

with $d_b = 2$ and $d_s = 1$.

Matrix elements useful for the calculation of various terms in the Watsonian are provided in Table X, with l, the signed quantum number of the vibrational angular momentum, related to v_b:

$$l = v_b, v_b - 2, \ldots, -v_b + 2, -v_b \quad (3.88)$$

or, using $k = \sum_b l_b$ (with $K = |k|$) $\quad (3.89)$

$$k = \sum_b v_b, \sum_b v_b - 2, \ldots, -\sum_b v_b + 2, -\sum_b v_b \quad (3.90)$$

The matrix elements are given in second quantization form in Table XI.

TABLE X
Vibrational Matrix Elements for Two-Dimensional Degenerate Vibrations

$\langle v,l \,|\, q_\pm, i\hat{p}_\pm | v+1, l \mp 1 \rangle = [(v \mp l + 2)/2]^{1/2}$

$\langle v,l \,|\, q_\pm, -i\hat{p}_\pm | v-1, l \mp 1 \rangle = [(v \pm l)/2]^{1/2}$

$\langle v,l \,|\, q_\pm q_\mp, -\hat{p}_\pm\hat{p}_\mp | v+2, l \rangle = \frac{1}{2}[(v+l+2)(v-l+2)]^{1/2}$

$\langle v,l \,|\, q_\pm q_\mp, \hat{p}_\pm\hat{p}_\mp | v,l \rangle = (v+1)$

$\langle v,l \,|\, q_\pm q_\mp, -\hat{p}_\pm\hat{p}_\mp | v-2, l \rangle = \frac{1}{2}(v^2 - l^2)^{1/2}$

$\langle v,l \,|\, q_\pm^2, -\hat{p}_\pm^2 | v+2, l \mp 2 \rangle = \frac{1}{2}[(v \mp l + 2)(v \mp l + 4)]^{1/2}$

$\langle v,l \,|\, q_\pm^2, \hat{p}_\pm^2 | v, l \mp 2 \rangle = [(v \mp l + 2)(v \pm l)]^{1/2}$

$\langle v,l \,|\, q_\pm^2, -\hat{p}_\pm^2 | v-2, l \mp 2 \rangle = \frac{1}{2}[(v \pm l - 2)(v \pm l)]^{1/2}$

$\langle v,l \,|\, q_\pm \hat{p}_\mp, \hat{p}_\pm q_\mp | v+2, l \rangle = \frac{-1}{2}i[(v+l+2)(v-l+2)]^{1/2}$

$\langle v,l \,|\, q_\pm \hat{p}_\mp, -\hat{p}_\pm q_\mp | v,l \rangle = \mp i(l \mp 1)$

$\langle v,l \,|\, q_\pm \hat{p}_\mp, \hat{p}_\pm q_\mp | v-2, l \rangle = \frac{1}{2}i(v^2 - l^2)^{1/2}$

With $q_\pm = q_{b1} \pm iq_{b2}$ and $\hat{p}_\pm = \hat{p}_{b1} \pm i\hat{p}_{b2}$

TABLE XI
Vibrational Matrix Elements in Second Quantization Form for
Two-Dimensional Degenerate Vibrations[a]

$$\langle v_b - 1, l_b - 1 | a_{b_+}^+ + hc | v_b, l_b \rangle = [\tfrac{1}{2}(v_b + l_b)]^{1/2}$$

$$\langle v_b + 1, l_b + 1 | a_{b_+}^- + hc | v_b, l_b \rangle = [\tfrac{1}{2}(v_b + l_b + 2)]^{1/2}$$

$$\langle v_b - 1, l_b + 1 | a_{b_-}^+ + hc | v_b, l_b \rangle = [\tfrac{1}{2}(v_b - l_b)]^{1/2}$$

$$\langle v_b + 1, l_b - 1 | a_{b_-}^- + hc | v_b, l_b \rangle = [\tfrac{1}{2}(v_b - l_b + 2)]^{1/2}$$

With

$$a_{b_+}^+ = 1/\sqrt{2}(a_{b_1}^+ - i a_{b_2}^+)$$
$$a_{b_-}^+ = 1/\sqrt{2}(a_{b_1}^+ + i a_{b_2}^+)$$
$$a_{b_+}^- = 1/\sqrt{2}(a_{b_1}^- - i a_{b_2}^-)$$
$$a_{b_-}^- = 1/\sqrt{2}(a_{b_1}^- + i a_{b_2}^-)$$

and

$$a_{b_1}^+ = 1/\sqrt{2}(q_{b_1} - i \hat{p}_{b_1})$$
$$a_{b_2}^+ = 1/\sqrt{2}(q_{b_2} - i \hat{p}_{b_2})$$
$$a_{b_1}^- = 1/\sqrt{2}(q_{b_1} + i \hat{p}_{b_1})$$
$$a_{b_2}^+ = 1/\sqrt{2}(q_{b_2} + i \hat{p}_{b_2})$$

[a] hc = Hermitian conjugate.

One also defines

$$\hat{N}_{b_1} = a_{b_1}^- a_{b_1}^+, \qquad \hat{N}_{b_2} = a_{b_2}^- a_{b_2}^+, \qquad \hat{N}_{b_\pm} = a_{b_\pm}^- a_{b_\pm}^+ \qquad (3.91)$$

leading to, for instance

$$\tilde{h}_{40} \equiv a_{b_+}^+ a_{b_-}^- a_{b'_+}^- a_{b'_-}^+ + h.c. \qquad (3.92)$$

2. l SUBLEVELS. The symmetry properties of the vibrational wavefunctions are directly connected to the value of $|k|$, with $k = \sum_b l_b$. The k/l components of the lower bending levels are detailed in Table XII, using labels $v_b^{|l_b|}$ for linear triatomic molecules, thus with only one such two-dimensional degenerate mode, and $v_b^{|l_b|} v_{b'}^{|l_{b'}|}$ for linear four atomic molecules that have two of them.

In the case of two bending vibrations, for each specific vibrational excitation, one must perform all possible angular momentum vector

TABLE XII

l Sublevels for Two-Dimensional Degenerate Vibrational Excitations in Linear Tops

Triatomic			Four-atomic		
$v_b^{\|l_b\|}$	$\|k\|$	Species	$v_b^{\|l_b\|} v_{b'}^{\|l_{b'}\|}$	$\|k\|$	Species
0^0	0	Σ^+	$0^0\,0^0$	0	Σ^+
1^1	1	Π	$1^1\,0^0$	1	Π
			$0^0\,1^1$	1	Δ
2^2	2	Δ	$2^2\,0^0$	2	Δ
2^0	0	Σ^+	$2^0\,0^0$	0	Σ^+
			$0^0\,2^0$	0	Σ^+
			$0^0\,2^2$	2	Δ
			$1^1\,1^1$	2	Δ
			$1^1\,1^1$	0	Σ^+
			$1^1\,1^1$	0	Σ^-
3^3	3	Φ	$3^3\,0^0$	3	Φ
3^1	1	Π	$3^1\,0^0$	1	Π
			$0^0\,3^1$	1	Π
			$0^0\,3^3$	3	Φ
			$2^2\,1^1$	3	Φ
			$2^2\,1^1$	1	Π
			$2^0\,1^1$	1	Π
			$1^1\,2^0$	1	Π
			$1^1\,2^2$	1	Π
			$1^1\,2^2$	3	Φ
4^4	4	Γ	etc.	etc.	etc.
4^2	etc.	etc.	etc.	etc.	etc.
etc.	etc.	etc.	etc.	etc.	etc.

additions $\vec{l}_b + \vec{l}_{b'}$ to generate all $\vec{k}$ vectors whose absolute value is listed in Table XII. For each combination $(v_b v_{b'})$ there is therefore a $\Pi_b(v_b + 1)$ degeneracy. Within the harmonic oscillator level of approximation, indeed, the energy is independent of l or k. This degeneracy factor includes a factor of 2 arising for each sublevel with $|k| \neq 0$, from $+k$ and $-k$. As we shall indicate later, the degeneracy of sublevels with different values of $|k|$ is removed by diagonal anharmonic terms, while the one between $+k$ and $-k$ components is affected when considering vibration–rotation interactions.

All levels are presented Table XII with their species in $C_{\infty v}$. In $D_{\infty h}$, one needs to add the adequate u/g subscript, corresponding to the symmetry of the wavefunction with respect to the inversion point group symmetry operation (with $u \times u = g \times g = g$, and $u \times g = g \times u = u$).

The two sublevels with different symmetries (Σ^+ and Σ^-) for the $(v_b^{|l_b|} v_{b'}^{|l_{b'}|}) = (1^1\,1^1)$ combination level can be distinguished using the labels

$(v_b^{l_b} v_{b'}^{l_{b'}}) = (1^{+1} 1^{-1})$ and $(1^{-1} 1^{+1})$, thus adopting the convention that, for $k = 0$, the combination with positive (negative) l_b value, that is, of the bending mode printed first, leads to the Σ^+ (Σ^-) level. The two levels can be thought of as corresponding to two different physical situations, with different signed l contributions to k arising from the different bending vibrations. The wavefunctions of the resulting Σ^+ and Σ^- sublevels in the case of $(v_b^{l_b} v_{b'}^{l_{b'}}) = (1^1 1^1)$ need to be defined according to the following transformation:

$$\frac{1}{\sqrt{2}} \left(|v_s, v_b, l_b, v_{b'}, l_{b'}\rangle \pm (-1)^k |v_s, v_b, -l_b, v_{b'}, -l_{b'}\rangle \right) \quad (3.93)$$

thus leading to a $+$ and a $-$ sublevel. Terms in the Hamiltonian, to be introduced in Section III.D, split the related energy levels and define e and f sublevels, as currently used in the literature nowadays. If the double vibrational excitation occurs in the same mode, for example, $(v_b^{l_b} v_{b'}^{l_{b'}}) = (2^0 0^0)$, there is no such ambiguity. There is only one sublevel with $k = 0$, which is of Σ^+ symmetry, that is, the sigma symmetry contained in the symmetric part of the direct product of the normal-mode representations $(\pi \otimes \pi)$. Each double excitation also generates a sublevel of Δ symmetry: $(v_b^{l_b} v_{b'}^{l_{b'}}) = (1^1 1^1)$, $(v_b^{l_b} v_{b'}^{l_{b'}}) = (2^2 0^0)$, and $(v_b^{l_b} v_{b'}^{l_{b'}}) = (0^0 2^2)$. Each of those sublevels carries a two-dimension degeneracy, with all signed l values either positive or negative, thus corresponding to $k = \pm 2$. This degeneracy is removed only when introducing rotation, as already pointed out and further detailed in Section III.D. This scheme can be extended to higher excitation levels with one or two degenerate modes of vibration. It needs to be carefully adapted when there are more than three such modes.

3. DIAGONAL ANHARMONIC TERMS. Diagonal vibrational terms arising from the transformed Hamiltonian $(H_{20} + \hat{h}_{40})$ give rise to the Dunham expansion for a linear top:

$$
\begin{aligned}
G(v) = {} & \sum_{i=1}^{2n-3} \tilde{\omega}_i \left(v_i + \frac{d_i}{2} \right) + \sum_{1 \le j=1}^{2n-3} x_{ij} \left(v_i + \frac{d_i}{2} \right) \left(v_j + \frac{d_j}{2} \right) \\
& + \sum_{1 \le j \le k=1}^{2n-3} y_{ijk} \left(v_i + \frac{d_i}{2} \right) \left(v_j + \frac{d_j}{2} \right) \left(v_k + \frac{d_k}{2} \right) \\
& + \sum_{b \le b'=1}^{n-2} g_{bb'} l_b l_{b'} + \sum_{b \le b'=1}^{n-2} y_b^{bb} \left(v_b + \frac{d_b}{2} \right) l_b^2 + \cdots - B_v k^2
\end{aligned}
\quad (3.94)
$$

The energy with respect to the ground vibrational level is [472,505,534,535]:

$$
\begin{aligned}
G_0(v) &= \sum_{i=1}^{2n-3} \tilde{\omega}_i^0 v_i + \sum_{1 \leq j=1}^{2n-3} x_{ij}^0 v_i v_j + \sum_{1 \leq j \leq k=1}^{2n-3} y_{ijk}^0 v_i v_j v_k \\
&+ \sum_{b \leq b'=1}^{n-2} g_{bb'} l_b l_{b'} + \sum_{b \leq b'=1}^{n-2} y_b^{bb} v_b l_b^2 + \cdots - B_v k^2
\end{aligned}
\tag{3.95}
$$

The term including the principal rotational constant B, although arising from rotation-type matrix elements to be developed later, is J-independent and is therefore included in the pure vibrational contribution. It is critical to include this term when retrieving the pure vibrational energy from some measured band center, specifically G_c or v_0, if the upper or lower level in the transition analyzed has $k \neq 0$. Thus

$$
G_0'(v) = G_0''(v) + G_c + B_v' k'^2 - D_v' k'^4 - B_v'' k''^2 + D_v' k'^4
\tag{3.96}
$$

The contribution in $g_{bb'}$ arises from the same effective operator ($\tilde{h}_{40}$) as the anharmonicity parameters x_{ij}. For a linear molecule with two 2-degenerate bending modes, a relevant diagonal contribution is

$$
\langle v_b, l_b, v_{b'}, l_{b'}, k | g_{bb'} (q_{b_+} \hat{p}_{b_-} + q_{b_-} \hat{p}_{b_+})(q_{b'_+} \hat{p}_{b'_-} + q_{b'_-} \hat{p}_{b'_+}) | v_b, l_b, v_{b'}, l_{b'}, k \rangle
\tag{3.97}
$$

Expressions for the diagonal anharmonicity constants, x_{ij} and $g_{bb'}$, in terms of the potential parameters are given, for example, by Mills [373] and Papousek and Aliev [47] for symmetric tops. Lehmann [528] (see also Jonas [536]) has provided the detailed local-mode expressions for some of them, applied to acetylene (see also Ref. 455):

$$
\begin{aligned}
x_{bb} &= \frac{1}{16} \phi_{bbbb} - \frac{1}{16} \phi_{sbb}^2 \left[\frac{3\tilde{\omega}_s^2 - 8\tilde{\omega}_b^2}{\tilde{\omega}_s(\tilde{\omega}_s^2 - 4\tilde{\omega}_b^2)} \right] \\
g_{bb} &= -\frac{1}{48} \phi_{bbbb} + \frac{1}{16} \phi_{sbb}^2 \left[\frac{\tilde{\omega}_s}{\tilde{\omega}_s^2 - 4\tilde{\omega}_b^2} \right]
\end{aligned}
\tag{3.98}
$$

which, to the limit $\tilde{\omega}_s^2 \gg \tilde{\omega}_b^2$, lead to the relation $x_{bb} = -3 g_{bb}$. As discussed by Lehmann and already mentioned in Section III.B.3.c.2, this relationship, as well as other x–K relations derived for the bends, is, however, not fulfilled in the example of acetylene, because of close cancellation between some terms [528].

Expression (3.95) has a term of the form $(g_{bb} - B)k^2$ that contributes to removal of the $\Pi_b(v_b + 1)$ degeneracy between sublevels with different $|k|$ values arising from the same combination of bending excitations. Thus, for $(g - B) > 0$ (< 0), the sublevels with higher $|k|$ values have higher (lower) energy. This is illustrated in the Section III.B.4.c.3.

c. *Off-Diagonal Terms* $(\hat{H}_v^{\text{anhar(lin)}^{\text{off}-\text{diag}}})$

1. l RESONANCE. Extra, J-independent contributions further remove the l degeneracy within the same vibrational level. It should be remembered that l is related to the $\hat{\pi}$ and $\hat{L}$ operators defined previously, which mix the rotational $(\hat{J}^2)$ and vibrational $(q, \hat{p})$ contributions. We nevertheless include the so-called l resonance in the present chapter devoted to the treatment of vibration. That resonance obeys the following l-off-diagonal, v- and k-diagonal interaction matrix elements [362,363,472,534,535,537]:

$$\langle v_b, l_b, v_{b'}, l_{b'}, k | \tilde{h}_{40} | v_b, l_b \pm 2, v_{b'}, l_{b'} \mp 2, k \rangle$$
$$= \tfrac{1}{4} r_{bb'} \{ (v_b \mp l_b)(v_b \pm l_b + 2)(v_{b'} \pm l_{b'})(v_{b'} \mp l_{b'} + 2) \}^{1/2} \quad (3.99)$$

with relevant terms in the effective operator of the form

$$q_{b_+}^2 q_{b'}^2 + q_{b_-}^2 q_{b'_+}^2 \quad (3.100)$$

and, for l- and v-off-diagonal, k-diagonal interactions:

$$\langle v_b, l_b, v_{b'}, l_{b'}, k | \tilde{h}_{40} | (v_b - 2), (l_b \mp 2), (v_{b'} + 2), (l_{b'} \pm 2), k \rangle$$
$$= \tfrac{1}{16} (r_{bb'} + 2g_{bb'}) [(v_{b'} \pm l_{b'} - 2)(v_b \pm l_b) \quad (3.101)$$
$$\times (v_{b'} \pm l_{b'} + 2)(v_{b'} \pm l_{b'} + 4)]^{1/2}$$

Other terms off-diagonal in k are discussed in the literature [47,534,538]. Related J-dependent matrix elements terms [539] are presented in Section III.D.

The $r_{bb'}$ parameter can itself be expanded into a series including v and J dependences [540]:

$$r_{bb'} = r_{bb'}^0 + r_{bb'}^J J(J+1) + r_{bb'}^{JJ} J^2(J+1)^2 + r_{bbb'}(v_b + 1)$$
$$+ r_{bb'b'}(v_{b'} + 1) + r_{bb'}^k k^2 \quad (3.102)$$

At the present stage of the theoretical development, thus without considering rotation, the structure of the $(v_b, v_{b'}) = (1, 1)$ l matrix for a linear molecule with 2D degenerate vibrations is

$\lvert v_b^{l_b}, v_{b'}^{l_{b'}} \rangle$	$\lvert 1^{+1}, 1^{+1} \rangle$	$\lvert 1^{-1}, 1^{-1} \rangle$	$\lvert 1^{+1}, 1^{-1} \rangle$	$\lvert 1^{-1}, 1^{+1} \rangle$
$\lvert 1^{+1}, 1^{+1} \rangle$	D^+			
$\lvert 1^{-1}, 1^{-1} \rangle$		D^+		
$\lvert 1^{+1}, 1^{-1} \rangle$			D^-	$r_{bb'}$
$\lvert 1^{-1}, 1^{+1} \rangle$			$r_{bb'}$	D^-

$$(3.103)$$

with

$$D^{\pm} = \tilde{\omega}_b^0 + \tilde{\omega}_{b'}^0 + x_{bb} + x_{b'b'} + x_{bb'} + g_{bb} + g_{b'b'} \pm g_{bb'} + \cdots - Bk^2$$

$$(3.104)$$

whose diagonalization separates the $k = 0$, Σ^+, and Σ^- sublevels:

$$G_0(1^{+1}, 1^{-1})^+ = G_0(1, 1, \Sigma^+) = D^- + r_{bb'}$$
$$G_0(1^{-1}, 1^{+1})^- = G_0(1, 1, \Sigma^-) = D^- - r_{bb'}$$

$$(3.105)$$

using the convention that $l_b > 0$ labels the Σ^+ sublevel, as stated previously. Whenever there is more than one level of the same symmetry generated by the same vibrational combination, the higher-energy one is conventionally labeled with an I, the next down in energy with a II, and so on.

As a general example, Table XIII lists some of the bending levels in $^{12}C_2H_2$. The energies in that table include additional contributions arising from various anharmonic resonances, as detailed in Section III.C. Labels such as I and II are not indicated.

The l-resonance terms we have introduced add a new dimension in the MIME as l matrices need thus to be set and diagonalized within the global structure we have built in the previous sections of this chapter.

2. ANHARMONIC RESONANCES. The lower-order anharmonic resonances in linear tops arise, as for asymmetric tops, from the v off-diagonal contributions in the H_{30} and H_{40} terms in the Watsonian. The anharmonic resonance matrix elements for the coupling between levels with excitation of nondegenerate vibrations only are similar to those listed for asymmetric tops in Section III.B.3. In the case of degenerate vibrations, some l dependence is introduced.

TABLE XIII

Observed or Predicted ($*$) Vibrational Energy^a of the first bending l sublevels in $^{12}C_2H_2$ [450]. Only the (v_4,v_5) combinations with even k values leading to g symmetry level (with v_4 the *trans*- and v_5 the *cis*-bending modes) are listed. The predicted energies account for the various anharmonic couplings introduced in the Cluster model of Section III.C.^{a}

$(v_4,v_5)_g$	$k=0$	$k=2$	$k=4$	$k=6$	$k=8$	$k=10$
(2,0)	1230.4	1233.5				
(0,2)	1449.1	1463.0				
(4,0)	2487.0*	2489.3*	2496.2*			
(2,2)	2648.0	2666.2	2715.0*			
	2683.8*	2692.7*				
(0,4)	2880.2	2894.1	2934.9*			
(6,0)	3767.0*	3769.4*	3776.4*	3788.1*		
(4,2)	3884.0*	3880.4*	3922.9*	3996.2*		
	3906.2*	3911.5*	3967.1*			
	3940.3*	3947.4*				
(2,4)	4060.0*	4076.0*	4127.4*	4208.9*		
	4086.9*	4099.7*	4164.9*			
	4124.6*	4135.1*				
(0,6)	4293.1*	4306.8*	4347.9*	4416.4*		
(8,0)	5068.5	5071.0*	5078.4*	5090.7*	5107.7*	
(6,2)	5143.0*	5141.0*	5138.4*	5207.5*	5305.1*	
	5172.9*	5178.0*	5190.2*	5268.7*		
	5216.5*	5222.7*	5240.7*			
(4,4)	5266.3*	5282.6*	5328.5*	5404.8*	5510.9*	
	5279.2*	5311.5*	5371.2*	5459.8*		
	5303.0*	5347.6*	5420.5*			
	5338.8*	5395.6*				
	5387.1*					
(2,6)	5455.4*	5470.6*	5517.0*	5600.2*	5714.7*	
	5499.9*	5512.7*	5550.7*	5648.6*		
	5554.0*	5564.9*	5597.1*			
(0,8)	5689.53*	5703.1*	5743.8*	5811.7*	5907.2*	
(10,0)	6386.7	6389.4*	6397.8*	6411.4*	6430.5*	6453.6*
(8,2)	6423.4	6422.5*	6421.9*	6424.4*	6519.6*	6640.3*
	6456.8*	6462.2*	6476.0*	6495.4*	6595.8*	
	6509.6	6515.1*	6532.0*	6559.2*		
(6,4)	6506.5*	6499.1*	6540.1*	6610.8*	6710.8*	6839.7*
	6527.8*	6532.5*	6586.0*	6668.4*	6779.8*	
	6561.2*	6568.4*	6635.6*	6731.2*		
	6606.9*	6614.3*	6695.2*			
	6665.8*	6673.2*				
(4,6)	6644.9*	6661.3*	6711.1*	6791.0*	6900.8*	7040.1*
	6675.7*	6690.3*	6753.6*	6845.6*	6966.9*	
	6713.6*	6724.7*	6801.9*	6907.2*		
	6763.7*	6773.2*	6863.0*			
	6826.6*					
(2,8)	6836.7*	6850.3*	6894.7*	6970.2*	7085.0*	7232.4*
	6903.2*	6914.2*	6951.6*	7013.8*	7144.3*	
	6973.1*	6984.1*	7017.0*	7071.0*		
(0,10)	7071.5*	7084.8*	7124.6*	7191.4*	7285.6*	7407.5*

^{a} Corresponding to hypothetical $J = 0$ levels when $k \neq 0$ with $k = l_4 + l_5$.

TABLE XIV
Some Anharmonic Resonance Matrix Elements for Linear Species

$$\langle v_s, v_b, l_b | \tilde{h}_{s/bb} | v_s - 1, v_b + 2, l_b \rangle = \tfrac{1}{2} K_{s/bb} \{ [(v_b + 2)^2 - l_b^2] v_s \}^{1/2}$$

$$\langle v_s, v_{s'}, v_b, l_b, k | \tilde{h}_{s/s'bb} | (v_s + 1), (v_{s'} - 1), (v_b - 2), l_b, k \rangle$$
$$= \tfrac{1}{4} K_{s/s'bb} [(v_s + 1) v_{s'} (v_b^2 - l_b^2)]^{1/2}$$

$$\langle v_s, v_{s'}, v_b, l_b, v_{b''}, l_{b'}, k | \tilde{h}_{s/s'bb'} | (v_s + 1),$$
$$(v_{s'} - 1), (v_b - 1), (l_b \pm 1), (v_{b'} - 1), (l_{b'} \mp 1), k \rangle =$$
$$\tfrac{1}{4} K_{s/s'bb''} [(v_s + 1) v_{s'} (v_b \mp l_b)(v_{b'} \pm l_{b'})]^{1/2}$$

$$\langle v_s, v_{s'}, v_b, l_b, v_{b'}, l_{b'}, k | \tilde{h}_{sb/s'b'} | (v_s + 1), (v_{s'} - 1), (v_b + 1),$$
$$(l_b \mp 1), (v_{b'} - 1), (l_{b'} \pm 1), k \rangle =$$
$$\tfrac{1}{4} K_{sb/s'b'} [(v_s + 1) v_{s'} (v_b \mp l_b + 2)(v_{b'} \mp l_{b'})]^{1/2}$$

$$\langle v_b, l_b, v_{b'}, l_{b'}, k | \tilde{h}_{bb/b'b'} | v_b - 2, l_b, v_{b'} + 2, l_{b'}, k \rangle$$
$$= \tfrac{1}{4} K_{bb/b'b'} \{ [(v_b^2 - l_b^2)][(v_{b'} + 2)^2 - l_{b'}^2] \}^{1/2}$$

$$\langle v_b, l_b, v_{b'}, l_{b'}, k | \tilde{h}_{bb/b'b'} | v_b - 2, l_b \mp 2, v_{b'} + 2, l_{b'} \pm 2, k \rangle$$
$$= \tfrac{1}{16} (r_{bb'} + 2g_{bb'}) [(v_b \pm l_b - 2)(v_b \pm l_b)(v_{b'} \pm l_{b'} + 2)(v_{b'} \pm l_{b'} + 4)]^{1/2}$$

Various relevant matrix elements are presented in Table XIV. The notation for the coupling constants is the same as for Table VIII. Higher-order terms for four-atom species are discussed by Perevalov et al. [539]. One should notice that the numerical coefficients in front of the K resonance parameters are sometimes partly included in those parameters. One should be careful for consistency when defining and using the coupling matrix elements.

Anharmonic resonances only affect levels of the same symmetry. Only those l sublevels of one symmetry may thus participate in the same interaction scheme. The resonances, in that case, further contribute to the removal of degeneracy of the l sublevels. This statement strictly holds in the absence of J-dependent interactions, which are considered later in this chapter only.

As an example, Fig. 29 presents the bending levels observed by Huet et al. in $^{12}C_2D_2$ through far infrared and infrared transitions [505]. Although the pattern of levels is, in principle, identical to the one in $^{12}C_2H_2$, because of different vibrational frequencies, a strong quartic anharmonic resonant coupling of Darling–Dennison type already occurs between the very first overtone levels. For example, it strongly couples the pairs of levels $2v_4/2v_5$, $3v_4/v_4 + 2v_5$ and $2v_4 + v_5/3v_5$. It is also observed in $^{12}C_2H_2$, but at much

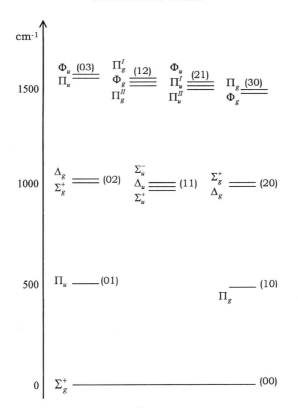

Figure 29. Bending levels observed in $^{12}C_2D_2$, with ν_4 and ν_5 the *trans* (π_g)- and *cis* (π_u)-bending modes, respectively. The vibrational levels are identified as $(\nu_4\nu_5)$ and their associated k sublevels are identified by their symmetry. The magnitude of the k splitting is greatly exaggerated.

higher vibrational excitations, when the related levels get closer in energy because of the relative vibrational frequencies and anharmonicities involved.

3. ANHARMONIC AND l RESONANCES. For those anharmonic resonances involving two-dimensional degenerate vibrations, as pointed out by Jonas [536], one must be careful not to first prediagonalize the l matrices, before considering the anharmonic resonances. The full matrix must be built, including l and anharmonic resonances, before diagonalization. Some of the numerical coefficients are otherwise affected.

As an example, let us consider the anharmonic resonance term developed in Table XIV coupling the $(\nu_s, \nu_{s'}, \nu_b, \nu_{b'}) = (1, 0, 1, 1)$ and (0,1,0,0) levels, thus also involving the $(\nu_b, \nu_{b'}) = (1, 1)$ l matrix discussed previously.

H:

$\left|v_s,v_{s'},v_b^{l_b},v_{b'}^{l_{b'}}\right\rangle$	$\left|0,1,0^0,0^0\right\rangle$	$\left|1,0,1^1,1^{-1}\right\rangle$	$\left|1,0,1^{-1},1^1\right\rangle$
$\left|0,1,0^0,0^0\right\rangle$	$G_0(0,1,0^0,0^0)$	$-\tfrac{1}{4}K_{s'/sbb'}$	$-\tfrac{1}{4}K_{s'/sbb'}$
$\left|1,0,1^1,1^{-1}\right\rangle$	$-\tfrac{1}{4}K_{s'/sbb'}$	$G_0(1,0,1^{+1},1^{-1})$	$r_{bb'}$
$\left|1,0,1^{-1},1^1\right\rangle$	$-\tfrac{1}{4}K_{s'/sbb'}$	$r_{bb'}$	$G_0(1,0,1^{-1},1^{+1})$

O:

$$\begin{vmatrix} 1 & 0 & 0 \\ 0 & \tfrac{1}{\sqrt{2}} & -\tfrac{1}{\sqrt{2}} \\ 0 & \tfrac{1}{\sqrt{2}} & \tfrac{1}{\sqrt{2}} \end{vmatrix}$$

O^THO:

$\left|v_s,v_{s'},v_b^{l_b},v_{b'}^{l_{b'}}\right\rangle$	$\left|0,1,0^0,0^0\right\rangle$	$\left|1,0,1^1,1^{-1}\right\rangle$	$\left|1,0,1^{-1},1^1\right\rangle$
$\left|0,1,0^0,0^0\right\rangle$	$G_0(0,1,0^0,0^0)$	$-\tfrac{\sqrt{2}}{4}K_{s'/sbb'}$	0
$\left|1,0,1^1,1^{-1}\right\rangle$	$-\tfrac{\sqrt{2}}{4}K_{s'/sbb'}$	$G_0(1,0,1^{+1},1^{-1})+r_{bb'}$	0
$\left|1,0,1^{-1},1^1\right\rangle$	0	0	$G_0(1,0,1^{-1},1^{+1})-r_{bb'}$

Figure 30. Matrix Hamiltonian and similitude transformation needed to represent the coupled anharmonic and l resonances in a linear species such as acetylene.

The matrix Hamiltonian and the required similitude transformation are defined in Fig. 30, leading to properly define the $\Sigma^+(e)$ and $\Sigma^-(f)$ sublevels. The factor $\sqrt{2}$ in the off-diagonal elements, neglected in the acetylene literature for a long period of time, results from this transformation. The e and f characterization of the levels [541] is defined in Section III.D.

The Δ sublevels were not discussed here because they are unaffected by the anharmonic resonance term selected in this example, strictly speaking. However, they have to be included for a faithful description of the spectrum because they are connected to the e and f levels of the Σ components through J-dependent terms also considered in Section III.D.

The sequence of couplings we have discussed leads to define the relative energy of those levels as presented, for a specific case, in Fig. 31, where the couplings are illustrated in the sequence of (1) the Dunham energy without

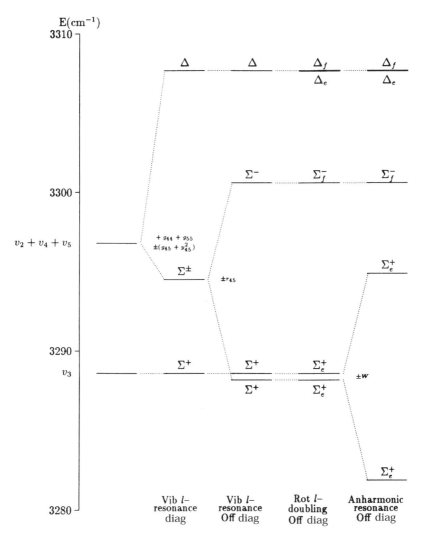

Figure 31. Scheme of the quartic anharmonic resonance interaction between the levels v_3 and $v_2 + v_4 + v_5$ in $^{12}C_2H_2$, including l effects, with modes 1 (σ_g^+) and 3 (σ_u^+) the symmetric and antisymmetric CH stretches, respectively; mode 2 (σ_g^+), the CC stretch; and modes 4 (π_g) and 5 (π_u), the *trans* and *cis* bends, respectively (adapted from Vander Auwera et al. [451]).

introduction of the vibrational degeneracy; (2) the diagonal l-resonance interaction separating sublevels with different $|k|$ values, thus accounting for the term in $\sum_{b \leq b'}(g_{bb'}^0 l_b l_{b'} - B_v k^2)$; (3) the off-diagonal l-resonance terms, with r_{45}, pushing apart the Σ^+ and Σ^-, $l = 0$ sublevels; (4) the rotational l

doubling (see Section III.D.2.c.1.) defining the *elf* character of the rotational levels; and (5) the 3/245 anharmonic resonance between v_3 (Σ^+) and the Σ^+ component of $v_2 + v_4 + v_5$. The levels should be successively labeled with (1) (v_2, v_3, v_4, v_5), (2) $(v_2, v_3, v_4^{|l_4|}, v_5^{|l_5|})$, (3) $(v_2, v_3, v_4^{l_4}, v_5^{l_5})$, (4) $(v_2, v_3, v_4^{l_4}, v_5^{l_5})$, and (5) a mixing of both zeroth-order labelling, weighted using the eigenvectors from the anharmonic matrix diagonalization. The present sequential description illustrates the various coupling cases for didactic purposes. According to the procedure defined in present section, the level energies ought to be represented directly from diagonalization of the MIME.

C. Level Clustering

1. Vibrational Polyads

a. One-Resonance Polyads. Let us assume a system with two oscillators ($|v_1, v_2\rangle$), coupled by one F resonance ($K_{1/22}$).

The diagonal and off-diagonal elements in the MIME are given by the following, restricting the development to terms in x_{ij}:

$$\left\langle v_1, v_2 \left| \frac{\hat{H}_v}{hc} \right| v_1, v_2 \right\rangle = G(v) \equiv D$$
$$= \tilde{\omega}_1^0 v_1 + \tilde{\omega}_2^0 v_2 + x_{11} v_1^2 + x_{12} v_1 v_2 + x_{22} v_2^2 \tag{3.106}$$

$$\left\langle v_1, v_2 \left| \frac{\hat{H}_v}{hc} \right| v_1 - 1, v_2 + 2 \right\rangle \equiv F = \frac{1}{2\sqrt{2}} K_{1/22} [v_1(v_2+1)(v_2+2)]^{1/2} \tag{3.107}$$

with, in equation (3.106), $\hat{H}_v \equiv \hat{H}_v^{\text{diag}} + \hat{H}_v^{\text{anh diag}} = H_{20} + \tilde{h}_{40}^{\text{diag}}$ and, in equation (3.107), $\hat{H}_v \equiv \tilde{h}_v^{\text{Fermi}}$.

The structure of the MIME is given in Fig. 32, for $v_i = 0, 1, 2$. The full MIME can be block-diagonalized into so-called polyads [108,486,542–544], as also demonstrated in that figure with rows and columns up to $v_i = 6$ taken into account.

b. Multiresonance Polyads. We can repeat the procedure for a system with three vibrational degrees of freedom ($|v_1 v_2 v_3\rangle$) coupled by two cubic resonances, of the type, for example, $K_{1/22}$ (F_1) and $K_{11/3}$ (F_2) thus enforced by near degeneracies $\tilde{\omega}_3 \approx 2\tilde{\omega}_1 \approx 4\tilde{\omega}_2$. The MIME can be built in two successive steps, as achieved above for the two-oscillator system, with the final result indicated in Fig. 33.

$|v_1,v_2\rangle$	$|0,0\rangle$	$|1,0\rangle$	$|2,0\rangle$	$|0,1\rangle$	$|1,1\rangle$	$|2,1\rangle$	$|0,2\rangle$	$|1,2\rangle$	$|2,2\rangle$
$|0,0\rangle$	D								
$|1,0\rangle$		D					F		
$|2,0\rangle$			D					F	
$|0,1\rangle$				D					
$|1,1\rangle$					D				
$|2,1\rangle$						D			
$|0,2\rangle$		F					D		
$|1,2\rangle$			F					D	
$|2,2\rangle$									D

$|v_1,v_2\rangle$	$|0,0\rangle$	$|1,0\rangle$	$|0,2\rangle$	$|2,0\rangle$	$|1,2\rangle$	$|0,4\rangle$	$|3,0\rangle$	$|2,2\rangle$	$|1,4\rangle$	$|0,6\rangle$
$|0,0\rangle$	D									
$|1,0\rangle$		D	F							
$|0,2\rangle$		F	D							
$|2,0\rangle$				D	F					
$|1,2\rangle$				F	D	F				
$|0,4\rangle$					F	D				
$|3,0\rangle$							D	F		
$|2,2\rangle$							F	D	F	
$|1,4\rangle$								F	D	F
$|0,6\rangle$									F	D

Figure 32. Vibration interaction matrix (upper) and level clustering (lower) for a two-mode system connected by a Fermi anharmonic resonance of the type $nv_1/2nv_2$.

$|v_1 v_2 v_3\rangle$	$|0,0,0\rangle$	$|0,1,0\rangle$	$|0,2,0\rangle$	$|1,0,0\rangle$	$|0,3,0\rangle$	$|1,1,0\rangle$	$|0,4,0\rangle$	$|1,2,0\rangle$	$|2,0,0\rangle$	$|0,0,1\rangle$
$|0,0,0\rangle$	D									
$|0,1,0\rangle$		D								
$|0,2,0\rangle$			D	F_1						
$|1,0,0\rangle$			F_1	D						
$|0,3,0\rangle$					D	F_1				
$|1,1,0\rangle$					F_1	D				
$|0,4,0\rangle$							D	F_1		
$|1,2,0\rangle$							F_1	D	F_1	
$|2,0,0\rangle$								F_1	D	F_2
$|0,0,1\rangle$									F_2	D

Figure 33. Vibrational interaction matrix and level clustering for a three-mode system connected by two Fermi anharmonic resonances, of the type $nv_1/2nv_2$ (F_1) and $2nv_1/nv_3$ (F_2).

The point of interest is, again, the formation of polyads, now with two anharmonic resonances. The notion of polyad in the spectroscopic literature concerns a specific set of interacting vibrational levels in a well-defined energy range, thus extends over the whole energy pattern, in the selected example. We also use, although polyad is the conventional denomination, "vibrational or V-cluster" to label each block of interacting levels. Our aim is to build the full MIME in a way allowing for new concepts, such as chaos, to naturally fit in. Interesting comments are given on this subject by Zhilinskii [112], Kellman [493], and Quack [544], among several other authors.

The present block diagonalization of the full vibration Hamiltonian into polyads or V-clusters is consistent with the existence of a few leading anharmonic resonance terms that, in normal-mode coordinates, shape the potential energy hypersurface.

According to the investigations on N_2O and acetylene illustrated later in this chapter, the clusters demonstrate the following features:

1. All vibration levels—in the range so far experimentally investigated— in the potential well do belong to one of the V-clusters.
2. Each level belongs to one V-cluster only.
3. All levels in a V-cluster are characterized by the same pseudo– quantum numbers of vibration.
4. There are rules allowing such numbers to be deduced for a specific molecule, once the dominant anharmonic resonances are known from the spectral analysis or from ab initio calculations.

Points 1 and 2 need to be empirically checked thus, so far. Points 3 and 4 are discussed in the next section.

Actually, the polyads picture in the selected example of Fig. 33 holds as long as the relation $\tilde{\omega}_3 \approx 2\tilde{\omega}_1 \approx 4\tilde{\omega}_2$ is fulfilled. In other words, if the respective anharmonicity constants x_{ij} of the vibrations are too different, then, as we stated in the classical picture presented in Section III.B.1, there is an increasing energy mismatch between the interacting levels and possibly a closer energy degeneracy between other sets of levels. In a simple second-order perturbation picture dealing with the off-diagonal anharmonicity resonance terms, the effects of the couplings follow that trend. The clustering may thus evolve with the excitation energy, although this evolution is usually smooth, given the respective orders of magnitude of $\tilde{\omega}$ and x values. In larger molecules, one can expect some resonances to replace others or, as we discuss later, extra resonances to enlarge the polyad size, in highly excited vibrational ranges.

c. Pseudo Quantum Numbers

1. CONSTANTS OF THE MOTION. In the one resonance system described in Section III.C.1.*a*, a $K_{1/22}$ coupling, thus a 1:2 resonance, was involved. The polyads with $N = 0, 1, 2, 3$ of the MIME were detailed. It appears that all levels such that $(v_1/1) + (v_2/2) = \text{constant} = N$, belong to the same polyad. Following the presentation from Sitja [545], the scheme can be easily expressed more generally in terms of a *A:B* resonance, thus the resonance such that levels $|v_1, v_2\rangle$ are coupled to levels $|v_1 - A, v_2 + B\rangle$ using the anharmonic resonance operator $\hat{K}_{AB} = K_{A_1/B_2}(a_1^{+A} a_2^{-B} + a_1^{-A} a_2^{+B})$. All levels such that

$$\frac{v_1}{A} + \frac{v_2}{B} = N \tag{3.108}$$

belong to the same cluster, of dimensions $(N + 1) \times (N + 1)$. The v_1 and v_2 quantum numbers do not hold anymore because of the mixing induced by the coupling matrix element. The two quantum numbers are replaced by the single N pseudo–quantum number. One may then refer to the ordering of the levels by increasing or decreasing energy, within the set of levels characterised by N, as calculated from some zeroth-order model. One should be careful when comparing this ordering from different sources in the literature.

The extension to the two resonance system, $K_{1/22}$ (F_1) and $K_{11/3}$ (F_2), presented in Section III.C.1.*b* can be performed by simple extrapolation. A look at the structure of the clusters indeed shows that all levels such that

$$2v_1 + v_2 + 4v_3 = N \tag{3.109}$$

belong to the same cluster, with an index number of $M = v_1 + v_2 + v_3$. The generalization of this double interaction scheme (coupling levels $|v_1, v_2, v_3\rangle$ and $|v_1 - A, v_2 + B, v_3\rangle$ on one hand and levels $|v_1, v_2, v_3\rangle$ and $|v_1 - C, v_2, v_3 + D\rangle$ on the other hand) defines the resonance quantum number:

$$\frac{Cv_1}{A} + \frac{v_2}{B} + Dv_3 = N \tag{3.110}$$

2. GENERAL PROCEDURE. A general procedure was presented by Kellman [493,546,547] (see also Refs. 548–551). He demonstrated a method to find the constants of motion related to the remnant vibrational quantum numbers

for a mth multiresonant system with m_{int} vibrational degrees of freedom. In this method, one associates a vector with each resonance operator. Some n of these m vectors are linearly independent and form a basis set defining a vector space of dimension t. The $(m_{\text{int}} - t)$ remaining vectors are orthogonal to the vector space and define as many constants of the motion, with which quantum numbers can be associated. The number of quantum numbers is thus also defined, with respect to the number of degrees of freedom and the number of linearly independent couplings. The latter concept is illustrated in the next section.

Let us illustrate here this method for a species with three normal modes, 1, 2, and 3, and two anharmonic resonances, $1/22$ and $11/3$. Thus, $m_{\text{int}} = 3$ and $m = 2$. The resonances define two linearly independent vectors:

$$\begin{aligned}
(\Delta v_1, \Delta v_2, \Delta v_3) &= (1, -2, 0) \\
(\Delta v_1, \Delta v_2, \Delta v_3) &= (2, 0, -1)
\end{aligned} \tag{3.111}$$

In this case, n, the number of linearly independent resonances is 2. There is therefore $(m_{\text{int}} - t)$, that is, 1 remnant constant of the motion. It corresponds to a vector perpendicular to the resonance vectors, which can be found using relations (3.111), with

$$\begin{aligned}
\Delta v_1 - 2\Delta v_2 &= 0 \\
2\Delta v_1 - \Delta v_3 &= 0 \\
\Rightarrow 2\Delta v_1 &= 4\Delta v_2 = \Delta v_3
\end{aligned} \tag{3.112}$$

Thus, any combination of v_i values such that $2pv_1 = pv_2 = 4pv_3$, with p a positive integer is a satisfactory associated quantum number. It is obviously related to the 2:1:4 ratio of the zeroth-order normal-mode vibrational frequencies. We select $p = 1$, defining as polyad or V-cluster quantum number:

$$N = 2v_1 + v_2 + 4v_3 \tag{3.113}$$

A new constant of the motion, N, is defined, which accounts for the vibrational clustering induced by the active set of anharmonic resonances, with several possible equivalent representations in terms of the combination of the zeroth-order normal-mode quantum numbers. This new constant of motion is directly related to the intramolecular dynamical properties of the molecule [493,501]. Although this is not the purpose of the present review to detail the latter concepts (see, e.g., Refs. 481 and 552), we briefly illustrate some of the related features in the next section.

At this stage, one can therefore build the MIME very systematically, polyad by polyad, by gathering all zeroth-order vibrational levels whose adequate combination of normal-mode quantum numbers leads to the same value of N. An index number allows the levels to be identified within a polyad.

The only prerequisite for applying such a procedure is to identify the dominant anharmonic resonances. This remains unfortunately, a major challenge, but for those structural resonances defined in Section III.B.3.a.1. One should aim at some kind of a priori identification of the dominant anharmonic resonances, using a procedure as systematic as those nowadays currently available for characterizing the normal modes of vibration in a species. Meanwhile, this crucial step can be achieved only either from the extensive spectral analysis or from large-scale ab $initio$ calculations. It still remains most difficult to achieve reliably, in particular for larger molecules. One should mention the apparent universal F resonance evidenced by Quack [544] in the sp^3 CH chromophores in species of the type CHX_3 (X = D, F, Cl, Br, CF_3 etc.). The universality of some structural DD resonances was introduced earlier, in Section III.B.

This concept of clustering should also help counting the number of levels, as demonstrated by Sadovskii and Zhilinskii [553]. Level density is an essential concept in unimolecular reaction dynamics. One should also mention the existence of so-called vibration–rotation clusters in molecules [554,555], although not directly considered in the present review.

2. Examples

a. Introduction. Detailed examples validating the cluster picture are provided in the literature by dedicated studies of nitrous oxide (N_2O) and of different isotopomers of acetylene (C_2H_2), which have actually supported the developments just presented and are detailed in this section. There are other species for which the quality of the experimental data allowed or ought to allow a global picture of the vibrational levels to be put forward. Most concerned species are listed hereafter with some, nonexhaustive, references. They include the triatomic species CO_2 [144,145,556–558], CS_2 [559–561], HCN [562–564], NO_2 [565], H_2S [566], H_2Se [567], OCS [568,569], and HOCl [570]. Besides acetylene, $^{12}C_2H_2$ and $^{12}C_2D_2$, the cluster picture could possibly be tested against other few four-atom species, such as $^{13}C_2H_2$ [571], C_2HF [510,572–577], C_2HBr [578], and H_2CO [579–582], for which a fair number of relevant data are available today. Concerning larger species, methane (CH_4) [159,550,583–588] and ethylene (C_2H_4) [376,589–591] are among the very few molecules for which enough precise information is today available to test global-type models. Some larger molecules are worth checking for any relevant trends for generalized clustering effects. It must be realized that the analysis of high-resolution vibration–rotation data is a prerequisite to any reliable information in this context.

In the present review, we consider the dedicated literature concerning $^{14}N_2\,^{16}O$, $^{12}C_2H_2$, and $^{12}C_2D_2$, to illustrate various features around vibrational level clustering. None of the methane literature, of very high interest to the global analysis concept, is reviewed here and the reader is referred to papers dedicated to this topic [e.g., 159,550,583–588]. We illustrate the preliminary results concerning ethylene in terms of vibrational clustering, according to still unpublished results from ULB [592]. The extension of the vibrational cluster picture to systematically include couplings of rotation nature is also highlighted later. Among all isotopomers of acetylene studied up to the overtone range, monodeuteroacetylene, $^{12}C_2HD$, presents less interest in the context of clustering effects, because of its asymmetry, leading to mismatch between otherwise resonant levels [444,593,594]. Rather, the related spectroscopic data prove relevant for discussing unperturbed overtone band intensities, as in Section III.E.

It is of interest to point out that, for nitrous oxide, a most accurate global fit, including rotation, was performed in the literature by Teffo et al. [595] before the cluster picture was introduced [596]. This was not the case for those isotopomers of acetylene demonstrating several efficient anharmonic couplings and studied in the literature: $^{12}C_2H_2$ and $^{12}C_2D_2$. Previous attempts of global fits in acetylene were achieved using force field and variational methods [52,143], local-mode [151,597–599] and algebraic [438,446,448,489,600,601] approaches, in particular (see also Refs. 602 and 603). However, they considered only a fraction of the vibrational degrees of freedom, usually excluding the bending vibrations, and achieved standard deviations on the level energies of the order of several reciprocal centimeters. Using the MIME model including the present cluster picture, Herman and co-workers reproduced almost all observed levels in $^{12}C_2H_2$ [450] and in $^{12}C_2D_2$ [525] up to the visible range, within better than $1\,cm^{-1}$. A similar study is actually planned on $^{13}C_2H_2$ [604]. The Cluster model, as we named it for acetylene, and earlier versions of it [605–607] already allowed high-resolution spectroscopy and intramolecular dynamics to be linked by various authors [1,493,552,608,609] and the apparition of chaos in acetylene to be discussed [112,481,482,525,533,610] (see, however, comments by Quack and Lewerenz [544,611]).

b.   Nitrous Oxide

1.   LEVEL CLUSTERING.   The model we discussed in Section III.C.1.*b* dealing with a triatomic species with $\tilde{\omega}_3 \approx 2\tilde{\omega}_1 \approx 4\tilde{\omega}_2$, and leading to two linearly

independent resonance vectors $(\Delta v_1, \Delta v_2, \Delta v_3)$ = (1,2,0) and (2,0,1), applies to nitrous oxide, $^{14}N_2{}^{16}O$. This is a linear species (NNO) with four vibrational degrees of freedom. Two of them are degenerate, defining the bending mode, mode 2. Modes 1 and 3 are, primarily, the NO and NN stretching vibrations, respectively. There are thus $m_{int} - t = 2$ constants of the motion. One of them has been identified above as $N = 2v_1 + v_2 + 4v_3$. The other corresponds to $|k|$ or $|l_2|$, the vibrational angular momentum quantum number. All levels with the value of N and the same vibrational symmetry are gathered in the same polyad or V-cluster. As an example, the polyad $\{N, k\} = \{12, 0\}$ is detailed in Table XV. The shift induced by the anharmonic resonances in the range considered in that table can be as high as $115 \, cm^{-1}$, for the highest energy level in the polyad. All predicted values correspond to the observed ones within less than $1 \, cm^{-1}$. Results detailed by Campargue et al. [596] demonstrate that the cluster picture in N_2O seems to hold up to the visible range, with the level at $14934.267 \, cm^{-1}$ the highest energy vibrational level so far reported experimentally, using the ICLAS technique described in Section IV.C.

The vibrational dependence of the rotational parameters was demonstrated to also fit the cluster picture [596,612]. The related procedure is presented in Section III.D.4.d, for acetylene.

2. BAND INTENSITIES. Let us detail the analysis of the relative absorption band intensities in the case of the $\{12, 0\}$ polyad we have just discussed. The bands observed by Fourier transform spectroscopy access 10 out of the 16 levels in this polyad [596].

Bands leading to levels with excitation of modes 1 and 3 are expected to carry the strongest intensity and define the zeroth-order bright states, as defined in Section III.C.3.a.3. There are four such levels involved in $\{12, 0\}$: $(v_1 v_2^{l_2} v_3)$ = (00^03), (20^02), (40^01), and (60^00). The composition of the eigenvectors resulting from the V-cluster matrix diagonalization shows that there is little mixing between those four zeroth-order levels. Each of their wavefunction is however spread into the one of several close by levels, defining sub-polyads. In a 2×2 interaction scheme, assuming that the other levels are zeroth-order dark levels, the band intensity should correspond to the square of the fraction of the zeroth-order bright state they contain, in agreement with the simple model detailed in Section III.C.3.a.3. The measured relative absorption band intensities within each sub-polyad are compared to those predictions in Table XVI. The agreement is rather satisfactory. It is even more convincing if one takes into account that the transitions unobserved in the polyad ($I_{rel} = 0$) correspond to those levels that carry less than 1% of any of the zeroth-order bright states.

TABLE XV
Vibrational Levels and Their Dominant Composition in the $\{N,k\} = \{12,0\}$ Vibrational Polyad in $^{14}N_2\,^{16}O$[a]

60^00	40^01	20^02	00^03
52^00	32^01	12^02	
44^00	24^01	04^02	
36^00	16^01		
28^00	08^01		
110^00			
012^00			

Rank	$\left(v_1 v_2^{l_2} v_3\right)^0$	$(G_{0\,pre.})^0$	$G_{0\,pre.}$	Dominant Contribution
1	00^03	6581.24	6580.87	00^03 (100%)
2	04^02	6667.30	6630.39	04^02 (74%)/12^02 (23%)
3	12^02	6755.53	6719.69	16^01 (42%)/08^01 (41%)/24^01 (13%)
4	20^02	6846.50	6768.38	12^02 (57%)/04^02 (23%)
5	08^01	6848.13	6849.85	110^00 (38%)/28^00 (30%)
6	16^01	6917.65	6868.57	20^02 (79%)/12^02 (19%)
7	24^01	6991.21	6882.80	08^01 (39%)/24^01 (39%)
8	32^01	7069.95	7024.15	32^01 (42%)/16^01 (30%)
9	012^00	7114.33	7029.75	012^00 (33%)/36^00 (31%)
10	40^01	7155.02	7137.08	40^01 (48%)/24^01 (27%)/16^01 (17%)
11	110^00	7164.01	7194.40	44^00 (31%)/012^00(28%) 28^00 (17%)/52^00(14%)
12	28^00	7219.02	7214.80	40^01 (41%)/32^01(36%)
13	36^00	7280.52	7340.97	52^00 (34%)/110^00(22%)
14	44^00	7349.65	7464.21	60^00 (33%)/44^00 (17%)
15	52^00	7427.57	7556.26	60^00 (51%)/28^00 (15%) 52^00 (14%)/36^00 (12%)
16	60^00	7515.42	7640.85	44^00 (30%)/36^00 (24%) 52^00 (22%)/28^00 (12%)

[a] The levels are identified in terms of the zeroth-order normal modes $\left(v_1 v_2^{l_2} v_3\right)^0$, with modes 1–3 primarily corresponding to the NO stretch, the bend, and the NN stretch, respectively. All levels in a row and in a column, in the first part of the table, are connected, respectively, through the $\Delta v_1 = -2$, $\Delta v_3 = 1$, and $\Delta v_1 = -1$, $\Delta v_2 = 2$ anharmonic resonances. All levels in the second part of the table are charterized by their energy rank in the polyad (i), their zeroth-order labeling $\left(v_1 v_2^{l_2} v_3\right)^0$, and their calculated energy in absence $(G_{0\,pre.})^0$ and in presence $(G_{0pre.})$ of the anharmonic resonances. The fraction of the zeroth-order states in the eigenstates, rounded to unity, is mentioned for the four dominant ones whenever they are above 10%, in terms of the zeroth-order energy levels labeling. It corresponds to the squared coefficients in the eigenvector resulting from the cluster matrix diagonalization (all constants used in the calculation are from Teffo et al. [595], and all details of the procedure are provided in Campargue et al. [596]).

TABLE XVI
Observed and Predicted Relative Band Intensities in the $\{N,k\} = \{12,0\}$
Vibrational Polyad in $^{14}N_2{}^{16}O^a$

Rank	$\tilde{\nu}_0(cm^{-1})$	Bright State	$I_{rel}^{(obs)}$	$I_{rel}^{(calc)}$
1	6580.83	00^03	100	100
2	6630.41	20^02	5	1
3			0	0
4	6768.48		19	20
5			0	0
6	6868.53		76	79
7			0	0
8	7024.07	40^01	11	10
9			0	0
10	7137.10		45	48
11			0	0
12	7214.65		44	41
13			0	0
14	7463.96	60^00	34	33
15	7556.11		50	51
16	7640.45		16	9

^{a} Modes 1–3 correspond primarily to the NO stretch, the bend, and the NN stretch, respectively. The sum of the observed relative intensities within each subpolyad is fixed to 100. Only observed band centers ($\tilde{\nu}_0$) are printed (adapted from Campargue et al. [596]). Each bright state is identified using its zeroth-order labeling.

c. Acetylene

1. ANHARMONIC RESONANCES IN $^{12}C_2H_2$. The gathering of all dominant anharmonic resonances in acetylene proved to be, as expected, a much more difficult task than in nitrous oxide. There are indeed seven vibrational degrees of freedom and thus, both a significantly increased number of possible couplings to consider as well as a larger amount of experimental and analysis work to achieve. The spectroscopic literature on ground-state acetylene is listed in Appendix B. Nine relevant resonant coupling mechanisms were identified in $^{12}C_2H_2$ from the experimental data, which are gathered in Table XVII. Only the coupling parameter is indicated in each case to represent the resonance, referring to the interaction mechanism and the related matrix elements presented in Sections III.B.3 and III.B.4. The existence of some of them was deduced from experimental evidence in related molecules, $^{13}C_2H_2$ [613] and $^{12}C_2HF$ [573] in particular. Those found to be efficient in $^{12}C_2D_2$ are also listed in Table XVII. All normal modes in acetylene were identified in Section II (Fig. 13).

TABLE XVII
Anharmonic Resonant Couplings, Including the Vibrational
l-Resonance, Used to Fit Vibrational Energy Levels in
$^{12}C_2H_2$ [450] and in $^{12}C_2D_2$ [525], up to Visible Range in
Both Cases

$^{12}C_2H_2$		$^{12}C_2D_2$
	Stretching DD	
*	$K_{11/33}$	*
	Bending DD	
*	$K_{44/55}$	*
	Stretch–Stretch	
	$K_{12/33}$	*
	Stretch–Bend	
*	$K_{3/245}$	
*	$K_{1/255}$	
*	$K_{1/244}$	*
*	$K_{15/34}$	
*	$K_{14/35}$	
	l Resonance	
*	r_{45}	*

It is interesting to note that no cubic resonance term is put forward in acetylene from the experimental data. The relevant contributions are almost exclusively quartic terms. This trend makes a significant difference with most other triatomic or larger molecules studied in the literature.

2. CLUSTER STRUCTURE IN $^{12}c_2H_2$. Using the set of resonances presented in Table XVII, one can build any V-cluster, starting from any of the vibrational levels, thus containing any number of excitation of each mode and of any symmetry. As an example, starting from the zeroth-order level $(v_1, v_2, v_3, v_4^{l_4}, v_5^{l_5}) = (0, 0, 3, 0^0, 0^0)$, and building the V-cluster using the set of resonances listed in Table XVII, one defines the set of zeroth-order levels mentioned in Table XVIII. The related V-cluster is thus made out of 16 zeroth-order levels. The way to build the matrix has been detailed previously, in Section III.B.

As pointed out before, one can empirically check that each vibrational level appears in only one V-cluster. It is to be emphasized that the clustering of levels is obviously a feature independent of any experimental technique used to monitor the levels. The V-cluster characteristics are linked only to the nature of the anharmonic couplings.

Applying Kellman's procedure [493,546,547], one associates a vector $V_{ab/cd}$ with each anharmonic resonance listed in Table XVII. There happens

TABLE XVIII

Cluster of Levels (Printed in Boldface) Around the Zeroth-order Level $(v_1, v_2, v_3, v_4^{l_4}, v_5^{l_5})^0 = (0, 0, 3, 0^0, 0^0)^a$

	0305^11^{-1}	0214^00^0	1112^00^0	0030^00^0	0303^30^{-3}	1203^11^{-1}	0303^13^{-1}	0121^11^{-1}	2010^00^0	0212^22^{-2}	0212^02^0	2101^11^{-1}	1110^02^0	1201^13^{-1}	0301^15^{-1}	021^004
$\mathbf{0305^11^{-1}}$		3/245														
$\mathbf{0214^00^0}$			1/244		44/55											
$\mathbf{1112^00^0}$					15/34	1/255										
$\mathbf{0030^00^0}$					3/245	15/34	44/55									
$\mathbf{0303^30^{-3}}$						**	1/244									
$\mathbf{1203^11^{-1}}$							1/255	14/35	1/244							
$\mathbf{0303^13^{-1}}$								3/245	14/35	44/55						
$\mathbf{0121^11^{-1}}$									3/245	3/245	1/244	44/55				
$\mathbf{2010^00^0}$										3/245	3/245	11/33	15/34			
$\mathbf{0212^22^{-2}}$											**	3/245	11/33			
$\mathbf{0212^02^0}$												**	3/245	1/244	44/55	
$\mathbf{2101^11^{-1}}$													1/244	14/35	1/255	
$\mathbf{1110^02^0}$														3/245	14/35	1/255
$\mathbf{1201^13^{-1}}$															1/255	14/35
$\mathbf{0301^15^{-1}}$																3/245
$\mathbf{021^004}$																

^{a}Defined by the set of quartic anharmonic resonances found to be dominant in $^{12}C_2H_2$, identified (in italic) using the subscript ab/cd related to the label $K_{ab/cd}$. Modes 1 and 3 are the symmetric and antisymmetric CH stretches; mode 2, the CC stretch; and modes 4 and 5, the *trans* and *cis* bends, respectively. All quartic anharmonic resonances are listed in the upper triangle of the matrix, including the vibrational *l*-resonance interaction denoted by two asterisks (**). A set of linearly independent anharmonic resonances is selected in the lower triangle of the matrix (see text).

153

TABLE XIX
Dominant Resonance Vectors in $^{12}C_2H_2$ [a]

	$V_{ab/cd}$	$(\Delta v_1,$	$\Delta v_2,$	$\Delta v_3,$	$\Delta v_4,$	$\Delta l_4,$	$\Delta v_5,$	$\Delta l_5,$	$\Delta k)$	
V-cluster	$V_{44/55}$	(0,	0,	0,	−2,	0,	2,	0,	0)	**a**
	$V^{3/245}$	(0,	1,	−1,	1,	1,	1,	−1,	0)	**b**
	$V^{11/33}$	(−2,	0,	2,	0,	0,	0,	0,	0)	**c**
	V^{l-res}_{vib}	(0,	0,	0,	0,	2,	0,	−2,	0)	**d**
	$[V^{1/244}$	$= -a/2 + b + c/2 - d/2]$									
	$[V^{1/255}$	$= a/2 + b + c/2 - d/2]$									
	$[V^{14/35}$	$= a/2 + +c/2 - d/2]$									
	$[V^{15/34}$	$= -a/2 + +c/2 - d/2]$									
	$[V^{l}_{44/55}$	$= a + + +d]$									
V/l-cluster	V^{l-doub}_{rot}	(0,	0,	0,	0,	±2,	0,	0,	±2)	**e**
V/l/C-cluster	$V_{2/444}$	(0,	−1,	0,	3,	±1,	0,	0,	±1)	**f**
	$[V_{2/455a}$	(0,	−1,	0,	1,	±1,	2,	0,	±1)	$=(a+f)]$
	$[V_{2/455b}$	(0,	−1,	0,	1,	∓1,	2,	±2,	±1)	$=(a-d+f)]$
	$[V_{22/35}$	(0,	−2,	1,	0,	0,	1,	±1,	±1)	$=(a-b+f)]$
	$[V_{22/14}$	(1,	−2,	0,	1,	±1,	0,	0,	±1)	$=(a/2 - b - c/2 + d/2 +f)]$

[a] Anharmonic and l resonances are considered in the subset V cluster; l-doubling effects are considered next, to define the V/l-cluster subset, and Coriolis couplings are listed in the subset $V/l/C$-cluster. A set of linearly independent resonance vectors is highlighted with labels **a–f**. Linear combinations for the other relevant resonances are detailed (see text). Modes 1 and 3 are the symmetric and antisymmetric CH stretches; mode 2, the CC stretch; and modes 4 and 5, the *trans* and *cis* bends, respectively.

to exist four linearly independent of them, with an arbitrarily selected choice listed in Table XIX, specifically, the vectors **a–d** in the subset labeled V-cluster. Any of the other resonance vectors is indeed a combination of those listed, as exemplified with all other relevant resonance vectors in Table XIX, in the same subset. The important feature is that these linearly independent resonances define the polyad size and content. The additional resonances provide additional links only between the levels. They therefore contribute to modify the strength of the couplings but not the size or content of the V-cluster. This is demonstrated in the lower interaction triangle in Table XVIII, which lists only the four selected linearly independent resonance couplings. It can be confirmed that the same V-cluster—containing exactly the same levels—is obtained. There are other possible selections of linearly independent resonances, whose role is, however, less demonstrative than the one we have selected.

Acetylene has $3n - 5 = 7$ vibrational degrees of freedom. Therefore, there remains $7 - 4 = 3$ constants of the motion. One can set them using the

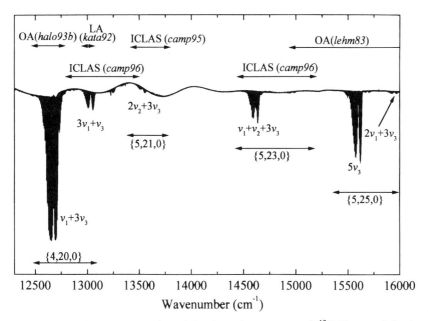

Figure 34. Higher-energy region of the absorption spectrum of $^{12}C_2H_2$ recorded using Fourier transform spectroscopy (from Campargue and co-workers [614], reproduced with permission from Taylor & Francis). The double arrows in the lower part define the energy spread of the clusters accessible through one-photon absorption experiments, identified using the $\{N_s, N_r, k\}$ label. The double arrows in the upper part indicate the ranges investigated by different authors using various laser techniques, all detailed in Section IV.C of the present review and revealing many extra spectral features. The related labels reference to the following papers: halo93b [615], kata92 [616], camp95 [617], camp96 [614], and lehm83 [618].

procedure described for N_2O, to

$$N_s = v_1 + v_2 + v_3$$
$$N_r = 5v_1 + 3v_2 + 5v_3 + v_4 + v_5 \qquad (3.114)$$
$$N_k = k = l_4 + l_5$$

The V-cluster presented in Table XVIII thus corresponds to the $\{N_s, N_r, k, u/g^{\pm}\} = \{3, 15, 0, u^{+}\}$, in which the u/g character is specified, as well as the $\pm$ symmetry, in case $k = 0(\Sigma^{+}$ or $\Sigma^{-})$ levels are concerned. The latter property is, by convention, related to the convention on the sign of l_4, as discussed in Section III.B.4.

An overview of the absorption spectrum of $^{12}C_2H_2$ at higher energy is presented in Fig. 34, to illustrate the clustering process.

3. PARAMETERS IN $^{12}C_2H_2$. Some 122 known experimental vibrational energy levels in $^{12}C_2H_2$ up to 12,000 cm^{-1} were gathered by Abbouti Temsamani and Herman [468] and a least-squares fitting procedure was carried on using a MIME pre-block-diagonalized into V-clusters. Some 35 parameters were fitted, using one constraint, only: $K_{1/244} = K_{1/255}$. The 122 levels were fitted within better than 1 cm^{-1}. As an example, Table XX gathers the results of the fit for the $\{3, 15, 0, u^+\}$ cluster. The resulting global parameters proved to be accurate enough to allow the visible range of the spectrum to be assigned [614,619]. The parameters are detailed in the literature [468], but not listed here. They are indeed being superseded, as several research groups, including Herman and co-workers [450], are currently investigating this problem. Their latest results allow some 219 energy levels observed up

TABLE XX
Levels in the V-Cluster $\{N_s, N_r, k, u/g\}^{+/-} = \{3, 15, 0, u\}^+$ in $^{12}C_2H_2$ [a]

Rank	$(v_1,$	$v_2,$	$v_3,$	$v_4^{l_4},$	$v_5^{l_5})^0$	$G_{0\,obs.}$	$G_{0\,calc.}$	%
1	0	3	0	5^{-1}	1^{-1}	—	9505.09	81|1⟩
2	0	2	1	4^0	0^0	—	9571.41	84|2⟩
3	1	1	1	2^0	0^0	9639.85	9639.44	64|4⟩
4	0	0	3	0^0	0^0	9668.13	9668.03	74|3⟩
5	0	3	0	3^3	3^{-3}	—	9695.61	32|7⟩
6	1	2	0	3^1	1^{-1}	—	9723.00	64|6⟩
7	0	3	0	3^1	3^{-1}	9741.62	9740.45	36|7⟩
8	0	1	2	1^1	1^{-1}	9744.55	9745.53	37|8⟩
9	2	0	1	0^0	0^0	9787.40	9787.41	44|10⟩
10	0	2	1	2^2	2^{-2}	—	9807.41	56|11⟩
11	0	2	1	2^0	2^0	9835.16	9834.58	68|9⟩
12	2	1	0	1^1	1^{-1}	9909.89	9907.70	59|13⟩
13	1	1	1	0^0	2^0	—	9940.73	62|12⟩
14	1	2	0	1^1	3^{-1}	9955.32	9955.91	60|14⟩
15	0	3	0	1^1	5^{-1}	—	9986.55	67|15⟩
16	0	2	1	0^0	4^0	—	10047.78	92|16⟩

[a] All levels are characterized by their energy rank in the cluster, their zeroth-order labeling, and their observed and fitted energy (in cm^{-1}). The fraction of the zeroth-order states in the eigenstates, rounded to unity, is also mentioned for the dominant contribution, in percent, in terms of the rank of the level as listed in the first column. It corresponds to the related squared coefficient in the eigenvector resulting from the V-cluster matrix diagonalization (adapted from Abbouti Teinsamani, and Herman [468]). Modes 1 and 3 are the symmetric and antisymmetric CH stretches; mode 2, the CC stretch; and modes 4 and 5, the *trans* and *cis* bends, respectively.

to $18,000\,\mathrm{cm}^{-1}$ to be fitted with a standard deviation better than $0.8\,\mathrm{cm}^{-1}$ (1σ), using 39 parameters. Although the number of constants is increased in that new investigation [450], with respect to the one we have just detailed at the beginning of this section [468], all basic trends of the cluster picture still fully hold. Actually, most of the information reported in the present review concerning the vibration clustering effect in $^{12}C_2H_2$ is taken from that new study, still to appear [450]. Only Table XX is adapted from the earlier study [468]. We have used the new parameters from [450] to calculate a set of vibrational levels in $^{12}C_2H_2$. It is listed in Appendix C, with additional information.

Checks on the band intensities were performed, as described for N_2O in a previous section. One of them is detailed in Section III.C.3. Relative band intensities could be nicely reproduced in all cases, independently of the symmetry (u/g) of the zeroth-order bright state [453], provided only one bright state appears in the V-cluster. Otherwise, the relative zeroth-order band intensity of each zeroth-order bright state is to be known a priori, and typical problems with the signs of the contribution have to be solved. Intensity features were discussed in Section II.C.2.b.4 and are further detailed in Section III.E.

Checks on the rotational constants are described in section III.D.4.d.

4. $^{12}C_2D_2$. It is relevant to stress the case of $^{12}C_2D_2$, for which the same global study was achieved in the literature, as explained above for the main molecule, by Herman and co-workers [525]. The set of active anharmonic resonances defining the V-clusters was given in Table XVII. The related vectors are developed in Table XXI. All five resonance vectors are linearly independent and, therefore, only two constants of the motion are defined in $^{12}C_2D_2$, for the V-clusters. The same global fitting procedure as in $^{12}C_2H_2$, also including levels up to the visible range, led to similar success [525].

TABLE XXI

Dominant Resonance Vectors in $^{12}C_2D_2{}^a$

	$V_{ab/cd}$	$(\Delta v_1,$	$\Delta v_2,$	$\Delta v_3,$	$\Delta v_4,$	$\Delta l_4,$	$\Delta v_5,$	$\Delta l_5,$	$\Delta k)$	
V cluster	$V_{44/55}$	(0,	0,	0,	$-2,$	0,	2,	0,	0)	a
	$V_{11/33}$	($-2,$	0,	2,	0,	0,	0,	0,	0)	b
	$V_{12/33}$	($-1,$	$-1,$	2,	0,	0,	0,	0,	0)	c
	$V_{1/244}$	($-1,$	1,	0,	2,	0,	0,	0,	0)	d
	$V_{\mathrm{vib}}^{1-\mathrm{res}}$	(0,	0,	0,	0,	2,	0,	$-2,$	0)	e

^{a} Modes 1 and 3 are the symmetric and antisymmetric CD stretches; mode 2, the CC stretch; and modes 4 and 5, the $trans$ and cis bends, respectively [525].

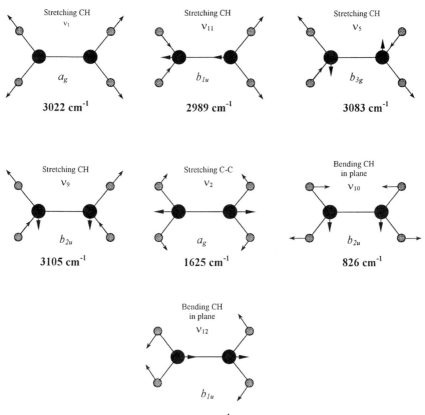

Figure 35. Selection of normal modes of vibration in $^{12}C_2H_4$ (symmetry labels are different from those usually reported in the literature). See [592].

d. Ethylene. We briefly present here results from a still unpublished study on ethylene ($^{12}C_2H_4$) [592]. A set of relevant normal modes of vibration in ethylene is schematically represented in Fig. 35. Although not very important in the present context, one should mention that the symmetry properties in that figure are slightly modified compared to most of the previous literature (see Table XXVIII).

Ethylene, C_2H_4, has four identical CH bonds giving rise to four normal modes of vibration whose label and symmetry are: $v_1(a_g)$, $v_5(b_{3g})$, $v_9(b_{2u})$, and $v_{11}(b_{1u})$. They lead to the following structural anharmonic resonances, with the term *structural* defined in Section III.B.1, with i, j, k, l as those four modes:

$$\left\langle v_i, v_j, v_k, \left| \frac{\hat{H}}{hc} \right| v_i + 2, v_j - 2, v_k \right\rangle$$

$$= \frac{1}{4} K_{ii/jj} [(v_i + 1)(v_i + 2)v_j(v_j - 1)]^{1/2}$$

$$\left\langle v_i, v_j, v_k, v_l \left| \frac{\hat{H}}{hc} \right| v_i + 1, v_j + 1, v_k - 1, v_l - 1 \right\rangle \qquad (3.115)$$

$$= \frac{1}{4} K_{ij/kl} [(v_i + 1)(v_j + 1)v_k v_l]^{1/2}$$

as already given in Table VIII. In addition, the following resonances were pointed out in the literature [620–624]:

$$\left\langle v_2, v_{10}, v_i \left| \frac{\hat{H}}{hc} \right| v_2 - 1, v_{10} + 2, v_i \right\rangle$$

$$= K_{10,10/2} \left[\frac{v_2(v_{10} + 1)(v_{10} + 2)}{8} \right]^{1/2} \qquad (3.116)$$

$$\left\langle v_2, v_{11}, v_{12}, v_i \left| \frac{\hat{H}}{hc} \right| v_2 + 1, v_{11} - 1, v_{12} + 1, v_i \right\rangle$$

$$= K_{2,12/11} \left[\frac{v_{11}(v_2 + 1)(v_{12} + 1)}{8} \right]^{1/2} \qquad (3.117)$$

The notation $K_{a/b,c}$ stands, as in Section III.B.3.a.2, for the potential-energy interaction coefficient (in cm^{-1}) between levels nv_a and $nv_b + nv_c$, as defined using rectilinear normal-mode coordinates.

According to this model, only one of the fundamental CH levels, v_{11}, is thus perturbed by one of the previously mentioned interactions, specifically, the 2,12/11 coupling just described. That same coupling scheme accompanies all levels involving some excitation in v_{11}.

The resulting polyad of vibrational levels involving the first CH overtone levels is presented in Fig. 36. The following pseudo–quantum number can be associated:

$$N = 3v_1 + 3v_5 + 3v_9 + 3v_{11} + 2v_2 + v_{10} + v_{12} \qquad (3.118)$$

The block represented in Fig. 36 thus corresponds to the $\{N, sym\} = \{6, a_g\}$ polyad. The picture can be easily extended to the full pattern of vibrational energy levels. Compared to the more complete picture demonstrated to fit

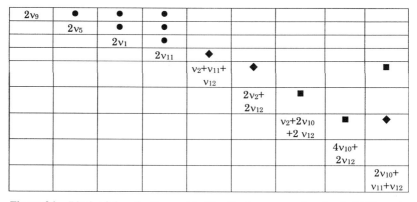

Figure 36. Block of the vibration matrix Hamiltonian representing the first CH overtone levels in $^{12}C_2H_4$ (the lower, symmetric triangle is not detailed). Diagonal elements are in the form of a Dunham expansion; off-diagonal elements are in the form of (3.116) (■), (3.117) (◆), and (3.115) (●). This corresponds to the $\{N, sym\} = \{6, a_g\}$ polyad. Adapted from Ref. [592].

acetylene and nitrous oxide, one should, however, remain most careful in the present case as the experimental data available do not allow one to check the validity of the polyad picture above the first CH overtone range. Information about the latter spectral region was recently significantly improved [589] using jet-cooled FT experiments, an experimental technique described in Section IV.B.

3. Giant Clusters

Acetylene is kept as the leading example in the present section, intended to open up the V-cluster picture to deal with the full MIME and quote with unusual and most interesting situations in high-resolution molecular spectroscopy, namely, chaotic behav' 'r.

a. V/l-Clusters

1. QUANTUM NUMBERS. Rotation, J-dependent terms, are considered in Section III.D. We need to point out here that the introduction of the rotation degrees of freedom in the MIME for linear species, adds the l-doubling interaction, in particular, leading to include matrix elements off-diagonal in k, with $\Delta k = \pm 2$. Such contributions thus lead to couple, for instance, l-substates of Σ and Δ symmetries arising from the same vibrational excitation. The consequences on the cluster size and quantum numbers are dramatic but consistent with the cluster picture. This extra coupling term indeed also leads to a resonance vector, which is indicated in the second

subset of Table XIX. It is linearly independent of the vectors defining the V-clusters, listed in that table. There are actually various types of such extra coupling terms, of similar nature, all listed for $^{12}C_2H_2$ by Perevalov and co-workers [540]. They all correspond to linear combinations of the vectors labeled **a–e** in Table XIX, not detailed here. The introduction of the l-doubling interaction thus defines new, larger clusters, which we label Vll-clusters [610]. There are now, taking into account the l-doubling interaction, $7 - 5 = 2$ remnant constants of the motion, instead of 3 for the V-clusters.

$$N_s = v_1 + v_2 + v_3$$
$$N_r = 5v_1 + 3v_2 + 5v_3 + v_4 + v_5$$
(3.119)

Returning to the example of the $\{3, 15, 0, u^+\}$ V-cluster, this means that now all clusters $\{3, 15, p, u^+\}$, with p even are gathered to define the $\{3, 15, u^{(+)}\}$ Vll-cluster, of larger size. This extension still perfectly fits in the cluster picture, with the same well-defined rules as for the V-clusters.

2. SPECTRA. Various examples of Vll-clusters are encountered in acetylene, leading to observe zeroth-order forbidden $\Delta - \Sigma^+$ bands in the spectra, as discussed in Section III.D.4.b.2. The most striking example of such a Vll-clustering is provided with the very highly excited bending levels observed by Field and co-workers [609,609a,609b] using dispersed laser-induced fluorescence. In each spectral portion they recorded, they expect to access only two zeroth-order rovibrational bright states, given their experimental conditions inducing very restrictive selection rules. The observed spectrum, in one of those portions, is presented in the upper part of Fig. 37. It consists in an impressive fractionation of the zeroth-order bright lines. It can be clearly reproduced using the cluster model and parameters, as demonstrated in the lower portion of Fig. 37. Each stick corresponds to a level in the Vll-clusters involved, with their height given by the fraction of the corresponding zeroth-order bright state. A model slightly more refined than the one previously discussed here, including some additional matrix elements to better describe the bend–bend DD coupling, was used to fit lower-energy bending levels, and its results are extrapolated in the region of interest [469]. The cluster picture thus accounts for the very complex spectrum observed. It is actually an essential feature in the MIME, in those very highly excited vibrational ranges.

b. Vll/C-Clusters. One can further include rotation by taking into account Coriolis-type couplings, to be introduced in Section III.D. The related matrix elements, from Perevalov et al. [540], are all detailed in the third subset in

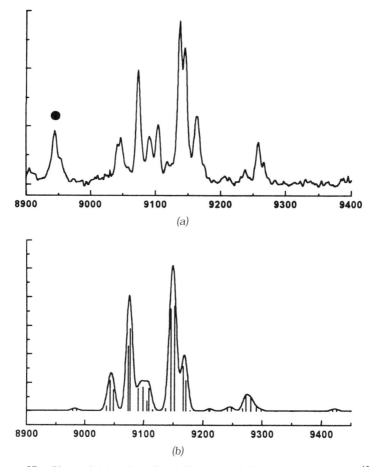

Figure 37. Observed (*a*) and predicted (*b*) dispersed fluorescence spectra of $^{12}C_2H_2$ reaching the $\{0, 14, 0, g\}$ and $\{0, 14, 2, g\}$ *V*-clusters, each composed of one zeroth-order bright state only (adapted from Abbouti Temsamani et al. [469]). The peak identified with a dot involves another *V*-cluster of levels, not taken into account in the simulation. Each vertical line indicates a dispersed induced fluorescence transition reaching a state of *g* symmetry, with its height proportional to the intensity (energy scale in cm^{-1}).

Table XIX. They correspond to interactions in $\Delta k = \pm 1$. One can further apply Kelman's procedure to define an extra, sixth linearly independent resonance vector, **f**. As indicated in Table XIX, all other Coriolis coupling vectors are a combination of the basis set, **a–f**. There remains therefore now only $7 - 6 = 1$ constant of the motion. The Coriolis couplings link clusters

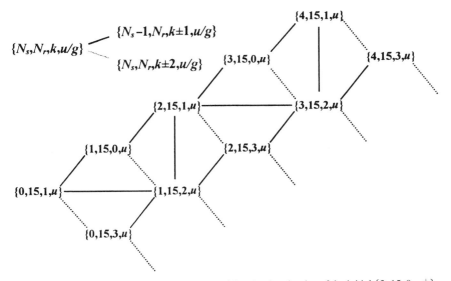

Figure 38. Scheme of the $V/l/C$-cluster model, enlarging the size of the initial $\{3, 15, 0, u^+\}$ V cluster in $^{12}C_2H_2$. The dashed lines represent the inter-V-cluster couplings via the l-doubling interactions and define the V/l-clusters. The solid lines correspond to the Coriolis interactions and define the $V/l/C$-cluster (adapted from Abbouti Temsamani and Herman [610]).

with $\Delta N_s = \pm 1$ and the only remaining quantum number is

$$N_r = 5v_1 + 3v_2 + 5v_3 + v_4 + v_5 \qquad (3.120)$$

The full picture presented in Fig. 38 defines the $\{Nr, u/g^{\pm}\} = \{15, u^+\}$ $V/l/$ C-cluster associated with the initial $\{3, 15, 0, u^+\}$ V-cluster. The important feature is that, again, the extension toward Coriolis coupling is naturally performed within the cluster picture. As a result, an unique set of parameters, including the zeroth-order vibrational–Dunham constants, the anharmonic and l-resonance parameters, and the l-doubling and Coriolis coupling constants, holds for all $V/l/C$-clusters. Their knowledge would thus lead to predict the full vibration–rotation energy levels in the electronic ground state of the molecule, using the clustered MIME. This model obviously holds as long as no coupling other than those we have considered is acting. In the case of acetylene, in particular, interaction with close-by triplet electronic states and with the vinylidene isomer (e.g., see Refs. [625–629] and references in Appendix B) can be expected to further perturb the present picture. A more general discussion of the role of Coriolis couplings within anharmonic resonances is provided by Carrington and Krishnan [630].

The next extension of the cluster picture for $^{12}C_2H_2$, following definition of the $V/l/C$-clusters, is quite obvious. Any extra linearly independent coupling, arising for instance, because of an extra near degeneracy of the levels occurring at high vibrational excitation, would lead to the suppression of all constants of the motion. The resulting absence of quantum numbers, but J, is a possible definition of chaos. The final cluster would then correspond to the full MIME. In fact, the related spectrum would then be close to showing the full density of levels, as no selection rules would remain. Still, other rules such as the energy mismatch would reduce the number of vibrational levels observed on the spectrum, around each zeroth-order bright state. Such data were already observed by Nesbitt and McIlroy [631] and by Lehmann and co-workers [632,633]. Figure 39 presents two such so-called clumps of lines, for which the conventional rules in high-

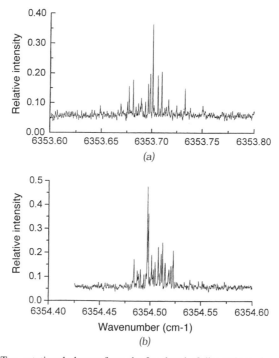

Figure 39. Two rotational clumps from the $3\nu_1$ band of diacetylene-d_1 revealing the role of extensive zeroth-order bright/dark coupling in modifying the conventional spectral fine structure [Ref. 633, $(a) J' = 0; (b) J' = 2$. Reprinted from *Chemical Physics*, Vol. 190, Gambogi et al., "Intramolecular vibrational dynamics of diacetylene and diacetylene-d_1 via eigenstate-resolved overtone spectroscopy," pp. 191–206, copyright (1995), with permission from Elsevier Science and from the authors].

resolution molecular spectroscopy need thus to be revised (see also, e.g., Ref. 634), in agreement with the present developments (see also, e.g., Refs. 552 and 635 for a more general discussion about overtone spectroscopy).

D. Vibration–Rotation Terms

1. General Picture

Classically, the components of the angular momentum and the rotational kinetic energy in the MAS defined previously, are

$$P_\alpha = I_\alpha \bar{\omega}_\alpha$$

$$E_r = \frac{1}{2} \bar{\omega}.I.\bar{\omega} = \sum \frac{1}{2} I_{\alpha\alpha} \bar{\omega}_\alpha^2$$

$$= \sum \frac{1}{2} I_\alpha \bar{\omega}_\alpha^2 \qquad (3.121)$$

with

$$P_\alpha = \frac{\partial E_r}{\partial \bar{\omega}_\alpha}$$

$I_{\alpha\alpha} = I_\alpha$ the three principal moments of inertia

When no torque is applied, the square of the total angular momentum $P^2 = P_x^2 + P_y^2 + P_z^2$ in the MAS is a constant. So is the rotational energy, whose classical expression is

$$E_r = \frac{1}{2} \left(\frac{P_x^2}{I_x} + \frac{P_y^2}{I_y} + \frac{P_z^2}{I_z} \right) = \text{const} \qquad (3.122)$$

In angular momentum space, the equation $P^2 =$ constant is a sphere with radius P, and E_r corresponds to an ellipsoid with principal semiaxes $(2I_\alpha E_r)^{1/2}$. Since the values of P_α must satisfy both equations, the solutions for P correspond to the intersection of both solids, and P is between the minimum and maximum axes of the ellipsoid. Striking three-dimensional representations of the various angular momenta related to the rotation degrees of freedom are provided by Harter [636].

One distinguishes

Spherical tops		$I_x = I_y = I_z$
Symmetric tops	Cigar-shaped and prolate:	$I_x = I_y > I_z$
	Disk shaped and oblate:	$I_z > I_x = I_y$
	Linear:	$I_x = I_y > I_z = 0$
Asymmetric tops		$I_x \neq I_y \neq I_z$

A classification is made for asymmetric tops using the ratio

$$\kappa = \frac{2B - A - C}{A - C} \qquad (3.123)$$

called the *Ray asymmetry parameter* [637]

Prolate symmetric top: $A > B = C$ and $\kappa = -1$
Oblate symmetric top: $A = B > C$ and $\kappa = +1$
Asymmetric top: $A > B > C$ and $-1 < \kappa < +1$

with the principal rotational constants defined (in cm^{-1}) as $B_\alpha = h/(8\pi^2 c I_{\alpha\alpha})$, with $B_\alpha = A, B, C$ and, by convention: $A > B > C$, and $I_a < I_b < I_c$.

In quantum terms, to evaluate the rotation contribution in the initial Watsonian, one must consider the term, from the initial Hamiltonian:

$$\sum_{\alpha\beta} \hat{J}_\alpha \mu_{\alpha\beta} \hat{J}_\beta \qquad (3.124)$$

in which, as we did for the potential energy, one develops the reciprocal inertia tensor in series [456]:

$$\mu_{\alpha\beta} = (I_{\alpha\beta})^{-1}\left[I_\alpha \delta_{\alpha\beta} - \sum_{i=1}^{3n-6} a_i^{\alpha\beta} Q_i + \tfrac{3}{4}\sum_{i,j,\gamma} a_k^{\alpha\gamma} Q_i I_\gamma^{-1} a_j^{\gamma\beta} Q_j - \dots \right](I_\beta)^{-1} \qquad (3.125)$$

The first, Q-independent term leads to the rigid rotor contribution and the next terms in the development generate the non–rigid rotor contributions. Both are evaluated in the next section.

2. Rigid Rotor Hamiltonian ($\hat{H}_r^{rig}$)

a. Symmetric Tops. The rigid rotor contribution thus arises from the H_{02} term in the initial Hamiltonian:

$$\frac{H_{02}}{hc} = \sum_\alpha B_\alpha^e \hat{J}_\alpha^2 \qquad (3.126)$$

with

$$B_\alpha^e \equiv A^e, B^e, C^e = \frac{\hbar^2}{2hc}\mu_{\alpha\alpha}^e = \frac{h}{8\pi^2 c I_{\alpha\alpha}^e} \qquad (3.127)$$

There are different ways to correlate the axis systems:

	I^r	I^l	II^r	II^l	III^r	III^l
x	b	c	c	a	a	b
y	c	b	a	c	b	a
z	a	a	b	b	c	c

with the I^r and III^r ("r" for right-handed) representations commonly used for quasi-prolate and quasi-oblate asymmetric tops, respectively. As demonstrated later, such mappings lead to lower the weight of the off-diagonal elements in the matrix Hamiltonian.

The quantum Hamiltonian and the related matrix elements are presented in Table XXII, in I^r representation ($a \equiv z, b \equiv x, c \equiv y$), still keeping the x,y,z labels for the operators to allow easier generalization. Only one equation is provided for all operators, separated by commas, giving the same eigenvalues. Condon and Shortley phase conventions [638] are followed. We also, as usual, follow Van Vleck [639], and use angular momenta referred to as molecule-fixed axes, which obey "anomalous" commutation relations $[\hat{J}_x, \hat{J}_y] = -i\hat{J}_z$. For completeness, the role of $\hat{J}_Z$ defining the projection of J on the Z axis in the LAS, specifically, M, is also included in Table XXII.

Using the second form of $\hat{H}_r$ in Table XXII, one can calculate the rotational energies. For symmetric tops, only the diagonal part of the Hamiltonian, in $\hat{J}^2$ and $\hat{J}_z^2$, is needed, as $B = C$. The energy levels are labeled using the J,k quantum numbers, or using J,K, with $K = |k|$ since only k^2 appears in the energy formulation:

$$\begin{aligned}
F(J,K) &= BJ(J+1) + (A-B)K^2 & \text{prolate} \\
F(J,K) &= BJ(J+1) + (C-B)K^2 & \text{oblate} \\
F(J,K) &= B\left[J(J+1) - K^2\right] & \text{linear}
\end{aligned} \tag{3.128}$$

The energy of the K sublevels is thus increasing or decreasing with increasing values of K, for prolate or oblate tops, respectively.

For linear tops, the v-independent term in BK^2 is to be included in the vibrational contribution to the energy, G_v, as we already pointed out in Section III.B.4.b.2. One should also notice the relations existing between the principal rotation constants A, B, and C in planar molecules [366,640,641].

b. *Asymmetric Tops.* For asymmetric tops, the problem is more complicated, as the off-diagonal elements generated by the $\hat{J}^{\pm^2}$ operators do not vanish. The quantum number J and the pseudo–quantum numbers K_a and K_c, which correspond to the unsigned projections of J on the corresponding principal axis of inertia in the MAS, are used. Knowing J, K_a, and K_c, one can also defined K_b.

TABLE XXII

Hamiltonian and Matrix Elements for Rotation (Constants in cm^{-1}), in Prolate Top
Representation

$\hat{H}_r/hc = (A\hat{J}_z^2 + B\hat{J}_x^2 + C\hat{J}_y^2)$

$\hat{H}_r/hc = \{[(B+C)/2]\hat{J}^2 + [A - (B+C)/2]\hat{J}_z^2 + [(B-C)/4][(\hat{J}_m^+)^2 + (\hat{J}_m^-)^2]\}$

$\hat{J}_m^\pm \equiv \hat{J}_x \pm i\hat{J}_y$

with

$\langle J,k,M \,|\, \hat{J}^2 \,|\, J,k,M \rangle = J(J+1)$

$\langle J,k,M \,|\, \hat{J}_z \,|\, J,k,M \rangle = k \qquad -J \leq k \leq +J$

$\langle J,k,M \,|\, \hat{J}_Z \,|\, J,k,M \rangle = M \qquad -J \leq M \leq +J$

In addition:

$\langle J, k \mp 1, M \,|\, \hat{J}^\pm \,|\, J,k,M \rangle = [J(J+1) - k(k \mp 1)]^{1/2}$

$\langle J, k-2, M \,|\, \hat{J}^{+^2} \,|\, J,k,M \rangle = \{[J(J+1) - (k-1)(k-2)][J(J+1) - k(k-1)]\}^{1/2}$

$\langle J, k+2, M \,|\, \hat{J}^{-^2} \,|\, J,k,M \rangle = \{[J(J+1) - (k+1)(k+2)][J(J+1) - k(k+1)]\}^{1/2}$

$\langle J, k \mp 1, M \,|\, \hat{J}_x \,|\, J,k,M \rangle = \frac{1}{2}[(J \pm k)(J \mp k + 1)]^{1/2}$

$\langle J, k \mp 1, M \,|\, i\hat{J}_y \,|\, J,k,M \rangle = \mp\frac{1}{2}[(J \pm k)(J \mp k + 1)]^{1/2}$

$\langle J, k, M \,|\, \hat{J}_x^2, \hat{J}_y^2 \,|\, J,k,M \rangle = \frac{1}{2}[J(J+1) - k^2]$

$\langle J, k \mp 2, M \,|\, \hat{J}_x^2, -\hat{J}_y^2 | J,k,M \rangle =$

$\frac{1}{4}[J(J+1) - k(k \mp 1)]^{1/2}[J(J+1) - (k \mp 1)(k \mp 2)]^{1/2}$

A matrix Hamiltonian has to be built for each J value, using initially the J,K representation. It is further block-diagonalized, making the following combination defining blocks $+$ and $-$, which is the so-called Wang transformation [640,642]:

$$\frac{1}{\sqrt{2}}[\,|\,J, +K > \pm |\, J, -K >] \tag{3.129}$$

which one obtains by performing the similarity transformation $U_W \hat{H}_r^{rig} U_W$, using the matrix U_W defined in Table XXIII. The procedure is outlined in that table, for $J = 2$. It is shown how the Hamiltonian matrix can be reorganized to define E^+, E^-, O^+, and O^- blocks, with E/O (i.e., even/odd) the parity of K. The resulting energy levels are labeled using J, K and K_c, selected according to the symmetric top level to which they correlate, with the diagonal (Λ_K), and off-diagonal ($\Lambda_{KK'}$) elements used in Table XXIII to be calculated using (3.128). The Wang transformation leads to define

TABLE XXIII

Steps of Operations Leading to Definition of E^-, O^-, O^+, and E^+ Block-Diagonal Hamiltonian Matrices Relevant for the Level Labeling and Energy Calculation of the Rotation Levels in Asymmetric Tops, for $J = 2$ (See Text)

$$
U_{\mathrm{w}} = 1/\sqrt{2} \times
\begin{vmatrix}
\cdot & \cdot & \cdot & \cdot & \cdot & \cdot & \cdot \\
\cdot & -1 & \cdot & \cdot & \cdot & 1 & \cdot \\
\cdot & \cdot & -1 & \cdot & 1 & \cdot & \cdot \\
\cdot & \cdot & \cdot & \sqrt{2} & \cdot & \cdot & \cdot \\
\cdot & \cdot & 1 & \cdot & 1 & \cdot & \cdot \\
\cdot & 1 & \cdot & \cdot & \cdot & 1 & \cdot \\
\cdot & \cdot & \cdot & \cdot & \cdot & \cdot & \cdot
\end{vmatrix}
$$

$|J,k\rangle$	$|2,-2\rangle$	$|2,-1\rangle$	$|2,0\rangle$	$|2,+1\rangle$	$|2,+2\rangle$
$|2,-2\rangle$	Λ_2		Λ_{20}		
$|2,-1\rangle$		Λ_1		Λ_{11}	
$|2,0\rangle$	Λ_{20}		Λ_0		Λ_{20}
$|2,+1\rangle$		Λ_{11}		Λ_1	
$|2,+2\rangle$			Λ_{20}		Λ_2

$|J,k\rangle^\pm$	$|2,2\rangle^-$	$|2,1\rangle^-$	$|2,0\rangle^+$	$|2,1\rangle^+$	$|2,2\rangle^+$
$|2,2\rangle^-$	Λ_2				
$|2,1\rangle^-$		$\Lambda_1 - \Lambda_{11}$			
$|2,0\rangle^+$			Λ_0		$\sqrt{2}\,\Lambda_{20}$
$|2,1\rangle^+$				$\Lambda_1 + \Lambda_{11}$	
$|2,2\rangle^+$			$\sqrt{2}\,\Lambda_{20}$		Λ_2

$|J,k\rangle^\pm$	$|2,2\rangle^-$	$|2,1\rangle^-$	$|2,1\rangle^+$	$|2,0\rangle^+$	$|2,2\rangle^+$
$|2,2\rangle^-$	Λ_2				
$|2,1\rangle^-$		$\Lambda_1 - \Lambda_{11}$			
$|2,1\rangle^+$			$\Lambda_1 + \Lambda_{11}$		
$|2,0\rangle^+$				Λ_0	$\sqrt{2}\,\Lambda_{20}$
$|2,2\rangle^+$				$\sqrt{2}\,\Lambda_{20}$	Λ_2

TABLE XXIV
Rotation Energies for $J = 2$, K Sublevels in Nearly Prolate (Left) and Nearly Oblate (Right) Asymmetric Tops

	I^r($B \cong C$)				off diag	III^r($A \cong B$)			
J	$K_a K_c$			$F(J, K_a K_c)$		$F(J, K_a K_c)$			J
2 2	2	1	E^-	$4A + B + C$		$4C + A + B$	E^-	2	1
2 1	1	1	O^+	$A + 4B + C$		$C + 4A + B$	O^+	2	2
2 1	1	2	O^-	$A + B + 4C$		$C + A + 4B$	O^-	2	1
2 0	0	2	E^+	$3B + 3C$	*	$3A + 3B$	E^+	2	2
2 2	2	0	E^+	$4A + B + C$	*	$4C + A + B$	E^+	2	0

*Off-diagonal terms are provided in the text.

$+$ and $-$ combinations. Reordering the wavefunctions, one defines the E^-, O^-, O^+, and E^+ blocks.

The resulting energies, directly calculated from the matrix elements in Table XXII, are given in Table XXIV, with the even/odd character of $\tau = K_a - K_c$ identifying the $+/-$ combination and the even/odd character of K_a (in I^r representation) or K_c (in III^r representation) identifying the E/O parity.

The off-diagonal terms required for the $J = 2(E^+)$ levels thus result from the following combination of the $|J, K\rangle$ symmetric top basis functions:

$$|2, 2, E^+\rangle = \frac{1}{\sqrt{2}}(|2, -2\rangle + |2, +2\rangle)$$
$$|2, 0, E^+\rangle = |2, 0\rangle$$

(3.130)

They are, for a prolate top in I^r representation:

	$|2, 0, E^+\rangle$	$|2, 2, E^+\rangle$
$|2, 0, E^+\rangle$	$3(B + C)$	$\sqrt{3}(B - C)$
$|2, 2, E^+\rangle$	$\sqrt{3}(B - C)$	$4A + B + C$

(3.131)

whose eigenvalues are, with $F(J, K_a, K_c) = F(2, E^+) = F(J_{K_a K_c})$:

$$F(2_{20}) = 2(A + B + C) + [3(B - C)^2 + (2A - B - C)^2]^{1/2}$$
$$F(2_{02}) = 2(A + B + C) - [3(B - C)^2 + (2A - B - C)^2]^{1/2}$$

(3.132)

The off-diagonal matrix element in (3.131) is small, as $B \approx C$, supporting the selection of the I^r representation for a prolate top.

For oblate tops, one usually associates the x,y,z axes in the MAS respectively to a,b,c. As a consequence, A and C are interchanged everywhere, with adequate changes in signs. The off-diagonal matrix element in (3.131) becomes $\sqrt{3}(A - B)$; thus it is again small and the III^r representation is indeed appropriate in the case of oblate tops.

The asymmetric top levels are labeled with the J quantum numbers and the K_a, K_c pseudo–quantum numbers which correlate to the K symmetric top quantum numbers, as defined in Fig. 40.

c. Linear Tops

1. l-DOUBLING. For linear tops, one must include the l-doubling matrix terms to obtain a faithful representation of the rotation level energies. The l-doubling is the rotational equivalent to the l-resonance discussed in Section III.B.4. In addition to the terms introduced via the l-resonance, one must therefore take into account the contributions arising from the following term in the Watsonian for linear molecules [643]:

$$-2B_e(\hat{J}_b \hat{\pi}_b + \hat{J}_c \hat{\pi}_c)$$

(3.133)

As usual, the introduction of the vibrational angular momentum l introduces a mixing between the rotation and vibrational degrees of freedom. The phenomenon of l-doubling occurs in bending vibrations of linear species from interaction with the stretching modes [359,360,367]. In the effective Hamiltonian, the result of this coupling is very similar to the K splitting occurring in asymmetric tops, and we therefore include its treatment in the present section, devoted to pure rotation, although it is truly of vibration–rotation Coriolis nature. It can be taken care of in a very pragmatic way, by simply replacing k by l in the relevant matrix elements presented in Table XXII.

Inserting appropriate forms of the $\hat{\pi}$ operators and using second-order perturbation theory leads to

$$\langle v_b, l_b \pm 2, J, k \pm 2 | \tilde{h} | v_b, l_b, J, k \rangle = \tfrac{1}{4} q_b^0 \{ [(v_b + 1)^2 - (l_b \pm 1)^2]$$
$$[J^2 - (k \pm 1)^2][(J + 1)^2 - (k \pm 1)^2] \}^{1/2}$$

(3.134)

with $k = \sum_b l_b$.

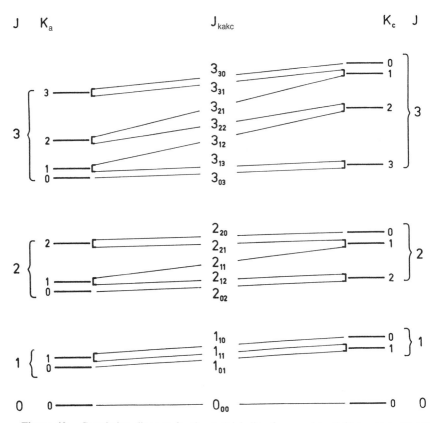

Figure 40. Correlation diagram for the J, K labeling from prolate (left) to oblate (right) symmetric top cases, through the asymmetric top levels.

Perturbation calculations are reported in the literature for the q constant, the so-called l-doubling constant, in symmetric tops [367,644,645]. In the case of triacetylene [646], for instance, it becomes

$$|q_b| \approx \frac{2B_0^2}{\tilde{\omega}_b} \sum_s \frac{3\tilde{\omega}_b^2 + \tilde{\omega}_s^2}{\tilde{\omega}_s^2 - \tilde{\omega}_b^2} \zeta_{sb}^2 \qquad (3.135)$$

where the sign indicated by Yamada et al. [647] is deliberately changed. A different convention is indeed adopted for the sign of q in this review, which is discussed later (see also Ref. 648).

The same expression holds by the way for acetylene, with the simplification from the sum rules on the Coriolis zeta constants given in the next section [536]:

$$q_b \approx \frac{2B_0^2}{\tilde{\omega}_b}\left(1 + 4\sum_s \frac{\zeta_{sb}^2\tilde{\omega}_b^2}{\tilde{\omega}_s^2 - \tilde{\omega}_b^2}\right) \tag{3.136}$$

Higher-order terms in q_b, resulting from $\tilde{h}_{22}$ and $\tilde{h}_{24}$ contributions, may have to be included, as detailed by Watson [372,643]:

2. e/f LEVELS

$$q_b = q_b + q_b^J J(J+1) + q_b^K(k \pm 1)^2 \tag{3.137}$$

For given v_b and J, the size of the matrix is determined by the conditions $|l_b| \le v_b$ and $|l_b| \le J$. In other words, the stack of rotational levels starts with the level with $J = |l_b|$. There is thus a rotational level with $J = 0$ only when $|l_b| = 0$. For $v_b > 0$ the matrix can be factorized in two blocks, corresponding to e and f parity classification of the rotational levels [541]. In even-electron molecules the e and f levels have parities $(-1)^J$ and $(-1)^{J+1}$, respectively. Using the previously selected phase convention [394]

$$E^*|v_b, l_b, J\rangle = (-1)^{J-l_b}|v_b, -l_b, J\rangle$$

the e and f functions are, for $l_b > 0$

$$|v_b^{l_b}, J\rangle^e \equiv |v_b, l_b, J, e\rangle = \frac{1}{2\sqrt{2}}[|v_b, l_b, J\rangle + (-1)^{l_b}|v_b, -l_b, J\rangle]$$

$$|v_b^{l_b}, J\rangle^f \equiv |v_b, l_b, J, f\rangle = \frac{1}{2\sqrt{2}}[|v_b, l_b, J\rangle - (-1)^{l_b}|v_b, -l_b, J\rangle] \tag{3.138}$$

The function $|v_b^{l_b}, J\rangle = |0^0, J\rangle$ belongs to the e series. This transformation, detailed by Pliva [472,535] and Winnewisser and Winnewisser [534], is similar to the Wang transformation used in the previous section for the asymmetric tops.

The matrix Hamiltonian for $v_b = 1$ is, before transformation to parity basis set

$\|v_b^{l_b}, J\rangle$	$\|1^{+1}, J\rangle$	$\|1^{-1}, J\rangle$
$\|1^{+1}, J\rangle$	D	$q_b J(J+1)$
$\|1^{-1}, J\rangle$	$q_b J(J+1)$	D

and, after transformation, omitting $|J\rangle$ in the notation:

$$
\begin{array}{c|cc}
|v_b^{l_b}\rangle^{e,f} & |1^{+1}\rangle^e & |1^{-1}\rangle^f \\
\hline
|1^{+1}\rangle^e & D(1^{+1}) - \frac{1}{2}q_b J(J+1) & \\
|1^{-1}\rangle^f & & D(1^{-1}) + \frac{1}{2}q_b J(J+1)
\end{array}
\tag{3.139}
$$

with $D = \tilde{\omega}_b^0 v_b + x_{bb}^0 v_b^2 + (g_{bb} - B)k^2 + BJ(J+1)$.

Extra terms, in particular those from centrifugal distortion, need to be included in the diagonal elements, as discussed later.

The transformation to the parity basis thus brings elements in q_b in the diagonal elements for $l_b = 1$, leading to the l-doubling of the $v_b = 1$, Π level. Effectively, the role of the l-doubling is to modify the value of B into $B \mp \frac{1}{2}q_b$, with the upper and lower signs respectively associated with the now split e and f rotational sublevels.

The wavefunctions have also changed through the Wang transformation and are now related to e and f linear combinations of the initial wavefunctions. The e and f subscripts accompany the wavefunctions in (3.139), for didactic purposes. They are usually missing in the literature. The distinction is, however, implicitly present through the use of the convention that the e levels are those with $l_b > 0$. It makes q_b positive, thus with the e sublevels having lower energy than the f ones. This definition is in agreement with the most commonly encountered situation that the bending vibration is lower in energy than the stretching ones. The l-doubling in the first bending state ($v_b = 1$) can be thought of as resulting from the interaction between its J levels of e symmetry with those of the higher-energy first stretching state, also of e symmetry. They are thus pushed toward lower energy, while the f levels of the first bending state are unaffected, leading to $q_b > 0$.

The e and f sublevels resulting from the present definitions correspond to c and d used by Pliva [472,535] and s and a of Winnewisser and Winnewisser [534], although there are sign differences in some of the matrix elements. Let us illustrate the procedure by building the matrix Hamiltonian for $v_b = 2$. It becomes, after transformation and omitting $|J\rangle$ in the notation

$$
\begin{array}{c|ccc}
 & \multicolumn{2}{c}{e} & f \\
|v_b^{l_b}\rangle & |2^{+2}\rangle & |2^{+0}\rangle & |2^{-2}\rangle \\
\hline
|2^{+2}\rangle & D(2^{+2}) & q_b[M(M-2)]^{\frac{1}{2}} & \\
|2^{+0}\rangle & q_b[M(M-2)]^{\frac{1}{2}} & D(2^{+0}) & \\
|2^{-2}\rangle & & & D(2^{-2})
\end{array}
\tag{3.140}
$$

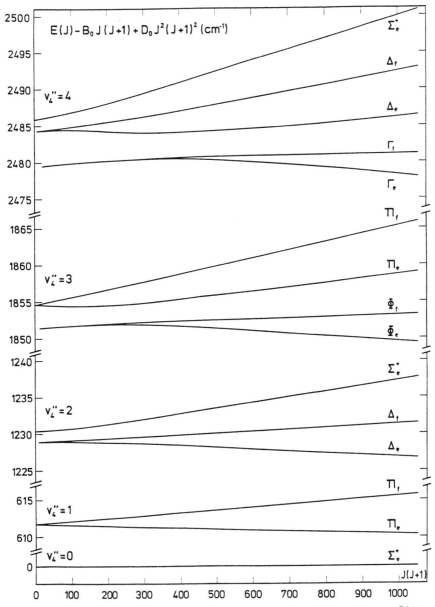

Figure 41. Rotational energy levels of $^{12}C_2H_2$ as a function of $J(J+1)$, in the $\tilde{X}^1\Sigma_g^+$ state for the l components of the *trans*-bending vibrational levels $v_4'' = 0$–4 (first part), and in the $\tilde{A}^1A_u$ state for the $K_a = 0$–5 sublevels of the *trans* bending vibrational levels $v_3' = 2$ (second part) (from Watson et al. [394]).

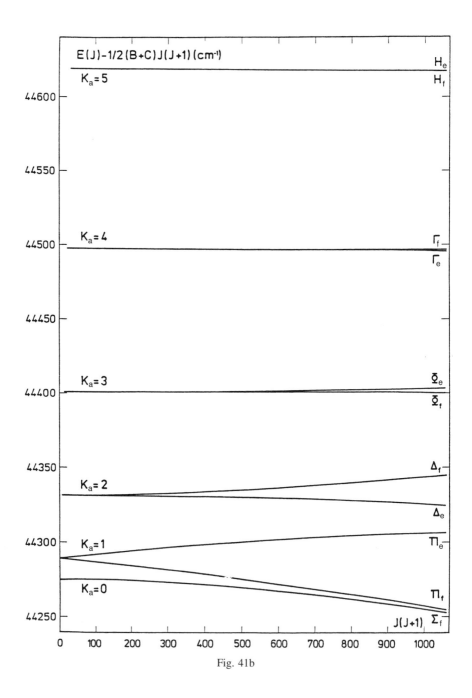

Fig. 41b

176

Thus, for the $v_b = 2$ level, there are two l sublevels, $\Delta(|k| = 2)$ and $\Sigma(k = 0)$. The l doubling leads to a splitting of the J rotational levels in the $l_b = 2\,(\Delta)$ sublevel, which, in this effective model, arises because of the matrix element coupling the $l_b = 2$ and $l_b = 0$ levels, within the e stack. The Δ_e and Σ_e^+ sublevels repel each other, while the Δ_f sublevel is not affected by the interaction. The matrix element (3.134) thus contributes to the l-doubling *and* l-resonance effects in $v_b > 1$ levels. A picture of the l-interaction mechanisms, merging l doubling and l resonance, was presented in Fig. 31.

The splitting arising from the l-doubling, for $v_b > 1$, is to be compared to the one whose size depends on the constant $(g\text{-}B)$ (section III.B.4.b.3). In the case of mode 4 in $^{12}\text{C}_2\text{H}_2$, the *trans* bending, this parameter is small $(-0.4\,\text{cm}^{-1})$. The l-sublevels arising from the same vibrational excitation are thus close in energy and the l doubling is important. It becomes so important at high J values, as it is $J(J + 1)$-dependent, that this doubling takes over the l-resonance effect and destroys the significance of l. Although l-doubling and K-splitting effects, in linear and asymmetric tops, respectively, are very similar in nature, a new regularity is introduced at high J values in cases similar to mode 4 *trans*-bend in $^{12}\text{C}_2\text{H}_2$, which makes a difference between both effects. These different behaviors are illustrated in Fig. 41.

TABLE XXV
Set of Matrix Operations Leading to Defintion of e and f Levels in a Linear Species with Two 2-Degenerate Vibration Modes

$|v_b^{l_b}, v_{b'}^{l_{b'}}\rangle$	$|1^{+1}, 1^{+1}\rangle$	$|1^{-1}, 1^{-1}\rangle$	$|1^{+1}, 1^{-1}\rangle$	$|1^{-1}, 1^{+1}\rangle$
$|1^{+1}, 1^{+1}\rangle$	D^+	—	$q_{b'}^0 J^2(J+1)^2$	$q_b^0 J^2(J+1)^2$
$|1^{-1}, 1^{-1}\rangle$	—	D^+	$q_b^0 J^2(J+1)^2$	$q_{b'}^0 J^2(J+1)^2$
$|1^{+1}, 1^{-1}\rangle$	$q_{b'}^0 J^2(J+1)^2$	$q_b^0 J^2(J+1)^2$	D^-	$r_{bb'}$
$|1^{-1}, 1^{+1}\rangle$	$q_b^0 J^2(J+1)^2$	$q_{b'}^0 J^2(J+1)^2$	$r_{bb'}$	D^-

with

$$D^{\pm} = \tilde{\omega}_b v_b + \tilde{\omega}_{b'} v_{b'} + x_{bb}^0 v_b^2 + x_{b'b'}^0 v_{b'}^2 + x_{bb'}^0 v_b v_{b'} + (g_{bb} + g_{b'b'} \pm g_{bb'} - B)k^2 + BJ(J+1)$$

$|v_b^{l_b}, v_b^{l_b}\rangle$	e		f	
	$|1^{+1}, 1^{+1}\rangle$	$|1^{+1}, 1^{-1}\rangle$	$|1^{-1}, 1^{-1}\rangle$	$|1^{-1}, 1^{+1}\rangle$
$|1^{+1}, 1^{+1}\rangle$	D^+	$\frac{1}{2}(q_b + q_{b'})$ $\times [M(M-2)]^{1/2}$		
$|1^{+1}, 1^{-1}\rangle$	$\frac{1}{2}(q_b + q_{b'})$ $\times [M(M-2)]^{1/2}$	$D^- + r_{bb'}$		
$|1^{-1}, 1^{-1}\rangle$			D^+	$\frac{1}{2}(q_{b'} - q_b)$ $\times [M(M-2)]^{1/2}$
$|1^{-1}, 1^{+1}\rangle$			$\frac{1}{2}(q_{b'} - q_b)$ $\times [M(M-2)]^{1/2}$	$D^- - r_{bb'}$

The complete l-matrix in the case of two 2D degenerate modes takes the form indicated in Table XXV, for $v_b = v_{b'} = 1$, before transformation to parity basis set and omitting $|J\rangle$ in the notation. Extra terms, in particular those from centrifugal distortion, need to be included in the diagonal elements, which are introduced in the next section (III.D.3). The l matrix after transformation is also printed in Table XXV. Matrix Hamiltonians corresponding to higher vibrational excitations in acetylene are explicitly given in Section III.D.4.a.

3. Vibration–Rotation Hamiltonian

a. General Picture. We have already dealt with vibration–rotation contributions in the MIME at various occasions. The main terms we are still left with are those arising from the non–rigid rotor rigid contributions ($\hat{H}_r^{nr}$), including the centrifugal distortion and the vibrational dependence of the rotational constants, as well as the Coriolis-type contributions ($\hat{H}_{rovib}^{Cor}$). They arise from parts in the initial Hamiltonian whose form is, respectively, $H_{12} + H_{22}$ and H_{21}, considering lower-order terms only. Their content has been indicated in Section III.A, where the dependence of the rotational tensors $\mu_{\alpha\alpha}$ in the vibrational coordinates that generates the nonrigid contributions, and the coupling between different vibrational states through rotation that is responsible for the Coriolis coupling were discussed. The latter term exists independently of the former. The related parameters, $a_i^{\alpha\beta}$ and ζ_{ij}^{α}, respectively, were also introduced in Section III.A.

Each of those partial Hamiltonian terms, just as the full Hamiltonian, needs to be totally symmetric. This condition therefore applies to the operator acting in the relevant term, such as $q_i \hat{J}_\alpha \hat{J}_\beta$ in H_{12}. Group theory therefore helps predicting when the coefficients $a_i^{\alpha\beta}$ and ζ_{ij}^{α} vanish in a specific molecule, as discussed by Bunker [46].

One needs to use perturbation theory to derive the useful effective forms of the vibration–rotation interactions. The relevant first and second-order contributions are, according to our previous developments (3.24):

$\tilde{h}$	1st Order	2nd Order	
$\tilde{h}_{02}$	H_{02}		
$\tilde{h}_{04}$		$H_{12}H_{12}$	(3.141)
$\tilde{h}_{12}$	H_{12}		
$\tilde{h}_{21}$	H_{21}		
$\tilde{h}_{22}$	H_{22}	$H_{30}H_{12}, H_{21}H_{21}, H_{21}H_{02}$	

The procedure is fully illustrated for the nonrigid terms in the next section, for diatomic species, then extended to linear polyatomic and asymmetric top species, in the next sections. Information concerning Coriolis coupling is eventually presented. Interesting vibration–rotation relations derived within the local mode limit, the so-called α-relations, can be found in the literature [649–653].

b. Diatomic Species. For a linear top, as presented in Section III.A.2.*a*

$$\hat{H}^{\text{lin}} = \frac{1}{2}\frac{\hbar^2}{I'}[\hat{J}_b - \hat{\pi}_b)^2 + (\hat{J}_c - \hat{\pi}_c)^2] + \frac{1}{2}\sum_{r=1}^{3n-5}\hat{P}_r^2 + V^{(Q)} \tag{3.142}$$

$$I''_{\alpha\beta} = I''_{\beta\alpha} = I^e_{\alpha\beta} + \frac{1}{2}\sum_k \left(\frac{\partial I_{\alpha\beta}}{\partial Q_k}\right)_e Q_k = I^e_{\alpha\beta} + \frac{1}{2}\sum_k a_k^{\alpha\beta}Q_k \tag{3.143}$$

$$I' = \frac{(I'')^2}{I^e} \tag{3.144}$$

Developing this expression for a diatomic species, and following the presentation by Kroto [641]; we have

$$\hat{\pi} = 0, \qquad I'' = I^e + \frac{1}{2}aQ, \qquad \text{and}$$

$$I' = I^e\left[1 + \frac{aQ}{I^e} + \frac{(aQ)^2}{4(I^e)^2}\right] = I^e\left(1 + \frac{1}{2}\vartheta_Q Q\right)^2 \tag{3.145}$$

$$(I')^{-1} = (I^e)^{-1}\left[1 - \vartheta_Q Q + \frac{3}{4}(\vartheta_Q Q)^2 - \cdots\right]$$

with: $\vartheta = a/I^e$.

Using the following form of development in series

$$(I')^{-1} = (I^e)^{-1}\left[1 - \vartheta_Q Q + \frac{3}{4}(\vartheta_Q Q)^2 - \cdots\right] \tag{3.146}$$

with the instantaneous moment of inertia

$$I' = mr^2 = m(r^e + s)^2 = (\sqrt{I^e} + Q)^2 = I^e + 2\sqrt{I^e}Q + Q^2 \tag{3.147}$$

and therefore

$$a = \left(\frac{\partial I'}{\partial Q}\right)_e = 2\sqrt{I^e} \tag{3.148}$$

Using $q = \gamma^{1/2}Q = (2\pi c\tilde{\omega}/\hbar)^{1/2}Q$, as defined in Section III.A.2.*a*

$$\vartheta_Q Q = \left(\frac{a}{I^e}\right)Q = \vartheta_q q = \left(\frac{2\pi c\tilde{\omega}}{\hbar}\right)^{-1/2}\left(\frac{a}{I^e}\right)q = \left(\frac{8B_e}{\tilde{\omega}}\right)^{1/2}q \tag{3.149}$$

Inserting the development of $(I')^{-1}$ in the Hamiltonian and using $\hat{\pi} = 0$ for a diatomic species, we see that the first term in $(I')^{-1}$ is just the rigid rotor contribution that we have already dealt with. The other two terms give the relevant contributions to the Hamiltonian to calculate the vibration–rotation terms, with $(I^e)^{-1}$ to be replaced by B_e after including all factors (viz., $\frac{1}{2}\hbar^2 \times (1/hc)$):

$$-B_e\vartheta_q q\hat{J}^2 + \tfrac{3}{4} B_e\vartheta_q^2 q^2\hat{J}^2 + \tfrac{1}{6} \phi_3 q^3 \qquad (3.150)$$

in which we have also kept the vibrational contribution, excluding the harmonic oscillator contribution, which we have also already dealt with, and excluding all noncubic anharmonic terms. This Hamiltonian corresponds to $H_{12} + H_{22} + H_{30}$, respecting the order of the terms.

Using second-order perturbation theory, according to the contribution listed in (3.141), one calculates

$$\tilde{h}_{04}(hc) = \frac{H_{12} \times H_{12}}{\Delta E_v^0} \qquad (3.151)$$

$$\langle v, J|\tilde{h}_{04}|v, J\rangle = (-B_e\vartheta_q)^2 \left[\frac{\langle v|q\hat{J}^2|v+1\rangle\langle v+1|q\hat{J}^2|v\rangle}{(E_{vJ}^0 - E_{v+1,J}^0)/hc} \right.$$

$$\left. + \frac{\langle v|q\hat{J}^2|v-1\rangle\langle v-1|q\hat{J}^2|v\rangle}{(E_{vJ}^0 - E_{v-1,J}^0)/hc} \right]$$

$$= (-B_e\vartheta_q)^2 \left[-\frac{J^2(J+1)^2}{2\tilde{\omega}} \right] \qquad (3.152)$$

Therefore, the centrifugal distortion constant appearing in the usual development:

$$F(v, J) = B[J(J+1)] - D[J(J+1)]^2 \qquad (3.153)$$

is, using (3.149)

$$D = \frac{4B_e^3}{\tilde{\omega}^2} \qquad (3.154)$$

Using again first and second-order perturbation contributions from (3.141), one can calculate $\tilde{h}_{22}$. The second order contribution from $H_{21} \times H_{02}$ and $H_{21} \times H_{21}$ are discarded here, as $\hat{\pi}$ and therefore H_{21} are zero for a diatomic species. Thus

$$\langle v,J|\tilde{h}_{22}|v,J\rangle = \frac{3}{4}B_e\vartheta_q^2\left(v+\frac{1}{2}\right)J(J+1) + \frac{1}{2}\left(\frac{B_e\vartheta_q\phi_3}{\tilde{\omega}}\right)\left(v+\frac{1}{2}\right)J(J+1)$$

$$(3.155)$$

in which we have omitted all J-independent terms.

Those terms provide the vibrational correction to the principal rotation constant B, which one should therefore write as

$$B_v = B_e - (v+\tfrac{1}{2})$$

with

$$\alpha = -\frac{6B_e^2}{\tilde{\omega}}(1+a_1)$$

$$a_1 = \frac{1}{6}\phi_3\left(\frac{2}{\tilde{\omega}B_e}\right)^{1/2} = \frac{1}{6}\frac{\phi_3 a}{\sqrt{\tilde{\omega}}}$$

$$(3.156)$$

Similarly, a vibrational dependence of D can be generated from higher-order terms:

$$D_v = D_e + \beta(v+\tfrac{1}{2})$$

$$(3.157)$$

c. *Linear Polyatomic Species.* In the case of linear polyatomic molecules, one must account for $\hat{\pi} \neq 0$. This operator takes, for example the following form for $^{12}C_2H_2$:

$$\hat{\pi}_\alpha = \sum_{s=1-3}\sum_{b=4,5}\zeta_{sb}^\alpha\left[\left(\frac{\tilde{\omega}_b}{\tilde{\omega}_s}\right)^{1/2}q_s\hat{p}_{b\beta} - \left(\frac{\tilde{\omega}_s}{\tilde{\omega}_b}\right)^{1/2}q_{b\beta}\hat{p}_s\right]$$

$$(3.158)$$

The various expressions just derived can be extended to the general case of linear tops by summing over all modes of vibrations and adding higher-order terms, [e.g., 654]:

$$B_v = B_e - \sum_i \alpha_i^B\left(v_i+\frac{d_i}{2}\right)$$

$$(3.159)$$

and, for the bending vibration of an unsymmetric triatomic molecule, with b the bending vibration (mode 2) and s (mode 1) and s' (mode 3), the stretching vibrations [641]:

$$\alpha_b = \frac{B_e^2}{\tilde{\omega}_b}\left[1 + \frac{4\zeta_{bs}^2\tilde{\omega}_b^2}{\tilde{\omega}_s^2 - \tilde{\omega}_b^2} + \frac{4\zeta_{bs'}^2\tilde{\omega}_b^2}{\tilde{\omega}_{s'}^2 - \tilde{\omega}_b^2}\right] - (2B_e)^{3/2}\left[\frac{\zeta_{bs'}\phi_{sbb}}{2\tilde{\omega}_s^{3/2}} - \frac{\zeta_{bs}\phi_{s'bb}}{2\tilde{\omega}_{s'}^{3/2}}\right]$$

$$(3.160)$$

As an example of the expression of the energy of a vibration–rotation level, we reproduce the one used to deal with experimental data of $^{12}C_2D_2$ by Huet et al., given with respect to the ground state [655]:

$$F(v,J) = \left[B_0 - \sum_i \alpha_i v_i + \sum_{1 \leq j} \gamma_{ij} v_i v_j + \sum_{1 \leq j \leq k} \gamma_{ijk} v_i v_j v_k\right.$$

$$\left. + \sum_{b \leq b'} \gamma^{bb'} l_b l_{b'} + \sum_b \sum_{b' \leq b''} \gamma_b^{b'b''} v_b l_{b'} l_{b''}\right][J(J+1) - k^2]$$

$$- \left[D_0 + \sum_i \beta_i v_i + \sum_{i \leq j} \beta_{ij} v_i v_j + \sum_{b \leq b'} \beta_{bb'} v_b v_{b'}\right][J(J+1) - k^2]^2$$

$$+ \left[H_0 + \sum_i H_i v_i\right][J(J+1) - k^2]^3 \qquad (3.161)$$

We should remember that one is required to include the J-independent term $(-Bk^2 + Dk^4 \cdots)$ in the vibrational contribution, to calculate vibrational-level energies from spectroscopic band centers.

The α parameter thus describes the vibrational dependence of the principal rotation constant. The role of α for a stretching vibration is to decrease the principal rotation constant in an excited stretching vibrational level, corresponding to an increase of the corresponding principal moment of inertia, specifically, to an increase of the average value of the stretching coordinate between the turning points in the classical potential-energy curve, as expected from the role of anharmonicity. For bending vibrations, α_b usually acts in the opposite way, thus increasing the principal rotation constant in the excited bending levels. This trend may be understood as if the mean structure were shrinking on bending rather than extending, leading to smaller average length of the molecule projected on the reference linear structure, and therefore to a smaller value of B. The present comments are refined in various papers (see, e.g., Refs. 59,515, and 641). Observed vibrational dependencies of rotational constants deserve further comments, as we illustrate with acetylene in Section III.D.4.d.

Table XXVI provides the values of α for different isotopomers in acetylene. Values for β, the constant for the vibrational dependence of D, are given for $^{12}C_2H_2$ in Ref. 610.

TABLE XXVI
Zeroth-order α (cm^{-1} × 10^3) Values for $^{12}C_2H_2$ and $^{12}C_2D_2$ (See, Refs. 610 and 525, Respectively, and References Cited Therein)

Mode	1	2	3	4	5
$^{12}C_2H_2$	6.9043	6.1814	5.8818	− 1.353535	− 2.232075
$^{12}C_2D_2$	5.585	3.133	4.488	− 2.08127	− 2.15884

d. Asymmetric Tops. Similar developments as those performed for the diatomic and linear species lead to the detailed formulas for the centrifugal distortion and the vibrational dependence of the principal rotational constants in asymmetric tops [e.g., 47]:

$$\alpha_i^{B_\alpha} = -\frac{2(B_\alpha^e)^2}{\tilde{\omega}_i}\left[\sum_\alpha \frac{3(a_i^{b\alpha})^2}{4I_\alpha} + \sum_j (\zeta_{ij}^b)^2 \frac{3\tilde{\omega}_i^2 + \tilde{\omega}_j^2}{\tilde{\omega}_i^2 - \tilde{\omega}_j^2} + \pi\left(\frac{c}{\hbar}\right)^{1/2}\sum_j \phi_{iij}a_j^{bb}\frac{\tilde{\omega}_i}{\tilde{\omega}_j^{3/2}}\right]$$

(3.162)

The problem of the centrifugal distortion constants is somewhat more complicated and was eventually solved by Watson [369,370,456,656]. The terms $\tilde{h}_{04},\tilde{h}_{06},\tilde{h}_{08}$, [456], whose form vary with the molecular symmetry point group, generate contributions to the Hamiltonian corresponding to the quartic, sextic, and octic centrifugal terms, respectively. We only point out here that those terms can be reduced in two appropriate forms [642]. In the so-called asymmetric top or A reduction, transformations of the initial Hamiltonian are made in such a way that matrix elements with $|\Delta k| > 2$ are eliminated. It is therefore a rigid asymmetric top Hamiltonian with higher-order term matrix elements of the form given in Table XXVII, accounting for $\tilde{h}_{02},\tilde{h}_{04},\tilde{h}_{06}$. They are to be dealt with in the same way as explained for the rigid rotor Hamiltonian, in Section III.D.2.*b*.

The other convenient reduced form of the Hamiltonian is the S form, which presents off-diagonal elements with Δk up to 6. The procedure remains similar to the one described previously, but the tridiagonal form of the matrix is lost. The matrix elements are also given in Table XXVII, in the second part. With the reduction procedure, the principal rotation constants $B_{x,y,z}^{(A),(S)}$ are not identical to those defined previously. The relationship, involving centrifugal distortion constants, is provided by Watson [642].

TABLE XXVII

Matrix Elements for Asymmetric Top Rotors with Centrifugal Distortion Terms

$$\langle J, k | \tilde{h}_r | J, k \rangle^A = \frac{1}{2} [B_x^{(A)} + B_y^{(A)}] J(J+1) + \{B_z^{(A)} - \frac{1}{2}[B_x^{(A)} + B_y^{(A)}]\} k^2$$

$$- \Delta_J J^2 (J+1)^2 - \Delta_{JK} J(J+1) k^2 - \Delta_K k^4$$

$$+ \Phi_J J^3 (J+1)^3 + \Phi_{JK} J^2 (J+1)^2 k^2 + \Phi_{KJ} J(J+1) k^4 + \Phi_K k^6$$

$$\langle J, k \pm 2 | \tilde{h}_r | J, k \rangle^A = \{\frac{1}{4}[B_x^{(A)} - B_y^{(A)}] - \delta_J J(J+1) - \frac{1}{2}\delta_K[(k \pm 2)^2 + k^2]$$

$$+ \Phi_J J^2 (J+1)^2 + \frac{1}{2}\Phi_{JK} J(J+1)[(k \pm 2)^2 + k^2] + \frac{1}{2}\Phi_K[(k \pm 2)^4 + k^4]\}$$

$$\{[J(J+1) - k(k \pm 1)][J(J+1) - (k \pm 1)(k \pm 2)]\}^{1/2}$$

$$\langle J, k | \tilde{h}_r | J, k \rangle^S = \frac{1}{2}[B_x^{(S)} + B_y^{(S)}] J(J+1) + \{B_z^{(S)} - \frac{1}{2}[B_x^{(S)} + B_y^{(S)}]\} k^2$$

$$- D_J J^2 (J+1)^2 - D_{JK} J(J+1) k^2 - D_K k^4$$

$$+ H_J J^3 (J+1)^3 + H_{JK} J^2 (J+1)^2 k^2 + H_{KJ} J(J+1) k^4 + H_K k^6$$

$$\langle J, k \pm 2 | \tilde{h}_r | J, k \rangle^S = \{\frac{1}{4}[B_x^{(S)} - B_y^{(S)}] + d_1 J(J+1) + h_1 J^2 (J+1)^2\}$$

$$\times \{[J(J+1) - k(k \pm 1)][J(J+1) - (k \pm 1)(k \pm 2)]\}^{1/2}$$

$$\langle J, k \pm 4 | \tilde{h}_r | J, k \rangle^S = [d_2 + h_2 J(J+1)]\{[J(J+1) - k(k \pm 1)]$$

$$[J(J+1) - k(k \pm 1)(k \pm 2)][J(J+1) - (k \pm 2)(k \pm 3)][J(J+1) - (k \pm 3)(k \pm 4)]\}^{1/2}$$

$$\langle J, k \pm 6 | \tilde{h}_r | J, k \rangle^S = h_3\{[J(J+1) - k(k \pm 1)][J(J+1) - (k \pm 1)(k \pm 2)]$$

$$[J(J \pm 1) - (k \pm 2)(k \pm 3)][J(J+1)] - (k \pm 3)(k \pm 4)]$$

$$[J(J+1) - (k \pm 4)(k \pm 5)][J(J+1) - (k \pm 5)(k \pm 6)]$$

e. Coriolis Coupling

1. ASYMMETRIC TOPS. We restrict the present section to first-order terms, with brief reference to second-order contributions. Useful references are 47,456, and 657–665.

Figure 42 illustrates first order Coriolis coupling in ethylene with the interaction between modes 10 and 4 induced through rotation around the b axis, as in Herzberg [515]. One considers initially the individual displacement vectors corresponding to normal mode 10 in the molecule (see Fig. 35). Using classical mechanics, one can readily calculate the

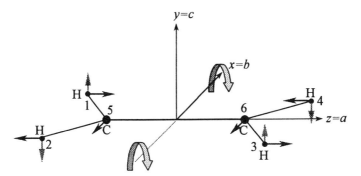

Figure 42. Mechanism for first-order Coriolis interaction between modes 10 (left part) and 4 (right part) in $^{12}C_2H_4$, induced by rotation around the b axis (adapted from Herzberg [515]).

Coriolis forces brought onto these vectors by a rotation around b, resulting in a set of modified displacements that is presented in the second part of the figure. The set of arrows one has generated exactly corresponds to mode 4. In this classical picture, the vibrational energy is, however, unchanged and thus still the one of mode 10. It is thus required that both modes have similar energy to make the resonance active, which is the case in $^{12}C_2H_4$.

The corresponding terms in the Hamiltonian arise from the contribution of $\tilde{h}_{21}$, which is, within a factor $1/hc$, identical to H_{21}:

$$\tilde{h}_{21} = -2 \sum_\alpha B_\alpha^e \sum_{i,j>i} \zeta_{ij}^\alpha \left[\left(\frac{\tilde{\omega}_j}{\tilde{\omega}_i} \right)^{1/2} q_i \hat{p}_j - \left(\frac{\tilde{\omega}_i}{\tilde{\omega}_j} \right)^{1/2} q_j \hat{p}_i \right] \hat{J}_\alpha \qquad (3.163)$$

As already pointed out, the direct product of the symmetry representations of the relevant operators, in the molecular symmetry group must be totally symmetric. For ethylene, the relevant information is provided in Table XXVIII [589]. Second-order Coriolis rotational operators are indicated as well in that table, for completeness. We also provide in Table XXVIII the symmetry of the components of the projections of the induced electric dipole moment on the molecular axis system, introduced in Section II and used further in Section III.E. The label of the modes in ethylene is the usual one [515], while the axis system and the point group symmetries are selected to be those in Bunker [46], in agreement with I^r representation ($a = z$, $b = x$, $c = y$). Using that table, one can thus easily predict the nonvanishing Coriolis constants. The related matrix elements of first order can be calculated from the general expressions of the action of the vibrational ($q_i \hat{p}_j$) and rotational operators, ($\hat{J}_x, i\hat{J}_y, \hat{J}_z$), provided in Tables VII and XXII, respectively.

TABLE XXVIII

Symmetry of Electric Dipole Transition Moments and Total Angular Momentum Operators ($\mu_\alpha, \hat{J}_\alpha$) of Rotation Wavefunctions (e = even, o = odd), and Normal Modes of Vibration (q_i), in $^{12}C_2H_4$ (D_{2h}) and $^{12}C_2H_2$ ($D_{\infty h}$)

D_{2h} [$z = a, y = c, x = b$]	q_i	$K_a K_c$	$\mu_\alpha, \hat{J}_\alpha$
A_g	1,2,3	ee	$\hat{J}^2, \hat{J}_b^2 - \hat{J}_c^2, \hat{J}_a^2$
A_u	4	—	μ_z
B_{1u}	11,12	—	μ_a
B_{2u}	9,10	—	μ_b
B_{3u}	7	—	μ_c
B_{1g}	—	eo	$\hat{J}_a, \{\hat{J}_b, i\hat{J}_c\}$
B_{2g}	8	oo	$\hat{J}_b, \{i\hat{J}_c, \hat{J}_a\}$
B_{3g}	5,6	oe	$\hat{J}_c, \{\hat{J}_a, \hat{J}_b\}$

$D_{\infty h}$ [$z = a, (x, y) = (b, c)$]	q_i	K	$\mu_\alpha, \hat{J}_\alpha$
Σ_g^+	1,2	0 (J even)	—
Σ_g^-	—	0 (J odd)	$\hat{J}_\alpha$
Σ_u^+	3	—	μ_α
Σ_u^-	—	—	—
Π_g	4	± 1	$(\hat{J}_b, \hat{J}_c)$
Π_u	5	—	(μ_b, μ_c)
Δ_g	—	± 2	—
$\cdots$	—	$\cdots$	—

The rotation matrix elements relevant for vibration–rotation transitions depend on the director cosines we have introduced in the Section II. Those have symmetry behaviors identical to those of the total angular momenta projection operators, $\hat{J}_\alpha$, active in the first order Coriolis couplings. Coriolis coupling and electric dipole transitions of the same α type thus connect rotation wavefunctions with identical selection rules. The various α-type selections rules to deal with electric-dipole-induced transitions are provided in Section III.E and can therefore be used here as well. The selection rules for first-order Coriolis couplings are usually called *Jahn's rules* [666].

In the example of modes 10 and mode 4 of C_2H_4, the selection rules for the interaction are of b type, namely, rotational levels with both ΔK_a and ΔK_c odd are connected through the couplings.

One should notice that this similarity with the electric dipole transition rotation selection rules does not extend further. Indeed, considering the example we have selected, the Coriolis interaction couples vibrational states in ethylene both having u symmetry, which cannot be connected through electric dipole transition.

2. ACETYLENE. Coriolis interactions in acetylene were already introduced while discussing the giant cluster picture, in Section III.C.3. Coriolis resonance vectors were presented at that occasion in Table XIX, for acetylene [540]. They actually correspond to higher-order Coriolis coupling schemes, which are only briefly referred to here. They are required for $^{12}C_2H_2$ because the levels coupled by low-order interactions are too distant in energy. Such lower-order terms might, however, be of some importance to deal with vibration–rotation intensities and, possibly, in very specific symmetry breaking perturbations [514]. We focus hereafter on first-order terms, only, as revealed explicitly during the analysis of overtone transitions [516].

The Coriolis constants of first order in symmetric acetylene and the related sum rules are, for symmetric isotopomers, ζ_{14}, ζ_{24}, and ζ_{35}, with the components on the principal axes b and c identical, and $(\zeta_{14})^2 + (\zeta_{24})^2 = 1$, and $(\zeta_{35})^2 = 1$, in agreement with the general form

$$\sum_{s,b} \zeta_{sb}\zeta_{sb'} = \delta_{bb'} \qquad (3.164)$$

The relevant symmetry information is provided in Table XXVIII. It can be checked that the first-order selection rules for the interaction are of perpendicular type: $\Delta K = \pm 1$.

TABLE XXIX
First-Order Coriolis Hamiltonian and Matrix Elements for Acetylene

---

$$\hat{H}^{Cor}_{rovib}/hc = -2B_e\left(\hat{J}_x\hat{\pi}_x - \hat{J}_y\hat{\pi}_y\right)$$

$$\hat{\pi}_x = \sum_{s=1,2,3;b=4,5} \zeta^x_{sb}\left[\left(\frac{\tilde{\omega}_b}{\tilde{\omega}_s}\right)^{1/2} q_s\hat{p}^y_b - \left(\frac{\tilde{\omega}_s}{\tilde{\omega}_b}\right)^{1/2}\hat{p}_s q^y_b\right]$$

$$\langle v_s - 1, (v_b + 1)^{l_b+1}, J, k+1 \| v_s, v_b b^{l_b}_b, J, k\rangle = B_e\zeta^y_{sb}\Theta_{sb}[v_s(v_b + l_b + 2)]^{1/2}[J(J+1)$$
$$- k(k+1)]^{1/2}$$

$$\langle v_s + 1, (v_b - 1)^{l_b+1}, J, k+1 \| v_s, v^{l_b}_b, J, k\rangle = -B_e\zeta^y_{sb}\Theta_{sb}(v_s + 1)^{1/2}(v_b - l_b)^{1/2}$$
$$[J(J+1) - k(k+1)]^{1/2}$$

with

$$k = l_b$$

$$\Theta_{sb} = \frac{1}{2}\left[\left(\frac{\tilde{\omega}_b}{\tilde{\omega}_s}\right)^{1/2} + \left(\frac{\tilde{\omega}_s}{\tilde{\omega}_b}\right)^{1/2}\right]$$

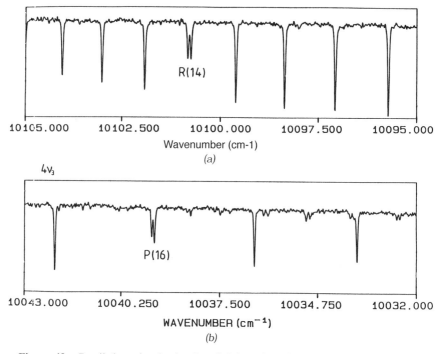

Figure 43. Detailed portion in the R and P branches of the same band in the high-resolution absorption spectrum of $^{12}C_2HD$ demonstrating the Coriolis coupling affecting the $J = 15$ rotation level in the upper vibrational level (from Abbouti Temsamani et al. [594], reproduced with permission from Taylor & Francis).

A complete treatment is presented by Halonen, and co-workers [667], which we have adapted in Table XXIX. The terms in that table are written using labels x and y for the axes b and c, and labels s and b for stretching and bending modes, respectively. All $\hat{\pi}_y \hat{J}_y$ matrix elements are identical to the $\hat{\pi}_x \hat{J}_y$ ones, listed in the table. As an example of Coriolis interaction in acetylene, we provide in Fig. 43 a portion of the vibration–rotation spectrum in $^{12}C_2HD$ demonstrating Coriolis coupling. As discussed in Section III.B.3.a.3 for vibrational bright and dark states, the zeroth-order wavefunctions of the now Coriolis coupled vibration–rotation levels mix, leading to light up a zeroth-order forbidden or extremely weak transition.

4. Vibration–Rotation Fits in Acetylene

a. MIME for Lower Bending Levels in $^{12}C_2H_2$. We illustrate here the theory provided in the present chapter by dealing with vibration–rotation structures in acetylene. Additional matrix elements [534] are sometimes required. We start in the present section with the full Hamiltonian used to fit the bending vibration–rotation energy levels in acetylene ($^{12}C_2H_2$), from Herman et al. [655], which is provided in Tables XXX and XXXI. It thus concerns two bending vibrations, with modes 4 (*trans*-bend) and 5 (*cis*-bend). The Hamiltonian matrices are given with each block appearing twice, once for e rotational levels (upper sign), once for the f rotational levels (lower sign). Some of the numerical coefficients are corrected compared to those in Ref. 655. All matrices are symmetric with respect to their diagonal and only the upper triangle is presented. A Hamiltonian is to be built for each different J-value. This model was built empirically and later confirmed theoretically by Perevalov and Sulakshina [539] who used contact transformation procedure to calculate all effective operators.

A fit to 1189 precise data belonging to selected bands recorded using Fourier transform spectroscopy in the far and mid infrared ranges, was performed using the matrix Hamiltonian provided in Table XXXI by Kabbadj et al. [504]. They determined the parameters reproduced in Table XXXII, which have updated those of the literature [452,472,535,668–670] (see also Ref. 671).

The root mean square error on the data was $0.000107\,cm^{-1}$.

A fit with extra data of the same quality, including information from more observed bands, was also reported by Kabbadj et al. [504]. The root mean square error on the 2307 data was $0.000439\,cm^{-1}$. Such an increase, compared to the other fit mentioned, calls for some extension of the model. It is probably required to include stretching–bending anharmonic and vibration–rotation interactions [604], and possibly higher-order bending terms, as being attempted by some research groups.

The parameters in Table XXXII were reproduced almost within their experimental uncertainty by Martin et al. [210], who extracted them from a potential calculated using ab initio means, as detailed in Section II. Only r_{45} was calculated a factor 2 lower in that paper. This discrepancy is, however, most probably apparent only because it resulted from another numerical definition of this constant in the ab initio approach.

A more extended series of bending levels in v_4, including highly excited states observed using stimulated emission pumping and laser-induced fluorescence spectroscopies, was reconsidered recently in the literature and all experimental vibrational bending energies refitted [469,609b]. That study considered only vibrational energies. Additional anharmonicity and vibrational dependencies were included in the model, to account, in particular, for the DD bend–bend anharmonic resonance. That interaction becomes stronger at higher vibrational excitation in $^{12}C_2H_2$ indeed, because of the different signs of x_{44} and x_{55}, which, combined with the relative ordering of modes 4 and 5 vibrational frequencies, make the pure mode 4 and mode 5

TABLE XXX
Matrix Elements Required to Fit Bending Vibration–Rotation Energy Levels in $^{12}C_2H_2$

Diagonal elements $[v_4^{l_4} v_5^{l_5} J]^k$

$$G_0(v_4^{l_4} v_5^{l_5}) = \sum_b \tilde{v}_b^0 v_b + \sum_{b \leq b'} (x_{bb'}^0 v_b v_{b'} + g_{bb'}^0 l_b l_{b'})$$
$$+ \sum_{b \leq b' \leq b''} y_{bb'b''}^0 v_b v_{b'} v_{b''} + \sum_b \sum_{b' \leq b''} y_b^{b'b''} v_b v_{b'} v_{b''}$$

$$F(v_4^{l_4} v_5^{l_5}) = (B_v + \sum_{b \leq b'} \gamma^{bb'} + \sum_b \sum_{b' \leq b''} \gamma_b^{b'b''} v_b l_{b'} l_{b''})(M - k^2)$$
$$- (D_v + \sum_{b \leq b'} \beta^{bb'} l_b l_{b'})(M - k^2)^2 + H_v(M - k^2)^3$$

$M = J(J + 1), k = l_4 + l_5$ and $b, b', b'' \equiv 4, 5$

$$B_v = B_0 - \sum_b \alpha_b v_b + \sum_{b \leq b'} \gamma_{bb'} v_b v_{b'}$$

$$D_v = D_0 + \sum_b \beta_b v_b + \sum_{bb'} \beta_{bb'} v_b v_{b'}$$

$$H_v = H_0 + \sum_b H_b v_b$$

$$r_{45} = r_{45}^0 + r_{45}^J J(J+1) + r_{45}^{JJ} J^2(J+1)^2 + r_{445}(v_4+1) + r_{455}(v_5+1)$$

$$q_b = q_b^0 + q_{bb} v_b + q_{bb'} v_{b'} + q_b^J J(J+1) + q_b^{JJ} J^2(J+1)^2 + q_b^k (k \pm 1)^2$$

$$\rho_b = \rho_b^0 + \rho_{bb} v_b + \rho_b^{b'} v_{b'} + \rho_b^J J(J+1)$$

with $\pm$ associated with $\Delta k = \pm 2$ matrix elements; off-diagonal elements

$$\langle v_4^{l_4}, v_5^{l_5}, k \parallel v_4^{l_4 \pm 2}, v_5^{l_5 \mp 2}, k \rangle = \tfrac{1}{4} r_{45}[(v_4 \mp l_4)(v_4 \pm l_4 + 2)(v_5 \pm l_5)(v_5 \mp l_5 + 2)]^{1/2}$$

$$\langle v_b^{l_b}, k, J \parallel v_b^{l_b \pm 2}, k \pm 2, J \rangle = \frac{1}{4} q_b\{(v_b \mp l_b)(v_b \pm l_b + 2)$$
$$[M - k(k \pm 1)][M - (k \pm 1)(k \pm 2)]\}^{1/2}$$

$$\langle v_b^{l_b}, k, J \parallel v_b^{l_b \pm 4}, k \pm 4, J \rangle = \frac{1}{16} \rho_b\{(v_b \mp l_b)(v_b \pm l_b + 2)(v_b \mp l_b - 2)(v_b \pm l_b + 4)$$
$$\times [M - k(k \pm 1)][M - k(k \pm 1)(k \pm 2)][M - k(k \pm 2)(k \pm 3)]$$
$$\times [M - k(k \pm 3)(k \pm 4)]\}^{1/2}$$

$$\langle v_4^{l_4}, v_5^{l_5}, k, J \parallel v_4^{l_4 \pm 2}, v_5^{l_5 \pm 2}, k \pm 4, J \rangle = \frac{1}{16} \rho_{45}\{(v_4 \mp l_4)(v_4 \pm l_4 + 2)(v_5 \mp l_5)(v_5 \pm l_5 + 2)$$
$$\times [M - k(k \pm 1)][M - (k \pm 1)(k \pm 2)]$$
$$\times [M - (k \pm 2)(k \pm 3)][M - (k \pm 3)(k \pm 4)]\}^{1/2}$$

$M = J(J+1), k = l_4 + l_5, b \equiv 4, 5$

Source: Adapted From Herman et al. [655].

190

(*continued*)

TABLE XXXI

Hamiltonian Used to Fit Bending Vibration–Rotation Energy Levels in $^{12}\mathrm{C}_2\mathrm{H}_2$ up to $\displaystyle\sum_b v_b = 4$

$[[1^1 J]^1 \mp q_b M]]$

$[2^2 J]^2 \pm \frac{1}{2}\rho_b M(M-2)$ $\qquad q_b[M(M-2)]^{1/2}$
$[2^0 J]^0$

$[3^3 J]^3$ $\qquad (\sqrt{3}/2)(q_b \mp \rho_b M)[(M-2)(M-6)]^{1/2}$
$[3^1 J]^1 \mp q_b M$

$[4^4 J]^4$ $\qquad q_b[(M-6)(M-12)]^{1/2}$ $\qquad \sqrt{3}\rho_b[M(M-2)(M-6)(M-12)]^{1/2}$
$[4^2 J]^2 \pm \frac{3}{2}\rho_b M(M-2)$ $\qquad \sqrt{3}q_b[M(M-2)]^{1/2}$
$\qquad\qquad\qquad\qquad [4^0 J^0]$

$[1^1 1^1 J]^2 \pm \frac{1}{4}\rho_{45}M(M-2)$ $\qquad \frac{1}{2}(q_5 \pm q_4)[M(M-2)]^{1/2}$
$[1^1 1^{-1} J]^0 \pm r_{45}$

$[1^1 2^2 J]^3$ $\qquad \frac{1}{2}(q_4 \mp \rho_5 M)[(M-2)(M-6)]^{1/2}$ $\qquad \sqrt{\frac{1}{2}[q_5 \mp \frac{1}{2}\rho_{45}M][(M-2)(M-6)]^{1/2}}$
$[1^{-1} 2^2 J]^1$ $\qquad \sqrt{2}(r_{45} \mp \frac{1}{2}q_5 M)$
$[1^1 2^0 J]^1 \mp \frac{1}{2}q_4 M$

TABLE XXXI (*continued*)

$[1^13^3J]^4$	$\frac{1}{2}q_4[(M-6)(M-12)]^{1/2}$ $[1^13^3J]^2$	$(\sqrt{3}/2)q_5[(M-6)(M-12)]^{1/2}$ $\sqrt{3}[r_{45}\pm\frac{1}{2}\rho_5 M(M-2)]$ $[1^13^1J]^2\pm\frac{1}{2}\rho_{45}M(M-2)$	$(\sqrt{3}/2)(\frac{1}{2}\rho_{45}\pm\rho_5)[M(M-2)(M-6)(M-12)]^{1/2}$ $\sqrt{3}/2q_5[M(M-2)]^{1/2}$ $[\frac{1}{2}q_4\pm q_5][M(M-2)]^{1/2}$ $[1^{-1}3^1J]^0\pm 2r_{45}$
$[3^31^1J]^4$	$\frac{1}{2}q_5[(M-6)(M-12)]^{1/2}$ $[3^31^1J]^2$	$(\sqrt{3}/2)q_4[(M-6)(M-12)]^{1/2}$ $\sqrt{3}[r_{45}\pm\frac{1}{2}\rho_4 M(M-2)]$ $[3^11^1J]^2\pm\frac{1}{2}\rho_{45}M(M-2)$	$(\sqrt{3}/2)(\frac{1}{2}\rho_{45}\pm\rho_4)[M(M-2)(M-6)(M-12)]^{1/2}$ $(\sqrt{3}/2)q_4[M(M-2)]^{1/2}$ $[\frac{1}{2}q_5\pm q_4][M(M-2)]^{1/2}$ $[3^11^{-1}J]^0\pm 2r_{45}$
$[2^22^2J]^4$	$\sqrt{\tfrac{1}{2}}q_5[(M-6)(M-12)]^{1/2}$ $[2^22^0J]^2\pm\frac{1}{2}\rho_4 M(M-2)$	$\sqrt{\tfrac{1}{2}}q_4[(M-6)(M-12)]^{1/2}$ $2r_{45}\pm\frac{1}{2}\rho_{45}[M(M-2)]$ $[2^02^2J]^2\pm\frac{1}{2}\rho_{45}M(M-2)$	$\frac{1}{2}(\rho_5\pm\rho_4)[M(M-2)(M-6)(M-12)]^{1/2}$ $\sqrt{\tfrac{1}{2}}q_5[M(M-2)]^{1/2}$ $\pm\sqrt{\tfrac{1}{2}}q_4[M(M-2)]^{1/2}$ $[2^22^{-2}J]^0$	$\sqrt{\tfrac{1}{2}}\rho_{45}[M(M-2)(M-6)(M-12)]^{1/2}$ $q_4[M(M-2)]^{1/2}$ $q_5[M(M-2)]^{1/2}$ $2\sqrt{2}r_{45}$ $[2^02^0J]^0$

Source: Adapted from Herman et al. [655].

192

TABLE XXXII
Vibration−Rotation Parameters (in cm^{-1}) Determined for Bending Levels in $^{12}C_2H_2$ the
Parameters are Defined in Eq. (3.161)

$\tilde{\omega}_4^0 = 608.985196(14)$		$\tilde{\omega}_5^0 = 729.157564(10)$
$x_{44} = 3.1049824(82)$	$x_{45} = -2.2878296(82)$	$x_{55} = -2.300723(56)$
$g_{44} = 0.7815203(90)$	$g_{45} = 6.6046746(79)$	$g_{55} = 3.4760741(63)$
$r_{45} = -6.238720(11)$		$r_{45}^J = 1.95353(63)10^{-4}$
$B_0 = 1.17664632(18)$	$D_0 = 1.62710(27)10^{-6}$	$H_0 = 1.60(11)10^{-12}$
$\alpha_4 = -1.353535(86)10^{-3}$		$\alpha_5 = -2.232075(40)10^{-3}$
$\gamma_{44} = 9.00(48)10^{-7}$	$\gamma_{45} = -2.3716(20)10^{-5}$	$\gamma_{55} = 1.8699(19)10^{-5}$
$\gamma^{44} = -6.578(11)10^{-5}$	$\gamma^{45} = -2.25627(49)10^{-4}$	$\gamma^{55} = -1.09922(41)10^{-4}$
$\beta_4 = 3.2581(43)10^{-8}$		$\beta_5 = 2.4528(16)10^{-8}$
$q_4^0 = 5.24858(12)10^{-3}$		$q_5^0 = 4.66044(12)10^{-3}$
$q_{44} = -1.783(11)10^{-5}$	$q_{45} = 7.915(42)10^{-5}$	$q_{55} = 3.798(12)10^{-5}$
	$q_{54} = 1.1012(46)10^{-4}$	
$q_4^J = -3.9301(65)10^{-8}$		$q_5^J = -3.8477(23)10^{-8}$
$\rho_4 = -3.68(57)10^{-9}$	$\rho_{45} = -1761(30)10^{-8}$	$\rho_5 = -6.86(15)10^{-9}$

Source: Adapted from Kabbadj et al. [504].

bend vibrational levels get closer at high v_b. This evolution requires the DD coupling to be included in the Hamiltonian matrix. Such terms were discussed in Section III.B.4.c.3. All bends observed below 10,000 cm^{-1} were recently refitted, together with the full set of experimentally observed vibrational levels in $^{12}C_2H_2$ [450].

b. Vibration−Rotation Wavefunctions in $^{12}C_2H_2$

1. LOWER ENERGY BENDING ENERGY LEVELS. The model and constants just reported, in Tables XXXI and XXXII, were used by Herman et al. [655] to calculate the J-dependent mixing between the bend vibration−rotation wavefunctions. The mixing of the functions of the zeroth-order basis set is given by the squared coefficients in the eigenvectors resulting from the each J-matrix Hamiltonian diagonalization. It is different from J to J level because some of the matrix elements are J-dependent. As a consequence (1) the off-diagonal matrix elements have different numerical values for different J values and (2) the zeroth-order values of the principal rotation constants, B_v, are different for the different interacting vibrational levels and, therefore, the energy difference between interacting levels also varies with J.

Figure 44 presents this mixing, for the Π_{1u} levels arising from $|v_4 v_5\rangle^e = |12\rangle^e$. As a result of l-type doubling, the combination with $l = 3$ is to be taken into account, in addition to the two $l = 1$ sublevels. The

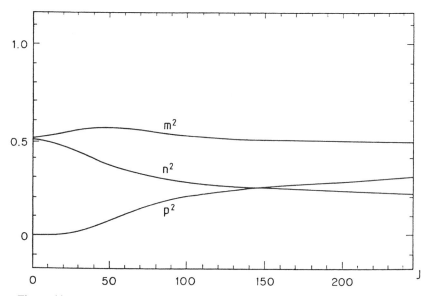

Figure 44. Evolution with J of the squared coefficients in the eigenvectors of the levels corresponding to $|v_4, v_5, J\rangle^e = |2, 1, J\rangle^e$ in $^{12}C_2H_2$; see text for definition of m, n, and p (from Herman et al. [655], reproduced with permission from Taylor & Francis).

wavefunction is (with $k > 0$ associated with e levels):

$$\Psi(12; \Pi_I^e) = \sum_i c_i |v_4^{l_4} v_5^{l_5}, J, k\rangle = m|2^0 1^1, J, 1\rangle$$
$$+ n|2^2 1^{-1}, J, 1\rangle + p|2^2 1^1, J, 3\rangle \qquad (3.165)$$

and $m^2 + n^2 + p^2 = 1$.

The mixing is very efficient between both Π sublevels at low J and decreases at higher J, while the Φ sublevel comes in. The same figure for the $|2, 1, J, k\rangle$ sublevels, presented in [655], shows different effects, with the Π sublevels less mixed at low J values. The off-diagonal elements connecting the two Π_e sublevels, of the form $[r_{45} - \frac{1}{2}qJ(J + 1)]$, are about the same size for both $|v_4 v_5\rangle^e = |21\rangle^e$ and $|12\rangle^e$ vibrational excitations, as $q_4 \approx q_5$. It is therefore the energy difference between the two Π sublevels, arising from the diagonal elements, which differentiates the two behaviors.

The mixings we have illustrated, and their difference from one set of levels to another, have dramatic influence on the vibration–rotation band intensities. Indeed, the two resulting wavefunctions are of the form

$$m|1^1 2^0\rangle + n|1^{-1} 2^2\rangle \qquad \text{and} \qquad n|1^1 2^0\rangle - m|1^{-1} 2^2\rangle \qquad (3.166)$$

and, because of the almost 1:1 mixing demonstrated in Fig. 44, $m \approx n$ and transitions reaching the state with the minus combination are about 30 times weaker than those reaching the combination with the plus sign, in agreement with the intensity scheme developed in Section III.B.3.a.3. This behavior is fully illustrated in Ref. 655, with, however, some confusion in the coherence of the state labelling in the equations and figures.

Complementary investigation on the role of the l-resonance on the vibration–rotation intensities in acetylene was achieved by Weber and co-workers [672–675].

2.   FUNDAMENTAL CH STRETCHING VIBRATION, v_3. Another example of vibration–rotation mixing was worked out for the fundamental v_3 level in $^{12}C_2H_2$ [451]. The scheme of the various resonances in the energy range, involving the $v_2 + v_4 + v_5$ level, were presented in Fig. 31. Thus the vibration–rotation Hamiltonian must include anharmonic resonance, l-resonance and l-doubling terms. The resulting J-dependent mixing, for the e levels, is presented in Figure 45. A 1:1 mixing occurs for the two Σ^+ states, right from $J = 0$. The matrix elements are also such that the $l = 2$, Δ sublevel is significantly mixed with the two $l = 0$ components through l doubling. This occurs for much lower J values than in the $l = 1,3$ substates, in Fig. 44. As a result, that zeroth-order Δ dark state borrows enough Σ^+ character to show up on the spectrum, as further discussed in Section III.E.8.c.3 and illustrated in Fig. 66.

c.   Lower Bending Levels in $^{12}C_2D_2$. The need for introducing the DD anharmonic coupling for dealing with the bends in acetylene was demonstrated with C_2D_2 [505], initially from the investigation of the $\tilde{A} - \tilde{X}$ electronic band system [676]. The first bend levels in that isotopomer were depicted in Fig. 29, illustrating the near degeneracy of the DD resonant pairs.

Some 1608 extremely precise microwave (MW) data and Fourier transform (FT) data recorded in the far-infrared and low-infrared ranges were gathered in the literature for $^{12}C_2D_2$ [505,677,678], and dealt with using matrix elements of the form given in Table XXX, introducing the following dependencies:

$$K_{44/55} = K^0_{44/55} + K^J_{44/55}J(J+1) + K_{445}v_4 + K_{455}(v_5 + 2)$$
$$r_{45} = r^0_{45} + r^J_{45}J(J+1) + r^{JJ}_{45}[J(J+1)]^2 + r_{445}v_4 + r_{455}(v_5 + 2)$$

$$(3.167)$$

The data were successfully fitted, within 3 times a dimensionless root mean squared value of 1.02. However, Perevalov and Sulakshina, later theoreti-

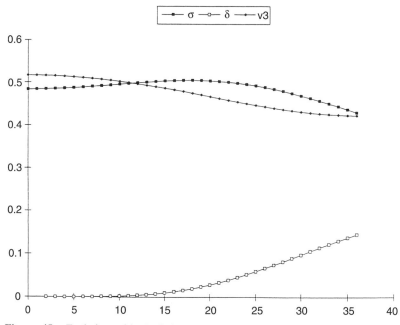

Figure 45. Evolution with J of the squared coefficients of the eigenvectors of the Hamiltonian matrix adapted to the $v_3/v_2 + v_4 + v_5$ anharmonic resonance in $^{12}C_2H_2$ in the wavefunction of $v_3 = 1$. Modes 3, 2, 4, and 5 are the asymmetric CH stretch, the CC stretch, and the *trans* and *cis* bends, respectively (from Vander Auwera et al. [451]).

cally checked this effective Hamiltonian and performed two successive contact transformations to remove as much ambiguity as possible between constants [539]. They reached the conclusion that additional, k off-diagonal terms need to be included to properly describe the very efficient DD coupling in $^{12}C_2D_2$, of the form

$$
\begin{aligned}
&\langle v_4, l_4, v_5, l_5, J, k \parallel v_4 - 2, l_4, v_5 + 2, l_5 \pm 2, J, k \pm 2 \rangle \\
&= \tfrac{1}{2} D^{l_5}[(v_4^2 - l_4^2)(v_5 \pm l_5 + 2)(J \pm k + 2) \\
&\quad (J \mp k - 1)(J \pm k + 1)(J \mp k)]^{1/2} \\
&\langle v_4, l_4, v_5, l_5, J, k \parallel v_4 - 2, l_4 \pm 2, v_5 + 2, l_5, J, k \pm 2 \rangle \\
&= \tfrac{1}{2} D^{l_4}\{[(v_5 + 2)^2 - l_5^2](v_4 \mp l_4) \\
&\quad (J \pm k + 2)(J \mp k - 1)(J \pm k + 1)(J \mp k)]\}^{1/2}
\end{aligned}
\tag{3.168}
$$

and provided explicit forms of all constants in their paper. They simultaneously concluded that one can remove the q_b^k constant from their

Hamiltonian. In addition, they suggested the use of a vibrational and J dependence of $(r_{45} + 2g_{45})$ taken as one parameter altogether. The data were so far not refitted using that modified Hamiltonian. The two extra matrix elements are not expected to play a significant role at low J value and could therefore be neglected if one cares only about the vibrational problem.

This example allows us to stress the well-known problems that is encountered when dealing with effective, empirical Hamiltonians, including a large number of constants to be fitted. Although the results of the fit seemed to converge satisfactorily during the initial work by Herman and co-workers, it turned out that the Hamiltonian was not fully adequate. Nevertheless, although this problem affects the higher-order constants, the initial experimental study turned out not only to provide extensive vibration–rotation assignments but also to stimulate theoreticians and provide useful background when later dealing with $^{12}C_2H_2$ at higher bend vibrational excitations. It therefore turned out to be a fruitful investment, in the end.

d.   Rotational Constants in $^{12}C_2H_2$ Overtone Levels.   As a final illustration of the MIME, we come back to the clustering process developed in Section III.C and consider the vibrational dependencies of the rotational constants in $^{12}C_2H_2$. In Section III.C.3.*b*, we derived the expected vibration–rotation dependence, α, of the principal rotation constant B, and briefly introduced β, the corresponding parameter for the centrifugal distortion constant, D. Values for α in acetylene were presented in Table XXVI. The zeroth-order predictions one makes for excited vibrational levels using such constants hold only in the absence of anharmonic resonances. Otherwise, the predictions need to be weighted by the squared coefficients of the eigenvectors resulting from the adequate Hamiltonian matrix diagonalization, as presented in (3.63). The procedure holds for 2×2 anharmonic interactions [515,679], and can be extended to larger couplings [517]. The vibrational clustering process occurring in acetylene gives higher dimension to this problem. Herman and co-workers applied the same procedure to calculate the B_v values of highly excited levels in $^{12}C_2H_2$ and $^{12}C_2D_2$ [450,468,516,525,614]. They used the coefficients in the eigenvectors produced by the diagonalization of the V-clusters to weight the predicted zeroth-order B_v constants and succeeded in reproducing the observations. Such predictions are highly valuable to help assigning the bands in the spectrum.

Abbouti Temsamani attempted to study the effect of the clustering on the centrifugal distortion constants, using a refined J-dependent procedure [533,610]. Some V/l-clusters were built and diagonalized for exactly all J values experimentally probed, constraining the set of vibration–rotation parameters from the literature in the calculation [468], including zeroth-

198 THE BACKWARD TRIP

TABLE XXXIII
Sample of Experimental Rotational Constants (in cm^{-1}) in $\{3, 15, u, J, e\}$ V/l-Cluster of
$^{12}C_2H_2$ and Corresponding Predictions, as Described in Text

G_v	B_v(obs)	B_v(calc)	D_v(obs) $\times 10^6$	D_v(calc) $\times 10^6$	H_v(obs) $\times 10^9$	H_v(calc) $\times 10^9$
9639.87	1.1583719	1.1586	1.7602	1.71	0.0192	0.01
9664.42	1.159988	1.1596	6.51	6.85	1.971	2.34
9668.16	1.161309	1.1615	-5.49	-5.54	-4.03	-4.04

Source: Adapted from Abbouti Temsamani and Herman [610].

order α and β constants to build the J-dependent diagonal elements [610]. The resulting calculated upper vibration–rotation level energies were then used to determine effective rotational constants (B, D, and H), using conventional procedures. A sample of the results is presented in Table XXXIII. The convincing agreement between the observed and predicted parameters thus highlights the origin of the random-like behavior of the rotational constants in that V-cluster, which present in some cases opposite signs for one level to another. One should also mention variational calculations reported by Lehmann, which led to precise predictions, although limited to B_V values [445].

E. Molecular Vibration–Rotation Spectra

Now that the basic theory governing the vibration–rotation energy levels structure of molecules has been developed, we are ready to consider the interaction between an electromagnetic radiation and these vibration–rotation levels.

After the definition of a few useful quantities and a brief introduction of transition probabilities, the absorption (and briefly the emission) of light is discussed. The expression of the integrated absorption coefficient associated with a vibration–rotation transition in terms of the electric dipole moment matrix elements or transition moment is then developed. From there, selection rules governing vibration and vibration–rotation transitions as well as intensities of the corresponding absorption features are discussed. This is done first in the harmonic oscillator–rigid rotor approximation. Then anharmonicity and vibration–rotation interactions are introduced and their influence on vibration–rotation spectra are discussed.

1. Spectral Intensity

Before discussing line intensities, it is necessary to define precisely what is meant by the term "intensity" [680]. For that purpose, we need to start from the so-called radiant energy.

The *radiant energy* W (measured in Joules (J)) refers to the total amount of energy emitted by a light source, transferred through a surface or collected by a detector. The *radiant power* Φ (measured in Watts (W)) is the radiant energy per second. The radiant power incident on unit detector area is called the *irradiance* I and is measured in W/m^2. The irradiance is the "intensity" to which the spectroscopic literature refers. It is related to the *radiant energy density* $\rho = W/V$ (expressed in J/m^3) of a plane wave by $I = c_0\rho$, where c_0 is the velocity of light in the vacuum. These four quantities—W, Φ, I, and ρ—refer to the total radiation integrated over the whole spectrum.

In many spectroscopic applications, the spectral irradiance and radiant energy density are used. The *spectral intensity* $I(\tilde{v}) = dI/d\tilde{v}$, also known as the *spectral irradiance* or (again) *intensity*, is the irradiance within a wavenumber interval $d\tilde{v}$ and is measured in W/m. The *spectral radiant energy density* $\rho(\tilde{v}) \equiv \rho_{\tilde{v}} = d\rho/d\tilde{v}$, measured in J/m^2, is the radiant energy density for a wavenumber interval $d\tilde{v}$. In the present chapter, we use the term "intensity" indifferently to mean the irradiance and the spectral irradiance.

2. Transition Probabilities

According to classical electrodynamics, any motion of a molecular system that is connected with a change of its electric dipole moment leads to the emission or absorption of radiation. During the vibration and/or rotation motion of a molecule, the charge distribution undergoes a periodic change, and therefore in general the dipole moment changes periodically at the frequencies of the molecular motions excited. In quantum theory, the frequencies of the absorption or emission features are determined by the energy differences of the vibration−rotation levels between which the transitions take place. In order to find out which transitions occur with what intensity, it is necessary to calculate the transition probabilities.

Let us assume that a molecular system with energy levels E_1 and E_2, such that $E_1 < E_2$, is in equilibrium with radiation at the temperature T. The spectral radiant energy density is given by Planck's radiation law [681]:

$$\rho(\tilde{v}) = 8\pi hc_0\tilde{v}^3 \frac{1}{\exp(hc_0\tilde{v}/kT) - 1} \qquad (3.169)$$

where h and k are Planck's and Boltzmann's constants. A transition from one level to the other will be accompanied by absorption or emission of radiation of wavenumber:

$$\tilde{v}_{21} = E_2 - E_1 (\text{in cm}^{-1}) \qquad (3.170)$$

This is the Bohr frequency rule. The probability per second, dP_{21}/dt, that the molecule in the lower level of energy E_1 absorbs a photon of energy $\tilde{v}_{21}$ is proportional to the number of photons having that energy per unit of volume. It can be expressed in terms of the radiant energy density $\rho(\tilde{v}_{21})$ of the radiation field in the wavenumber interval $[\tilde{v}_{21}, \tilde{v}_{21} + d\tilde{v}]$ as (see, for example, Ref. 681):

$$\frac{dP_{21}}{dt} = B_{21}\rho(\tilde{v}_{21}) \qquad (3.171)$$

The factor B_{21} is the *Einstein coefficient for absorption* from level 1 to level 2.

The radiation field can also induce molecules in the excited level E_2 to make a transition down to the lower level E_1 with simultaneous emission of a photon of energy $\tilde{v}_{21}$. This process is called *stimulated emission*. The probability per second that a molecule emits one induced photon is given by

$$\frac{dP_{21}}{dt} = B_{12}\rho(\tilde{v}_{21}) \qquad (3.172)$$

The constant B_{12} is the *Einstein coefficient for stimulated emission*.

An excited molecule in level E_2 can also *spontaneously* convert its excitation energy into an emitted photon of energy $\tilde{v}_{21}$. The probability per second that such a phenomenon occurs is independent of the external radiation field. It is equal to

$$\frac{dP_{12}}{dt} = A_{12} \qquad (3.173)$$

where A_{12} is the *Einstein coefficient for spontaneous emission* and is often called the *spontaneous emission transition probability*. It is the inverse of the lifetime of the levels involved.

The concepts developed in the next sections 3 and 4 apply to vibration–rotation transitions as well as to transitions between different electronic states.

3. Light Absorption

a. Absorption Induced by Electric Dipole Interactions. This section briefly outlines the semiclassical treatment of the interaction of electromagnetic radiation with molecules using time-dependent perturbation theory. The time-dependent Schrödinger equation for the system is

$$i\hbar\frac{\partial\Psi(t)}{\partial t} = (\hat{H}_0 + V)\Psi(t) \qquad (3.174)$$

where $\hat{H}_0$ is the time-independent Hamiltonian for the free molecule, the solution of which yields the set of stationary states $|1\rangle, |2\rangle, \ldots, \Psi(t)$ represents the time-dependent wavefunctions of the perturbed molecular system, which are linear combinations of the unperturbed wavefunctions. V represents in this section a time-dependent electric dipole interaction potential. If the radiation incident on the molecule is described by a classical electromagnetic wave of wavenumber $\tilde{\nu}$ propagating in the direction $\vec{u}$ at time t

$$\vec{E}(\vec{\mu}, \tilde{\nu}, t) = \vec{E}_0 \cos(\vec{k}\vec{u} - 2\pi c_0 \tilde{\nu}t) \qquad (3.175)$$

then V can be written as

$$V = -\sum_{\Xi} \mu_{\Xi} E_{\Xi}^0 \cos(2\pi c_0 \tilde{\nu}t) \qquad (3.176)$$

assuming that the space-dependent part of the cosine can be neglected because the phase of the electromagnetic wave does not change much within the volume of a molecule (see, e.g., Refs. 681 and 682 for details). In that latter expression, μ_{Ξ} and E_{Ξ}^0 are the projections on the LAS ($\Xi = X, Y, Z$) of the electric dipole moment of the molecule and the amplitude of the electromagnetic wave, respectively.

Solving that problem for the probability per second that a molecule initially in the lower level E_1 is found in the upper level E_2 yields (in the wavenumber domain):

$$\frac{dP_{21}}{dt} = \frac{1}{4\hbar^2 c_0} \sum_{\Xi} |\langle 2|\mu_{\Xi} \cdot E_{\Xi}^0|1\rangle|^2 \qquad (3.177)$$

Remembering that for an unpolarized plane electromagnetic wave, the density of radiant energy in the wavenumber interval $[\tilde{\nu}_{21}, \tilde{\nu}_{21} + d\tilde{\nu}]$ is equal to

$$\rho(\tilde{\nu}_{21}) = \sum_{\Xi} \varepsilon_0 \overline{|E(\tilde{\nu}_{21}, t)|^2} = \frac{3}{2} \varepsilon_0 E_0^2 \qquad (3.178)$$

where $\overline{|E(\tilde{\nu}_{21}, t)|^2}$ represents the average of the modulus square of the electric field strength associated to the incident radiation of wavenumber $\tilde{\nu}_{21}$, ε_0 is the electric permitivity of vacuum and $E_0^2 \equiv (E_X^0)^2 = (E_Y^0)^2 = (E_Z^0)^2$. Equating (3.177) and (3.171) yields, taking (3.178) into account (again when wavenumbers are considered)

$$B_{21} = \frac{2\pi}{3\hbar^2 c_0} \frac{1}{4\pi\varepsilon_0} \sum_{\Xi} |\langle 2|\mu_{\Xi}|1\rangle|^2 \qquad (3.179)$$

This relation, together with (3.171), indicates that the transition from the lower level $|1\rangle$, of energy E_1, to the upper level $|2\rangle$, of energy E_2, occurs only if the matrix elements of at least one component of the space-fixed electric dipole moment, the so-called transition moment, is nonzero.

Because $|\langle 2|\mu_\Xi|1\rangle|^2 = |\langle 1|\mu_\Xi|2\rangle|^2$, the Einstein coefficient for absorption is equal to that for stimulated emission. More generally, if the combining levels are degenerate, then this relation becomes

$$g_1 B_{21} = g_2 B_{12} \qquad (3.180)$$

where g_1 and g_2 are the degeneracy (statistical weight) of the levels, detailed in Section III.E.3.e.3. Furthermore, it can be shown that for a system at thermodynamical equilibrium, the following relation holds [681]:

$$A_{12} = 8\pi h c_0 \tilde{v}_{21}^3 B_{12} \qquad (3.181)$$

b. Absorption Coefficient: Beer's Law. Let us consider a radiation going through an optically thin gaseous medium. The intensity of the radiation can be attenuated at every wavenumber as a result of absorption by the molecules. That variation of intensity is characterized by the *absorption coefficient* $\alpha(\tilde{v})$. This quantity represents the fraction of intensity dI of the incident radiation of intensity I at the wavenumber $\tilde{v}$ absorbed along the pathlength dx, such that

$$dI(\tilde{v}) = -\alpha(\tilde{v})I(\tilde{v})dx \qquad (3.182)$$

In the usual experimental conditions, that is, involving a homogeneous medium and a low-intensity radiation that does not perturb the Boltzmann equilibrium distribution of the population of the energy levels, the integration of (3.182) over the whole absorption pathlength yields the well-known Beer's law:

$$I(\tilde{v}) = I_0(\tilde{v})e^{-\alpha(\tilde{v})\ell} = I_0(\tilde{v})e^{-\sigma(\tilde{v})Nl} \qquad (3.183)$$

where I_0 is the intensity of the incident light beam and ℓ is the length of the path in the absorbing medium. Usually, ℓ is expressed in centimeters; the absorption coefficient is thus given in reciprocal centimeters (cm^{-1}). The second equality expresses the intensity in terms of the absorption cross section σ, given in square centimeters (cm^2). It is linked to the absorption coefficient by

$$\alpha(\tilde{v}) = \sigma(\tilde{v})N \qquad (3.184)$$

where N is the particle density (or concentration) of the absorbing gas, expressed in molecule $\times$ cm^{-3} or cm^{-3}. In standard conditions of pressure

(1013.2458 hPa) and temperature (273.15 K), the particle density is given by Loschmidt's number, $n_L = 2.687 \times 10^{19}\,\text{cm}^{-3}$.

On the basis of Beer's law, the following three quantities can be defined to characterize the absorption by a gaseous medium (neglecting scattering):

Transmittance:

$$\tau(\tilde{\nu}) = \frac{I(\tilde{\nu})}{I_0(\tilde{\nu})} \qquad (3.185)$$

Naperian absorbance:

$$A_e(\tilde{\nu}) = -\ln\left(\frac{I(\tilde{\nu})}{I_0(\tilde{\nu})}\right) = \alpha(\tilde{\nu})\ell \qquad (3.186)$$

Absorption:

$$A(\tilde{\nu}) = 1 - \tau(\tilde{\nu}) \qquad (3.187)$$

c. *Absorption Line Shape.* Consider again a transition between a lower level $|1\rangle$ of energy E_1 and an upper level $|2\rangle$ of energy $E_2 : E_2 \leftarrow E_1$. As stated above, it can lead to the observation of a corresponding absorption line in the spectrum. Because of various phenomena, that absorption line is not strictly monochromatic; it presents a spectral distribution of finite width around the central wavenumber $\tilde{\nu}_{21}$, given by (3.170). Figure 46 schematizes the corresponding evolution of the absorption coefficient with the wavenumber. It also indicates the various quantities used to characterize the profile of that line. In addition to the line position, they are the *half-width at half-maximum*, which is equal to $\gamma = |\tilde{\nu}_b - \tilde{\nu}_a|/2$, the *amplitude*, that is, the absorption coefficient at $\tilde{\nu}_{21}$, and the *intensity*. That latter quantity is the area subtended by the line profile:

$$\alpha_{21} = \int_{\text{line}} \alpha(\tilde{\nu})d\tilde{\nu} \qquad (3.188)$$

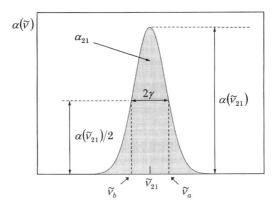

Figure 46. Schematic representation of an absorption line and parameters characterizing its shape.

or integrated absorption coefficient. The line intensity can also be defined as the integrated absorption cross section:

$$\sigma_{21} = \int_{\text{line}} \sigma(\tilde{\nu})d\tilde{\nu} \qquad (3.189)$$

That latter quantity is the line intensity appearing, for instance, in the Hitran database [683]. The integrated absorption coefficient is expressed in cm^{-2} and the integrated absorption cross section in cm^2 molecule^{-1} cm^{-1} or cm $\times$ molecule^{-1}. Finally, the integrated absorption coefficient is usually given for a sample pressure of one atmosphere (1 atm). It is the so-called absolute intensity;

$$S_{21} = \frac{\alpha_{21}}{P} \qquad (3.190)$$

where P is the pressure of the gaseous sample.

On the basis of the notion of line intensity, the dependence with the wavenumber of the absorption coefficient or the cross section can be expressed in the following way:

$$\begin{aligned} \alpha(\tilde{\nu}) &= \alpha_{21} \cdot g(\tilde{\nu} - \tilde{\nu}_{21}) \\ \sigma(\tilde{\nu}) &= \sigma_{21} \cdot g(\tilde{\nu} - \tilde{\nu}_{21}) \end{aligned} \qquad (3.191)$$

The function g is the so-called line profile function. It follows from the preceding expressions that g is normalized, that is

$$\int_{\text{line}} g(\tilde{\nu} - \tilde{\nu}_{21})d\tilde{\nu} = 1 \qquad (3.192)$$

In molecular spectroscopy in the gas phase, the lineshape function $g(\tilde{\nu} - \tilde{\nu}_{21})$ usually results from the simultaneous effects of Doppler broadening (leading to a Gaussian lineshape g_D) and pressure broadening (leading to a Lorentzian lineshape g_L) [681]. The convolution of these profiles leads to the well-known Voigt line profile:

$$g_V(\tilde{\nu} - \tilde{\nu}_{21}) = \int_{-\infty}^{+\infty} g_L(\tilde{\nu} - \tilde{\nu}_{21} - \tilde{\nu}')g_D(\tilde{\nu}')d\tilde{\nu}' \qquad (3.193)$$

The convolution integral (3.193) cannot be evaluated in closed form and must therefore be computed numerically. This has led to the development of a wide variety of algorithms and tabulations, [e.g., 684,685].

Systematic deviations from the Voigt profile are observed experimentally. They may arise from the reduction of the Doppler broadening due to the averaging effect of velocity-changing collisions (Dicke narrowing; see, e.g., 686–688). They may also be explained by the speed dependence of broadening and shifting parameters for systems in which the perturber mass is larger than the absorber mass (see, e.g., Refs. 689 and 690). Line mixing effects also lead to discrepancies from the Voigt lineshape (see the review by Lévy and co-workers [691]). The Van Vleck–Weisskopf [692] and related profiles are also worth mentioning. They are used, for instance, to describe absorption in the far linewings of H_2O–CO_2 mixtures observed in the millimetre and infrared ranges (see Ref. 693 and references cited therein).

To correctly describe the observed lineshape, the degradation of the preceding molecular profile by the spectrometer should be taken into account. That effect is usually modelled by the introduction of an "instrument function" that is convoluted with the molecular spectrum (3.183). The instrument function is the response of the instrument to an infinitely narrow input line. It represents, more or less, the variations of the flux received by the detector as a function of the wavenumber. For a discussion of the instrument function of a Fourier transform spectrometer (see, e.g., Refs. 694–696).

d. Line Intensity. Equation (3.182) shows that the radiant energy density absorbed per second within the wavenumber interval $[\tilde{\nu}, \tilde{\nu} + d\tilde{\nu}]$ is equal to

$$\frac{d\rho(\tilde{\nu})}{dt} = \alpha(\tilde{\nu})I(\tilde{\nu})d\tilde{\nu} \tag{3.194}$$

The radiant energy density absorbed per second on the transition $E_2 \leftarrow E_1$ is then

$$\frac{d\rho_{21}}{dt} = \int_{\text{line}} \alpha(\tilde{\nu})I(\tilde{\nu})d\tilde{\nu} \tag{3.195}$$

where the integration extends over all wavenumbers which contribute to this transition. If the incident radiation $I(\tilde{\nu})$ does not change much within the wavenumber range of the absorption profile (which means roughly within the absorption half-width), $I(\tilde{\nu}) = I(\tilde{\nu}_{21})$ can be assumed to be constant and equation (3.195) becomes

$$\frac{d\rho_{21}}{dt} \cong I(\tilde{\nu}_{21}) \int_{\text{line}} \alpha(\tilde{\nu})d\tilde{\nu} = I(\tilde{\nu}_{21}) \cdot \alpha_{21} \tag{3.196}$$

According to (3.171), the probability per second that one molecule undergoes the transition $E_2 \leftarrow E_1$ is proportional to the Einstein coefficient B_{21}.

Then, for N_1 molecules per unit volume in the lower level, the radiant energy absorbed per second on the transition $E_2 \leftarrow E_1$ is equal to $N_1 B_{21} \rho(\tilde{v}_{21}) \cdot hc_0 \tilde{v}_{21}$. Because of stimulated emission, radiant energy density is also emitted. Neglecting spontaneous emission, the net radiant energy density absorbed per second is, assuming that there are N_2 molecules per unit volume in the upper level:

$$\frac{d\rho_{21}}{dt} = (N_1 B_{21} - N_2 B_{12})\rho(\tilde{v}_{21})hc_0\tilde{v}_{21} \qquad (3.197)$$

Since the intensity $I(\tilde{v})$ of a plane wave is proportional to the energy density $\rho(\tilde{v})$ (see Section III.E.1), comparison of (3.196) and (3.197) yields, including the link between the Einstein coefficients for absorption and stimulated emission (3.180)

$$\alpha_{21} = h\tilde{v}_{21}\left(\frac{N_1}{g_1} - \frac{N_2}{g_2}\right)g_1 B_{21} \qquad (3.198)$$

Replacing the Einstein coefficient for absorption by its expression (3.179) yields the following form of the integrated absorption coefficient associated with the transition $E_2 \leftarrow E_1$:

$$\alpha_{21} = \frac{8\pi^3}{3hc}\frac{1}{4\pi\varepsilon_0}\tilde{v}_{21}g_1\left(\frac{N_1}{g_1} - \frac{N_2}{g_2}\right)\sum_{\Xi}|\langle 2|\mu_\Xi|1\rangle|^2 \qquad (3.199)$$

The general expression (3.199) is correct for all transitions as long as spontaneous emission can be neglected.

Although it is beyond the scope of this section, it is interesting to consider briefly the case of electronic transitions. The integrated intensity of an electronic transition $|2\rangle \leftarrow |1\rangle$ is often expressed in terms of the *oscillator strength* f_{21}, which is dimensionless. The oscillator strength is proportional to the Einstein transition probability B_{21} [681]:

$$f_{21} = \frac{(4\pi\varepsilon_0)m_e hc_0^2\tilde{v}_{21}}{\pi e^2}B_{21} \qquad (3.200)$$

where $m_e \approx 9.109 \times 10^{-28}$ g and $e \approx 1.602 \times 10^{-19}$ C are the electron rest mass and charge, respectively.

We now deal in some details with the population factors involved in (3.199). The last—and perhaps most important—term, the transition moment, is discussed later, in Section III.E.5.

e. Population Factors

1. THERMAL DISTRIBUTION OVER THE ENERGY LEVELS. As appears from (3.199), the difference between the population densities of the two levels involved in the transition determines the intensity of the corresponding absorption line. If it is assumed that the system under investigation is at thermal equilibrium, the Boltzmann distribution law states that the number of molecules per unit volume N_1 in the level of energy E_1 and degeneracy g_1 is given by

$$\frac{N_1}{g_1} = \frac{N}{Q(T)} \exp\left(-\frac{hc_0E_1}{kT}\right) \qquad (3.201)$$

In that expression, T is the absolute temperature (in Kelvin), $Q(T)$ is the total internal partition function of the gas (the translation motion is considered to be "removed" through the use of a suitably chosen system of axis, as discussed in Section II.B.1.c), and N is the total number of molecules per unit volume. The latter quantity is often expressed in terms of Loschmidt's number n_L; thus, in the perfect gas approximation

$$N = n_L P \frac{T_0}{T} \qquad (3.202)$$

where $T_0 = 273.15\,\text{K}$.

The population factor appearing in (3.199) can then be reformulated taking (3.201) and (3.170) into account:

$$\frac{N_1}{g_1} - \frac{N_2}{g_2} = \frac{N}{Q(T)} \exp\left(-\frac{hc_0E_1}{kT}\right) \cdot \left\{1 - \exp\left(-\frac{hc_0\tilde{v}_{21}}{kT}\right)\right\} \qquad (3.203)$$

2. TOTAL INTERNAL PARTITION FUNCTION. The total internal partition function appearing in the Boltzmann distribution law (3.201) determines how molecules in thermodynamic equilibrium are distributed among the various energy levels at a given temperature. It is defined as a sum over all levels of the molecule with all degeneracy factors included:

$$Q(T) = \frac{1}{\sigma} \sum_n g_n \exp\left(-\frac{hcE_n}{kT}\right), \qquad (3.204)$$

where g_n is the degeneracy of the level n discussed below (see Section III.E.3.e.3) and σ is equal to 2 for homonuclear diatomics and 1 otherwise.

Various methods have been devised to evaluate $Q(T)$. The most accurate is direct summation. In this case, the energy of all levels below a given cutoff is computed and the partition sum is calculated with appropriate degeneracy factors. This approach has been used only rarely because of the vast number of energy levels that must be calculated and stored, assuming that a complete set of parameters is available, which is seldom the case. The fact that the summation must be done separately at each temperature of interest is also a limiting factor.

Assuming separability of the electronic, vibrational and rotational energies of a molecule, $Q(T)$ can be approximated by the product:

$$Q(T) = \frac{1}{\sigma} Q_e(T) Q_v(T) Q_r(T) \qquad (3.205)$$

where $Q_e(T)$, $Q_v(T)$, and $Q_r(T)$ are the electronic, vibration, and rotation partition sums, respectively. They have the same form as in (3.204). Each of these terms can be evaluated by direct summation, classical formula [697], approximate analytical methods [698], or using polynomial expansions in the temperature [699]. Except in a very few cases, the energy of the excited electronic states are such that their contribution to the electronic partition function can be neglected: $Q_e(T) = 1$.

A detailed discussion of the calculation of the total internal partition function using various methods can be found in Ref. 699.

3.   DEGENERACY OF THE LEVELS. The discussion of the degeneracy factor g_n of the level $|n\rangle$ appearing in (3.199) and (3.204) given here is limited to the case of linear or asymmetric top molecules. The degeneracy factor can be expressed as the following product:

$$g_n = g_v(2S+1)g_j g_i, \qquad \text{with} \qquad g_j = \prod_j (2I_j + 1) \qquad (3.206)$$

The $(2J+1)$ factor arising from the J spatial degeneracy (see Table XXII) is not included in this expression because it is usually put in the line strength factor (see below).

In (3.206) g_v represents the degeneracy of the vibrational wavefunction. It is always equal to 1 for an asymmetric rotor. For a linear molecule, it is equal to 1 for levels with $k = 0$. For the $k \neq 0$ levels, $g_v = 2$ except when the k-degeneracy is removed by l-type resonance, as detailed in Section III.B.4.c.1. The factor $(2S+1)$ is the spatial degeneracy in the electron spin. The last two terms correspond to the nuclear spin degeneracy factor. The first of these, g_j, represents the nuclear spin degeneracy for unpaired nuclei: each unpaired nucleus j, of spin I_j, contributes a factor $(2I_j + 1)$, which is constant for all levels and can be taken out of the summation involved in the partition function. Because that factor is constant and many applications take the ratio of $Q(T)$ at different temperatures, it is usually omitted. The second factor, g_i, represents the statistical weight resulting from the coupling of symmetrically equivalent nuclear pairs with the rotational wavefunctions. It is discussed in the next section.

4. STATISTICAL WEIGHT OF THE ROVIBRONIC LEVELS. The statistical weight of a rovibronic level $\Psi_{evr} = \Psi_e\Psi_v\Psi_r$ is the number of nuclear spin wavefunctions Ψ_{ns} compatible with the symmetry of the total molecular wavefunction $\Psi_{tot} = \Psi_e\Psi_v\Psi_r\Psi_{es}\Psi_{ns}$:

$$\Gamma(\Psi_{tot}) = \Gamma(\Psi_e) \otimes \Gamma(\Psi_v) \otimes \Gamma(\Psi_r) \otimes \Gamma(\Psi_{es}) \otimes \Gamma(\Psi_{ns}) \qquad (3.207)$$

where $\otimes$ represents the direct product of the irreducible representations Γ (see Section III.E.6) and Ψ_{es} stands for the electron spin wavefunction. The Pauli exclusion principle stipulates that the total molecular wavefunction is symmetric under the permutation of two identical nuclei of integral spin (bosons) and antisymmetric under that of half-integral spin particles (fermions).

Determination of the statistical weight of the levels thus relies on molecular symmetry. The rovibronic levels of linear molecules can be classified in the extended molecular symmetry group $C_{\infty v}(EM)$ and $D_{\infty h}(EM)$ [46]. In the present context, only molecules belonging to the latter are of interest because they possess a centre of symmetry. In the case of acetylene, the permutation of identical nuclei is performed by the operation $(12)(34)_\varepsilon$ of $D_{\infty h}(EM)$ [46]. Therefore, the symmetry of the total molecular wavefunction must be Σ_g^- or Σ_u^+ for $^{12}C_2H_2$ and $^{13}C_2D_2$ and Σ_g^+ or Σ_u^- for $^{13}C_2H_2$ and $^{12}C_2D_2$. Given the symmetry of the rovibronic [46] and the nuclear spin wavefunctions, the statistical weight of the rovibronic levels can be determined. Table XXXIV outlines the process for $^{12}C_2H_2$.

Table XXXV summarizes the statistical weight for various symmetric isotopomers of acetylene.

The statistical weight of the levels obviously affects the intensity of the lines, through (3.199) and (3.206). Figure 47 shows an example of this effect by comparing spectra of C_2H_2, C_2HD, and C_2D_2.

f. Integrated Absorption Coefficient and Cross Section. The final form of the expression of the integrated absorption coefficient of an absorption line corresponding to a transition $|2\rangle \leftarrow |1\rangle$ is obtained by insertion of (3.203) into (3.199):

$$\alpha_{21} = \frac{8\pi^3}{3hc_0}\frac{\tilde{v}_{21}}{4\pi\varepsilon_0}\frac{N}{Q(T)}g_1\exp\left(-\frac{hcE_1}{kT}\right)\left\{1 - \exp\left(-\frac{hc\tilde{v}_{21}}{kT}\right)\right\}\sum_{\Xi}|\langle 2|\mu_\Xi|1\rangle|^2$$

$$(3.208)$$

TABLE XXXIV
Determination of Statistical Weight of Rovibronic Levels of $^{12}C_2H_2$ in Ground Electronic State $[\Gamma(\Psi_e) = \Sigma_g^+]$

$\Gamma(\Psi_v)$	k	Parity	J	$\Gamma(\Psi_{evr})$	$\Gamma(\Psi_{ns})$	$\Gamma(\Psi_{tot})$	g_i
$\Sigma_g^+, \Sigma_g^-, \ldots$	$=0$	e	Even	Σ_g^+	$3\Sigma_g^+ + \Sigma_u^+$	Σ_g^- or Σ_u^+	1
			Odd	Σ_g^-			3
$\Pi_g, \Delta_g, \ldots$	$\neq 0$	e	Even	Σ_g^+	$3\Sigma_g^+ + \Sigma_u^+$	Σ_g^- or Σ_u^+	1
			Odd	Σ_g^-			3
		f	Even	Σ_g^-			3
			Odd	Σ_g^+			1
$\Sigma_u^+, \Sigma_u^-, \ldots$	$=0$	e	Even	Σ_u^+	$3\Sigma_g^+ + \Sigma_u^+$	Σ_g^- or Σ_u^+	3
			Odd	Σ_u^-			1
$\Pi_u, \Delta_u, \ldots$	$\neq 0$	e	Even	Σ_u^+	$3\Sigma_g^+ + \Sigma_u^+$	Σ_g^- or Σ_u^+	3
			Odd	Σ_u^-			1
		f	Even	Σ_u^-			1
			Odd	Σ_u^+			3

TABLE XXXV
Statistical Weight of Rovibronic Levels of Various Isotopomers of Acetylene in Ground Electronic State $[\Gamma(\Psi_e) = \Sigma_g^+]$

$\Gamma(\Psi_v)$	Parity	J	Statistical Weight g_i			
			$^{12}C_2H_2$	$^{12}C_2D_2$	$^{13}C_2H_2$	$^{13}C_2D_2$
$\Sigma_g^+, \Sigma_g^-, \Pi_g, \ldots$	e	Even	1	3	6	15
		Odd	3	6	10	21
	f	Even	3	6	10	21
		Odd	1	3	6	15
$\Sigma_u^+, \Sigma_u^-, \Pi_u, \ldots$	e	Even	3	6	10	21
		Odd	1	3	6	15
	f	Even	1	3	6	15
		Odd	3	6	10	21

From (3.184), it can be easily shown that the following similar expression can be obtained for the corresponding integrated absorption cross section σ_{21}:

$$\sigma_{21} = \frac{8\pi^3}{3hc_0} \frac{\tilde{\nu}_{21}}{4\pi\varepsilon_0} \left\{ 1 - \exp\left(-\frac{hc\tilde{\nu}_{21}}{kT} \right) \right\} \sum_\Xi |\langle 2|\mu_\Xi|1\rangle|^2 \qquad (3.209)$$

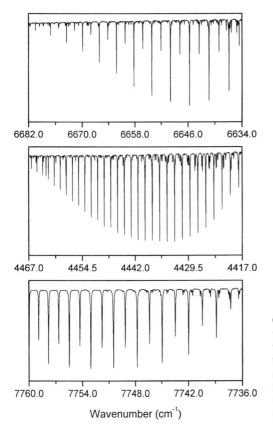

6682.0   6670.0   6658.0   6646.0   6634.0

4467.0   4454.5   4442.0   4429.5   4417.0

7760.0   7754.0   7748.0   7742.0   7736.0

Wavenumber (cm^{-1})

Figure 47. Spectra of C_2H_2, C_2HD, and C_2D_2 showing the varying effect of the statistical weight on the line intensities. These three R branches are observed in the single-beam absorption spectrum of the same C_2D_2-enriched sample (unpublished results from Herman and co-workers, ULB).

4. Emission

We briefly outline here the general expression of the intensity of emission lines, to allow for comparison with the absorption process.

Similarly as for absorption, we consider an optically thin gaseous medium of some species in which only two levels $|1\rangle$ and $|2\rangle$ of respective energy E_1 and E_2 are considered. As a result of some excitation process, N_2 molecules per unit volume are found in the upper level. The spectral intensity (or spectral irradiance) emitted per unit of solid angle by the sample over a pathlength ℓ as a result of spontaneous emission on the transition $|2\rangle \rightarrow |1\rangle$ is

$$I_{12}(\tilde{\nu}) = \frac{1}{4\pi} N_2 \ell hc_0 \tilde{\nu}_{21} A_{12} g(\tilde{\nu} - \tilde{\nu}_{21}) \qquad (3.210)$$

In this expression, the function g is the line profile function introduced in (3.191). The factor 4π (steradians) is the total solid angle around a point. $I_{12}(\tilde{\nu})$ is thus expressed in $W\,cm^{-1}\,sr^{-1}$. Replacing N_2 by the Boltzmann distribution (3.201) and A_{12} by its expression from (3.181), taking (3.180) and (3.179) into account, yields the intensity emitted as a result of the spontaneous transition from the upper level $|2\rangle$ to the lower level $|1\rangle$:

$$I_{12}(\tilde{\nu}) = \frac{16\pi^3 c_0 \ell}{3} \frac{1}{4\pi\varepsilon_0} \tilde{\nu}_{21}^4 \frac{N}{Q(T)} g_1 \exp\left(-\frac{hcE_2}{kT}\right)$$
$$\times \sum_{\Xi} |\langle 2|\mu_\Xi|1\rangle|^2 g(\tilde{\nu} - \tilde{\nu}_{21})$$

(3.211)

5. Dipole Moment Matrix Elements for Vibration–Rotation Transitions

A look at expressions (3.208), (3.209), and (3.211) shows that, in either absorption or emission, the intensity of a line, and thus the occurrence of the corresponding transition, is determined by the dipole moment matrix element or transition moment $\langle 2|\mu_\Xi|1\rangle$.

The general expression of the transition moment for a rovibronic transition in the rigid rotor–harmonic oscillator approximation (the latter corresponding to the neglect of mechanical anharmonicity) is derived in Section II. The discussion presented in this part builds on (2.38), in which the components $\mu_\xi(e, e)$ of the electric dipole moment of the molecule in the ground electronic state have been expanded in terms of the dimensionless normal coordinates of vibration q_i about their value at the nuclear equilibrium configuration of that state (2.71). Considering vibration–rotation transitions within the same electronic state, the transition moment can be written as follows for the Ξth component of the electric dipole moment in the LAS:

$$\langle \Psi_e \Psi_v' \Psi_r' | \mu_\Xi | \Psi_e \Psi_v'' \Psi_r'' \rangle = \sum_\xi \langle \Psi_r' | \lambda_{\Xi\xi} | \Psi_r'' \rangle \left[\sum_i m_i^\xi \langle \Psi_v' | q_i | \Psi_v'' \rangle \right.$$
$$\left. + \frac{1}{2} \sum_{i,j} m_{ij}^\xi \langle \Psi_v' | q_i q_j | \Psi_v'' \rangle + \cdots \right]$$

(3.212)

In this equation, the electronic (Ψ_e), vibrational (Ψ_v), and rotational (Ψ_r) parts of the wavefunction associated with a given level are explicitly written, taking into account the fact that only one electronic state is involved:

$$|1\rangle = |\Psi_e \Psi_v'' \Psi_r''\rangle \quad \text{and} \quad |2\rangle = |\Psi_e \Psi_v' \Psi_r'\rangle$$

(3.213)

The term $m_0^{\xi}\langle \Psi_v' | \Psi_v'' \rangle$, involving the components of the permanent electric dipole moment of the molecule m_0^{ξ} [see (2.71)], does not appear in (3.212) because the integral $\langle \Psi_v' | \Psi_v'' \rangle$ is equal to zero since the transitions occur within the same electronic state. The electron and nuclear spin parts of the wavefunction are not explicitly written here. Their contribution to the transition moment, together with the orthogonality of the wavefunctions, leads to selection rules of the form $\Delta S = 0$ and $\Delta I = 0$, respectively [46]. $\lambda_{\Xi\xi}$ are the direction cosine elements, given in Table I. The dots in (3.212) indicate elements involving higher-order terms of the power series expansion of the electric dipole moment in the normal coordinates of vibration.

The zero or nonzero nature of the transition moment defines restrictions on the transitions between pairs of levels. When they result from the approximations yielding (3.212) to which the neglect of electric anharmonicity [i.e., neglecting all except the first term in the bracketed expression appearing in (3.212)] is added, these restrictions are known as the *selection rules*. When the transition moment is zero, as it is in the vast majority of cases, the transition is said to be *forbidden* and, according to (3.208), (3.209), and (3.211), the associated line has no intensity. On the opposite side, *allowed transitions* correspond to a nonzero value of the transition moment (3.212).

On the basis of (3.212), the remainder of this section first discusses the selection rules governing transitions between vibrational levels of a nonrotating molecule and the intensity of the corresponding "lines." A similar approach is then followed for vibration–rotation transitions. In both cases, the discussion starts within the rigid rotor–harmonic oscillator approximation including the neglect of electric anharmonicity and then considers the effects of the introduction of mechanical and electric anharmonicities as well as vibration–rotation interaction.

All the aspects developed below are for a rigid and a semirigid molecule. The effect of molecular nonrigidity such as the torsional motion (see, e.g., Ref. 700) and quasi-linearity [701] are not considered.

6. Molecular Symmetry

Most of the discussion in this section, especially the aspects associated with the determination of the selection rules, relies on the use of molecular symmetry. For a detailed presentation and discussion of the subject, the reader is referred to Ref. 46, for example. Here, we only provide some definitions of symbols used to denote the symmetry of the levels, the wavefunctions and the various operators, as well as some useful relations, for didactic purposes.

A molecule possesses structural symmetry that can be described either in terms of rotations about axes and reflections through planes of its vibronic variables or in terms of permutations of identical nuclei and inversion of the molecular coordinates in the center of mass. The former corresponds to the use of the molecular point group; the latter, to the molecular symmetry group, to label or classify the various molecular motions and properties. The behavior of a given molecular property with respect to the application of the elements of the group of symmetry to which the molecule belongs is expressed by the irreducible representations. In the present review, irreducible representations are denoted by the symbol Γ.

One of the uses of symmetry is to allow the easy determination of whether two given energy levels can interact with each other as a result of the addition of some unconsidered term $\hat{H}'$ of the molecular Hamiltonian. The matrix element

$$\langle \Psi_0' | \hat{H}' | \Psi_0'' \rangle \tag{3.214}$$

where Ψ_0' and Ψ_0'' are the unperturbed wavefunctions associated with the two levels, does not vanish if the direct product of the irreducible representations characterizing the symmetry of the three terms involved contains the totally symmetric irreducible representation $\Gamma^{(s)}$:

$$\Gamma(\Psi_0') \otimes \Gamma(\hat{H}') \otimes \Gamma(\Psi_0'') \supset \Gamma^{(s)} \tag{3.215}$$

This is the *vanishing integral rule*, which is used extensively, for example, when dealing with the selection rules. Two symmetry behaviors that are useful and that will be used below are, in the molecular symmetry group [46]

$$\Gamma(\mu_\xi) = \Gamma(T_\xi) \tag{3.216}$$

and

$$\Gamma(\lambda_{\Xi\xi}) = \Gamma(J_\xi) \tag{3.217}$$

where $\xi = x, y, z, T_\xi$ are the translational coordinates and J_ξ are the components of the total angular momentum on the molecule-fixed frame (MAS).

7. Vibrational Spectra

a. Harmonic Selection Rules

1. NONDEGENERATE VIBRATIONS. Vibrational transitions within a given electronic state are controlled by the terms appearing in the square brackets

in (3.212). Indeed, the zero or nonzero value of these terms determines selection rules to which the vibrational transitions obey. When nonzero, these terms lead to the observation of bands of different kind. In the following, we examine the selection rules generated by several of these terms and apply them to bands associated to transitions originating from the ground level. Such bands are sometimes called *cold bands*. *Hot* or *difference bands* correspond to absorption transitions starting from an excited vibrational level. Those bands obey the same selection rules as cold bands, adapted for the symmetry of the lower excited level.

1. *Fundamental Bands*. For the first term in (3.212) to be nonvanishing, we must have

$$\langle \Psi'_v | q_i | \Psi''_v \rangle \neq 0 \tag{3.218}$$

and, from (3.215)

$$\Gamma(\Psi'_v) \otimes \Gamma(\Psi''_v) \supset \Gamma(T_\xi) \tag{3.219}$$

That latter expression includes the following relation, obtained by taking (2.71) and (3.216) into account:

$$\Gamma(q_i) = \Gamma(\mu_\xi) = \Gamma(T_\xi) \tag{3.220}$$

Because the ground vibrational level is always totally symmetric, it appears from (3.219) that transitions from that level to levels with $\Gamma(\Psi_v) \supset \Gamma(T_\xi)$ are allowed. In the harmonic oscillator approximation, relation (3.218) yields the following selection rule (see Table VII)

$$\Delta v_i = \pm 1 \tag{3.221}$$

where, from (3.219), the only allowed vibrational transitions are those for which the normal coordinate transforms like a translation. Vibrational transitions originating from the ground level and obeying the preceding selection rules, give rise to *fundamental bands*. In the *double harmonic approximation*, that is, for a harmonic oscillator for which all except the first term in the square brackets in (3.212) are not considered (i.e., electrical anharmonicity is also neglected), the only allowed transition are those that obey the selection rule $\Delta v_i = \pm 1$.

2. *Overtone and Combination Bands*. The second and next terms appearing in the square brackets in (3.212), and thus associated with the electrical anharmonicity, can lead to $\Delta v_i \neq 1$ transitions. Those arising from the ground state give rise to *overtone* or *combination bands*, depending on

whether they involve the excitation of a single or of several degrees of freedom of vibration. Overtone bands are thus transitions characterized by

$$|\Delta v_i| = n, \text{ with } n \geq 2 \qquad (3.222)$$

They can be observed if

$$\langle \Psi_v' | q_i^n | \Psi_v'' \rangle \neq 0 \qquad (3.223)$$

and, similar to (3.218),

$$\Gamma(\Psi_v') \otimes \Gamma(\Psi_v'') \supset [\Gamma(T_\xi)]^n \qquad (3.224)$$

The latter condition shows that molecules with one totally symmetric translational coordinate T_ξ have all their overtone bands allowed. Linear molecules of $C_{\infty v}$ symmetry fall into that category. On the contrary, when the molecule belongs to a group of symmetry lacking totally symmetric translational coordinates, then overtone bands corresponding to the excitation of an even number of quanta of vibration are forbidden. These forbidden bands are in fact observed in the Raman spectrum, not considered here. For example, linear molecules of $D_{\infty h}$ symmetry and asymmetric tops of C_{2h} symmetry exhibit such behavior.

Combination bands correspond to the simultaneous change of more than one vibrational quantum number:

$$\Delta v_i = n, \Delta v_j = m, \ldots \qquad (3.225)$$

with $n \neq 0$, $m \neq 0$, and so on, and correspond in the harmonic oscillator approximation to nonvanishing matrix elements in (3.212) of the form

$$\langle \Psi_v' | q_i^n q_j^m | \Psi_v'' \rangle \neq 0 \qquad (3.226)$$

As pointed out above, such bands are forbidden in the double harmonic approximation.

2.   DEGENERATE VIBRATIONS.   As pointed out in Section III.B.4.a, a linear polyatomic molecule consisting of n atoms has $(n-1)$ nondegenerate stretching modes of vibration and $(n-2)$ two-dimensional degenerate bending vibrations. The energy and wavefunction associated with each twofold degenerate vibration are usually expressed in terms of the quantum numbers v_b and l_b instead of v_{b_1} and v_{b_2}, which are associated to each pair of degenerate vibrations q_{b_1} and q_{b_2} see, e.g., Ref. 46). In the same way,

these normal coordinates are expressed using the coordinates introduced in (3.86): $q_\pm = q_{b_1} \pm q_{b_2}$.

Similar to the nondegenerate case, fundamental, overtone, combination, and hot (difference) bands can be shown to occur depending on which term appearing in the square brackets in (3.212) is considered. The present discussion is limited to two-dimensional degenerate vibrations, which are those found in linear molecules.

Fundamental bands and hot bands corresponding to the single excitation of one degenerate mode of vibration are controlled by the first term in the square brackets in (3.212). This term does not vanish only if

$$\langle \Psi'_{v,k} | q_\pm | \Psi''_{v,k} \rangle \neq 0 \qquad (3.227)$$

In that expression, it is taken into account that the wavefunctions depend not only on the vibration quantum number v but also on the total vibration angular momentum quantum number k [see (3.90)]. In the present case, we consider only one degenerate mode b, and $k \equiv l_b$, with l_b the vibration angular momentum associated to that mode. Again, symmetry conditions apply. They lead, in the harmonic oscillator approximation, to the following selection rule, which applies together with (3.219):

$$\Delta v_b = \pm 1 \qquad \text{and} \qquad \Delta l_b = \pm 1 \qquad (3.228)$$

The selection rules for overtone and combination bands of degenerate vibrations can be obtained in a similar way as for the nondegenerate modes, by application of molecular symmetry. It can be shown that overtone bands of degenerate vibrations are governed by the following selection rules in the harmonic approximation:

$$\Delta v = \pm n \qquad \text{with} \qquad \Delta k = 0, \pm 1 \qquad (3.229)$$

with $n \geq 2$, $v = v_b$, and $k = l_b$. For combination bands, the same selection rules apply with $v = \sum_b v_b$ and $k = \sum_b l_b$. Again, the symmetry of the upper and lower levels must be consistent with that of the electric dipole moment components. In the double harmonic approximation, allowed transitions obey the selection rules (3.228), only.

b. Vibrational Bands in the Harmonic Oscillator Approximation. As a result of the above-mentioned selection rules, bands associated with the transitions between the vibrational levels are thus observed. We discuss here the intensity of these bands.

The intensity or integrated absorption coefficient of a vibrational band, known as the "band strength" S_v, is often defined (see, e.g., Ref. 702) by [derived from (3.208)]:

$$S_v = \frac{8\pi^3}{3hc_0} \frac{\tilde{v}_0}{4\pi\varepsilon_0} \frac{N}{Q_v(T)} g_v \exp\left(-\frac{hc_0 E_v''}{kT}\right) |R_v|^2 \qquad (3.230)$$

In this expression, R_v represents the terms between brackets in (3.212) and is known as the *vibrational transition moment*. E_v'' is the energy of the lower vibrational level. All the other terms have been defined. S_v can also be obtained by summing values of α_{21} obtained from (3.208), provided one remembers that the stimulated emission correction term has been omitted from (3.230), so that

$$\sum_{\text{all lines}} \alpha_{21} = S_v \left[1 - \exp\left(\frac{-hc_0\tilde{v}_0}{kT}\right)\right] \qquad (3.231)$$

However, the two values are not identical for two reasons: (1), the band origin $\tilde{v}_0$ appears in (3.231), rather than the position of each line, and (2), and the Herman–Wallis factors are not accounted for (see below). If the latter are zero, agreement between the two methods of obtaining S_v is better than 0.5%.

For an allowed transition, the band strength thus depends on the actual value of the vibration factors such as (3.218), (3.223), or (3.227), in addition to the other terms appearing in (3.230). The values of these terms in the harmonic oscillator approximation are listed in Tables VII and X. The effect of such terms on the band strength is illustrated here in the case of the single excitation of nondegenerate and twofold degenerate vibrations.

For the single excitation of nondegenerate vibrations, the matrix elements detailed in Table VII show that

$$\langle v_i + 1 | q_i | v_i \rangle = \sqrt{\frac{v_i + 1}{2}}$$
$$\langle v_i - 1 | q_i | v_i \rangle = \sqrt{\frac{v_i}{2}} \qquad (3.232)$$

When a twofold degenerate vibration of a linear molecule is excited, (3.227) yields (from Table X)

$$\langle v_i + 1, l_i \pm 1 | q_\pm | v_i, l_i \rangle = \sqrt{\frac{(v_i \pm l_i + 2)}{2}}$$
$$\langle v_i - 1, l_i \pm 1 | q_\pm | v_i, l_i \rangle = \sqrt{\frac{(v_i \mp l_i)}{2}} \qquad (3.233)$$

TABLE XXXVI
Dipole Moment Derivatives Obtained from $^{12}C^{16}O_2$ Bands Near 15 μm[a]

$\tilde{\nu}_0(cm^{-1})$	Transition	$|R_v|^2$	$(\partial\mu_\pm/\partial q_\pm)^2$
667.380	$(01^10)-(00^00)$	0.03294	0.03294
667.752	$(02^20)-(01^10)$	0.06406	0.03203
668.115	$(03^30)-(02^20)$	0.09574	0.03191
720.805	$(10^00)-(01^10)$	0.01505 ⎫	0.03225
618.029	$(02^00)-(01^10)$	0.01721 ⎭	
741.724	$(11^10)-(02^20)$	0.01509 ⎫	0.03148
597.338	$(12^00)-(02^20)$	0.01639 ⎭	
791.447	$(11^10)-(02^00)$	0.00132 ⎫	
688.671	$(11^10)-(10^00)$	0.04714 ⎢	0.03283
647.062	$(12^00)-(02^00)$	0.04866 ⎢	
544.286	$(12^00)-(10^00)$	0.00136 ⎭	

[a] The square of the vibrational transition moment and the dipole moment derivatives are expressed in D^2 (dipoles squared).

Source: Adapted from Johns and Vander Auwera [702].

Let us consider a series of bands involving the same vibrational excitation Δv_i but in which each band corresponds to a transition that starts from a level that has one additional quantum of vibrational excitation. As can be seen from the expressions above, the transition moment associated to each band becomes larger than the previous one in a regular, predictable way. Thus, for example, for a stretching mode of vibration v_s, the vibration factor (3.232) is $\sqrt{5}$ larger for the band $5v_s - 4v_s$ than for the fundamental band v_s. On the other hand, the dipole moment derivatives m_s^ξ may be expected to be the same for all these bands. This is somewhat confirmed in the case of bending vibrations by measurements performed in the 15-μm region of the spectrum of $^{12}C^{16}O_2$ [702], some results of which are listed in Table XXXVI. The experimental dipole moment derivatives obtained for the first three bands indeed confirm the expectations. However, the remaining bands do not seem to agree with that zeroth-order picture. As discussed in Section III.B.3, the main reason for these discrepancies is the presence of anharmonic resonances that affect the levels involved by these bands. For those latter cases, it is necessary to sum the appropriate effective dipole moment derivatives [702].

c. Application to Linear Molecules. Table XII provides the symmetry of the vibrational levels in a linear molecule as a function of the excitation: it depends on the value of k associated with the levels. Accordingly, the transitions and bands are identified by these symmetry species in the upper

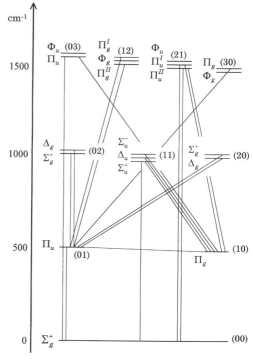

Figure 48. Lowest energy levels of C_2D_2 associated with the excitation of up to three quanta in the *trans* (v_4)- and *cis* (v_5)-bending modes and allowed transitions originating from the ground level and the $v_4 = 1$ and $v_5 = 1$ levels. The vibrational levels are identified as (v_4v_5) and their associated k sub-levels are identified by their symmetry. The magnitude of the k splitting is greatly exaggerated.

and lower levels. For instance, a Δ–Π transition involves an upper level with $k = 2$ and a lower level with $k = 1$.

According to (3.219), allowed vibrational transitions in symmetric linear molecules (point group $D_{\infty h}$) involve levels for which the direct product of their symmetry contains the irreducible representation Σ_u^+ (symmetry of μ_z) or Π_u [symmetry of (μ_x, μ_y)]. For an unsymmetric linear species, thus belonging to the $C_{\infty v}$ point group, the subscript g/u should be dropped.

Figure 48 illustrates allowed vibrational transitions in linear molecules. It presents the lowest vibrational levels of C_2D_2 associated with the excitation of up to three quanta in the two bending modes and the allowed transitions originating from the ground level and the $v_4 = 1$ and $v_5 = 1$ levels.

d. Vibrational Bands for the Anharmonic Oscillator

1. INFLUENCE OF MECHANICAL ANHARMONICITY. The introduction of mechanical anharmonicity affects thoroughly the vibrational energy-level structure (see Sections III.B–III.D). These modifications correspond to the introduction of diagonal as well as off-diagonal contributions to the Hamiltonian matrix or

MIME. Both contributions affect the energy of the vibrational levels and, in turn, the position of the bands corresponding to transitions involving these levels.

As the discussion of the evolution of the intensities of overtones associated with the excitation of the CD and CH stretching modes of $^{12}C_2HD$ given in Section II.C.2.b.4 shows, mechanical anharmonicity also affects overtone intensities. Indeed, because of the presence of off-diagonal terms in the Hamiltonian matrix, the vibrational levels are not pure anymore; their wavefunctions are a linear combination of harmonic oscillator basis functions. As a result, the transition moment of overtone bands involves different orders of electrical anharmonicity instead of only one in the harmonic oscillator approximation (see Fig. 19). Ab initio calculations of overtone intensities, reported for various species including H_2O, H_2O_2, NH_3, CH_4, H_2CO, CHD_2F, and HCN [334,703–705], confirm that general picture. The experimentally measured intensities of overtone bands (see, e.g., Refs. 596 and 706) show an evolution that roughly involves a decrease by about one or two orders of magnitude from the fundamental to the first overtone and each successive overtone band. Such observation is illustrated in Fig. 18, which presents the evolution of the square of the transition moment associated to stretching vibrations of C_2HD [444]. Lehmann and Smith studied the factors that influence overtone intensity using realistic potential and dipole moment functions that model the acetylenic stretching vibrations [707]. Their conclusions were that mechanical and to a lesser extent electrical anharmonicity are responsible for the observed overtone intensities. Moreover, they found that overtone intensities are extremely sensitive to small changes in the potential, especially those affecting its inner wall.

Overtone stretching spectra and intensities have also been measured and studied theoretically for larger species. Recent examples can be found in Refs. 708–715.

2. RESONANCES AND INTENSITY BORROWING. The off-diagonal contributions of mechanical anharmonicity to the MIME are at the basis of the so-called anharmonic resonances and lead to the formation of polyads, or V-clusters [468], of interacting levels (see the extensive discussion in Section III.C).

In addition to modifying the energy of the levels, the anharmonic resonances mix their zeroth-order (harmonic oscillator) wavefunctions. As a result, the intensity of both the allowed and forbidden vibrational bands associated with transitions involving these mixed levels is also modified. Such alteration can be a transfer of intensity of an allowed transition to forbidden ones or an averaging of the intensity of allowed transitions (see Section III.B.3.a.3).

Teffo and co-workers [462,463,716] studied the effect of anharmonic resonances on band intensities in triatomic linear molecules, with application to the CO_2 and N_2O molecules. They developed an effective Hamiltonian and effective dipole moment approach for the global quantitative analysis and prediction of line positions and intensities of infrared cold and hot bands, including forbidden transitions. That model relies on the approximate resonances $\tilde{\omega}_3 \approx 2\tilde{\omega}_1 \approx 4\omega_2$, as described in Section III.C.2.b. For example, the model and parameters obtained by Lyulin et al. [463] explained quantitatively the observed relative band intensities, the vibrational term values, and the principal rotational parameters derived from spectra of $^{14}N_2{}^{16}O$ recorded between 6500 and 15,000 cm^{-1} using both FT spectroscopy and ICLAS [596] (see also the discussion in Section II.C.2.b.4).

Clustering of the vibrational levels is also observed to be dominant in intermediate-size molecules such as acetylene (see, e.g., Refs. 468,513,571, and 717) and haloacetylenes [718]. A detailed discussion of the "Cluster model" and its application to the understanding of relative intensities of bands associated with the excitation of stretching and bending vibrations of various isotopomers of acetylene is given in Section III.C.2.b.2 (see also Fig. 37).

8. Vibration–Rotation Spectra

a. Rigid Rotor Selection Rules. Rotational transitions occur simultaneously to the vibrational transitions discussed above. The selection rules governing these vibration–rotation transitions are determined neglecting the vibration–rotation interactions. As a result, they are obtained by combining those for vibrational transitions given above (see Section III.E.7.a) with the rules for pure rotation transitions. The latter correspond to a nonvanishing value of the matrix element involving the direction cosines in (3.212). That is, they must satisfy

$$\langle \Psi'_r | \lambda_{\Xi\xi} | \Psi''_r \rangle \neq 0 \qquad (3.234)$$

Before discussing of the selection rules applicable to vibration–rotation transitions, we must determine when the condition (3.234) is fulfilled. We know that a molecule possesses a total angular momentum J defined quantum-mechanically by the vector operator $\hat{J}$ and its projections $\hat{J}_X, \hat{J}_Y, \hat{J}_Z$ on the LAS (X,Y,Z) and $\hat{J}_x, \hat{J}_y, \hat{J}_z$ on the MAS (x,y,z). Because $\hat{J}^2$, $\hat{J}_z$, and $\hat{J}_Z$ commute with each other, they define a common system of eigenfunctions, which are denoted $|J, k, M\rangle$, where J, k, and M are the quantum numbers respectively associated with these operators (see Table XXII). These functions are also eigenfunctions of the Hamiltonian of the rigid rotor as the latter is a function of these operators (see Table XXII):

$$\Psi_r \equiv |J, k, M\rangle \tag{3.235}$$

In the absence of external fields, the $2J + 1$ eigenvalues of $\hat{J}_Z$ are degenerate. Furthermore, the three space-fixed directions are equivalent. We can then consider any-one of the three matrix elements of (3.212) $\langle \Psi_e \Psi_v' \Psi_r' | \mu_\Xi | \Psi_e \Psi_v'' \Psi_r'' \rangle$ with $\Xi = X$, Y, or Z to determine the selection rules for the rigid rotor. In the following discussion, we consider the Z space-fixed component of the electric dipole matrix elements. Following Papousek and Aliev [47], it is convenient to write (2.38) in the form

$$
\begin{aligned}
\langle \Psi_e \Psi_v' \Psi_r' | \mu_Z | \Psi_e \Psi_v'' \Psi_r'' \rangle = {} & \langle \Psi_v' | \mu_z(e,e) | \Psi_v'' \rangle \langle \Psi_r' | \lambda_{Zz} | \Psi_r'' \rangle \\
& + \tfrac{1}{2} \langle \Psi_v' | \mu_x(e,e) - i\mu_y(e,e) | \Psi_v'' \rangle \langle \Psi_r' | \lambda_{Zx} \\
& + i\lambda_{Zy} | \Psi_r'' \rangle \\
& + \tfrac{1}{2} \langle \Psi_v' | \mu_x(e,e) + i\mu_y(e,e) | \Psi_v'' \rangle \langle \Psi_r' | \lambda_{Zx} \\
& - i\lambda_{Zy} | \Psi_r'' \rangle
\end{aligned}
\tag{3.236}
$$

to derive the selection rules on the rotation quantum numbers J and k. It can be shown that [47]

$$\langle \Psi_r' | \lambda_{Zz} | \Psi_r'' \rangle \neq 0 \quad \text{if} \quad \Delta k = k' - k'' = 0 \tag{3.237}$$

$$\langle \Psi_r' | \lambda_{Zx} \pm i\lambda_{Zy} | \Psi_r'' \rangle \neq 0 \quad \text{if} \quad \Delta k = \mp 1 \tag{3.238}$$

with $\Delta J = 0, \pm 1$ holding for all three matrix elements. These optical selection rules are general.

b. Vibration–Rotation Transitions. The general selection rules governing rotation transitions were described in the previous section. To discuss the selection rules governing the vibration–rotation transitions, we need to explicitly define the class of molecule under consideration that is spherical, symmetric, asymmetric, or linear top (see Section III.D.1). In the following discussion, we focus on symmetric tops excited in nondegenerate vibration levels, asymmetric tops and linear molecules (For the excitation of degenerate vibrations, see, e.g., Ref. 719).

1. SYMMETRIC TOP. As defined in Section III.D.1, the symmetric top has two identical principal moments of inertia. The remaining axis, associated with the z axis of the MAS, is the symmetry axis of the top. The fact that a molecule is a symmetric top comes almost always from its structural

symmetry and an n-fold (with $n \geq 3$) axis of symmetry coincides with the top axis. The only exceptions are "accidental" symmetric tops, that is, molecules of lower symmetry or having no symmetry at all for which two of the principal moments of inertia happen to have the same value.

1. *Selection Rules.* The symmetry properties of a "true" symmetric top generates two types of allowed transitions leading to the following types of bands [697]: parallel bands and perpendicular bands.

Parallel bands are associated with transitions induced by the electric dipole moment oscillating in a direction parallel to the top axis (z axis). For example, in the double-harmonic approximation, they correspond to (the dipole moment derivatives m_i^ξ introduced in (2.71) are written explicitly here for didactic purposes)

$$\left(\frac{\partial \mu_\pm(e,e)}{\partial q_\pm}\right)_0 = 0 \quad \text{and} \quad \left(\frac{\partial \mu_z(e,e)}{\partial q_i}\right)_0 \neq 0 \qquad (3.239)$$

with $\mu_\pm(e,e) = \mu_x(e,e) \pm i\mu_y(e,e)$ [see equation (3.236)]. As a result, the selection rules for the rotation quantum numbers depend on λ_{Zz}. They are thus, from (3.237):

$$
\begin{aligned}
\Delta K = 0, \quad & \Delta J = 0, \pm 1 \quad && \text{when} \quad K \neq 0 && (3.240) \\
\Delta K = 0, \quad & \Delta J = \pm 1 \quad && \text{when} \quad K = 0 && (3.241)
\end{aligned}
$$

all with $J' = 0 \leftrightarrow J'' = 0$ forbidden. These selection rules are expressed for convenience in terms of the quantum number $K = |k|$.

Perpendicular bands correspond to transitions associated with the electric dipole moment oscillating in a direction perpendicular to the top axis. Again, in the double-harmonic approximation, they involve

$$\left(\frac{\partial \mu_\pm(e,e)}{\partial q_\pm}\right)_0 \neq 0 \quad \text{and} \quad \left(\frac{\partial \mu_z(e,e)}{\partial q_\pm}\right)_0 = 0 \qquad (3.242)$$

The selection rules on the rotation quantum numbers then depend on λ_{Zx} and λ_{Zy} and are then, from (3.238):

$$\Delta K = \pm 1, \quad \Delta J = 0, \pm 1 \quad (\text{with } J' = 0 \leftrightarrow J'' = 0 \text{ forbidden}) \quad (3.243)$$

It is possible to observe bands for which, in the double harmonic approximation, all the dipole moment derivatives are nonzero:

$$\left(\frac{\partial \mu_\pm(e,e)}{\partial q_\pm}\right)_0 \neq 0 \quad \text{and} \quad \left(\frac{\partial \mu_z(e,e)}{\partial q_i}\right)_0 \neq 0 \qquad (3.244)$$

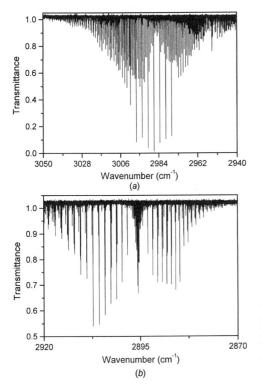

Figure 49. Spectrum of jet-cooled C_2H_6 around $3000\,cm^{-1}$, recorded at a resolution of $0.005\,cm^{-1}$ (unpublished results from M. Hepp and M. Herman, ULB), illustrating the structure of perpendicular (*a*) and parallel (*b*) bands of a symmetric top.

These bands are known as *hybrid bands*. Such transitions can be observed in asymmetric tops, in accidental symmetric tops and in high overtone and combination bands of true symmetric tops. They have the features of both parallel and perpendicular bands, and thus correspond to the superposition of these two types of bands. The selection rules are then given by (3.240), (3.241), and (3.243). Figure 49 presents the absorption spectrum of supersonic jet-cooled C_2H_6 in the 3.45-μm region recorded at a resolution of $0.005\,cm^{-1}$ using Fourier transform spectroscopy (unpublished results from M. Hepp and M. Herman, ULB). The v_7 band at $2985\,cm^{-1}$, appearing in the upper panel, presents the typical structure of a perpendicular band with very strong Q branches spread over almost the entire band. The v_5 band at $2896\,cm^{-1}$ shown in the lower panel of the figure is a parallel band. Its Q branches remain in the vicinity of the origin and have about the same intensity as the P and R branches. Furthermore, the P and R lines involving the same value of K form stacks, leading to a branch structure, which resembles that of a diatomic molecule.

TABLE XXXVII
Values of Elements of Direction Cosine Matrices, Separated into Three Factors [103]

Matrix Element	Value of J'		
	$J+1$	J	$J-1$
$\langle J' \lvert \lambda_{\Xi\xi} \rvert J \rangle$	$[4(J+1)\sqrt{(2J+1)(2J+3)}]^{-1}$	$[4J(J+1)]^{-1}$	$[4J\sqrt{4J^2-1}]^{-1}$
$\langle J'K \lvert \lambda_{\Xi z} \rvert JK \rangle$	$2\sqrt{(J+1)^2-K^2}$	$2K$	$+2\sqrt{J^2-K^2}$
$\langle J'K+1 \lvert \lambda_{\Xi x} \rvert JK \rangle = \pm i \langle J'K+1 \lvert \lambda_{\Xi y} \rvert JK \rangle$	$\mp\sqrt{(J\pm K+1)(J\pm K+2)}$	$\sqrt{(J\mp K)(J\pm K+1)}$	$\pm\sqrt{(J\mp K)(J\mp K-1)}$
$\langle J'M \lvert \lambda_{Z\xi} \rvert JM \rangle$	$2\sqrt{(J+1)^2-M^2}$	$2M$	$+2\sqrt{J^2-M^2}$
$\langle J'M\pm1 \lvert \lambda_{X\xi} \rvert JM \rangle = \mp i \langle J'M\pm1 \lvert \lambda_{Y\xi} \rvert JM \rangle$	$\mp\sqrt{(J\pm M+1)(J\pm M+2)}$	$\sqrt{(J\mp M)(J\pm M+1)}$	$\pm\sqrt{(J\mp M)(J\mp M-1)}$

2. *Line Strength Factors.* The value of the matrix elements of the direction cosines (3.237) and (3.238) for a symmetric top can be expressed analytically. They may be broken up into three factors [103]:

$$\langle J'K'M'|\lambda_{\Xi\xi}|J''K''M''\rangle = \langle J'|\lambda_{\Xi\xi}|J''\rangle\langle J'K'|\lambda_{\Xi\xi}|J''K''\rangle\langle J'M'|\lambda_{\Xi\xi}|J''M''\rangle$$

$$(3.245)$$

Table XXXVII gives the values of the three factors involved in the elements of the direction cosines matrices. They are given in the phase convention of Condon and Shortley [638], adopted by Hougen and Watson [393] and already previously introduced. It is such that

$$E^*|J,K,M\rangle = (-1)^{J-K}|J,-K,M\rangle \qquad (3.246)$$

where E^* is the inversion of the spatial coordinates of all nuclei and electrons through the center of mass [46]. The direction cosines may be used to calculate the line strength, or Hönl–London, factors for symmetric top molecules in the rigid rotor approximation. For a given transition $J'_{K'} \leftarrow J''_{K''}$, the line strength factor is defined as the sum of the modulus square of the direction cosine matrix elements for the three space-fixed axes ($\Xi = X, Y, Z$) and all the degenerate components contributing to that transition. Because we consider a field-free space, the summation involves all M' and M'' components:

$$L_J^\xi = \sum_\Xi \sum_{M,M''} |\langle J'K'M'|\lambda_{\Xi\xi}|J''K''M''\rangle|^2 \qquad (3.247)$$

Because X, Y, Z are equivalent, the summation of the squared elements over Ξ can be accomplished by multiplying the squared elements for any given Ξ by 3. Thus

$$L_J^\xi = 3 \times |\langle J'|\lambda_{Z\xi}|J''\rangle|^2 |\langle J'K'|\lambda_{Z\xi}|J''K''\rangle|^2 \sum_{M',M''} |\langle J'M'|\lambda_{Z\xi}|J''M''\rangle|^2$$

$$(3.248)$$

L_J^ξ is the line strength factor of the transition with component ξ of the electric dipole moment. Their analytical expression for R-, Q- and P-branch lines of parallel and perpendicular bands of a symmetric top are listed in Table XXXVIII (for the perpendicular bands, the line strength factor involves the sum of the x and y contributions, see (3.236)).

TABLE XXXVIII

Line Strength Factors for Bands of a Symmetric Top in the Absence of an External Field ($M \equiv M'' = M'$)

Parallel Band

$\Delta J = +1$

$$L_J^z = 3 \times \sum_M |\langle J+1, K, M|\lambda_{Zz}|J, K, M\rangle|^2 = \frac{(J-K+1)(J+K+1)}{J+1}(2 - \delta_{K,0})$$

$\Delta J = 0$

$$L_J^z = 3 \times \sum_M |\langle J, K, M|\lambda_{Zz}|J, K, M\rangle|^2 = \frac{K^2(2J+1)}{J(J+1)}(2 - \delta_{K,0})$$

$\Delta J = -1$

$$L_J^z = 3 \times \sum_M |\langle J-1, K, M|\lambda_{Zz}|J, K, M\rangle|^2 = \frac{J^2 - K^2}{J}(2 - \delta_{K,0})$$

Perpendicular Band[a]

$\Delta J = +1$

$$L_J^{x,y} = 3 \times \sum_M \{|\langle J+1, K\pm 1, M|\lambda_{Zx}|J, K, M\rangle|^2 + |\langle J+1, K\pm 1, M|\lambda_{Zy}|J, K, M\rangle|^2\} = \frac{(J\pm K+2)(J\pm K+1)}{J+1}$$

$\Delta J = 0$

$$L_J^{x,y} = 3 \times \sum_M \{|\langle J, K\pm 1, M|\lambda_{Zx}|J, K, M\rangle|^2 + |\langle J, K\pm 1, M|\lambda_{Zy}|J, K, M\rangle|^2\} = \frac{(J\pm K+1)(J\mp K)(2J+1)}{J(J+1)}$$

$\Delta J = -1$

$$L_J^{x,y} = 3 \times \sum_M \{|\langle J-1, K\pm 1, M|\lambda_{Zx}|J, K, M\rangle|^2 + |\langle J-1, K\pm 1, M|\lambda_{Zy}|J, K, M\rangle|^2\} = \frac{(J\mp K-1)(J\mp K)}{J}$$

[a] When $K = 0$ and $\Delta K = +1$, the values given must be multiplied by 2.

2. ASYMMETRIC TOP. In this case, the principal axes of inertia are labeled a, b, and c, which are mapped to x, y, and z according to one of the representations given in Section III.D.2.a.

1. *Selection Rules.* The rotation selection rules in vibrational transitions of asymmetric top molecules are commonly derived from symmetry arguments. The rigid rotor Hamiltonian of an asymmetric rotor (Table XXII) is invariant to the Euler angle transformations caused by a twofold rotation about a, b, or c, which form the symmetry group D_2 [46]. The irreducible representations of that group can thus be used to classify the rotational wavefunctions of an asymmetric rotor and the direction cosines $\lambda_{\Xi\xi}$. The rotation selection rules are then derived from (3.234):

$$\Delta K_a = \text{even} \quad \Delta K_c = \text{odd} \quad \text{if } \xi = a$$
$$\Delta K_a = \text{odd} \quad \Delta K_c = \text{odd} \quad \text{if } \xi = b \qquad (3.249)$$
$$\Delta K_a = \text{odd} \quad \Delta K_c = \text{even} \quad \text{if } \xi = c$$

all with $\Delta J = 0, \pm 1$ (with $J'' = 0 \leftrightarrow J'' = 0$ forbidden) giving rise to a-, b- and c-type bands. If the molecule is a near-prolate top, then the selection rules on K_a are more restrictive, that is $\Delta K_a = $ even or odd can be replaced by $\Delta K_a = 0$ or ± 1, respectively. In the case of a near-oblate top, the same comment applies to ΔK_c. Table XXXIX links the types of bands in the prolate and oblate cases of the asymmetric top to those of the symmetric top molecule.

Figures 50–52 present the allowed transitions between a few rotational levels belonging to two different vibrational levels of a near-prolate asymmetric top. The corresponding spectra, generated using the constants listed in Table XL, are schematized in Fig. 53, 54, and 55 respectively (more examples can be found in Ref. 720). The rotational transitions of a near-prolate asymmetric top are labeled with $^{\Delta K_a \, \Delta K_c}\Delta J_{K_a''}(J'')$. However, when

TABLE XXXIX
Correlation between Types of Bands of Prolate and Oblate
Asymmetric Tops and Those of a Symmetric Rotor

Band Type		
Prolate Rotor	Oblate rotor	Symmetric Top
a	c	Parallel
b	b	Hybrid
c	a	Perpendicular

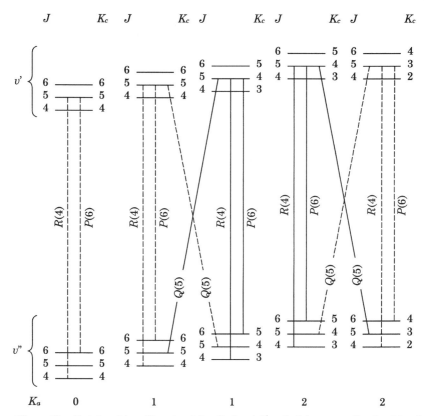

Figure 50. Rotational transitions involving the level $J' = 5$ of the upper vibrational level, satisfying the selection rules of an a-type band of a near-prolate asymmetric top. The effects of nuclear spin statistics are exemplified for *trans*-$C_2H_2O_2$ in its ground electronic state (C_{2h} symmetry); the solid lines represent transitions leading to absorption 3 times as strong as the dashed ones.

the two K_c components of J_{K_a} levels are not resolved, the labeling of the transitions is often simplified to $^{\Delta K_a}\Delta J_{K_a''}(J'')$. The same comments apply to near-oblate asymmetric tops, provided the indices "a" are replaced by "c". Figures 53–55 show that the rotational structure of the bands of an asymmetric top can be conveniently described by the superposition of "subbands." Each subband is made of R, Q, and P branches and is characterized by a constant value of K_a'' or K_c'' and the parity of the lower levels (see Section III.D.2.b).

Figures 56 and 57 present examples of the rotational structure of a-type bands of near-prolate and near-oblate asymmetric tops, respectively. Figure

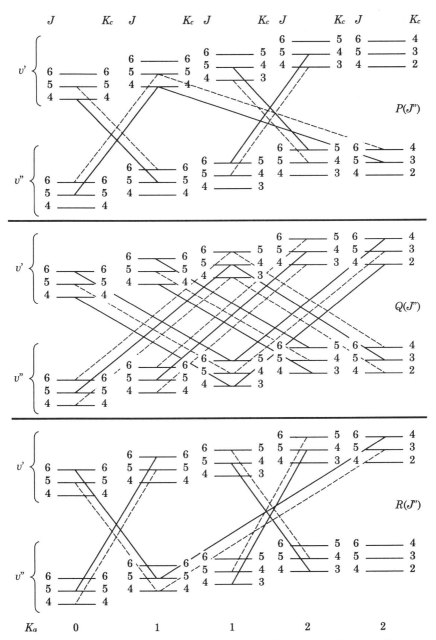

Figure 51. Rotational transitions involving the levels $J' = 4, 5, 6$ of the upper vibrational level, satisfying the selection rules of a b-type band of a near-prolate asymmetric top. The effects of nuclear spin statistics are exemplified for *trans*-$C_2H_2O_2$ in its ground electronic state (C_{2h} symmetry); the solid lines represent transitions leading to absorption 3 times as strong as the dashed ones.

231

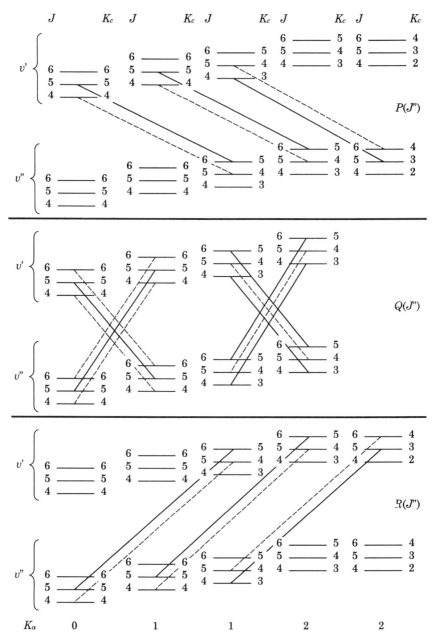

Figure 52. Rotational transitions involving the levels $J' = 4, 5, 6$ of the upper vibrational level, satisfying the selection rules of a c-type band of a near-prolate asymmetric top. The effects of nuclear spin statistics are exemplified for *trans*-$C_2H_2O_2$ in its ground electronic state (C_{2h} symmetry); the solid lines represent transitions leading to absorption 3 times as strong as the dashed ones.

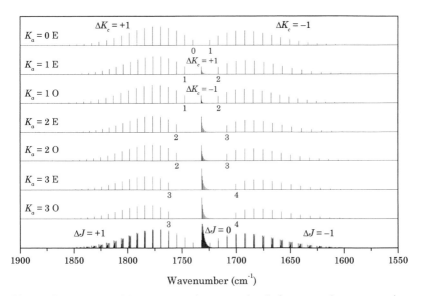

Figure 53. Structure of the spectrum of an a-type band of a near-prolate asymmetric top (calculated using the constants listed in Table XL), in the absence of the effects of nuclear spin statistics. The upper traces schematize the subbands originating from the levels with $K_a'' = 0$–3. The complete spectrum is shown in the lowest part. The line intensities were calculated for a temperature of 298 K. Symbols E and O indicate the parity, even or odd, of $J + K_a + K_c$.

56 shows the observed spectrum of the v_2 band of hypochlorous acid (HOCl), and Fig. 57 [721] presents the spectrum of the v_1 band of pyrrole (C_4H_5N). Because of the existence of two isotopes of chlorine, there are actually two fundamental v_2 bands observed in the spectrum of Fig. 56. Moreover, the absence of Q branches in these bands is apparent. As demonstrated in the lower part of the figure, they are indeed spread over the whole bands instead of remaining in the vicinity of the origin. Such behavior results from the relative values of the rotational constants in both vibrational levels.

2. *Line Strength Factors.* The line strength factors for asymmetric top molecules can be written as

$$L_J^\xi = \sum_\Xi \sum_{M',M''} |\langle J'K_a'K_c'M'|\lambda_{\Xi\xi}|J''K_a''K_c''M''\rangle|^2 \qquad (3.250)$$

where $\xi = a$, b, or c. It can be computed from those for the symmetric top, given in Table XXXVIII, because the wavefunctions of an asymmetric rotor

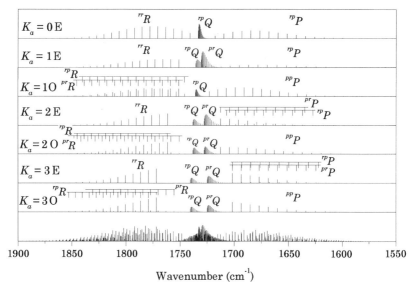

Figure 54. Structure of the spectrum of a *b*-type band of a near-prolate asymmetric top (calculated using the constants listed in Table XL), in the absence of the effects of nuclear spin statistics. The upper traces schematize the subbands originating from the levels with $K_a'' = 0–3$. The complete spectrum is shown in the lowest part. The line intensities were calculated for a temperature of 298 K. Symbols E and O indicate the parity, even or odd, of $J + K_a + K_c$.

are usually expressed as linear combinations of symmetric top wavefunctions, as detailed in Section III.D.2.*b*.

3. LINEAR MOLECULE. The vibration–rotation wavefunctions of linear molecules, including *l*-type doubling and resonance, were given in (3.135) in the case in which one degenerate bending vibration is excited. More generally, for an arbitrary number of stretching and bending vibrations, the wavefunctions can be written as

$$|v_s, v_b^{l_b}, k, J, e\rangle = \frac{1}{\sqrt{2}} [|v_s, v_b^{l_b}, k, J\rangle + (-1)^k |v_s, v_b^{-l_b}, -k, J\rangle]$$
$$|v_s, v_b^{l_b}, k, J, f\rangle = \frac{1}{\sqrt{2}} [|v_s, v_b^{l_b}, k, J\rangle - (-1)^k |v_s, v_b^{-l_b}, -k, J\rangle]$$
(3.251)

and when all $l_b = 0$,

$$|v_s, v_b^0, 0, J, e\rangle = |v_s, v_b^0, 0, J\rangle$$
(3.252)

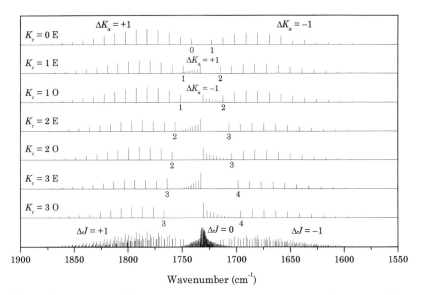

Figure 55. Structure of the spectrum of a c-type band of a near-prolate asymmetric top (calculated using the constants listed in Table XL), in the absence of the effects of nuclear spin statistics. The upper traces schematize the subbands originating from the levels with $K_c'' = 0$–3. The complete spectrum is shown in the lowest part. The line intensities were calculated for a temperature of 298 K. Symbols E and O indicate the parity, even or odd, of $J + K_a + K_c$.

TABLE XL

Values of Rotational Constants of Lower and Upper Vibrational Levels Used to Generate Spectra of Figs. 53–55 in the I^r Representation

Constant	Value (cm^{-1})
A''	5.241
B''	3.946
C''	3.878
A'	5.211
B'	3.870
C'	3.855
$\tilde{v}_0$	1732
κ	-0.9

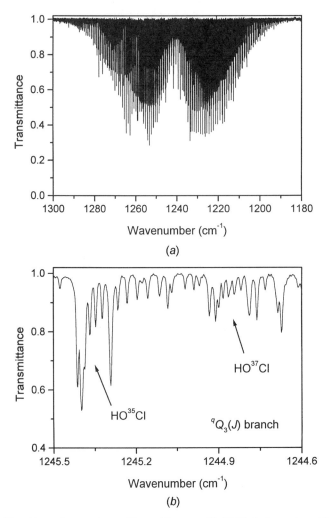

Figure 56. Absorption spectrum of hypochlorous acid, HOCl, in the region of the bending fundamental v_2 near $1238\,\mathrm{cm}^{-1}$: (a) structure of the whole band, which is an a-type band of a near-prolate asymmetric top ($\kappa = -0.9987$); (b) 0.9-cm^{-1} region of the spectrum showing the $^qQ_3$ branches of HO^{35}Cl and HO^{37}Cl (unpublished results from J. Vander Auwera, ULB).

In these expressions, v_s and v_b represent the vibration quantum number associated with the stretching and bending vibrations, respectively, as defined previously.

1. *Selection Rules.* The selection rules governing vibration–rotation transitions can be classified into the following three categories:

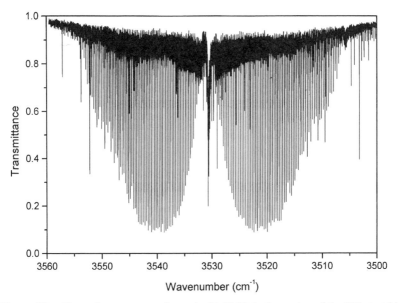

Figure 57. Absorption spectrum of pyrrole (C_4H_5N) in the region of the NH stretching mode. The band observed is an a-type band of the near-oblate top ($\kappa = +0.9438$; adapted from Ref. 721).

$$
\begin{array}{llll}
\Delta k = 0 \text{ with } k' = k'' = 0 : & \Delta J = \pm 1 & \text{with} & e \leftrightarrow e \\
& & & f \leftrightarrow f \\
\Delta k = 0 \text{ with } k' \neq 0, k'' \neq 0 : & \Delta J = 0 & \text{with} & e \leftrightarrow f \\
& \Delta J = \pm 1 & \text{with} & e \leftrightarrow e \\
& & & f \leftrightarrow f \\
\Delta k = \pm 1 : & \Delta J = 0 & \text{with} & e \leftrightarrow f \\
& \Delta J = \pm 1 & \text{with} & e \leftrightarrow e \\
& & & f \leftrightarrow f
\end{array}
\tag{3.253}
$$

where the transition $J' = 0 \leftrightarrow J'' = 0$ is always forbidden. The first two categories correspond to a nonzero electric dipole moment component along the molecule z axis. The latter involves nonzero components on the x and y axes. According to these selection rules, the following types of vibration–rotation bands can occur in linear molecules:

a. $\Delta k = 0$ *Bands, with* $k' = k'' = 0$. These are parallel bands for which $k = 0$ in both the lower and upper levels. In the case of unsymmetric linear species, thus belonging to the point group $C_{\infty v}$, this category includes vibration bands of the types $\Sigma^+ – \Sigma^+$ and $\Sigma^- – \Sigma^-$. For the

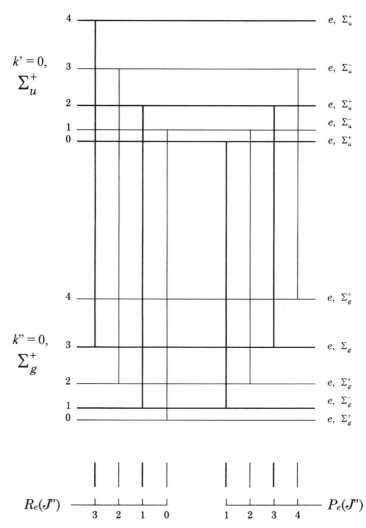

Figure 58. Energy-level diagram for $\Sigma_u^+ - \Sigma_g^+$ transitions of linear polyatomic molecules. The B values are assumed to be the same in the lower and upper levels. The resulting spectrum is schematized at the bottom. The intensity alternation represented by the heavy and light lines refers to molecules such as C_2H_2 (see Section III.E.3.*e*.4). For $C_{\infty v}$ molecules, the g/u subscripts and intensity alternation should be disregarded.

symmetric linear molecules (point group $D_{\infty h}$), the vibration bands are of the types $\Sigma_u^+ - \Sigma_g^+$, $\Sigma_g^+ - \Sigma_u^+$, $\Sigma_u^- - \Sigma_g^-$, and $\Sigma_g^- - \Sigma_u^-$. As stated in the selection rules, this type of bands does not possess a Q branch. Figure 58 presents the allowed transitions between the lowest energy rotational levels in the case of a band of type $\Sigma_u^+ - \Sigma_g^+$ and the resulting

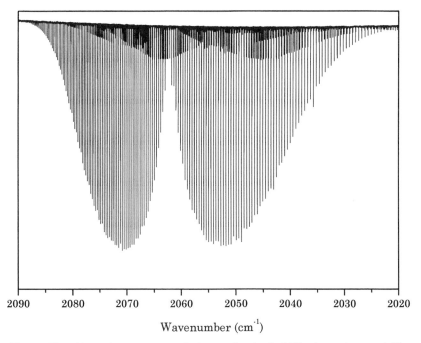

Figure 59. Absorption spectrum of the ν_3 band of OCS observed near 4.85 µm (unpublished results from J. Vander Auwera, ULB), illustrating the structure of a $\Sigma^+ - \Sigma^+$ band; ν_3 is the asymmetric stretching mode.

spectrum. This diagram also represents $\Sigma^+ - \Sigma^+$ transitions of $C_{\infty v}$ molecules if the subscript g/u and heavy/light lines are omitted. Figure 59, which presents the ν_3 fundamental of OCS, illustrates the $\Sigma^+ - \Sigma^+$ type of band. Two examples of $\Sigma_u^+ - \Sigma_g^+$ bands can be found in Fig. 70, which shows the $5\nu_3$ band of C_2H_2, and Fig. 81, which presents the $2\nu_1 + \nu_2 + 3\nu_3$ band of C_2D_2. These latter examples clearly demonstrate the influence of the statistical weight on the individual line intensities, detailed in Section III.E.3.e.4.

 b. $\Delta k = 0$ Bands, with $k' = k'' \neq 0$. These are parallel bands for which $k \neq 0$ in both the lower and upper levels. They are, for instance $\Pi_u - \Pi_g$, $\Pi_g - \Pi_u$, and $\Delta_u - \Delta_g$ bands (where again the subscript is dropped for $C_{\infty v}$ molecules). Contrary to the other type of parallel bands, these present a Q branch. Because k is nonzero in the lower level, no fundamental bands are of this type. Figure 60 shows a diagram of the energy levels and transitions allowed between them, according to the selection rules for a $\Pi_u - \Pi_g$ band. Because both the lower and upper

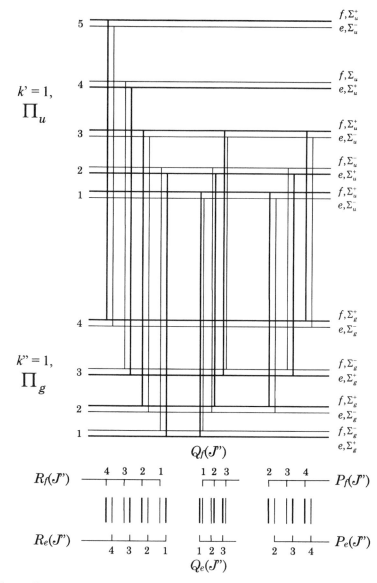

Figure 60. Energy-level diagram for $\Pi_u - \Pi_g$ transitions of linear polyatomic molecules. The B values are assumed to be the same in the lower and upper levels. The resulting spectrum is schematized at the bottom. The Q-branch lines are spread out even though they should coincide because $B' = B''$. The intensity alternation represented by the heavy and light lines refers to molecules such as C_2H_2 (see Section III.E.3.e.4). For $C_{\infty v}$ molecules, the g/u subscripts and intensity alternation should be disregarded.

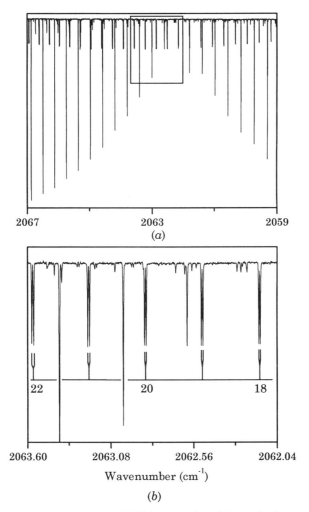

Figure 61. Absorption spectrum of OCS in the region of the v_3 fundamental showing a portion of the R branch of the $v_2 + v_3 - v_2$ band (unpublished results from J. Vander Auwera, ULB): (*a*) view showing the R branch relative to the origin of the fundamental v_3; (*b*) enlarged view of the region indicated in the upper panel. The lines are identified by J''. The e and f components of these lines of this $\Pi_u - \Pi_g$ band are clearly observed; v_2 and v_3 are the bending and asymmetric stretching modes, respectively.

levels have e and f components as a result of l-type doubling, all R-, P- and Q-branch lines exhibit two components. This is illustrated in Figure 61. For symmetric linear molecules, the nuclear spin statistics can induce an intensity difference between both components. This is clearly illustrated in Fig. 62, which shows the absorption spectrum of

242

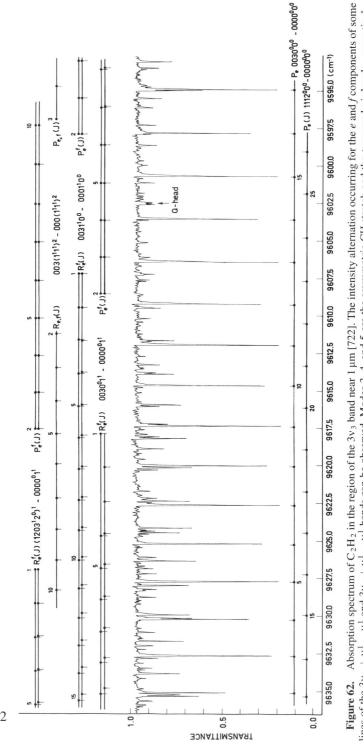

Figure 62. Absorption spectrum of C_2H_2 in the region of the $3\nu_3$ band near 1 μm [722]. The intensity alternation occurring for the e and f components of some lines of the $3\nu_3 + \nu_5^1 - \nu_5^1$ and $3\nu_3 + \nu_4^1 - \nu_4^1$ bands can be observed. Modes 3, 4, and 5 are the asymmetric CH stretch and the *trans* and *cis* bends, respectively.

C_2H_2 in the region of the $3\nu_3$ band [722]. It must be noted that when l-type doubling is negligible, that is when both e and f components overlap, the intensity alternation is not observed. This can be seen in the spectrum presented in Fig. 80.

$c.$ $\Delta k \neq 0$ *Bands.* These perpendicular bands arise from the excitation of the bending vibrations. The fundamental bands associated with these bending vibrations are of this type. For the unsymmetric linear species, perpendicular bands can be of the types $\Pi-\Sigma^+$, $\Pi-\Sigma^-$, $\Pi-\Delta$, or inversely. For $D_{\infty h}$ molecules, g and u subscripts have to be added according to the selection rule $u \leftrightarrow g$. Figure 63 shows an energy levels diagram, allowed transitions and associated spectrum corresponding to a $\Pi_u-\Sigma_g^+$ band. Such case is illustrated in Fig. 64, which shows the $\nu_2^1 + \nu_3$ band of N_2O near $2798\,\mathrm{cm}^{-1}$.

2. *Line Strength Factors.* The line strength factors for linear molecules are obtained from Table 38 by simply replacing K by k.

$c.$ *Vibration—Rotation Interaction.* In the absence of vibration—rotation interactions, the relative intensities of the rotation lines in allowed vibration bands are given by the line strength factors. The influence of vibration—rotation interactions is to introduce additional rotational dependence in the transition moment (3.236).

The discussion in this section outlines the effect of vibration—rotation interactions on the line intensities in a general way. It then focuses on linear molecules. Only global problems are introduced, in agreement with the strategy adopted in the present review. The role of accidental resonances is not detailed. An example of Coriolis-induced line intensity borrowing was illustrated in Fig. 43.

1. THE EFFECTIVE DIPOLE MOMENT OPERATOR. Let us recall the notion of effective Hamiltonians, discussed by Papousek and Aliev [47] and detailed in Sections III.A–III.E. The calculation of the vibration—rotation energy levels and wavefunctions is usually performed by combining perturbation and variational methods. This two-step process is currently the most effective way of analyzing experimental data. Both methods rely on a matrix representation of the Hamiltonian, usually built into the harmonic oscillator basis of wavefunctions. The perturbation method applied in the first step, such as those presented previously and the widely used contact transformation method (see, e.g., Refs. [47,456,723]), removes off-diagonal terms of the Hamiltonian that couple vibrational levels belonging to different previously defined polyads of interacting levels. That step leads to an *effective block-diagonal Hamiltonian* whose individual blocks describe the various polyads. The variational method applied in the second step

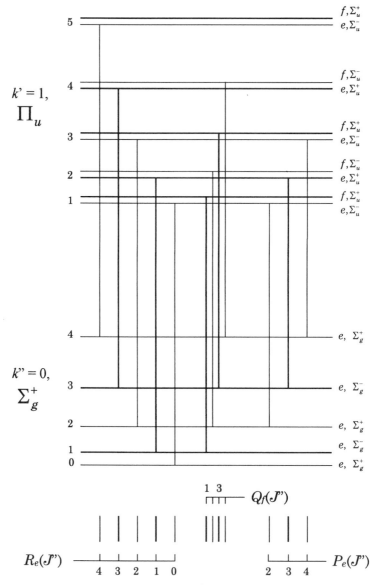

Figure 63. Energy-level diagram for $\Pi_u - \Sigma_g^+$ transitions of linear polyatomic molecules. The B values are assumed to be the same in the lower and upper levels. The resulting spectrum is schematized at the bottom. The Q-branch lines are spread out even though they should coincide because $B' = B''$. The intensity alternation represented by the heavy and light lines refers to molecules such as C_2H_2 (see Section III.E.3.e.4). For $C_{\infty v}$ molecules, the g/u subscripts and intensity alternation should be disregarded.

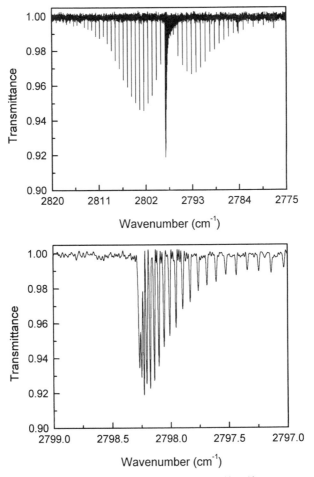

Figure 64. Absorption spectrum of the $\nu_2^1 + \nu_3$ band of $^{14}N_2{}^{16}O$ near $2798\,\text{cm}^{-1}$ (from Herman and co-workers [596]), illustrating the structure of a $\Pi-\Sigma^+$ band: (*a*) the whole band and (*b*) enlargement showing the structure of the *Q* branch. Modes 2 and 3 are the bend and primarily the NN stretch, respectively.

corresponds to diagonalization of the individual blocks to yield the desired energies and wavefunctions of the vibration–rotation levels. In almost all the treatments performed in molecular spectroscopy, the calculation of vibration–rotation energy levels and wavefunctions is thus performed using a Hamiltonian that is effective to some degree. Several examples were provided in Sections III.B and III.D.

Because the wavefunctions used to calculate the matrix elements of the dipole moment operator (or any physical quantity) result from the diagonalization of an effective—transformed—Hamiltonian, it is necessary to perform the same transformations on that physical quantity (see, e.g., Ref. [456] and references cited therein). The resulting effective dipole moment operator presents rotational dependencies, known as Herman–Wallis corrections (from the name of the authors who originally introduced this type of correction for diatomic molecules [724]). These correction terms add to the contribution from the direction cosine matrix elements. They are usually expressed as a factor multiplying the square of the transition moment and denoted by the letter F.

The theory of Herman–Wallis corrections for diatomic molecules and its extension to linear and nonlinear polyatomic molecules was developed in numerous papers (see, e.g., Ref. 725 and references cited therein). For example, for parallel and perpendicular bands of a linear molecule, the Herman–Wallis factors for P- and R-branch lines are [725]

$$F_{RP} = [1 + A_1 m + A_2^{RP} m^2]^2 \qquad (3.254)$$

with $m = -J$ and $J + 1$, respectively, while for Q-branch lines they are

$$F_Q = [1 + A_2^Q J(J + 1)]^2 \qquad (3.255)$$

The expression for the factors A_1, A_2^{RP}, and A_2^Q can be found in Ref. 725. As an example of the magnitude of these terms, $A_1 = -1.432(82) \times 10^{-4}$ and $A_2^{RP} = +7.9 \times 10^{-8}$ for the v_3 band of CO_2. The former is an experimental value [726], while the second was calculated using literature values for the parameters on which that coefficient depends [725]. Watson also derived the expression of Herman–Wallis corrections for symmetric and asymmetric top molecules [727]. In practical terms, these Herman–Wallis factors are considered as parameters, whose values are determined through fits to the experimental intensity data (see, e.g., Refs. [726 and 728–732] and references cited therein).

2. INFLUENCE OF VIBRATION–ROTATION RESONANCES ON INTENSITIES. Diagonalization of the blocks of the effective Hamiltonian matrix also affects the transition moment. This effect can be easily grasped with the following example, introduced in Section III.B.3.*a*.3, and repeated here for the sake of clarity.

Consider two vibrational levels, 1 and 2, of zeroth-order energy E_1^0 and E_2^0, which interact with each other through some off-diagonal matrix element W. Because of that interaction, the vibrational wavefunction associated with each level is a linear combination of the unperturbed wavefunction Ψ_1^0 and Ψ_2^0:

$$\begin{aligned} \Psi_1 &= a\Psi_1^0 - b\Psi_2^0 \\ \Psi_2 &= b\Psi_1^0 + a\Psi_2^0 \end{aligned} \qquad (3.256)$$

where a and b are the mixing coefficients. Now consider two transitions from an unperturbed lower level leading to these two levels and corresponding to parallel-type bands of a linear molecule. Let R_{10}^0 and R_{20}^0 represent the unperturbed vibronic transition moment associated with these two transitions, that is the transition moment of (3.236) excluding the line strength factor. The observed (perturbed) transition moment associated with the transition leading to level 1 is then given by

$$\begin{aligned} R_{10} &= \langle \Psi_1|\mu_z(e,e)|\Psi_0\rangle = a\langle \Psi_1^0|\mu_z(e,e)|\Psi_0\rangle - b\langle \Psi_2^0|\mu_z(e,e)|\Psi_0\rangle \\ &= aR_{10}^0 - bR_{20}^0 \end{aligned} \qquad (3.257)$$

where Ψ_0 is the wavefunction associated with the unperturbed lower level. Similarly, the vibronic transition moment associated with the transition to level 2 is

$$R_{20} = bR_{10}^0 + aR_{20}^0 \qquad (3.258)$$

Intensity Borrowing. This very simple example illustrates the effect of vibration–rotation interactions on the transition moment and thus on the intensity of vibrational bands. Because the intensity of the bands is proportional to the modulus square of the transition moment, the expressions (3.257) and (3.258) show that their initial intensity, proportional to $|R_{10}^0|^2$ and $|R_{20}^0|^2$, respectively, is distributed over them to the extent allowed by the mixing coefficients. If, for some reason, the transition from the lower level 0 to level 2 is forbidden, that is $R_{20}^0 = 0$, then the intensity of the corresponding band is proportional to the intensity of the allowed $1 \leftarrow 0$ band, with a proportionality factor b^2 [see (3.258)]. Band $2 \leftarrow 0$ thus borrows intensity from the allowed one.

Herman–Wallis Corrections. Let us introduce rotation; E_1^0, E_2^0, and eventually the interaction term W, are functions of molecular rotation. As a result, the mixing coefficients a and b will also be functions of molecular rotation. The observed transition moments associated to both bands will thus exhibit a rotation dependence, which adds to the

Hönl–London factor and the effect of the transformation of the Hamiltonian.

3. HERMAN–WALLIS EFFECTS IN LINEAR MOLECULES. There are many examples of experimental measurements of the intensity of rotation lines within vibrational bands of linear species in the literature, including the observation of Herman–Wallis effects. Many measurements were and are still performed using conventional absorption techniques with FT (see, e.g., Refs. 726,728–730,733, and 734) and tunable-difference frequency or diode laser spectroscopy [e.g., 732,735], among other techniques. More recently, absolute intensity measurements were reported using intracavity laser absorption spectroscopy (ICLAS) and cavity ringdown spectroscopy (CRDS) [e.g., 736]. An extensive review of such experimental techniques, dedicated to overtone spectroscopy, is provided in Section IV. Herman–Wallis effects were also observed in emission (see Ref. 737 and references cited therein and Ref. 738). Finally, information about the electric dipole moment function of diatomic species such as ArH^+, ClO, NH (see Ref. 737 and references cited therein) and OH [739,740] has been obtained from observed Herman–Wallis dependencies in vibration bands.

1. *Influence of Anharmonic Resonances.* In Section III.E.7.*d*, the occurrence of intensity borrowing through anharmonic resonances was discussed. However, as just highlighted, intensity borrowing and Herman–Wallis effects are closely coupled. Anharmonic resonances, which are purely vibrational interactions, can also lead to significant Herman–Wallis effects, because the rotation structure is usually different in different vibrational levels. Similar examples, in which vibration interactions affect the effective rotational constants, were discussed in Section III.D.4.*d*. We include below a few examples that show that such effects can in fact contribute significantly to the rotational dependence of line intensities in vibration bands of linear molecules.

2. *Herman–Wallis Effects in the* ν_3 *region of* $^{12}C_2H_2$. As described in Sections III.B.4.*c*.3 and III.D.4.*b*.2, the (0010^00^0) level in $^{12}C_2H_2$ (of Σ_u^+ symmetry) belongs to a polyad of interacting levels that also involves the two Σ_u^+ and Δ_u^e sublevels of the (0101^11^1) level. The block of the effective Hamiltonian matrix describing that polyad is given in Table IV of Ref. 451. The energy and wavefunction of these e levels, as well as of the f components observed from the $v_4 = 1$ level, were characterized from the analysis of the positions of vibration–rotation lines observed in the 3- and 3.7-μm regions [451]. The vibronic transition moments associated with the rotational lines of the ν_3 and $\nu_2 + \nu_4 + \nu_5$ (Σ_u^+–Σ_g^+) parallel bands, extracted using (3.208) and (3.236), are plotted as a function of m in Fig. 65

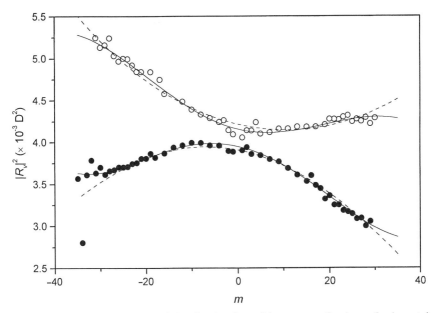

Figure 65. Variation with m of the vibrational transition moment for the ν_3 fundamental (bottom curve) and the $\nu_2 + \nu_4 + \nu_5$ ($\Sigma_u^+ - \Sigma_g^+$) combination band (top curve), observed in the 3-μm region of $^{12}C_2H_2$; the dots represent the experimental data and the dashed and solid curves, the rotational dependencies calculated at the end of the fit of both data sets to eqs. (3.259) and (3.260), respectively (adapted from Vander Auwera et al. [451]).

($m = -J''$ for P-branch lines and $J'' + 1$ for R-branch lines). The dashed line in the Figure corresponds to a fit of each data set to

$$|R_n|^2 = |R_n^0|^2 [1 + A_1^{RP} m + A_2^{RP} J'(J'+1)]^2 \qquad (n \equiv 3 \text{ or } 245) \quad (3.259)$$

in which R_n^0, A_1^{RP}, and A_2^{RP} were fitted as effective parameters. Because of the resonances between the upper levels, these transition moments R_n can be expressed as linear combinations of the zeroth-order transition moments associated to the allowed transitions. The situation is thus very similar to the discussion in Section III.E.7.*d*.2. The observed data were actually fitted to the following expressions, similar to (3.257) and (3.258):

$$\begin{aligned} |R_3|^2 &= |a(J)R_3^0 - b(J)R_{245}^0|^2 (1 + A_1^{RP} m)^2 \\ |R_{245}|^2 &= |b(J)R_3^0 + a(J)R_{245}^0|^2 (1 + A_1^{RP} m)^2 \end{aligned} \qquad (3.260)$$

taking the values of the mixing coefficient $a(J)$ and $b(J)$ from the analysis of the line positions (see Fig. 65). R_3 (R_3^0) and R_{245} (R_{245}^0) represent the observed (unperturbed) transition moments for the v_3 and $v_2 + v_4 + v_5$ ($\Sigma_u^+ - \Sigma_g^+$) bands, respectively. Figure 45 shows the composition of the wavefunction of the $v_3 = 1$ level as a function of J. Values of

$$R_3^0 = \pm 0.08907(3)D \text{ and } R_{245}^0 = \mp 2.63(3) \times 10^{-3}D \qquad (3.261)$$

were obtained, showing the strength of the intensity borrowing (as $|R_3|^2 \simeq |R_{245}|^2$). These results also show that it is possible with such treatment to determine the relative signs of the unperturbed transition moments. The Hamiltonian of Table IV of Ref. 451 accounts for the quadratic Herman–Wallis term. The linear term still present in equation

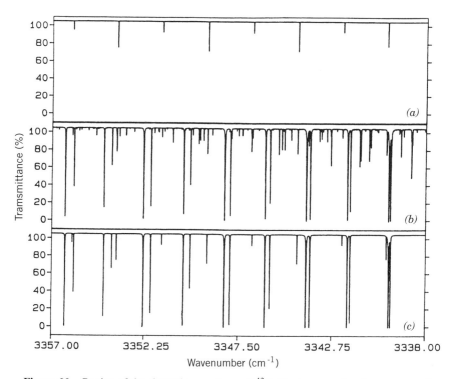

Figure 66. Portion of the absorption spectrum of $^{12}C_2H_2$ demonstrating the presence of the forbidden $v_2 + v_4 + v_5$ ($\Delta_u^e - \Sigma_g^+$) band: (*a*) calculated spectrum of the band; (*b*) observed spectrum; (*c*) calculated spectrum including all the subbands studied (from Vander Auwera et al. [451]). Modes 2, 4, and 5 are the CC stretch and the *trans* and *cis* bends, respectively.

(3.260), comes from vibration–rotation interactions not included in the model. Its value determined in both fits was actually the same, within uncertainties. Its presence puts forward the effective character of the Hamiltonian used here. As can be seen from Fig. 45, the $v_3 = 1$ level is mixed with the Δ_u^e component of the (0101^11^1) level at higher J values. As a result, the forbidden $v_2 + v_4 + v_5$ $(\Delta_u^e - \Sigma_g^+)$ band can be observed, as confirmed by the spectra presented in Fig. 66.

3. *Herman–Wallis Effects in Triatomic Species.* Similar examples can be found for triatomic molecules. In the case of CO_2, Watson [741] showed that the anharmonic interaction between the (10^01) and (02^01) levels accounts for the essential part of the magnitude of the A_2^{RP} Herman–Wallis factors observed for the $v_1 + v_3$ and $2v_2^0 + v_3$ bands near $2.7\,cm^{-1}$ [742]. The treatment also included *l*-type resonance with (02^21). However, its contribution to the Herman–Wallis effect was found to be much smaller [741]. Many examples of such effects can also be found in $^{16}O^{12}C^{32}S$ [569,730,743]. For instance, the anharmonic interaction between the $v_1 + 4v_2^0$ and $v_1 + v_3$ levels associated with $k_{2222/3}$ is responsible for the very strong Herman–Wallis dependence observed for the $v_1 + 4v_2^0$ band near $2937\,cm^{-1}$ and shown in Fig. 67. The mixing of these levels is larger at low J and decreases with increasing J as does the energy difference between these levels.

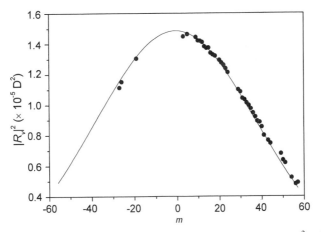

Figure 67. Evolution of the square of the vibronic transition moment (in D^2) with m in the P and R branches of the $v_1 + 4v_2^0$ band of $^{16}O^{12}C^{32}S$ near $2937\,cm^{-1}$. The dots indicate the experimental data, and the solid line shows the evolution given by the mixing coefficients. Almost no measurements are available for the P branch because of overlap with the much stronger $v_1 + v_3$ band at $2918\,cm^{-1}$ (adapted from Vander Auwera and co-workers [730]). Modes 1 and 3 are the symmetric and asymmetric stretches, respectively, and 2 is the bend.

4. *Contribution from l-Type Resonance.* As mentioned above, l-type resonance does also affect the rotational distribution of the line intensities within vibrational bands (see also, e.g., Refs. 472,504, and 744). Figure 66 showed an example of J-dependent intensity borrowing through such interaction observed in the 3-μm region of $^{12}C_2H_2$. Watson [741] showed that, similar to anharmonic interaction, l-type resonance can contribute to the quadratic Herman–Wallis term A_2 for the R, P, and Q branches of linear molecules. For example, using second-order perturbation theory, he explained the somewhat odd values of the A_2 coefficients measured by Johns [702] for the $\nu_1-\nu_2^1$, $2\nu_2^0-\nu_2^1$, and $2\nu_2^{2e}-\nu_2^1$ perpendicular bands of CO_2. Indeed, A_2 was found to be small for the P and R branches of all three bands and the Q-branch of $2\nu_2^{2e}-\nu_2^1$, while being quite significant for the Q branch of the remaining two bands. Watson showed that this was due to cancellations between contributions, due to the anharmonic interaction between ν_1 and $2\nu_2^0$ and the l-type resonance in all branches except these two Q branches [741]. Another example is provided by the observation of the forbidden $2\nu_2^{2e}$, $\nu_1 + 2\nu_2^{2e}$ and $2\nu_2^{2e} + \nu_3$ bands in a number of isotopomers of HCN [734].

5. *Other Contributions.* The intensity distribution of the rotation lines of vibration bands can also be modified through J-dependent intensity borrowing resulting, for instance, from Coriolis or centrifugal distortion interactions [456,725]. An illustrative example of such effects is provided by the ν_3 fundamental of $H^{12}C^{14}N$. That band presents an odd intensity distribution, especially in the R branch, and its transition moment is very small ($|R_3| = 1.362(12)10^{-3}$ D for $H^{12}C^{14}N$ [745]). Maki and co-workers showed that such behavior is the result of the combined effect of the very small transition moment R_3 and the large value of the linear Herman–Wallis coefficient A_1 for that parallel band. The origin of the large A_1 coefficient can be understood by looking at its expression (equation 8 of Ref. 745, derived from Ref. 725):

$$A_1 = \frac{1}{\sqrt{2}}\left[-4B(\tilde{\omega}_3\tilde{\omega}_2)^{1/2}(\tilde{\omega}_3^2 - \tilde{\omega}_2^2)^{-1}\frac{\zeta_{23}R_2}{R_3} - 2\left(\frac{2B}{\tilde{\omega}_3}\right)^{3/2}\frac{\zeta_{12}\mu_e}{R_3}\right]$$

$$(3.262)$$

where the constants have their usual meaning, defined in Section III.D. The first term describes intensity borrowing from the ν_2 fundamental by Coriolis coupling. The second one describes intensity stealing from the pure rotation spectrum by centrifugal interaction. Using known values of the constants involved (see Table V of Ref. 745), it can be easily shown that the large A_1 value results from the combined effect of the large permanent dipole moment of $H^{12}C^{14}N$ ($\mu_0 = 2.985$ D) and the very small transition moment R_3. Another example of the effect of Coriolis interaction on the intensity distribution within vibration bands is again provided by HCN. Indeed, forbidden Q-branch transitions $2\nu_2^{2f}$, $\nu_1 + 2\nu_2^{2f}$, and $2\nu_2^{2f} + \nu_3$ were recently observed for a number of isotopomers of HCN [734]. Because they are forbidden even when the effects of l-type resonance are considered, they must get their intensity from some other Coriolis interaction. The mechanism proposed by Maki and co-workers, and later confirmed by Watson [746], involves Coriolis intensity stealing mainly from the allowed ν_2 fundamental, enhanced by the large anharmonicity constant $k'_{122} = 1197\,\text{cm}^{-1}$. We observe a similar effect in the 7.5 μm region of $^{12}C_2H_2$ [746a].

IV. EXPERIMENTAL OVERTONE SPECTROSCOPY

A. Introduction

The dramatically low intensity and high density of vibration–rotation lines for molecules and spectral ranges of interest to overtone spectroscopy and dynamics precluded and sometimes misled the development of this field, for a long period of time. Its spectacular recent rebirth, which supports the present review, is to be attributed mostly to the explosion of instrumental and technical developments. It is today possible to record a number of overtone transitions in polyatomic species with the sensitivity, resolution, and simplification required to perform high-resolution investigations. A significant number of experimental state-of-the-art achievements were and are still reported in the literature in that respect.

The present section intends to provide some information of a more technical nature than the rest of the review on this specific aspect. It is, in many respects, complementary to, or updates the content of previous reviews, such as those of Quack [544] and Lehmann and co-workers [747]. We concentrate on selected techniques used for the study of overtone transitions. We restrict ourselves to high-spectral-resolution-type techniques, meaning those allowing the fine rotation structure to be accessed to, provided they are not broadened by intramolecular mechanisms. Laser techniques are by far dominant in this respect, and we review, in Section IV.C, on the basis of a list of references not intended to be exhaustive, the most recent developments in cavity ringdown spectroscopy (CRDS), intracavity laser spectroscopy (ICLAS), optoacoustic (OA), and optothermal (OT) spectroscopies. This review extends beyond pure vibrational overtone observations. Diode-laser absorption spectroscopy is also mentioned, for completeness. Double-resonance methods and ion-based detection techniques are also, briefly, dealt with because of their relevance for overtone investigations.

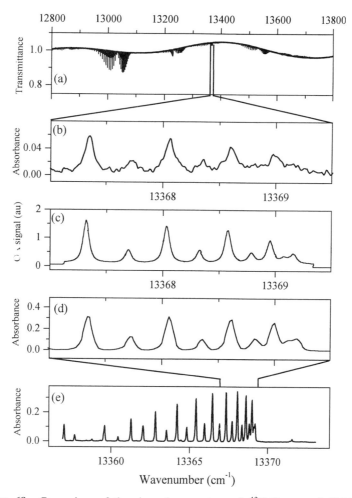

Wavenumber (cm^{-1})

Figure 68. Comparison of the absorption spectrum of $^{12}C_2H_2$ around 13,368 cm^{-1} recorded by (a, b) FTS associated with a multipass cell ($l = 49$ m) at a pressure of 328 hPa, (c) OA spectroscopy with a pressure of 39.5 hPa, and (d, e) ICLAS with an equivalent pathlength of about 33 km and a pressure of 35.5 hPa. This range corresponds to the band head of the Q branch of the $\Pi_u - \Pi_g$ band centered at 13,230.15 cm^{-1}. It is critical to notice that none of the spectra were initially recorded for the present comparison purposes. They could all be improved. Nevertheless, significant information can be obtained by comparing them, highlighted later. In these recordings a slightly better S/N ratio is obtained by OA detection than by ICLAS. However, ICLAS ensures a quick data acquisition due to the multiplex advantage, while OA detection requires a much longer time to scan the monomode laser. The respective recording times for the range presented were of the order of 30 s and 20 min. The FT spectrum took an even longer time to record, despite the multiplex advantage, because of the need for accumulating scans to reach an optimal S/N ratio. The FT spectrum was recorded by Herman at the "Laboratoire de Chimie Physique Moléculaire" (ULB), the ICLAS spectrum by Campargue at the Laboratoire de Spectrométrie Physique (U. J. Fourier, Grenoble), and the OA spectrum at "Laboratoire de Chimie Physique Moléculaire" (ULB) again, using the OA cell build by Hadj Bachir and Huet, at the "Laboratoire de Physique des Lasers, Atomes et Molécules" (USTLille) [758], and their co-workers, in all cases.

Fourier transform (FT) spectroscopy has also been used in the present general framework, and provided interesting case studies, some of which are mentioned hereafter, in Section IV.B. Particular emphasis is set on the recent FT–jet overtone investigations performed at ULB.

It is most relevant to highlight that, historically, overtone spectroscopy was first achieved using grating spectrometers equipped with long absorbing path cells, since the 1930s [748–757]. These apparatuses have provided high-quality absorption spectra in the photographic infrared of C_2H_2 [756,757], CHD_3 [752], and of heavier atom molecules such as CO [748], CO_2 [750], and N_2O [749]. As an example, the weakest overtone bands of CO_2 [750] were observed by means of a 22-m absorption cell equipped with a White mirror allowing up to 250 round trips corresponding to a pathlength of 5500 m. This required a sample of several cubic meters of gas to fill the cell, which therefore limits the use of such equipment to low-cost gases. However, the method proved to be so sensitive than some of these previous observations have not yet been reproduced by laser techniques.

As a snapshot of some of today's available experimental means, Fig. 68 [758] presents the same spectral range in acetylene, recorded using three different experimental setups.

B. Fourier Transform Spectroscopy (FTS)

1. Introduction

The development of high-resolution FT spectroscopy (HRFTS) started with the pioneering work of P. and J. Connes and the construction of homemade Michelson-type step-scanning instruments [759–765]. Such development became feasible because of large improvements in computing techniques, related essentially to the advent of the Cooley–Tukey algorithm [694,766,767] and the availability of more powerful computers. State-of-the-art results obtained in the field of astronomy [768,769] and metrology [770–772] significantly contributed to settle HRFTS in the scientific community. The next decisive step in HRFTS came with the availability of commercial fast-scanning instruments, dedicated to fundamental spectro-scopic research investigations. That step initiated a worldwide expansion of the technique.

Nowadays, high-resolution FTS either in absorption or in emission is used as a basic means of investigation over a wide spectral range. HRFTS in absorption offers many well-known combined advantages over most laser-type investigations, including broad spectral coverage, absolute intensity measurements, and energy calibration. State-of-the-art achievements are actually being reported using FTS from the far infrared (see, e.g., Refs. 505 and 773–783) to the ultraviolet (see, e.g., Refs. 784–788) and vacuum ultraviolet (see Ref. 789 and references cited therein) regions, including the infrared and near-infrared ranges (see some recent examples in the literature [453,571,730,734,790–816] and most references in this section). More specifically, individual line position measurements using FTS have been reported with a stated accuracy as high as $\pm 1 \times 10^{-6}\,cm^{-1}$ in the 10-μm region of N_2O [817]. Similarly, the best of the recent measurements of the strengths of individual absorption lines shows precisions that are often better than 1%, while estimates of their accuracy are in the 5% range ([e.g., 726,818] and the Hitran database [683]). The technique still evolves. Developments include, for example the introduction of time resolution (see, e.g., Refs. 819,820 and 821–825 and references cited therein) and spectral simplification using supersonic jets (see below), collisional cooling cells [e.g., 826–828] and absorption cells operating at varying temperatures below room temperature (see, e.g, Ref. 829 and references cited therein).

HRFTS is also used extensively in emission. Infrared, near-infrared, and visible emission spectroscopy of stable and transient molecules led to the analysis of electronic transitions (see review on laser-induced fluorescence studies [830] and recent examples including $CuCl_2$ [831], NbO [832], O_2 [833], TiCl [834], ZrCl [835], and HO_2 [836]) and vibration–rotation transitions (e.g., OH [740], CO_2, and CO [837,838], H_2O [839,840]). Pure rotation spectra have also been recorded using far infrared emission spectroscopy (e.g., H_2O [841], LiH and LiD [842]).

The expansion of HRFTS also stimulated the development of efficient algorithms dedicated to either the manipulation of the raw FT data (for example, *The Giessen Program Package* [843]) or the measurement of line parameters in the FT spectra. Computer programs belonging to that latter category include *Decomp* [844]; *Intmet* [702,845,846], which later evolved to *Intbat* [847]; and *Spectra* [848]. They all treat one spectrum at a time. Recently, Benner et al [849] developed an algorithm allowing the measurement of line parameters from the simultaneous fitting of multiple spectra, recorded in different experimental conditions.

The technique and instrumentation are reviewed in many books (see, e.g., Refs. 694 and 696) and have been extensively discussed in the literature. Examples include the study of the effects of phase errors [850], optical misalignments (see Refs. 851 and 852 and references cited therein), sample blackbody emission [853,854], and nonlinearity in the detecting system (see Ref. 855 and references cited therein). We intend here only to highlight some of the features related to overtone spectroscopy. In any case, as pointed out in the introduction to this section, the emphasis is set almost exclusively on *high-resolution*-type investigation involving as highly excited vibrational levels as possible. Yet, the literature coverage in this section (IV.B) is not meant to be exhaustive and the results obtained of ULB are highlighted, in particular.

2. Overtone Spectroscopy Using FTS

The most challenging aspect of overtone spectroscopy in absorption using Fourier transform spectrometers is the weakness of the targeted bands. Indeed, HRFTS lacks sensitivity compared to most laser-type investigations as exemplified in Fig. 68. As briefly discussed in the introduction to this section, FT spectrometers, just as conventional spectrographs, require the use of long absorption paths and high-pressure samples. Multiple reflection cells providing absorption paths up to several kilometers, usually of White type [856], have thus been added to Fourier transform spectrometers [e.g., 857–860]. The sensitivity that can be obtained with such equipment is illustrated in Fig. 69. It presents a portion of the FT vibration–rotation spectrum of H_2O, recorded using a 50-m baselength longpath multiple-reflection absorption cell set to 603-m total absorption pathlength [861]. The unapodized resolution was $0.06\,cm^{-1}$. Si and GaP diodes were used together with various optical filters to record the spectral range from 8000 to $27,000\,cm^{-1}$. The signal-to-noise ratio (S/N), achieved after some 12 h of recording time, is of the order of 2000, allowing very weak lines to be observed (the strongest absorption in the region presented in Fig. 69 is about 7%).

The improvement in S/N of transmittance FT spectra such as those presented in Fig. 69 is due to the multiplex advantage [694], which consists in measuring simultaneously N spectral elements as is done with FTs. It results in an increased sensitivity by a factor $\sqrt{N}$ compared to a sequential single-element scanner. A further means of increasing the sensitivity of HRFTS involves the reduction of the noise due to statistical fluctuations in the rate of arrival of the photons on the detector ("photon noise"). Indeed, for n photons reaching the detector in a given time interval, the photon noise is equal to $\pm\sqrt{n}$ [695]. So, longer observation times lead to better S/N ratios. Fast scanning FTs implement this practically by allowing interferograms to be accumulated. For example, the visible overtone spectrum of $^{12}C_2H_2$ presented in Figure 70 [862] is the result of the coaddition of 6900 interferograms, recorded over 96 h. As usual, filtering the absorption source to window the region of interest significantly helped the improvement of the quality of the spectrum.

With its high sensitivity, HRFTS in emission can be used to detect transitions arising from very high vibrational levels at high resolution, in the "normal" infrared range. In CO_2 excited in electric discharges or produced in $CH_4 + O_2$ flames, for instance, Bailly and co-workers was able to monitor $\Delta v_2 = 1$ and $\Delta v_3 = 1$ transitions involving vibrational levels as high as $v_2 = 4$ [863] and $(2v_1 + v_2) = 5$ [838]. High-resolution FT resolved emission was also recorded in vibrationally excited acetylene, leading to

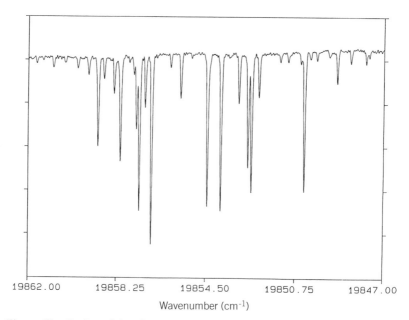

19862.00   19858.25   19854.50   19850.75   19847.00

Wavenumber (cm^{-1})

Figure 69. Portion of the vibration–rotation spectrum of H_2O recorded with the high-resolution portable Bruker IFS120M Fourier transform spectrometer of the Laboratoire de Chimie Physique Moléculaire of the Université Libre de Bruxelles and the 50-m baselength longpath multiple reflection absorption cell of the Groupe de Spectrométrie Moléculaire et Atmosphérique of the Université de Reims [861]. The H_2O pressure was 14 torr; the temperature, 291 K; the absorption pathlength, 603 m; and the unapodized resolution, 0.06 cm^{-1} (unpublished data reported with permission from the authors).

the observation of the fundamental band v_3 only [864]. A surprising lack of intensity alternation was observed in the latter spectra and attributed to emission/reabsorption cancellation effects. Emission was also reported more recently in acetylene using a high-resolution FT interferometer, following overtone levels excitation with a CW Ti : Sa laser [865]. This technique is most promising and should provide new insight in highly excited vibration–rotation levels previously experimentally inaccessible.

Going back to absorption spectroscopy using FTs, hot bands can provide another way to access vibrationally excited levels, otherwise forbidden through one-photon transitions. These hot bands can also bring relevant information on anharmonic resonances through the observation of zeroth-order dark states [602]. Hot bands are, however, often of the type $v_x + v_y - v_y$. In other words, they are weaker than and overlapped by the main band, v_x. Their high-resolution analysis is a problematic task in larger species. Herman and co-workers recently reported on two different

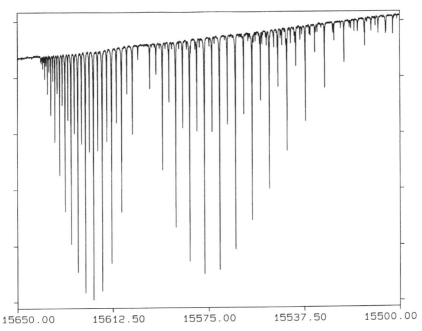

15650.00   15612.50   15575.00   15537.50   15500.00

Figure 70. Portion of the vibration–rotation spectrum of $^{12}C_2H_2$ in the visible range, recorded with a high-resolution FT (from Herman and co-workers, adapted from Ref. 862) (energy scale in cm^{-1}).

procedures to selectively highlight hot bands in a way that allowed their fine structure to be investigated. A first approach consisted into simulating the full fine structure of the main band, after performing its detailed rotation analysis, and dividing the simulated spectrum from the observed spectrum, also containing the hot bands. This method was applied with success to pyrrole (C_4H_5N) [866]. In other investigations, two experimental FT spectra recorded at different sample temperatures were divided from each other, selectively highlighting the hot bands. In that latter case, the experiments were achieved under supersonic jet conditions and drastic simplification was brought to the fine structure of the hot band. An example of the result of this procedure is provided in Fig. 71 [867]. In many respects, the achievements are comparable to those obtained by Reuss and co-workers in a jet, using a sophisticated and powerful collision-enhanced double resonance laser procedure [868–871].

The coupling of supersonic jets with FTS was achieved long ago, by Snavely et al. [872] and later promoted more systematically by Quack and co-workers [544]. The technique has now generated many contributions in

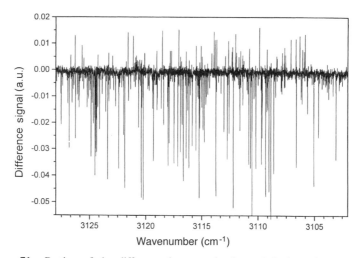

Figure 71. Portion of the difference between the jet-cooled absorption spectrum of $^{12}C_2H_4$ in the region of 3000 cm^{-1} recorded with and without a heated slit. The hot band points upward. Unpublished data from M. Hepp and M. Herman recorded under experimental conditions identical to those detailed in Ref. 867.

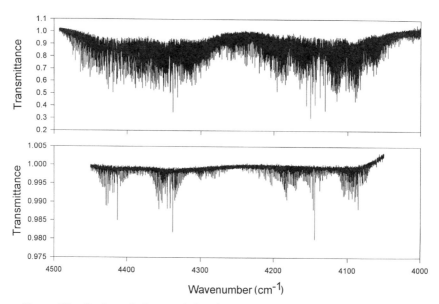

Figure 72. Portion of the near-infrared spectrum of ethane recorded under room temperature (upper) and under FT-jet cooled (lower) experimental conditions (reproduced from Hepp and Herman [894], with permission from Taylor & Francis).

260

the literature [544,586,721,872–916], as also summarized and discussed by Arno and Bevan [917]. The lack of sensitivity however confined the experiments to the study of fundamental bands, in most cases. The development of a supersonic jet cell allowing for longer absorption path, by Herman and co-workers, recently promoted the technique to the overtone range. The longer path was produced, either using a multinozzle [890] or a slit nozzle [891], totaling in both cases some 16 cm in absorption pathlength.

Despite the limited FT instrumental resolution, linewidths smaller than the room-temperature Doppler width could be recorded using the slit jet [895]. Using this experiment, overtone and combination transitions were recorded in various species, including methane [586], ethylene [891], and ethane [893,894], allowing the fine structure to be investigated. Typical rotational temperatures reached with this experimental setup are 30 to 50-K. As an example, Figure 72 presents the comparison between the room-temperature and the jet-cooled spectra of ethane in the range 4000–4500 cm^{-1}. Figure 73 details the highest energy band recorded so far using that equipment, in ethylene.

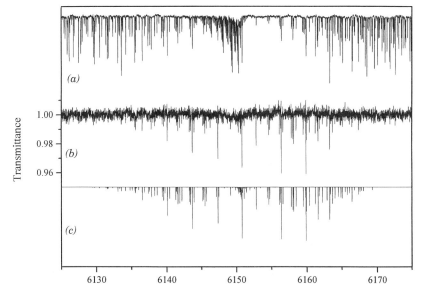

Figure 73. Highest-energy vibration–rotation band in ethylene recorded using FT jet-cooled experimental conditions (*b*), and comparison with room temperature (*a*) and simulated jet-cooled (*c*) spectra (reproduced from Georges et al. [891], with permission from Taylor & Francis) (energy scale in cm^{-1}).

C. Laser Overtone Spectroscopy

The decrease of overtone absorption intensity at high energy is a stringent limitation to the use of Fourier transform interferometers in the overtone range. Grating spectrometers equipped with long absorbing path cells have been developed since the 1930s [748–757], as already pointed out in the introduction to the present section.

The availability of lasers in the near-infrared and visible ranges makes them powerful tools to increase the detectivity and give access to very excited vibrational states. They appear ideally complementary to FT spectroscopy. In this section, we discuss the different laser-based techniques currently used to give access to the rovibrational energy levels of molecules in the gas phase. We consider in more detail the cavity ringdown spectroscopy (CRDS) and the intracavity laser absorption spectroscopy (ICLAS) as they are not yet as widely available in laboratories as other techniques such as frequency modulation diode laser (FMDL), optoacoustic (OA), and optothermal (OT) spectroscopy. The performances of the different techniques are illustrated in the present section by reviewing the results obtained in the field of overtone spectroscopy in the electronic ground state. Particular emphasis is put on the fine structure of overtone and combination bands in small molecules. Finally we provide some comparison between the different methods.

It is worth recalling that the minimum detectable absorption coefficient $\alpha(\tilde{\nu})$ is given by

$$\alpha_{\min}(\tilde{\nu}) = \frac{1}{\ell}\ln\left[\frac{I_0(\tilde{\nu})}{I(\tilde{\nu})}\right] \cong \frac{1}{\ell}\frac{\delta I}{I_0} \qquad (4.1)$$

where ℓ is the absorption pathlength, $I_0(\tilde{\nu})$ the reference intensity (i.e., the transmitted intensity in the absence of absorption) and $I(\tilde{\nu})$ the transmitted intensity. In order to increase the sensitivity (i.e., the S/N), the absorption pathlength (i.e., the signal) should be increased and the minimum value of the relative variation of the intensity (i.e., the noise) should be decreased. FMDL, CRDS, and ICLAS are methods, that use different ways to fulfill these two requirements.

However, we should underline the fact that the fundamental limitation of any absorption approach is determined by the quantum noise, specifically, the statistical fluctuation of the photons distribution. This common limitation is simply given by $\sqrt{n}/n$, where n is the number of photons incident on the detector during the measurement. This quantum noise limit has been achieved in some experiments, but instrumental noise generally limits the sensitivity of most of the setups currently used for spectroscopy. In a rough simplification, one can say that ICLAS and CRDS favor the increase of the signal while FMDL favors the minimization of the noise. Indeed, the high sensitivity of FMDL spectroscopy is mostly based on the frequency modulation of the laser diode, which allows a lock-in amplifier detection while CRDS and ICLAS are two highly sensitive techniques based on a virtual increase of the absorption pathlength. Similar to FT spectrometers, FMDL spectrometers are frequently associated with multipass cells. The maximum pathlength and hence the optimal sensitivity, which can be achieved by using a multipass cell, depends on several factors: the reflectivity of the mirrors, the length of the cell, and the noise levels of the light source and the detector. Currently, pathlengths of the order of several tens or hundreds of meters are achieved with multipass cells, while absorption equivalent pathlengths of several kilometers are currently obtained by CRDS and ICLAS using either a "passive" optical cavity injected by a laser beam (CRDS) or an "active" laser cavity (ICLAS). In both cases, photons trapped inside an optical cavity during a time interval Δt interact with the gas sample filling the cavity over an equivalent absorption pathlength $\ell_{eq} = c\Delta t$. For a typical value of $\Delta t = 50\,\mu s$, an absorption equivalent pathlengths of 15 km is achieved by both techniques. In CRDS, the photon lifetime in the cavity is increased by using very high-reflectivity mirrors while in ICLAS, the gain medium creates and maintains by laser amplification the photons in the cavity of a multimode laser.

Similar to conventional absorption, CRDS and ICLAS detect the attenuation of the transmitted radiation, namely, the difference δI of two large quantities, which is, in principle, not favorable for a high sensitivity. Optoacoustic and optothermal spectroscopies are based on the indirect detection of the absorbed photons. The excited molecules transfer by collision their energy to a microphone (OA) or to a bolometer (OT). These methods have the advantage of being, to a large extent, dark background methods and will be described below.

1. Frequency Modulation with Diode Lasers (FMDL).

a. Method. Frequency modulation is a standard method, in particular in microwave spectroscopy. The basic idea is to modulate the signal in a frequency domain where the instrumental noise is weak i.e., in the high-frequency domain; (in the megahertz range or more) and to use a phase-sensitive detector centered at the modulation frequency. In the case of diode lasers, the optical frequency is easily modulated through the modulation of the diode current. There are a large variety of approaches for frequency modulation that can be classified according to the amplitude and frequency of the modulation. Wavelength modulation spectroscopy uses large amplitude and relatively low-frequency modulations. The laser spectrum shows numerous close sidebands around the central wavelength that simultaneously sample the absorption line. In contrast, Frequency modulation spectroscopy uses low-amplitude and high-frequency (100 MHz–10 GHz) modulation, which leads to well separated sidebands. A detailed study of these techniques falls out of the scope of the present review, and the reader may refer, for instance, to Refs. 918–930 for more detail. The development of diode lasers in many spectral ranges has made this field of research particularly active for applications to atmospheric trace-gas monitoring [921,927–930].

As a consequence of the phase-sensitive detection, the noise level is subsequently reduced by limiting the detection to a narrow bandwidth centered at the frequency Ω of the laser modulation or at one of its overtone frequencies $n\,\Omega$. The detected signal obtained by scanning the central wavelength of the laser is then the first derivative of the absorption signal or the derivative of the nth order in the case of the detection at a frequency $n\,\Omega$. This derivative method, to a very large extent, gets rid of the laser intensity fluctuations and the smoothly frequency dependent absorption background. It has allowed shot-noise-limited measurements to be performed [918,926,931–935] even with noisy lasers [936,937]. It is also possible to use two close frequencies, Ω_1 and Ω_2, to modulate the laser frequency. The advantage of this two frequency modulation spectroscopy is that the absorption signal is detected at the frequency difference $\Omega_1 - \Omega_2$ [919,927,938,939] which falls in a frequency domain where the detection is easier.

b. Application of FMDL to Spectroscopy. As mentioned above, FMDL is a powerful tool for trace detection, and a number of detectors using FMDL are now commercially available. In absorption spectroscopy, the high sensitivity and high resolution of FMDL made it a powerful method that has provided an important contribution to high-resolution spectroscopy, mostly in the infrared and near-infrared ranges (see, e.g., Ref. 917). Tunable diode lasers, scanned by varying the temperature and frequency modulation, are routinely used associated with a multipass cell [939,940] and a reference cell. As an example, in the near-infrared range the Doppler-limited absorption spectrum of $^{12}C_2H_2$ has been investigated by FMDL using DFB [941,942] or GaAs laser diodes [943]. The same method was applied to the overtone spectrum of C_2D_2 near 0.82 μm [939].

The different high-sensitivity laser absorption techniques are compared below. In terms of sensitivity and spectral resolution, the quality of the FMDL measurements applied to overtone spectroscopy clearly illustrate the high performance of this technique. Among the limitations, the reduced scanning range allowed by laser diodes and the possible occurrence of mode hops during the recording must be mentioned.

Finally, it is worth noting that the frequency modulation technique has been applied to other types of lasers such titanium–sapphire lasers. The output beam is modulated by an electrooptic element in the megahertz (MHz) range, which is higher than the upper limit of the laser amplitude noise [944,945]. Compared to FMDL, this tone-burst spectroscopy takes advantage of the exceptionally wide tunability of the titanium–sapphire laser [946,947]. It has been associated with a cell cooled at 77 K with Stark modulation to rotationally assign the $3v_1 + v_3$ band of methane [945].

2. Cavity Ringdown Spectroscopy (CRDS)

a. Pulsed CRDS. Until recently, CRDS was performed by injecting a small fraction of the intense laser pulses through one of the high-reflectivity mirrors (typically $R = 0.99999$) of an optical cavity filled by the absorber.

Figure 74 describes the principle of pulsed CRDS, assuming a cavity with two mirrors of equally very high reflectivity facing each other, separated by a distance ℓ. When a short laser pulse, of typically a few nanoseconds (ns) in

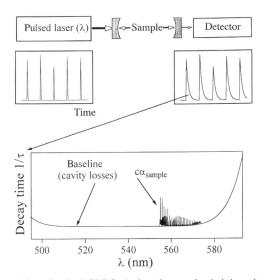

Figure 74. Principles of pulsed CRDS. A short laser pulse is injected through one of the high-reflectivity mirrors of the passive optical cavity filled with the absorber. The detector records the output signal, which is a series of laser pulses separated by the round-trip time. The envelope of this ringdown signal is an exponential decay depending on the cavity loss rate. When the wavelength of the pulsed laser is scanned, the sample absorption causes a rapid increase in the ringdown time which can be easily distinguished from the smooth baseline, due to the cavity losses (transmission of the mirrors, scattering...).

duration, is injected through one of the two mirrors, a fraction of the photons is trapped inside the cavity, making thousands of round trips. The output signal collected by a photomultiplier or a photodiode is then a periodic series of possibly interfering laser pulses separated by the round-trip time $2\ell/c$ (a few ns). The envelop of this periodic structure is an exponential decay with a time constant τ, called the *ringdown time constant*, fixed by the cavity loss rate over a round trip [948].

$$\frac{1}{\tau} = (T + L + \alpha_g(\tilde{v})\ell)\frac{c}{\ell} = \frac{1}{\tau_0} + c\alpha_g(\tilde{v}) \qquad (4.2)$$

where T is the transmittivity (of the order of 10^{-5} or less), L is the losses including absorption by the dielectric coating and scattering on the surface, and $\alpha_g(\tilde{v})$ is the sample absorption coefficient.

Note that the diffraction losses due to the mirrors are usually negligible as the light beam remains collimated along the cavity axis. The energy conservation of the injected energy requires

$$R + T + L = 1 \qquad (4.3)$$

where R is the reflectivity. Since the cavity ringdown time is usually much longer than the round-trip time, the exponential envelope is detected with a long detector time constant that smoothens the fast structure due to the round trips. By scanning the laser wavelength, the absorption coefficient $\alpha_g(\tilde{v})$ is simply obtained by dividing the decay rate $1/\tau$ by the velocity of light [see, (4.2)]. Indeed, the contribution of the mirror losses $1/\tau_0$ to the decay rate depends smoothly on the wavelength and gives the baseline of the absorption spectrum. For a rotationally resolved absorption spectrum, the narrow absorption features are easily distinguished against the baseline profile (see Fig. 74). In the case of broad unresolved absorption bands, the sample absorption may be obtained from the difference in decay rate measured with and without the sample in the cavity [949].

To our knowledge, the first experiments using cavity ringdown decays were performed in 1980 by Herbelin et al. [950] to measure the reflection coefficient of high-reflectivity mirrors. However, the first application to molecular spectroscopy was performed by O'Keefe and Deacon [951] in 1988. One of the key advantages of pulsed CRDS is that pulsed laser sources are available from the 10-μm infrared region (OPO lasers) [952] up to the UV range (200 nm by harmonic generation with pulsed lasers). However, the optimal sensitivity is achieved in the near-infrared and visible ranges where the highest reflectivity mirrors are available. A detection limit as low as 3×10^{-10}/cm was achieved with standard pulsed dye lasers [948]. For a complete understanding of CRDS, the cavity modes that act as a regularly spaced frequency filter, must be considered [948,953,954]. Indeed, a laser pulse usually excites several of the transmission modes of the cavity, and mode beating may generate oscillations in the ringdown time [955]. In principle, a Fourier transform–limited pulse could selectively excite one cavity mode, only if the spectral linewidth of the pulse is narrower than the free spectral range. This, in principle, would allow a resolution limited by the mode width but, in practice, requires an accurate mode matching and the controlled tuning of the cavity length when the pulsed laser is scanned. Some interesting solutions have been proposed very recently [954,956,957]. In the following section, we describe how to use similar principles and achieve very high resolution by using a single frequency CW laser (CW-CRDS).

A spectral resolution higher than that provided by pulsed lasers is generally needed for supersonic jet spectroscopy or, even for Doppler–limited spectroscopy. Meijer et al. [954] and Romanini et al. [949,958,959] have recently succeeded in coupling narrowband CW laser sources to ringdown cavities. The possibility of implementing low-cost CW diode laser sources for CW-CRDS is particularly promising for trace detection.

b.   CW-CRDS. When a single frequency CW laser is tuned to one of the cavity resonances, an important buildup of the intracavity power occurs and a strong ringdown signal may be recorded after a quick interruption of the injected laser beam. However, the frequency locking of the passive cavity to the incoming laser is extremely difficult to achieve, in particular because of the frequency jitter of the laser. A stabilization of a free-running laser diode to only a few MHz, in the presence of strong reflections of a ringdown cavity, was very recently achieved using an acoustooptic modulator [960]. Another more elegant solution was proposed by Romanini et al. [949,958,959] that is both simple and efficient. It is based on the periodic coincidences of the laser frequency with the cavity modes, when the length of the cavity is modulated. It is worth mentioning that the original idea can be found in the pioneering work of Anderson et al. [961], who exploited the accidental coincidences of a He–Ne laser with cavity modes to measure the mirrors reflectivity from the free ringdown signal. However, these authors could not scan the laser and did not mention the potentials of the method for spectroscopy.

Figure 75 presents a simplified scheme of the CW-CRDS setup which has been used with both a ring dye laser [949] and an external cavity tunable diode laser [958]. One of the high reflectivity mirrors was mounted on a piezoelectric transducer to modulate the cavity length, and therefore, the frequency of one of the cavity modes around the laser line. When sufficient buildup is transmitted by the cavity and measured by a photodiode, a threshold circuit interrupts the laser with the help of an acoustooptic

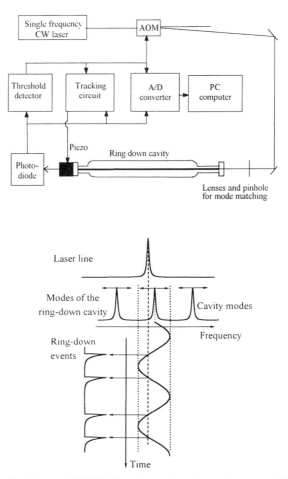

Figure 75. Principles of CW-CRDS (adapted from Romanini et al. [959]). A single-frequency CW laser is injected into a linear ringdown cavity with two supermirrors. An acoustooptic modulator (AOM) interrupts the laser beam when the threshold circuit detects a sufficient buildup. The laser is mode-matched to the TEM_{00} modes of the ringdown cavity. The cavity length is piezoelectrically modulated to cause the frequency of its modes to oscillate around the laser line. The modulation amplitude is less than the mode spacing, and ringdown events are observed each time the mode passes through the laser line: twice per period. A similar procedure as for pulsed-CRDS is then followed to obtain the absorption spectra from the ringdown signals.

modulator. The ringdown signals are digitized, recorded, and averaged. The laser wavelength is scanned with typical speeds of 10 GHz per minute [958]. Figure 76 shows the diode laser ringdown spectrum of the $6v_3$ band of N_2O recorded at a pressure of 39.5 hPa with a noise-equivalent absorption

LASER OVERTONE SPECTROSCOPY 269

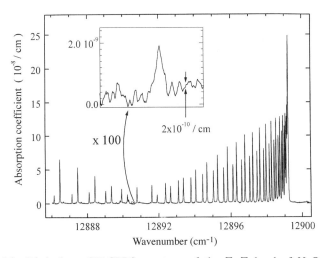

Figure 76. Diode laser CW-CRDS spectrum of the Σ–Σ band of N_2O centred at 12,891.15 cm^{-1} (adapted from Romanini et al. [958]). The pressure was 39.4 hPa. As shown by the zoom, the RMS noise-equivalent absorption coefficient is about 2×10^{-10}/cm. The upper state of the observed transition belongs to the $\{N, k\} = \{24, 0\}$ polyad, as defined in Section III.C [596,612], with a dominant fraction of 92% corresponding to the $(v_1 v_2^{l_2} v_3)^0 = (00\,^\circ 6)$ bright state, with modes 1–3 corresponding primarily to the NO stretch, the bend and the NN stretch, respectively.

coefficient of 2.10^{-10} cm^{-1} [958], which compares favorably to previous ICLAS results [596].

A detection scheme based on the measurement of the magnitude of the phase shift that an intensity-modulated CW laser experiences on passing through an unstabilized optical cavity was proposed by Engeln et al. [962] and tested on the weak $b(^1\Sigma_g^+)v = 2 - X(^3\Sigma_g^-)v = 0$ transition of $^{18}O_2$. However, it seems that up to now this interesting alternative to the aforementioned CW-CRDS method [949,958,959] has not provided comparable results.

The excellent performance obtained with low-power tunable diode lasers (of external cavity or distributed feedback types) shows that compact and affordable spectroscopic devices with extreme sensitivity are now available. The combination of CW-CRDS with diode lasers is very promising in the 1–2-μm range for applications in spectroscopy and trace detection. The very high resolution achieved by CW-CRDS was illustrated by the sub-Doppler spectroscopic investigation of NO_2 expanded in a supersonic slit jet expansion [958]. A residual Doppler width of 125 MHz (HWHM) in the jet was measured. CW-CRDS also appeared well suited for investigating very broad spectral features; broadband [949] and even continuum absorption intensities were measured by subtracting the zero absorption baseline obtained with an empty cavity since an excellent reproducibility of the baseline can be achieved using a mechanically stiff apparatus.

In the preceding description, we assumed a perfect mode matching, that is, no excitation of high-order transverse cavity modes that have shorter decay times. Mode matching prevents significant excitation of the transverse modes and this makes easier the analysis of the ringdown signal. This is not the case for the phase shift CRDS [962], which monitors the phase shift introduced by the cavity when CW amplitude-modulated laser light is coupled through it. The theoretical interpretation of the ringdown signal due to the high spectral density of transverse modes is not obvious.

c.   Capabilities of CRDS. CRDS has experienced a rapid growth recently in a wide range of applications as illustrated later, mainly with pulsed CRDS.

1.   SPECTROSCOPY. In the UV range, the absolute transition strength of the spin-forbidden Cameron band in CO near 206 nm has been measured by using the antistoke Raman shifting of a frequency doubled dye laser [963]. Pulsed CRDS has been performed with jet-cooled metallic clusters near 400 nm [964–968]. The $^2\Pi-\tilde{X}^2\Pi$ electronic transition of C_6H generated in a hollow cathode discharge was detected near 500 nm [969], while the kinetics of phenyl radical reactions has been studied in the same range [970,971]. New Herzberg bands of O_2 and improved spectroscopic data have been obtained from the CRDS spectrum near 250 nm [972,973], and the predissociation lifetimes of SH were studied from CRDS spectra near 300 nm [974]. To our knowledge the only extensive overtone CRDS studies reported so far were performed on the HCN molecule between 17,500 and 23,000 cm^{-1} [948,975–977] and on the $\tilde{A}^1A''-\tilde{X}^1A'$ transition of HNO between 16,100 and 18,500 cm^{-1} [978,979]. In the latter study, the predissociation dynamics of HNO were analyzed on the basis of the measurement of the lifetime-broadened linewidth.

2.   DIAGNOSTICS. As a diagnostics tool, CRDS has been successfully applied to measure low concentration of C_2H_2 in a cell [980]. In flames CRDS was used to monitor CH_3 near 3 μm using an OPO [952], H_2O [981] in the near infrared, HCO [982] in the visible range, and OH [961,983,984] and CH_3 [985] in the UV. In discharges, the absolute density of transient species such as molecular ions [986] and SiH_2 [987] has been measured. These latter studies demonstrate that CRDS can be successfully performed in corrosive environments. Dielectric mirrors are generally sufficiently resistant against corrosive agents and are usually protected from dirt deposits by flowing a buffer gas across their surface [961,987]. In the 3-μm region, infrared CRDS

was employed for direct measurement of water cluster concentration in a pulsed supersonic expansion [988]. Laser-desorbed diphenylamine was also detected by CRDS near 308 nm [989].

3. Intracavity Laser Absorption Spectroscopy (ICLAS)

a. Principles. When an absorption cell is inserted inside a single-mode laser cavity, the high intracavity power allows the detection sensitivity to be considerably enhanced when the absorbed photons are monitored by laser-induced fluorescence or OA spectroscopy (see the following section). Another possibility is to use the high sensitivity to intracavity losses of the single-mode laser output power when it is operated near the threshold. The reader is referred to Ref. 681 for calculation and discussion of sensitivity enhancement in single-mode operation.

We discuss here a different method based on the *multimode* oscillation of the laser, known as intracavity laser absorption spectroscopy (ICLAS). In this case, the increased sensitivity is a result of the competition between the modes sharing the same amplification medium in a cavity free of narrowband, frequency-selective optical elements. Any active medium with broad gain profile such as dyes, color centers, vibronic solid-state media (Ti:sapphire, Cr^{4+}:YAG (yttrium aluminum garnet), Cr^{4+}: LISAF, etc.), rare-earth-doped glass such as Nd:glass and to a lesser extent doped fibers and semiconductor amplification media, are well suited for ICLAS (see examples below). Since its introduction in the 1970s by Sviridenkov [990–993], Keller [994,995], Toschek [996], and Atkinson [997], ICLAS has been the subject of continuous interest, including for the study of the dynamics of a multimode laser spectrum [998–1003].

The most common experimental arrangement for ICLAS is presented in Fig. 77. The first acoustooptic modulator (AOM1) allows a square modulation of the pumping of the amplification medium (e.g., a dye) at a high repetition rate (about 1 kHz). After a given delay from the rising edge of the pump pulse, called the *generation time*, t_g, the second acoustooptic modulator deflects the laser beam during a short time interval $\Delta t (<< t_g)$ to the entrance slit of a high-resolution spectrograph. The laser beam is dispersed, and a portion of the absorption spectrum of the sample cell inserted in the laser cavity is observed superimposed on the broad spectrum of the laser spectrum, and recorded as a whole by a photodiode array or a charge-coupled device (CCD). The time-resolved intracavity spectrum is shown in Fig. 78 for different generation times.

The observed kinetics of a broadband laser can be satisfactorily accounted for by the results of a series of theoretical works (see, e.g., Refs. 991,999, and 1004–1009). After saturation of the gain medium and assuming that the gain profile centered at $\tilde{v}_0$, can be approximated by a parabolic function of

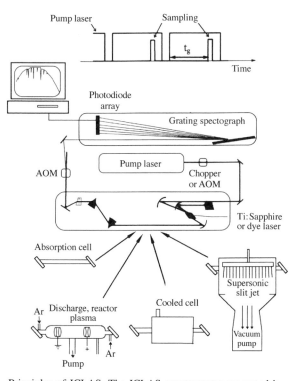

Figure 77. Principles of ICLAS. The ICLAS spectrometer presented here is based on a Ti:sapphire laser whose central wavelength can be adjusted by two prisms. As shown by the time sequence, two acoustooptic modulators are used to control the generation time. The first one (which can be replaced by a mechanical chopper) chops the pump laser with a typical time interval of 500 μs. The second one, AOM2, sends the laser beam into the spectrograph during a short time after a delay t_g from the beginning of the laser generation. A photodiode array (or a CCD) records a portion of the spectrum. The signal is then averaged over a few thousand spectra by the microcomputer. In the long arm of the laser cavity, different types of samples can be inserted, such as absorption cells (eventually cooled), reactors, and supersonic slit jet expansions.

bandwidth $\Delta \tilde{v}_0$, the following expression is obtained for the time evolution of the number of photons, M_q, in the mode q:

$$M_q(t) = M_q(0)\left(\frac{t_g}{\pi T}\right)^{1/2} \exp\left[-\left(\frac{\tilde{v}_q - \tilde{v}_0}{\Delta \tilde{v}_0}\right)^2 \frac{t_g}{T}\right] \exp\left[-\alpha(\tilde{v}_q)\frac{\ell}{L} c t_g\right] \quad (4.4)$$

where $M_q(0)$ is the initial number of photons seeded in the mode q, T is the mean photon lifetime, and $\alpha(\tilde{v}_q)$ is the absorption coefficient of the qth mode

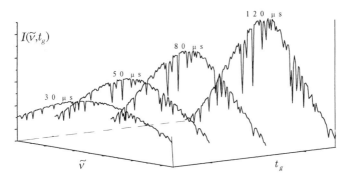

Figure 78. Time evolution of an ICLAS spectrum. The absorption lines due to the intracavity absorber appear superimposed on the laser spectrum. The baseline is nearly Gaussian with a width decreasing with the generation time t_g, and the absorption lines have an intensity increasing linearly with the generation time t_g. The total width of the laser spectrum is of the order of $100\,\mathrm{cm}^{-1}$ for a dye or Ti : sapphire laser at a generation time of $100\,\mu\mathrm{s}$.

due to the intracavity sample. This analytical expression has been deduced from a system of rate equations coupling the gain medium and the number of photons in the cavity. The second exponential is the Beer–Lambert absorption law with an equivalent absorption pathlength given by

$$\ell_{eq} = \frac{\ell}{L} c t_g \qquad (4.5)$$

where ℓ/L is the filling ratio of the sample inside the cavity, specifically, the optical length of the sample cell (ℓ) divided by that of the resonator (L). With a standing-wave cavity, typical generation times are of the order of $300\,\mu\mathrm{s}$. The equivalent absorption pathlength is then of the order of 45 km for a 50% filling ratio. If the noise on the baseline is around 1%, a line corresponding to an absorption coefficient $\alpha(\tilde{v})$ of $2\ 10^{-9}\,\mathrm{cm}^{-1}$ can be detected and measured.

The first exponential term in (4.4) describes the time evolution of the laser after saturation: the broad Gaussian baseline is narrowing as $1/\sqrt{t_g}$ while its amplitude increases as $\sqrt{t_g}$ since the total laser intensity has reached its stationary value. It is worth noting that for dyes the gain saturates very quickly (within a few microseconds or less) while for solid-state laser media such as Ti:Sa or Nd:glass, the excited-state relaxation is slower than the photon lifetime T, leading to large relaxation oscillations before a steady state of the total laser intensity is achieved. Even in this case, it has been shown that the linear dependence of the equivalent pathlength with the generation time is not affected [1000–1002]. Solid-state lasers may therefore be used for quantitative intensity measurement in ICLAS. Details concerning adequate experimental systems and data acquisition can be found in Refs. 570 and 1010–1014. Pulsed lasers have frequently been used, but accurate absorption measurements require the careful control of the delay between the beginning of the laser generation and the observation [1015], which can be achieved using CW lasers operated intermittently (long pulses) only.

b. Sensitivity Limitations. The agreement between line intensity measurements performed either by ICLAS or by conventional multipass techniques has been checked for very weak absorption lines of NH_3 [1016], H_2O [1014,1015], O_2 [1017–1019], and CH_4 [1014,1016,1020]. As shown by Stoeckel et al. [1015], the accuracy of such measurements by ICLAS relies on a careful control of the generation time, which is almost impossible when using a pulsed laser as the intracavity absorption is integrated over the total duration of the pulse. The line intensity is usually calculated from the slope of the Naperian absorbance plotted versus t_g [see (4.4) and Refs. 558,1015,1017,1018, and 1020–1025 for details].

Depending on the amplification medium, several effects may limit the sensitivity. The laser dynamics governs the range of generation times over which the equivalence with a multipass cell is valid. The linear relation between ℓ_{eq} and t_g occurs typically up to several hundred microseconds or a few milliseconds. For longer t_g, ℓ_{eq} is observed to increase more slowly until saturation. In an ideal case, the individual modes of the broadband laser are coupled only to the gain medium and are free of direct mutual interaction. The ultimate theoretical limit of sensitivity is fixed by the spontaneous emission of photons in the laser modes [1007,1008,1026], which have been neglected in the rate equations leading to (4.4). Extra spontaneous photons emitted during the laser generation induce a loss of spectral memory and a saturation of the spectral evolution [1000,1001,1008,1027]. When the laser is working just above threshold and in the absence of absorption, this steady state is predicted to be observed for t_g of the order of 1 s ($\ell_{eq} = 300,000$ km), for dye and for Ti:Sa lasers [1000,1001]. This fundamental limit has been approached [1028] using an unidirectional ring dye laser with compensated intracavity dispersion. Since the spectral width of the laser reduces to a few wavenumbers, such long generation times could, unfortunately, hardly be used in spectroscopy. Still, the acceleration of the saturation of the number of photons in the modes affected by absorption was recently experimentally observed in agreement with theory [1029].

More often, absorption growth is limited to shorter times, due to nonlinear mode coupling, and the laser reaches a steady state long before 1 s [1016,1030,1031]. Several mechanisms generate nonlinear mode couplings [1000,1030,1032]: stimulated Brillouin scattering, Rayleigh scattering for dyes, four-wave mixing [1019], spatial hole burning [1033], and Kerr-lens modulation of the broadband intracavity losses. In the case of a Ti:Sa ring laser, the linearity of the absorbance with t_g was observed up to a value for t_g of 3 ms and the mechanism responsible for nonlinear mode coupling was tentatively assigned [1000,1001] to mode locking due to Kerr-lens modulation of the broad intracavity losses. Furthermore, it has been recently established that, even in the absence of mode coupling mechanisms, the modes affected by absorption experience a quicker saturation than the others [1029].

Other technical limitations may arise from parasitic optical interference or birefringence giving fringes growing with t_g as the absorption features. These parasitic spectral modulations are observed especially with standing wave lasers where the fringes are due partly to spatially localized losses (dust over optical components or defect in solid-state amplification medium). Indeed, losses due to a diffusion center affect differently the modes having a node or a peak at their location. This type of parasitic spectral modulation can be eliminated using a traveling-wave unidirectional cavity [1028]. Ring configuration is also required to prevent nonlinear mode coupling by stimulated Brillouin scattering and spatial hole burning. Fringes may also arise from interference between beams partially reflected by intracavity optical surfaces. In practice, this type of parasitic modulation is reduced by using thick and wedged optical elements or by modulating the cavity length.

c. Seeding Noise. The rate equation model leading to (4.4) neglects random quantum forces [1007,1026] describing the random character of the fully random initial distribution [$M_g(0)$ in (4.4)] of spontaneous photons seeded in the cavity modes. At the beginning of the laser generation, the distribution of mode intensities is very noisy and it is required to average a great number of spectra to extract the absorption lines from the 100% mode fluctuations [1034]. In practice, several thousands of spectra, corresponding to the same generation time, are averaged within a few seconds to achieve a smooth spectrum.

For a given laser generation, the intensity of each mode follows the time evolution given by (4.4); thus the initial distribution is highly fluctuating from mode to mode, but each mode evolution keeps the memory of the spontaneous photons seeded at the beginning of the laser generation. This is the basis of a recent development proposed by Kachanov et al. in which the laser spectrum is sampled at two different times during the same generation [1035]. In this single-pulse ICLAS scheme, the two highly noisy spectra are thus recorded at t_g and $t_g + \Delta t_g$. The ratio of the second spectrum to the first, corresponding to an equivalent pathlength $\ell_{eq} = c\Delta t_g$, gives a spectrum free of the initial seeding noise. With this new detection technique, a noise

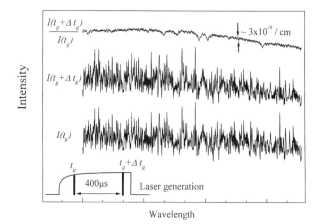

Figure 79. Two-time sampling ICLAS spectra from a single-laser pulse (adapted from Kachanov et al. [1035]). The spectra of a Ti:sapphire laser sampled at t_g and $t_g + \Delta t_g$ during the same generation are plotted with an offset for clarity. The ratio of the two highly noisy spectra removes the mode to mode fluctuation of the laser intensity. With $\Delta t_g = 400\,\mu s$, specifically, $l_{eq} = 12\,km$, a noise level of $2.10^{-9}\,cm^{-1}$ was achieved [1035].

level of $2.10^{-9}\,cm^{-1}$ was obtained (see Fig. 79) within $400\,\mu s$. Taking into account that 1000 spectral elements are recorded at the same time (multiplex advantage), a detection limit of $1.4 \times 10^{-12}\,cm^{-1}/\sqrt{Hz}$ was achieved [1035]. Highly sensitive spectroscopy or detection of species produced by a single event is thus made possible using single-pulse ICLAS.

d. Spectral Resolution and Calibration. The high-resolution spectrograph used to disperse the laser output determines the spectral resolution. At the exit of the spectrograph, the dispersed spectrum is detected by a multichannel detector (e.g., a CCD). The magnification of the optical system is chosen such that about 5 pixels sample the apparatus function. With a spectral resolution of $0.05\,cm^{-1}$ and a total number of pixels of 1024, for instance, a slice of $10\,cm^{-1}$ of the ICLAS spectrum is recorded as a whole within a few seconds. Grating spectrographs adapted to the spectral analysis of a laser beam and working in a high order of diffraction have been developed [570,1010–1012] and a practical resolving power as high as 1,300,000 has been achieved [1011]. Nevertheless, even with this high resolution, the laser modes are not resolved and the absorption lines appear on a smooth baseline resulting from the cavity modes broadened by the spectrograph instrumental function [1013].

   The central wavenumber of the laser spectrum is fixed/mainly by the gain profile of the amplification medium and the reflection profile of the laser mirrors. It can be adjusted over the laser gain profile by tilting a thin (several μm) intracavity beamsplitter acting as a Fabry–Pérot with low finesse [1010], by a set of intracavity prisms [1011] or by a Lyot filter. A detection technique based on a synchronized Fourier spectrometer has also been reported [1036,1037]. ICLAS with Fourier transform detection in principle provides all the advantages of FT instruments (high resolving power, direct wavenumber calibration, multiplex advantage) with, in addition, the possibility of removing parasitic interference modulation directly from the interferogram. However, the fluctuations of the laser intensity from one generation to the other may become a significant limitation in the overall performance.

In slit jet spectroscopy, the residual Doppler linewidth may be smaller than the laser mode spacing. If the mode frequencies fluctuate during the data acquisition time (a few seconds), the narrow absorption lines are only occasionally sampled by the laser modes, which, in turn, reduces the sensitivity [1013].

A recent study has, however, taken advantage of the high spectral purity of the laser modes, to achieve a resolution higher than that of the spectrograph [1038]. By inserting an intracavity etalon, only a subset of regularly spaced modes operate. The etalon thickness is chosen to ensure that the mode separation is well resolved by the spectrograph. The ICLAS spectrum is then sampled by a series of regularly spaced modes whose linewidth determines the spectral resolution. The whole spectrum is obtained by scanning the etalon step by step and reconstructed by interleaving these spectra [1038].

The wavenumber calibration may use several absorption markers superimposed on the sample spectrum according to the spectral range of interest. As air fills part of the laser cavity, atmospheric water vapor and oxygen lines [683] are currently used for that purpose in the visible range. In the regions where atmospheric lines are absent, the iodine spectrum from an external, possibly heated, cell may be used [770,771]. If only a few lines exist for a slice of the spectrum sampled by the spectrograph, the accuracy of the calibration may be improved by inserting an etalon whose fringes allow the wavenumber scale to be linearized. A precision of $0.01\,\mathrm{cm}^{-1}$ on the wavenumber scale results from the latter procedure.

e.   Capabilities of ICLAS. Two major fields of application are accessible to highly sensitive absorption techniques: the measurement of low concentration of absorbers and the spectroscopy of weak absorption bands. ICLAS, just as CRDS, is highly versatile and any optical system enclosed by clean windows at Brewster angle may be inserted in the laser cavity (see Fig. 77). ICLAS has been combined with, for instance, slit jet expansions

[1013,1039–1041] liquid-nitrogen-cooled cells [1021–1024,1042], different types of reactors [987,1043–1047], and atmospheric-pressure burners [1048–1050].

1. OVERTONE SPECTROSCOPY. From its earlier development, it became clear that ICLAS was an ideal tool for spectroscopy. In 1976, Krugel et al. pioneered the spectroscopic applications of ICLAS by reporting the low-resolution electronic spectrum of plutonium hexafluoride [1051]. ICLAS has since been applied for the extensive study of the rotationally resolved high overtone transitions of intermediate size molecules (3–6 atoms) in the ground electronic state. In the spectral region between 10,000 and 20,000 cm^{-1} covered by ICLAS spectrometers using Ti:Sa or one of the dye lasers, the absorption spectrum is dominated by the overtone transitions of the high frequency and highly anharmonic stretching modes such as the X–H (X = C, O, N, Si, S, Ge, etc.) stretching modes. The corresponding absorption features are, however, several orders of magnitude weaker than the fundamental bands, and ICLAS therefore appears ideally suited in this context. A review of the overtone spectroscopy using ICLAS was published in 1990 [1010]. The ICLAS spectrum of a number of intermediate size molecules has since been explored and analyzed between 11,000 and 18,500 cm^{-1}. Campargue and co-workers, in particular, investigated molecules with high local-mode character which are mainly free of stretch–bend interaction: GeH$_4$, SiH$_4$, and their deuterated derivatives (see Refs. 1052, 1053 for a recent compilation of the observed bands), and H$_2$S [1054]. The systematic investigation of the ICLAS spectrum of CHD$_3$ [1052,1053], C$_2$H$_2$ [516,614], C$_2$D$_2$ [525], and N$_2$O [596,612], made it possible to study the stretch–bend anharmonic resonances affecting the highly excited levels of these molecules. An example of the ICLAS spectrum of weak overtone transitions is given for C$_2$D$_2$ and C$_2$H$_2$ in Figs. 80 [1055] and 81, respectively. The ICLAS visible overtone spectrum of HOCl has recently been studied by Abel et al. with equivalent pathlengths of up to 300 km [1011,1054]. The couplings of the observed levels with background states were discussed in relation to intramolecular vibrational energy redistribution (IVR).

In the near infrared (0.9–1.6 µm) several solid-state amplification media with an homogeneously or inhomogeneously broadened gain profile have been used for ICLAS: Nd glass [1012,1056–1058], color centers such as F$_2$:LiF [1057,1059,1060], and KCl:Tl (1) [1061], and laser crystals such as Cr^{4+}:YAG [1062] and Cr^{4+}:LISAF [1063]. The high sensitivity of ICLAS was first demonstrated in the 1970s with this class of laser whose dynamics has since been extensively studied. Note that antireflection-coated diode lasers with an external cavity have been experimentally and theoretically

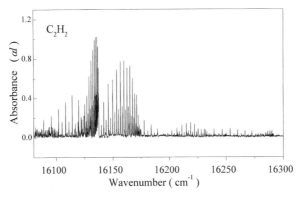

Figure 80. Example of ICLAS spectrum: the $\Pi_u - \Pi_g$ band at 16,139.10 cm^{-1} and close bands of $^{12}C_2H_2$ recorded at a pressure of 214 hPa (adapted from Campargue and co-workers [1055]). The generation time was 100 μs, which, taking into account a filling ratio of 47%, leads to an absorption equivalent pathlength of 14 km. Note the marked difference in intensity between the R and P branches.

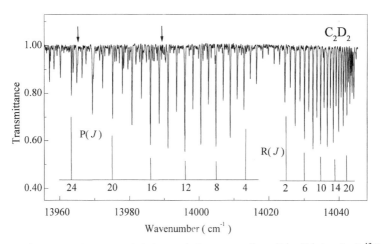

Figure 81. ICLAS spectrum of the weak $2v_1 + v_2 + 3v_3$, $\Sigma_u^+ - \Sigma_g^+$ band of $^{12}C_2D_2$ centered at 14019.72 cm^{-1} with modes 1–3 the CD symmetric stretch, the CC stretch, and the CD antisymmetric stretch (adapted from Campargue and co-workers [525]). The pressure was 196 hPa and the absorption equivalent pathlength was about 22 km. The two arrows indicate the band origin of two hot bands.

studied for ICLAS [1006] but the dynamics of this class of laser and the residual reflectivity of the facets have prevented it, up until now, from achieving a high sensitivity.

In spite of the high sensitivity (no saturation of the absorption is observed after several milliseconds), these near-infrared solid-state lasers have been less extensively used than the dye lasers for spectroscopic studies, possibly because of to their limited spectral coverage. Sinitsa and co-workers [1060,1064–1066], reported the ICLAS spectrum of deuterated water (HDO and D_2O) in the range 8200–10,800 cm^{-1} with pulsed F_2: LiF and Nd:glass lasers. In the range of the Nd:glass laser (9100–9500 cm^{-1}), ICLAS was used to record the spectrum of acetylene [1067–1069], nitrous oxide [1070], and ozone [1012]. A compilation of small polyatomic molecules studied by ICLAS between 1 and 1.25 µm can be found in Refs. 1057 and 1071.

Several studies were devoted to the measurement of the line intensities of weak transitions, especially for molecules of atmospheric interest, in particular, CO_2 and N_2O [558,1025], CH_4 [1021–1024], and O_2 [1018,1072]. Because of the importance of methane in planetology, line intensity measurements were performed for various temperatures, down to 77 K [1021–1024]. Such data are required because the highly congested near infrared and visible spectrum of methane has not yet been rotationally assigned, and the variation of the line intensity with temperature cannot be calculated.

The first association of ICLAS with supersonic beams concerned the electronic spectrum of ultracold large molecules [1039]. It was presented as an alternative method to fluorescence excitation for studying non- or weakly fluorescent transitions. Rotational cooling is particularly well suited to drastically simplify highly congested spectra. The combination of ICLAS with a 24-cm-long intracavity slit jet expansion was used to record the $\nu_{CH} = 4$ overtone transition of methane [1040,1041,1073] and the visible electronic spectrum of NO_2, which shows correlation properties characteristic of the vibronic chaos [1014]. The rotationally resolved visible spectrum of the O_2 dimer formed either in a supersonic expansion or in a cell cooled at liquid nitrogen temperature was also recently recorded by ICLAS [1042].

High-resolution time resolved ICLAS allows lineshape to be analyzed. The measurement of linewidths by ICLAS has been applied to the determination of pressure broadening coefficients [1014,1015,1017], translational temperature from Doppler-broadened linewidth [1043–1045], and excited-state lifetimes [1012,1074].

2. DIAGNOSTICS. Compared to the laser induced fluorescence (LIF) detection technique, frequently used for plasma diagnostics, ICLAS directly provides the absorption coefficient whereas LIF requires a separate calibration of the signal. However, LIF may offer a good spatial resolution while ICLAS (as CRDS) gives line-of-sight-averaged absorption. When line intensities of transient absorbing species are available, very small absolute concentration of these species may be calculated from the measured absorption coefficient.

Such investigations were performed to obtain the concentration profile of excited hydrogen ($n = 2$) in a hot-filament-assisted diamond deposition reactor [1043] or in RF plasmas used for diamond-like carbon films deposition [1044,1045]. Similar studies were undertaken in methane–oxygen–nitrogen flames to monitor atomic oxygen [1075], HCO [1048,1050], singlet CH_2 [1049], and in a DC discharge in argon with 5% SiH_4, to measure the density of SiH_2 [987,1046,1047]. The temperature in the flame was measured using the ICLAS spectrum of water vapor [1076].

4. Optoacoustic (OA) Spectroscopy

Since its discovery in 1880 [1077–1080], optoacoustic (OA) spectroscopy has been applied to a wide variety of fields and is now a well-established technique that has been described in several review articles [1081–1085]. The reader is invited to refer to these review papers for a more detailed description of the advantages, experimental arrangements and the possible applications of OA spectroscopy. The latter includes trace detection [1086–1091], nonlinear spectroscopy, and vibrational and rotational energy transfer investigation. We limit the presentation here to the basic principles and emphasize, through a number of examples, the applicability of OA to overtone spectroscopy.

a. Basic Principles. The gas sample under investigation is enclosed in a cell and excited by a laser beam modulated at a frequency of the order of several kilohertz. The excitation of the absorbing molecules may be transferred by collision into translational, rotational, or vibrational energy of all the molecules present in the sample. This relaxation process leads to a modulation of the thermal energy and, hence, of the pressure and temperature, at the same modulation frequency. The resulting acoustic wave is detected by a highly-sensitive microphone with lock-in detection. It is worth noting that photoacoustics can also use pulsed laser excitation which has the advantage to extend the class of tunable lasers suitable for OA spectroscopy [1092,1093]. The absorbance is proportional to the ratio of the OA signal to the laser power and the absorption spectrum is obtained by tuning the laser. The method requires that the collisional energy transfer from the excited state of the absorber to the translation be shorter than the spontaneous lifetime. This gives a low-pressure limit for the absorption spectroscopy involving levels with short lifetime. At low pressures, indeed, part of the absorbed energy may be lost by fluorescence. However, in the case of vibrationally excited molecules in the near-infrared and visible ranges, this limitation is not important as these states have a sufficiently long lifetime compared to the time of flight corresponding to pressures of some tens of hectopascals.

Several methods have been used to increase the sensitivity. One possibility, as suggested by Kamm [1094], is to optimize the geometry of the acoustical resonator and chop the radiation source at a resonant frequency of the cell [1081,1084,1095]. Another way to enhance the OA signal is to operate with a multipass configuration [1081,1096–1100]. As the signal is, to a large extent, proportional to the beam power, the sensitivity may also be increased by orders of magnitude by inserting the photoacoustic cell equipped with wedged Brewster angle windows inside the laser cavity. High-power tunable lasers such as color center, Ti:sapphire, and dye lasers are then suitable for an optimum sensitivity. These lasers are the same as those used for ICLAS. In ICLAS, however, the best sensitivity is obtained with the laser operating just above the threshold. A high laser power in ICLAS would increase the nonlinear intermode coupling and substantially decrease the sensitivity. This is an important difference between OA spectroscopy and ICLAS, which renders meaningless a global comparison of the sensitivity of the two methods.

The linewidth of the single-mode CW lasers is of the order of 1 MHz, when using active frequency control. However the required modulation of the laser beam in OA spectroscopy prevents the use of an external cavity to stabilize the laser frequency. The effective resolutions are therefore not better than some tens [1097,1098] or hundreds of megahertz [1101–1103]. In practice, this is still significantly narrower than the Doppler linewidth (1 GHz or more). As a pioneer example, we note the measurement of the pressure broadening of a rovibrational line of the $v_{CH} = 6$ overtone transition of methane performed by Gelfand et al. 20 years ago [1102,1103].

b. *Application to overtone spectroscopy.* OA spectroscopy is then well suited to record the rotationally resolved spectrum of intermediate size molecules with Doppler-limited resolution. We try here to highlight the performances of this technique by reviewing the OA overtone spectra obtained in the near-infrared and visible ranges with emphasis on the rotationally resolved spectra that are of particular interest for the present review. As the first example, we must mention the impressive set of absorption bands of a number of alkynes, in particular C_2H_2 and its isotopic derivatives observed for the first time, in 1984, by Hall between 12,000 and 24,000 cm^{-1} [1099,1104]. This experimental work, performed at high resolution with dye lasers over a very wide spectral range, reported the observation of the most intense bands of each Cluster (see Section III.C) of different isotopomers of acetylene up to 24,000 cm^{-1}. To our knowledge, this is still the only observation of overtone bands of acetylene at energies higher than 18,500 cm^{-1}. Lehmann et al. recorded and analyzed the high-resolution visible OA spectrum of HCN [1105], acetylene [618], methane

[1101], and ammonia [1106]. They, for example, showed that the linewidth of the fine structure in the $v_{CH} = 5$ overtone band of methane recorded at 77 K was Doppler-limited and not homogeneously broadened [1101]. Halonen et al. used a similar setup, based, however, on a ring laser configuration and recorded the high-resolution spectrum of acetylene [615,1107] of a series of monohalogenoacetylenes, C_2HX (X = F,I,Br) [577,578,1108] and of some local-mode molecules such as stannane [1109,1110] and stibine [1111].

The rotationally resolved OA spectra of some deuterated methane molecules [1112] and of trideuterated silane [1113] were investigated with a resolution of the order of $1 \, cm^{-1}$. For completeness, we mention the OA spectra of N_2O, CO_2 [1114], and OCS [1115] recorded in the range of the Ti:sapphire with a 1-cm^{-1} resolution. Reilly et al. [1116] associated a cryogenic OA cell cooled down to 100 K with a single-mode Ti:sapphire to study the Doppler-limited spectrum of the $v_{CH} = 4$ overtone transition of $^{12}CH_4$ [1117], $^{13}CH_4$ [1118], CH_3D [1119], and propyne [1120]. Other high-resolution studies have been devoted to the CH overtone spectra of formaldehyde [1121], the NH stretch overtone of pyrrole [1122], the OH stretch overtones of HDO [1123], H_2O_2 [1124], and NH_2OH [1125], and the $v_{AsH} = 6$ overtone of arsine [1126]. In the range 6000–6700 cm^{-1}, a highly sensitive OA spectrometer using a $NaCl:OH^-$ color ring laser was developed to record some overtone bands of C_2H_2 [1096], CS_2 [561,1127], and OCS [1127] with an optimal sensitivity of $\alpha_{min} = 2 \, 10^{-9} \, cm^{-1}$ and a laser linewidth narrower than 1 MHz. Takahashi et al. have used a pulsed tunable infrared laser derived from a tunable dye laser to observe three vibrational transitions of C_2H_2 between 3200 and 8500 cm^{-1} [1128].

OA detection has been extensively used to study the overtone spectra of larger molecules in gas and liquid phases. In most cases, the spectra are unresolved as a result of rotational congestion, overlap with hot bands, and an increased number of extra, zeroth-order forbidden, lines (see Section III). Lasers with narrow spectral width are not required for such investigations. An extensive review of the applications of OA spectroscopy to high-energy overtone transitions of large species falls out of the scope of the present review. We only mention the OA spectra of C_6H_6 obtained by Reddy and Berry up to $\Delta v_{CH} = 9$ in the visible range [1129–1131], which has stimulated a number of theoretical works devoted to the importance of IVR. The local-mode model was used by Sage [1132], Henry et al. [714,1085], Fang and Swofford [1133–1135], Crofton et al. [1136], Duncan et al. [526,1137,1138], and Wong and Moore [1139,1140] to account for the rotationally unresolved OA spectra of a series of alkanes and alkenes in the $\Delta v_{CH} = 5-9$ range of absorption. The structure and dynamics of the excited CH chromophore in trihalomethanes CHX_3 have been studied on the basis of

OA investigations of their overtone spectra [544,1141–1145] as well. Evidence of the homogeneous broadening responsible for the unresolved rovibrational structure was presented and discussed in the latter studies in relation to selective vibrational energy redistribution.

5. Optothermal (OT) Spectroscopy

The OA laser spectroscopy is based on the transfer of energy from the internal degrees of freedom of the molecules to the translation. It cannot be applied to molecular beams. Photothermal lensing effects of supersonic jet expansions have, however, been observed in the early expansion stages of a supersonic free jet of acetylene [1146].

Optothermal (OT) spectroscopy is a highly sensitive laser technique devoted to the spectroscopy of molecules cooled in a supersonic expansion. It was introduced 20 years ago by Gough, Miller and Scoles [1147–1149]. The basic idea is to illuminate molecules in a beam expanding in a high-vacuum chamber and detect the internal excitation by using a cryogenic detector (bolometer). When the radiative lifetime is longer than the flight time, and this is the case with overtone excitation, the optical energy deposited in the molecules in the beam is transferred by collision to the cooled bolometer. The resulting increase of the bolometer temperature then images the absorbance, and the spectrum is obtained by monitoring the temperature of the detector when the excitation laser is tuned. The sensitivity of the method is increased by chopping the laser beam and using a lock-in detection, by maximizing the absorption pathlength in the jet and by using very sensitive semiconductor detectors at low temperatures as bolometers. Technical details about the experimental setup can be found in Refs. 1124 and 1150–1155.

Optothermal detection is a powerful tool in high-resolution overtone spectroscopy. It naturally provides the additional advantages of a drastic reduction in both the Doppler linewidth and the spectral congestion. As very narrowband (a few MHz) CW lasers are available, OT detection allows highly sensitive Doppler-limited spectroscopy under beam conditions. Particularly in the near-infrared range, Reilly et al. studied the $v_{CH} = 4$ overtone manifold in H_2O_2 [1124] and CH_4 [1117,1154,1155]. The Stark effect in the $3v_1 + v_3$ band of methane was measured with a 7.5-MHz Doppler-limited resolution [1117]. A supersonic beam OT infrared spectro-meter was also developed by Bassi and co-workers to measure the $v_{CH} \leq 4$ overtones of benzene [1156,1157] and F-substituted benzenes [1158]. The observation of several unresolved vibrational bands for each v_{CH} manifold was described on the basis of calculations using the algebraic approach [1156–1158] (see Section II). It is worth noting that the low overtone spectrum of benzene was rotationally resolved by Page, Shen, and Lee, only by combining rotational cooling in a supersonic expansion and laser labeling with resonant two-photon ionization detection [1159,1160]. OT spectroscopy was also very recently applied to high-resolution measurements of vibronic transitions of *s*-tetrazine, dimethyl-*s*-tetrazine, and pyridine near 285 nm [1161].

In the range of the lower-energy overtones, the CH stretching fundamental of fluoroform expanded in a free jet has been recorded at sub-Doppler resolution by OT spectroscopy using a color center laser [1162]. The same approach [1163,1164] was used by Scoles, Lehmann, and co-workers to study the intramolecular vibrational redistribution of the $v_{CH} = 1-3$ stretch in molecules such as propyne [1163,1165], CF_3CCH [1166], and methylsilane [1167].

6. Comparison of FMDL, CRDS, ICLAS, and OA Methods

The present review of the highest-performance laser absorption techniques, with a particular emphasis on CRDS and ICLAS, gives the opportunity to discuss the advantages of the different highly sensitive laser techniques described above. Romanini recently published a comparison of the performances of ICLAS and CRDS [1168].

For spectroscopic applications, the spectral resolution, the scanning range, the speed of data acquisition, the possible combination of a technique with low-temperature cells or jet expansion, and, of course, the sensitivity need to be considered. It is meaningless to use the sensitivity reported in the literature as a final criterion. Indeed, the optimal experimental parameters are different for each technique. As an example, a high laser power increases the sensitivity of an OA system, while the performance of ICLAS or CRDS is unaffected or even reduced by an increased laser power. With the exception of FMDL, which is frequently associated with a multipass cell, the different methods we have considered all require a small amount of gas (1 mmol or less). This is a considerable advantage over the above-mentioned grating spectrometer techniques, especially for the study of unstable or expensive species, such as isotopically substituted molecules.

Diode lasers are available in any range from 1 to 2 μm and allow compact low cost systems, based on FMDL or on CW-CRDS approaches, to be built for trace detection and diagnostics. The ever-increasing spectral ranges accessible with diode lasers together with their low cost and easy operation, gives a key advantage on FMDL and CW-CRDS for trace detection and diagnostics. The use of ICLAS for diagnostic applications in the near infrared range is limited by the availability of homogeneously or inhomogeneously broadened amplification media in this region. Another characteristic of the ICLAS technique that has limited, up to now, its application to laboratory investigations, is the need of a high-resolution spectrograph, at least 1 m long.

FMDL, CRDS, CW-CRDS, and ICLAS provide absolute measurements of absorption coefficients. Similar measurements of absorption cross sections are difficult to obtain by OA spectroscopy requiring both a preliminary calibration of the OA signal and the knowledge of several parameters such as the laser power and the thermodynamical properties of the gas under investigation [1130,1139,1169,1170]. As an instructive example, the conflicting OA intensity measurements of some weak overtone bands of HCN [1171] were definitely resolved after CRDS measurements were reported by Romanini and Lehmann [948]. We note, however, that an excellent agreement between the cross sections of the $5\nu_{CH}$ and $6\nu_{CH}$ overtone bands of ethane and ethylene measured either by FT or by OA spectroscopy was reported after calibration of the OA signal against the well-known cross sections of HD overtone bands [1172]. In the case of unresolved absorption spectra, CW-CRDS and to a lesser extent CRDS and OA spectroscopy are probably best suited. Indeed, the zero absorption baseline can be obtained by FMDL only by comparison with a reference cell. In the case of ICLAS, unresolved features are observable only when their spectral width is less than the laser spectral width (some tens cm^{-1}). ICLAS is therefore not dedicated to the overtone spectroscopy of large molecules. Finally, CRDS and OA spectroscopy are well suited for the measurement of broadband losses even if scattered light complicates the determination of the zero absorption baseline in OA spectroscopy. As an example, the initial formation and growth of nanometersize particle when igniting a plasma have been followed by monitoring the scattering losses [987] measured by CRDS. In the case of the superposition of resolved features with a broadband spectrum (or a continuum), OA detection is probably most convenient. In such a case, indeed, ICLAS remains, to a large extent, insensitive (and then tolerant) to the broadband losses, and only the resolved spectrum will be highlighted. In CRDS, the cavity factor will be partly spoiled by the broadband losses and the detection sensitivity of discrete lines will be reduced.

Spectral multiplexing is probably the most important intrinsic advantage of ICLAS. As mentioned above, ICLAS is the only highly sensitive technique that provides several thousand spectral elements simultaneously. We note, however, that Engeln and Meijer [1173] have recently succeeded in using a time-resolved Fourier transform spectrometer to record as a whole the ringdown rates of a cavity injected by a broadband pulsed laser. As a consequence of the multiplex advantage, ICLAS allows rapid data acquisition compared to the other techniques, which require scanning the spectrum step by step with narrowband lasers.

In ICLAS, however, the multiple spectral elements are recorded after the dispersion of the laser spectrum by a spectrograph, which limits the spectral resolution. A resolving power of the order of 10^6 is at present the best that can be achieved for specifically built spectrographs. A resolution of a few MHz is accessible with the CW monomode laser sources used by FMDL, CW-CRDS, or OA spectroscopy, while a resolution of the order of 1 GHz is achieved with the standard pulsed CRDS. In most cases, this is sufficient to record overtone spectra at room temperature with a Doppler-limited resolution. It is, however, a severe sensitivity limitation for the study of jet-cooled molecules and especially when using a slit jet expansion where the residual Doppler width is reduced to some 100 MHz or even less. In that case, indeed, as the absorbance integrated over the line profile is roughly independent of the spectral resolution, the maximum of absorption at the line center is reduced by the ratio of the spectral resolution to the intrinsic linewidth. Therefore, the sensitivity of ICLAS and pulsed CRDS is considerably reduced. CW-CRDS thus appears to be the best-suited method, with OT detection, for jet-cooled absorption spectroscopy.

Optoacoustic detection has an intrinsic low-pressure limitation which prevents its application to jet spectroscopy and to spectroscopy of transient species generated in low-pressure environments such as that encountered in reactors and discharges. As for ICLAS, OA spectroscopy has been combined with cells cooled down to liquid nitrogen temperature. To our knowledge, CRDS or CW-CRDS of gases cooled to low temperature has not yet been reported. This is probably because the inability to insert windows inside the passive cavity makes it necessary to cool the entire CRDS cavity. The spectral range of interest often determines the optimal selection of a technique, although, in practice, it is usually determined by the available experimental equipment! From this point of view, pulsed CRDS appears to be more universal as it can use a large variety of laser sources, including new optical parametric oscillator (OPO) systems, which provide broad tunability in the 1–10-μm range and even free electron lasers [1174] for low-resolution investigations. The performance of CRDS and CW-CRDS are strongly dependent on the quality of the high-reflectance mirrors of the cavity and then decrease in the UV and IR ranges. ICLAS and OA spectroscopy have no such limitation. They suffer, however, from the absence of universal tunable CW lasers, which must be of high power for OA detection. The UV range is thus, not yet accessible, for both techniques. We note, however, that single-mode OPO systems have recently been used as narrow-bandwidth infrared sources for OA spectroscopy near 1 μm [1175]. Finally, we should mention that the ICLAS performances may be, especially in the near infrared range, limited by relatively strong atmospheric absorption bands. A buffer gas may be introduced in the laser cavity to reduce this effect.

The acetylene molecule provides an instructive case study to compare the possibilities of some of the considered techniques. Its absorption bands, which extend over the whole spectrum up to the blue region, show a high variation in intensity. Moreover, this molecule is frequently used as a test species to estimate the sensitivity of a given experimental setup. As an example, the performance of CW-CRDS with a ring dye laser at 17,500 cm^{-1} was compared with OA measurements [959].

A review of the transitions of $^{12}C_2H_2$ observed at energies higher than 11,000 cm^{-1} can be found in Refs. 516,614,1055 with the corresponding experimental techniques. Figure 68 (above) shows an overview of the low-energy region of the visible absorption spectrum recorded using Fourier transform spectroscopy. In the upper part of this figure, the highest performing absorption investigations are specified with the corresponding experimental technique. CRDS does not appear on this plot as it emerged very recently in the literature and has been used, so far, only to investigate over a wide spectral range the overtone spectra of HCN and HNO. Our ICLAS systematic investigation of the visible absorption spectrum of $^{12}C_2H_2$ illustrates the spectral multiplexing advantage of ICLAS, which makes it possible to rapidly explore large spectral regions at the level of rotational resolution with a high sensitivity. This is in contrast to FMDL investigations, which are limited by the reduced tuning range of the diode lasers. The spectra in Fig. 68 illustrate the high speed of data acquisition of ICLAS compared to OA spectroscopy. In terms of spectral resolution, the FMDL and OA investigations have provided higher-quality spectra when they were performed with single-mode lasers. In term of sensitivity, a wealth of new data have been obtained by ICLAS, in particular in the ranges previously explored by OA spectroscopy. It is, however, difficult to generalize. It appears, in particular, that the intracavity resonant cell developed by Halonen et al. [615,1097,1098] has provided some high-quality spectra in the visible range comparable to ICLAS in terms of sensitivity but better in terms of resolution.

7. Other Laser Investigations of Vibration–Rotation Levels

High-resolution spectra recorded at room temperature are too complex for analysis, especially for larger species and in the high overtone region where many rovibrational levels may interact. Supersonic jet cooling is a powerful method to reduce the rotational congestion and to a lesser extent hot-band congestion. As discussed above, FT, FMDL, CRDS, and ICLAS have been successfully associated to rotational cooling in a supersonic slit jet expansion, while OT spectroscopy with bolometric detection is a technique specifically devoted to jet-cooled molecules.

In this section, we review several other methods that were developed to probe the spectroscopy and dynamics of molecules at high vibrational excitation on their ground potential electronic surface. These experimental approaches, frequently associated with jet cooling or (and) double resonance excitation have given key insights into IVR and revealed the possibilities and challenges of bond specific unimolecular reactions (see, e.g., Refs. 544,1176, and 1177).

These methods may be less versatile than the direct absorption methods described above, but they may provide highly complementary information and are particularly well suited for study of IVR processes.

a. Laser Labeling. Because of the vibrational selection rules, and poor Franck–Condon access, only a very limited number of highly excited vibrational states are accessible by conventional absorption spectroscopy. Double-resonance excitation is suitable for both labeling some specific rovibrational levels and giving access to classes of levels that cannot be excited from the ground state. However, because of the weakness of the overtone transitions, double-absorption schemes using a first tunable laser to populate selected rovibrational levels and a second laser to probe the absorption from the excited rovibrational state have been used in a very limited number of studies. As an example, De Martino et al. studied the $N = 2$ and $N = 3$ stretching manifolds of silane [1178] and methane [1179, 1180] by this method.

Another approach proposed by Page, Shen, and Lee [1159,1160,1181] to study the rotationally resolved overtone transition ($v_{CH} \leq 3$) of benzene is to ionize the molecule by resonant absorption of two photons from the ground state (R2PI). The overtone spectrum is obtained by monitoring the decrease of the ionization signal when the overtone transition depopulates the ground state. This contribution established that the homogeneous contribution to the low CH stretching overtone transition of C_6H_6 is limited to a few cm^{-1} and inhomogeneous congestion dominates at room temperature.

The fluorescence signal from an electronic excited state can be used instead of the ionization signal to probe vibrational excited states populated by a first laser. A pulsed laser double-resonance scheme was developed by Utz et al. to study the weak state mixing of the $3v_{CH}$ manifold of acetylene [513,1182]. The combination of tunable infrared vibrational excitation and UV LIF via the $\tilde{A}^1 A_u$ electronic state was shown to be a sensitive probe of weak interactions that escape detection in conventional absorption experiment. Orr et al. used a comparable approach in the range of the $v_2 + 3v_3$ manifold of C_2H_2 and demonstrated symmetry-breaking perturbations [514,1183,1184].

b.   Photofragment   Spectroscopy.   Another   efficient   way   to   detect   the vibrationally excited molecules is to dissociate them selectively and to detect the photofragment. Indeed, the absorption spectrum may be faithfully reproduced by monitoring the photofragment yield when the excitation laser is tuned. In the case of molecules with low dissociation energy, a direct (one-photon) excitation of reactant overtone transitions may dissociate the molecules. This method was used in conjunction with both supersonic jet techniques and double-resonance labeling for detection of overtone transitions of HOOH [1185–1190], NH_2OH [1191], HN_3 [1192], and HOCl [1193]. The LIF detection of the OH or NH fragment was used to monitor the absorption spectrum of these molecules.

In the case of strongly bound molecules, the photodissociation may be assisted by a selective electronic UV excitation from the vibrationally excited state. This approach was used in particular by Crim et al. to study the bond selected breaking of HOD [1194–1198] after the vibrational excitation.

Hippler and Quack recently used multiphoton ionization to detect the fragments. This method called *overtone spectroscopy by vibrationally assisted dissociation and photofragment ionization* (OSVADPI), was applied both in flow cells [1199,1200] and in supersonic jets [1201], to molecules such as $CHCl_3$. A Ti:sapphire laser (eventually associated with a Raman cell) excites the $v_{CH} = 3$ or 4 stretch overtones while subsequent UV excitation dissociates the vibrationally excited molecules. The chlorine atom fragments are probed by multiphoton ionization with the same UV laser, and the $v_{CH} = 3$ or 4 overtone spectrum is obtained by monitoring the ionization yield when the IR laser is scanned. As the REMPI detection of fragments is extremely sensitive, this versatile technique is, in principle, generally applicable both to low-pressure flow cells and to free molecular jet expansions (see also Ref. 1202). A similar approach was used by Rosenwaks et al., who utilized the vibrational excitation of the fourth overtone CH stretch of acetylene followed by the electronic photodissociation of C_2H_2/C_2D_2 combined with REMPI detection of H/D fragment [1203–1206].

A recent innovation proposed by Rizzo et al. uses a vibrational rather than an electronic excitation to dissociate vibrationally excited molecules. This approach,   called   *infrared-laser-assisted   photofragment   spectroscopy* (IRLAPS) [1207,1208] was applied to study the IVR in jet-cooled CH_3OH [1208], and CHF_3 [1207]. It has also been combined with infrared–optical double-resonance excitation for rotational labeling [1209].

c.   Stimulated Emission Pumping and Dispersed Fluorescence.   Stimulated emission pumping (SEP) is a double-resonance technique largely comple-mentary to direct vibrational overtone excitation that has been used to study high vibrational levels of many polyatomic molecules in their ground electronic state [552,1210–1212]. A first laser promotes a molecule to a single rovibrational level of an electronically excited state, while a second laser stimulates the emission down to a highly excited vibrational state of the ground electronic state. The SEP spectrum is measured as a decrease of the side fluorescence when the dump beam causing the stimulated emission is scanned [1213,1214].

Because of improved Franck–Condon access for transition corresponding to an important change of molecular geometry, SEP gives access to a class of vibrational states involving a high bending excitation that are not easily excited by direct overtone excitation. In addition, in the case of molecules with an inversion center such as $^{12}C_2H_2$, while direct overtone spectroscopy samples mostly *ungerade* vibrational levels, the electronic emission samples *gerade* levels. As a result, at high vibrational excitation, direct absorption provides information about states involving high excitation of the CH stretch and low excitation of the bends whereas SEP primarily involves high excitation of bends and low excitation of CH stretches. Using high-resolution SEP, Field and co-workers have identified the first stages of IVR within the $\tilde{X}^1\Sigma_g^+$ ground state of C_2H_2 in the range 5000–18,000 cm^{-1} [469,473, 1215]. As a rule, it was shown that CC stretch excitation is necessary for stretch–bend coupling and that the quantum numbers obey some well-defined relations that restrict the possible couplings with the initial bright state during the first stages of IVR. Dispersed fluorescence (DF) is a suitable alternative to SEP when high resolution is not necessary. Indeed this spontaneous emission technique allows a much quicker coverage of large spectral regions. SEP and DF have been used together to study C_2H_2 [469,473,609,1215,1216], SO$_2$ [1217], HFCO [1218,1219], and HCO [1220], among others. The mode-specific dynamics of HFCO [1218,1219], HCO [1220,1221], and DCO [1222] above the dissociation limit were investigated by SEP, too.

The case of the NO$_2$ molecule is particularly interesting as quantum chaos within the vibrational levels is induced by couplings of the first excited state $\tilde{A}^2B_2$ with the high vibrational levels of the ground state. Jost, Delon, and co-workers obtained by DF with jet-cooled NO$_2$, the complete set of the 191 vibrational levels of the $\tilde{X}^2A_1$ ground state up to 11000 cm^{-1} [565]. This regular sequence of vibrational levels was satisfactorily reproduced by a Dunham expansion. The situation is different in the visible range; strong vibronic couplings between the vibrational levels of the $\tilde{X}^2A_1$ and $\tilde{A}^2B_2$ states induce strong mixing and statistical properties characteristic of quantum chaos [28,1013,1223]. The major progress in the understanding of this highly complex part of the spectrum was made by LIF spectroscopy under supersonic jet [1224–1227]. This high-resolution technique is roughly equivalent to direct absorption techniques. It has made it possible to analyze rovibronic interactions by the Zeeman effect and anticrossing experiments

[1228] and to determine accurately the NO_2 photodissociation threshold near 25,128.57 cm^{-1} [1229]. As the jet-cooled visible LIF is limited to the red range by the rapid decrease of the oscillator strength, ICLAS associated with a jet expansion was used to observe 249 $^2B_2$ vibronic bands between 11,200 and 16,150 cm^{-1} [1013]. This set of vibronic levels has recently been completed by new LIF measurements in this range [1223], leading to an almost complete set (89%) of the $^2B_2$ vibronic levels of NO_2 up to 19,300 cm^{-1}. Finally, we mention that the very high resolution allowed by the OT technique was recently used to study the hyperfine structure of one vibronic band of jet-cooled NO_2 recorded with a residual Doppler width of 12 MHz [1230].

ACKNOWLEDGMENTS

The present review could not have been prepared without the help of various friends and collaborators. We thank them all, warmly:

- Drs. Joyeux, Romanini and Kachanov, Prof. Stoeckel and Mr. Day (Univ. J. Fourier–Grenoble), Prof. Jost (LCMI, Grenoble), and Dr. Perrin (Univ. Paris–Sud), for a critical reading of some sections of the manuscript
- Prof. Di Lonardo (Univ. Bologna), Dr. Hurtmans (Institut d'Aéronomie Spatiale de Belgique), Dr. Kou (LPMA–Orsay), and Dr. Sauval (Observatoire Royal de Belgique), who helped us in preparing Appendix B
- Mr. El Idrissi (ULB–Belgium), who helped us in various ways in preparing the manuscript, including the compilation of Appendix C, Drs. Carleer and Hepp, Mr. Bach, Mr. Herregodts, and Mr. Mellouki (ULB–Belgium) for preparing some figures
- Profs. Botschwina (Univ. Gottingen), Crim (Univ. Madison, Wisconsin), Dai (Penn State Univ.), Lehmann (Princeton Univ.), Mills (Reading Univ.), Dr. Romanini (Univ. Fourier–Grenoble), and Dr. Watson (Steacie Institute for Molecular Sciences, NRCC), who provided us with a copy of figures and/or permission for reproducing some of their work, as well as all those who gave us a list of their publications on a specific subject
- Mrs. Leclercq (ULB–Belgium) for her help, which included typing most references
- Our families for their incredible patience

The ULB scientific achievements referred to in the present review are presently supported by the FNRS Belgium (FRFC) and by the ULB. MH and

AC are indebted to the CNRS and FNRS/CGRI for a collaborative research grant.

REFERENCES

1. R. W. Field, J. P. O'Brien, M. P. Jacobson, S. A. B. Solina, W. F. Polik, and H. Ishikawa, *Adv. Chem. Phys.* **CI**, 463 (1997).
2. M. Born and J. R. Oppenheimer, *Ann. Phys.* (Leipzig) [4] **84**, 457 (1927).
3. G. Herzberg, *Electronic Spectra of Polyatomic Molecules* Van Nostrand-Reinhold, Princeton, NJ, 1966.
4. R. Renner, *Z. Phys.* **92**, 172 (1934).
5. J. A. Pople and H. C. Longuet-Higgins, *Mol. Phys.* **1**, 372 (1958).
6. J. A. Pople, *Mol. Phys.* **3**, 16 (1960).
7. H. C. Longuet-Higgins, *Adv. Spectrosc.* **2**, 429 (1961).
8. J. M. Brown, *J. Mol. Spectrosc.* **68**, 412 (1977).
9. C. A. Mead, *Chem. Phys.* **49**, 23 (1980).
10. C. Jungen and A. J. Merer, *Mol. Phys.* **40**, 1 (1980).
11. C. Jungen, K.-E. Hallin, and A. J. Merer, *Mol. Phys.* **40**, 25 (1980).
12. J. M. Brown and F. Jorgenson, *Adv. Chem. Phys.* **52**, 117 (1983).
13. G. Duxbury, C. Jungen, and J. Rostas, *Mol. Phys.* **48**, 719 (1983).
14. G. Fischer, *Vibronic Coupling* Academic Press, London, 1984.
15. S. Carter and N. C. Handy, *Mol. Phys.* **52**, 1367 (1984).
16. S. Carter, N. C. Handy, P. Rosmus, and G. Chambaud, *Mol. Phys.* **71**, 605 (1990).
17. G. J. Atchity, S. S. Xantheas, and K. Ruedenberg, *J. Chem. Phys.* **95**, 1862 (1991).
18. B. T. Sutcliffe, *J. Chem. Soc. Faraday Trans.* **89**, 2321 (1993).
19. H. Muller and H. Koppel, *Chem. Phys.* **183**, 107 (1994).
20. T. Neuheuser, N. Sukumar, and S. D. Peyerimhoff, *Chem. Phys.* **194**, 45 (1995).
21. P. Jensen, M. Brumm, W. P. Kraemer, and P. R. Bunker, *J. Mol. Spectrosc.* **171**, 31 (1995).
22. G. Duxbury and B. D. McDonald, *Mol. Phys.* **89**, 767 (1996).
23. B. Kendrick and R. T. Pack, *J. Chem. Phys.* **104**, 7475 (1996).
24. B. Kendrick and R. T. Pack, *J. Chem. Phys.* **104**, 7502 (1996).
25. B. Kendrick and R. T. Pack, *J. Chem. Phys.* **106**, 3519 (1996).
26. B. Kendrick, *Phys. Rev. Lett.* **79**, 2431 (1997).
27. U. Manthe and H. Koppel, *J. Chem. Phys.* **93**, 1658 (1990).
28. R. Georges, A. Delon, and R. Jost, *J. Chem. Phys.* **103**, 1732 (1995).
29. E. Leonardi, C. Petrongolo, G. Hirsch, and R. J. Buenker, *J. Chem. Phys.* **105**, 9051 (1996).
30. B. Kirmse, A. Delon, and R. Jost, *J. Chem. Phys.* **108**, 6638 (1998).
31. M. Kolbuszewski, P. R. Bunker, W. P. Kraemer, G. Osmann, and P. Jensen, *Mol. Phys.* **88**, 105 (1996).
32. G. Duxbury, B. McDonald, M. Van Gogh, A. Alijah, C. Jungen, and H. Palivan, *J. Chem. Phys.* **108**, 2336 (1998).

33. G. Duxbury, A. Alijah, B. D. McDonald, and C. Jungen, *J. Chem. Phys.* **108**, 2351 (1998).

34. G. Duxbury, A. Alijah, and R. R. Trieling, *J. Chem. Phys.* **98**, 811 (1993).

35. P. Jensen, *J. Mol. Spectrosc.* **181**, 207 (1997).

36. G. Osmann, P. R. Bunker, and P. Jensen, *J. Mol. Spectrosc.* **186**, 319 (1997).

37. J. P. Gu, R. J. Buenker, G. Hirsch, P. Jensen, and P. R. Bunker, *J. Mol. Spectrosc.* **178**, 172 (1996).

38. P. Jensen, M. Brumm, W. P. Kraemer, and P. R. Bunker, *J. Mol. Spectrosc.* **172**, 194 (1995).

39. G. Osmann, P. R. Bunker, P. Jensen and W. P. Kraemer, *Chem. Phys.* **225**, 33 (1997).

40. C. Cohen-Tannoudji, B. Diu, and F. Laloe, *Quantum Mechanics* Wiley, New York 1977.

41. D. Heidrich, W. Kliesch, and W. Quapp, *Properties of Chemically Interesting Potential Energy Surfaces* Springer-Verlag, Heidelberg, 1991.

42. B. T. Sutcliffe, in *Methods in Computational Molecular Pysics* (S. Wilson and G. H. F. Diercksen, eds.), NATO ASI Ser. Ser., B: Phys., p. 19. Plenum, New York and London, 1992.

43. M. J. Bramley, W. H. Green, Jr., and N. C. Handy, *Mol. Phys.* **73**, 1183 (1991).

44. T. J. Lukka, *J. Chem. Phys.* **102**, 3945 (1995).

45. E. R. Wilson, J. C. Decius, and P. C. Cross, *Molecular Vibrations* McGraw-Hill, London, 1955.

46. P. R. Bunker, *Molecular Symmetry and Spectroscopy.* Academic Press, New York, 1979.

47. D. Papousek and M. R. Aliev, *Molecular Vibrational-Rotational Spectra.* Elsevier, Amsterdam, and Academia, Publishing House of the Czechoslovak Academy of Sciences, Prague, 1982.

48. C. Eckart, *Phys. Rev.* **47**, 552 (1935).

49. P. Jensen, in *Methods in Computational Molecular Physics* (S. Wilson and G. H. F. Diercksen, eds.), NATO ASI Ser., Ser. B: Phys. p. 423. Plenum, New York and London, 1992.

50. M. L. Sage, *J. Phys. Chem.* **98**, 3317 (1994).

51. A. R. Hoy, I. M. Mills, and G. Strey, *Mol. Phys.* **24**, 1265 (1972).

52. G. Strey and I. M. Mills, *J. Mol. Spectrosc.* **59**, 103 (1976).

53. M. L. Sage and J. Jortner, *Adv. Chem. Phys.* **47**, 293 (1981).

54. E. L. Sibert, III, J. T. Hynes, and W. P. Reinhardt, *J. Phys. Chem.* **87**, 2033 (1983).

55. M. S. Child and L. Halonen, *Adv. Chem. Phys.* **57**, 1 (1984).

56. C. R. Quade, *J. Chem. Phys.* **64**, 2783 (1976).

57. A. B. McCoy and E. L. Sibert, III, *J. Chem. Phys.* **95**, 3476 (1991).

58. A. Requena, A. Bastida, and J. Zúñiga, *Chem. Phys.* **175**, 255 (1993).

59. J. T. Hougen, P. R. Bunker, and J. W. C. Johns, *J. Mol. Spectrosc.* **34**, 136 (1970).

60. F. T. Smith, *Phys. Rev. Lett.* **45**, 1157 (1980).

61. T. R. Horn, R. B. Gerber, and M. A. Ratner, *J. Chem. Phys.* **91**, 1813 (1989).

62. M. S. Child and R. T. Lawton, *Faraday Discuss. Chem. Soc.* **71**, 273 (1981).

63. L. Halonen and M. S. Child, *Mol. Phys.* **46**, 239 (1982).

64. L. Halonen, *Adv. Chem. Phys.* **104**, 41 (1998).

65. B. T. Sutcliffe, in *Current Aspect of Quantum Chemistry* (R. Carbo, ed.) 99, Elsevier, Amsterdam, 1983.

66. A. Nauts and X. Chapuisat, *Mol. Phys.* **55**, 1287 (1985).
67. N. C. Handy, *Mol. Phys.* **61**, 207 (1987).
68. X. Chapuisat, A. Nauts, and J.-P. Brunet, *Mol. Phys.* **72**, 1 (1991).
69. H. Wei and T. Carrington, Jr., *J. Chem. Phys.* **107**, 2813 (1997).
70. H. Wei and T. Carrington, Jr., *J. Chem. Phys.* **107**, 9493 (1997).
71. C. Froese Fischer, T. Brage, and P. Jönsson, *Computational Atomic Structrure.* Institute of Physics Publishing, Bristol and Philadelphia, 1997.
72. A. Szabo and N. S. Ostlund, *Modern Quantum Chemistry: Introduction to Advanced Electronic Structure Theory.* Macmillan, New York, 1982.
73. K. P. Lawley, ed., *Ab initio Methods in Quantum Chemistry*, Part 1, Vol. 67. Advances in Chemical Physics, Wiley, New York, 1987.
74. K. P. Lawley, Ed., *Ab initio Methods in Quantum Chemistry*, Part 2, Vol. 67. Advances in Chemical Physics, Wiley, New York, 1987.
75. E. Clementi, ed., *Modern Techniques in Computational Chemistry: MOTECC-91.* ESCOM, Leiden, 1991.
76. R. McWeeny, *Methods of Molecular Quantum Mechanics.* Academic Press, San Diego, CA, 1992.
77. S. Aberg, H. Flocard, and W. Nazarewicz, *Annu. Rev. Nucl. Part. Sci.* **40**, 439 (1990).
78. G. Pacchioni, P. S. Bagus, and F. Parmigiani, *Cluster Models for Surfaces and Bulk Phenomenas*, Plenum, New York, 1992.
79. U. Kaldor, *Many-body Methods in Quantum Chemistry*, Springer-Verlag, Heidelberg, 1989.
80. E. A. Hylleraas and B. Undheim, *Z. Phys.* **65**, 759 (1930).
81. J. K. L. MacDonald, *Phys. Rev.* **43**, 830 (1933).
82. M. R. Godefroid, C. Froese Fischer, and P. Jönsson, *Phys. Scr.* **65**, 70 (1996).
83. J. Liévin, A. Delon, and R. Jost, *J. Chem. Phys.* **108**, 8931 (1998).
84. T. C. Thompson and D. G. Truhlar, *Chem. Phys. Lett.* **75**, 87 (1980).
85. H. Romanowski, J. M. Bowman, and L. B. Harding, *J. Chem. Phys.* **82**, 4155 (1985).
86. G. D. Carney and R. N. Porter, *J. Chem. Phys.* **60**, 4251 (1974).
87. G. D. Carney, L. L. Sprandel, and C. W. Kern, *Adv. Chem. Phys.* **37**, 305 (1978).
88. E. L. Sibert, III, *Int. Rev. Phys. Chem.* **9**, 1 (1990).
89. B. T. Sutcliffe and J. Tennyson, *Int. J. Quantum Chem.* **39**, 183 (1991).
90. N. C. Handy, *Int. Rev. Phys. Chem.* **8**, 275 (1989).
91. J. M. Bowman, ed., *Comput. Phys. Commun.*, Vol. 55, Spec. Issue (1988).
92. Z. Bačić and J. C. Light, *Annu. Rev. Phys. Chem.* **40**, 469 (1989).
93. A. G. Császár and I. M. Mills, *Spectrochim. Acta A* **53**, 1101 (1997).
94. J. R. Henderson, J. Tennyson, and B. T. Sutcliffe, *J. Chem. Phys.* **98**, 7191 (1993).
95. T. Carrington, Jr., in *Encyclopedia of Computational Chemisty*, P. von R. Schleyer, N. L. Allinger, T. Clark, J. Gasteiger, P. Kollmann, and H. F. Schaefer III, eds., in press, Wiley, New York, 1998.
96. S. Carter, N. C. Handy, and I. M. Mills, *Philos. Trans. R. Soc. London, Ser. A* **332**, 309 (1990).
97. D. J. Searles and E. I. von Nagy-Felsobuki, in *Vibrational Spectra and Structure* (J.R. Durig, ed.) p. 151. Elsevier, Amsterdam, 1991.

98. C. R. Le Sueur, S. Miller, J. Tennyson, and B. T. Sutcliffe, *Mol. Phys.* **76**, 1147 (1992).

99. P. N. Roy and T. Carrington, Jr., *J. Chem. Phys.* **103**, 5600 (1995).

100. D. Neuhauser, *J. Chem. Phys.* **102**, 8011 (1995).

101. B. H. Bransden and C. J. Joachain, *Physics of Atoms and Molecules.* Longman, London, 1983.

102. M. Weissbluth, *Atoms and Molecules.* Academic Press, London, 1978.

103. P. C. Cross, R. M. Hainer, and G. W. King, *J. Chem. Phys.* **12**, 210 (1944).

104. D. M. Brink and G. R. Satchler, *Angular Momentum.* Oxford University Press, Oxford, 1979.

105. I. W. M. Smith, in *Modern Gas Kinetics: Theory, Experiment, and Application* (M. J. Pilling and I. W. M. Smith, eds.), p. 55. Blackwell, Oxford, 1987.

106. J. I. Steinfeld, J. S. Francisco, and W. L. Hase, *Chemical Kinetics and Dynamics.* Prentice-Hall, Englewood Cliffs, NJ, 1989.

107. S. C. Farantos, *Time Dependent Quantum Mechanics: Experiment and Theory.* Plenum, New York, 1992.

108. B. I. Zhilinskii, *Chem. Phys.* **137**, 1 (1989).

109. Z. Li, L. Xiao, and M. E. Kellman, *J. Chem. Phys.* **92**, 2251 (1990).

110. M. S. Child, *Semiclassical Mechanics with Molecular Applications.* Clarendon Press, Oxford, 1991.

111. P. Gaspard, A. Alonso, and I. Burghardt, *Adv. Chem. Phys.* **90**, 105 (1995).

112. B. Zhilinskii, *Spectrochim. Acta, Part A* **52**, 881 (1996).

113. R. Schinke, *Photodissociation Dynamics.* Cambridge University Press, Cambridge, UK, 1993.

114. B. M. Garraway and K.-A. Suominen, *Rep. Prog. Phys.* **58**, 365 (1995).

115. C. Leforestier, R. H. Bisseling, C. Cerjan, M. D. Feit, R. Friesner, A. Guldberg, A. Hammerich, G. Jolicard, W. Karrlein, N. Meyer, N. Lipkin, O. Roncero, and R. Kosloff, *J. Comput. Phys.* **94**, 59 (1991).

116. T. J. Lee, Ed., "Ab initio and ab initio derived force fields: state of science", *Spectrochim Acta Part A*, Vol. 53, Spec. Issue (1997).

117. J. M. Martin and R. P. Taylor, *Spectrochim. Acta, Part A* **53**, 1039 (1997).

118. K. A. Peterson, *Spectrochim. Acta, Part A* **53**, 1051 (1997).

119. C. E. Dateo and T. J. Lee, *Spectrochim. Acta, Part A* **53**, 1065 (1997).

120. J. Koput and S. Carter, *Spectrochim. Acta, Part A* **53**, 1091 (1997).

121. C. Puzzarini, R. Tarroni, and P. Palmieri, *Spectrochim. Acta, Part A* **53**, 1123 (1997).

122. T. Schmelz, P. Rosmus, A. Berning, and H. J. Werner, *Spectrochim. Acta, Part A* **53**, 1133 (1997).

123. P. Palmieri, C. Puzzarini, R. Tarroni, and A. O. Mitrushenkov, *Spectrochim. Acta, Part A* **53**, 1139 (1997).

124. A. Berces, *Spectrochim. Acta, Part A* **53**, 1257 (1997).

125. J. F. Stanton and J. Gauss, *Spectrochim. Acta, Part A* **53**, 1153 (1997).

126. R. A. King, C. D. Sherrill, and H. F. Schaefer, III, *Spectrochim. Acta, Part A* **53**, 1163 (1997).

127. S. Carter, J. M. Bowman, and L. B. Harding, *Spectrochim. Acta, Part A* **53**, 1179 (1997).

128. Q. Wu and J. Z. H. Zhang, *Spectrochim. Acta, Part A* **53**, 1189 (1997).

129. A. J. R. da Silva, H. Y. Cheng, D. A. Gibson, K. L. Sorge, Z. Liu, and E. A. Carter, *Spectrochim. Acta, Part A* **53**, 1285 (1997).

130. P. Botschwina, A. Heyl, M. Oswald, and T. Hirano, *Spectrochim. Acta, Part A* **53**, 1079 (1997).

131. N. C. Handy and A. Willetts, *Spectrochim. Acta, Part A* **53**, 1169 (1997).

132. G. Fogarasi, *Spectrochim. Acta, Part A* **53**, 1211 (1997).

133. C. W. Bauschlicher and S. R. Langhoff, *Spectrochim. Acta, Part A* **53**, 1225 (1997).

134. R. Liu, X. Zhou, and H. Kasmai, *Spectrochim. Acta, Part A* **53**, 1241 (1997).

135. H. Sun and D. Rigby, *Spectrochim. Acta, Part A* **53**, 1301 (1997).

136. S. Dasgupta, K. A. Brameld, C. F. Fan, and W. A. Goddard, III, *Spectrochim. Acta, Part A* **53**, 1347 (1997).

137. A. A. Jarzecki, P. M. Kozlowski, P. Pulay, B. H. Ye, and X. Y. Li, *Spectrochim. Acta, Part A* **53**, 1195 (1997).

138. G. D. Smith, O. Borodin, M. Pekny, B. Annis, D. Londono, and R. L. Jaffe, *Spectrochim. Acta, Part A* **53**, 1273 (1997).

139. M. A. Glaser, N. A. Clark, E. Garcia, and D. M. Walba, *Spectrochim. Acta, Part A* **53**, 1325 (1997).

140. D. T. Colbert and E. L. Sibert, III, *J. Chem. Phys.* **91**, 350 (1989).

141. A. B. McCoy and E. L. Sibert, III, *J. Chem. Phys.* **97**, 2938 (1992).

142. A. B. McCoy and E. L. Sibert, III, in *Dynamics of Molecules and Chemical Reactions* (R. E. Wyatt and J. Z. H. Zhang, eds.), 151, Dekker, New York, 1995.

143. M. J. Bramley, S. Carter, N. C. Handy, and I. M. Mills, *J. Mol. Spectrosc.* **157**, 301 (1993).

144. M. Joyeux, *Chem. Phys.* **221**, 269 (1997).

145. M. Joyeux, *Chem. Phys.* **221**, 287 (1997).

146. S. Carter and N. C. Handy, *Comput. Phys. Rep.* **5**, 15 (1986).

147. J. Tennyson, *Comput. Phys. Rep.* **4**, 1 (1986).

148. D. Searles and E. von Nagy-Felsobuki, *Ab initio Variational Calculations of Molecular Vibrational-rotational Spectra*, Springer-Verlag, Heidelberg, 1993.

149. M. J. Bramley and T. Carrington, Jr., *J. Chem. Phys.* **99**, 8519 (1993).

150. M. S. Child and R. T. Lawton, *Faraday Discuss. Chem. Soc.* **71**, 273 (1981).

151. L. Halonen, M. S. Child, and S. Carter, *Mol. Phys.* **47**, 1097 (1982).

152. I. M. Mills and F. J. Mompean, *Chem. Phys. Lett.* **124**, 425 (1986).

153. J. E. Baggott, *Mol. Phys.* **62**, 1019 (1987).

154. L. Halonen, *J. Phys. Chem.* **93**, 3386 (1989).

155. A. B. McCoy and E. L. Sibert, III, *J. Chem. Phys.* **92**, 1893 (1990).

156. E. Kauppi and L. Halonen, *J. Chem. Phys.* **96**, 2933 (1992).

157. R. C. Mayrhofer and E. L. Sibert, III, *Theor. Chim. Acta* **92**, 107 (1995).

158. E. L. Sibert, III and A. B. McCoy, *J. Chem. Phys.* **105**, 469 (1996).

159. L. Halonen, *J. Chem. Phys.* **106**, 831 (1997).

160. M. S. Child and L. Halonen, *Adv. Chem. Phys.* **57**, 1 (1984).

161. J. S. Hutchinson, *Adv. Chem. Phys.* **73**, 637 (1989).

162. B. R. Henry, in *Vibrational Spectra and Structure* (J. R. Durig, ed.), p. 270. Elsevier, Amsterdam, 1981.

163. J. L. Dunham, *Phys. Rev.* **41**, 721 (1932).

164. G. Simon, R. G. Parr, and J. M. Finlan, *J. Chem. Phys.* **59**, 3229 (1973).

165. J. F. Ogilvie, *Proc. R. Soc. London, Ser. A* **378**, 287 (1981).

166. R. G. Parr and R. J. White, *J. Chem. Phys.* **49**, 1059 (1968).

167. G. D. Carney, *Mol. Phys.* **37**, 659 (1979).

168. G. J. Sexton and N. C. Handy, *Mol. Phys.* **51**, 1321 (1984).

169. S. Carter and N. C. Handy, *J. Chem. Phys.* **87**, 4294 (1987).

170. L. B. Harding and W. C. Ermler, *J. Comput. Chem.* **6**, 13 (1985).

171. L. B. Harding and W. C. Ermler, *QCPE Bull.* **6**, 25 (1986).

172. D. J. Searles, S. J. Dunne, and E. I. von Nagy-Felsobuki, *Spectrochim. Acta, Part A* **44**, 505 (1988).

173. F. Culot and J. Liévin, *Phys. Scr.* **46**, 502 (1992).

174. F. Culot, F. Laruelle, and J. Liévin, *Theor. Chim. Acta* **92**, 211 (1995).

175. W. Meyer, P. Botshwina, and P. G. Burton, *J. Chem. Phys.* **84**, 891 (1986).

176. P. Jensen, *J. Mol. Spectrosc.* **128**, 478 (1988).

177. P. Jensen, *J. Chem. Soc., Faraday Trans. 2* **84**, 1315 (1988).

178. P. Jensen, in *Molecules in the Stellar Environment*, (U.G. Jørgensen, ed.), p. 478. Springer-Verlag, Berlin, 1994.

179. P. M. Morse, *Phys. Rev.* **34**, 57 (1929).

180. J. Tennyson and A. van der Avoird, *J. Chem. Phys.* **76**, 5710 (1982).

181. J. Tennyson, *Comput. Phys. Commun.* **38**, 39 (1986).

182. S. Carter, I. M. Mills, and N. C. Handy, *J. Chem. Phys.* **93**, 3722 (1990).

183. V. S. Vasan and R. J. Cross, *J. Chem. Phys.* **78**, 3869 (1982).

184. B. T. Sutcliffe and J. Tennyson, *Mol. Phys.* **58**, 1053 (1986).

185. E. Kauppi, *Chem. Phys. Lett.* **229**, 661 (1994).

186. R. N. Zare, *Angular Momentum*. Wiley, New York, 1988.

187. T. C. Thompson and D. G. Truhlar, *J. Chem. Phys.* **77**, 3031 (1982).

188. R. Lefebvre, *Int. J. Quantum Chem.* **23**, 543 (1983).

189. N. Moiseyev, *Chem. Phys. Lett.* **98**, 233 (1983).

190. R. M. Roth, R. B. Gerber, and M. A. Ratner, *J. Phys. Chem.* **87**, 2376 (1983).

191. Z. Bačić, R. B. Gerber, and M. A. Ratner, *J. Phys. Chem.* **90**, 3606 (1986).

192. L. L. Gibson, R. M. Roth, and M. A. Ratner, *J. Chem. Phys.* **85**, 4538 (1986).

193. R. B. Gerber and M. A. Ratner, *Adv. Chem. Phys.* **70**, 97 (1988).

194. A. Hidalgo, J. Zúñiga, J. M. Frances, A. Bastida, and A. Requena, *Int. J. Quantum Chem.* **40**, 685 (1991).

195. J. Zúñiga, A. Bastida, and A. Requena, *J. Phys. Chem.* **96**, 9691 (1992).

196. W. D. Allen, A. G. Császár, V. Szalay, and I. M. Mills, *Mol. Phys.* **89**, 1213 (1996).

197. D. W. Schwenke, *J. Phys. Chem.* **100**, 2867 (1996).

198. H. F. Schaefer, III, ed., Modern Theoretical Chemistry, Vol. 4. Plenum, New York, 1977.

199. D. G. Truhlar, ed., *Potential Energy Surfaces and Dynamics Calculations*. Plenum, New York, 1981.

200. S. Wilson and G. H. F. Diercksen, *Methods in Computational Molecular Physics*. Plenum, New York, 1992.

201. P. von. R. Schleyer, N. L. Allinger, T. Clark, J. Gasteiger, P. Kollmann, and H. F. Schaefer III, eds., *Encyclopedia of Computational Chemistry*. Wiley, New York, 1998.

202. C. W. Bauschlicher and P. R. Taylor, *Theor. Chim. Acta* **71**, 263 (1987).

203. C. W. Bauschlicher and S. R. Langhoff, *Theor. Chim. Acta* **731**, 43 (1988).

204. H. Koch, G. E. Scuseria, C. Scheiner, and H. F. Schaefer, II, *Chem. Phys. Lett.* **149**, 43 (1988).

205. J. F. Stanton, W. N. Lipscomb, D. H. Magers, and R. J. Bartlett, *J. Chem. Phys.* **90**, 3241 (1989).

206. W. D. Allen, Y. Yamaguchi, A. G. Császár, D. A. Clabo, Jr., R. B. Remington, and H. F. Schaefer, III, *Chem. Phys.* **145**, 427 (1990).

207. J. R. Thomas, B. J. DeLeeuw, G. Vacek, and H. F. Schaefer, III, *J. Chem. Phys.* **98**, 1336 (1993).

208. J. L. Martin and P. R. Taylor, *Chem. Phys. Lett.* **225**, 473 (1994).

209. J. L. Martin, *J. Chem. Phys.* **100**, 8186 (1994).

210. J. M. L. Martin, T. J. Lee, and P. R. Taylor, *J. Chem. Phys.* **108**, 676 (1998).

211. K. A. Peterson and T. H. Dunning, Jr., *J. Chem. Phys.* **106**, 4119 (1997).

212. J. Olsen, P. Jørgensen, and J. Simons, *Chem. Phys. Lett.* **169**, 463 (1990).

213. A. O. Mitrushenkov, *Chem. Phys. Lett.* **217**, 559 (1994).

214. J. Andzelm, M. Klobukowski, E. Radzio-Andzelm, Y. Sakai, and H. Tatewaki, *Gaussian Basis Sets for Molecular Calculations*. Elsevier, New York, 1984.

215. R. Poirier, R. Kari, and I. G. Csizmadia, *Handbook of Gaussian Basis sets. Compendium for ab initio Molecular Orbital Calculations*. Elsevier, New York, 1985.

216. W. J. Hehre, L. Radom, P. R. von Schleyer, and J. A. Pople, *Ab initio Molecular Orbital Theory*, Wiley, New York, 1986.

217. K. Ruedenberg, M. W. Schmidt, M. M. Gilbert, and S. T. Elbert, *Chem. Phys.* **71**, 41 (1982).

218. B. O. Roos, *Adv. Chem. Phys.* **67**, 399 (1987).

219. I. Shavitt, in *Modern Theoretical Chemistry* (H. F. Schaefer, III, ed.), Vol. 3, p. 189. Plenum, New York, 1977.

220. R. J. Buenker, S. D. Peyerimhoff, and W. Butscher, *Mol. Phys.* **37**, 771 (1978).

221. R. J. Buenker, in *Current Aspect of Quantum Chemistry* (R. Carbo, ed.) p. 17, Elsevier, Amsterdam, 1983.

222. H.-J. Werner and P. J. Knowles, *J. Chem. Phys.* **89**, 5803 (1988).

223. P. J. Knowles and H.-J. Werner, *Chem. Phys. Lett.* **145**, 514 (1988).

224. S. R. Langhoff and E. R. Davidson, *Int. J. Quantum Chem.* **8**, 62 (1974).

225. J. Cizek, *Adv. Chem. Phys.* **14**, 35 (1969).

226. G. D. Purvis, III and R. J. Bartlett, *J. Chem. Phys.* **76**, 1910 (1982).

227. G. E. Scuseria and H. F. Schaefer, III, *J. Chem. Phys.* **90**, 3700 (1989).

228. G. E. Scuseria, C. L. Janssen, and H. F. Schaefer, III, *J. Chem. Phys.* **89**, 7382 (1988).

229. J. Paldus, in *Methods in Computational Molecular Physics*, (S. Wilson and G. H. F. Diercksen, eds.), NATO ASI Ser., Ser. B: Phys. p. 99. Plenum, New York and London, 1992.

230. K. Raghavachari, G. W. Trucks, J. A. Pople, and M. Head-Gordon, *Chem. Phys. Lett.* **157**, 479 (1989).

231. T. J. Lee and J. E. Rice, *Chem. Phys. Lett.* **150**, 406 (1988).
232. C. Hampel, K. A. Peterson, and H.-J. Werner, *Chem. Phys. Lett.* **190**, 1 (1992).
233. C. Møller and M. S. Plesset, *Phys. Rev.* **46**, 618 (1934).
234. J. A. Pople, R. Seeger, and R. Krishnan, *Int. J. Quantum Chem. Symp.* **11**, 149 (1977).
235. R. Krishnan and J. A. Pople, *Int. J. Quantum Chem.* **14**, 91 (1978).
236. R. Krishnan, M. J. Frisch, and J. A. Pople, *J. Chem. Phys.* **72**, 4244 (1980).
237. Y. El Youssoufi, M. Herman, J. Liévin, and I. Kleiner, *Spectrochim. Acta, Part A* **53**, 881 (1997).
238. E. S. Kryachko and E. V. Ludeña, *Energy Density Functional Theory of Many-electron Systems.* Kluwer Academic Publishers, Dordrecht, The Netherlands, 1990.
239. N. C. Handy, C. W. Murray, and R. D. Amos, *J. Phys. Chem.* **97**, 4392 (1993).
240. J. M. L. Martin, *Mol. Phys.* **86**, 1437 (1995).
241. M. W. Wong, *Chem. Phys. Lett.* **256**, 391 (1996).
242. F. De Proft, J. M. L. Martin, and P. Geerlings, *Chem. Phys. Lett.* **250**, 393 (1996).
243. E. Clementi, G. Corongiu, and O. G. Stradella, in *Modern Techniques in Computational Chemistry: MOTECC-91*, E. Clementi, Ed. p. 295. ESCOM, Leiden, 1991.
244. A. D. Becke, *Phys. Rev. A* **38**, 3098 (1988).
245. C. Lee, W. Yang, and R. G. Parr, *Phys. Rev. B* **37**, 785 (1988).
246. W. Kohn and L. J. Sham, *Phys. Rev. A* **140**, 1133 (1965).
247. B. Feuston, C. Lee, and E. Clementi, in *Modern Techniques in Computational Chemistry: MOTECC-91*, p. 279. ESCOM, Leiden, 1991.
248. P. G. Mezey, *Potential Energy Hypersurfaces.* Elsevier, Amsterdam, 1987.
249. J. N. Murrel, S. Carter, C. Farantos, P. Huxley, and A. J. C. Varandas, *Molecular Potential Energy Functions.* Wiley, Brisbane, 1984.
250. K. S. Sorbie and J. N. Murrel, *Mol. Phys.* **29**, 1385 (1975).
251. J. N. Murrel and S. Farantos, *Mol. Phys.* **34**, 1185 (1977).
252. J. N. Murrel, S. Carter, I. M. Mills, and M. F. Guest, *Mol. Phys.* **42**, 605 (1981).
253. A. J. C. Varandas and J. N. Murrel, *Chem. Phys. Lett.* **88**, 1 (1982).
254. S. Carter, I. M. Mills, J. N. Murrel, and A. J. C. Varandas, *Mol. Phys.* **45**, 1053 (1982).
255. J. N. Murrel, W. Craven, M. Vincent, and Z. H. Zhu, *Mol. Phys.* **56**, 839 (1985).
256. J. Ischtwan and M. A. Collins, *J. Chem. Phys.* **100**, 8080 (1994).
257. T. S. Ho and H. Rabitz, *J. Chem. Phys.* **104**, 2584 (1996).
258. M. Sana, *Int. J. Quantum Chem.* **19**, 139 (1981).
259. M. Sana, *QCPE Bull.* **19**, 117 (1981).
260. M. Sana, *Theor. Chim. Acta* **60**, 543 (1982).
261. M. Sana, *Rev. Quest. Sci.* **155**, 345 (1984).
262. G. E. P. Box, *Appl. Stat.* **6**, 81 (1957).
263. G. E. P. Box, *Biometrics* **10**, 16 (1954).
264. G. E. P. Box and J. S. Hunter, *Ann. Math. Stat.* **28**, 195 (1957).
265. G. E. P. Box and J. S. Hunter, *Technometrics* **3**, 449 (1961).
266. D. H. Doehlert, *Appl. Stat.* **19**, 231 (1970).
267. K. G. Roquemore, *Technometrics* **18**, 419 (1976).
268. J. Liévin, *J. Mol. Spectrosc.* **156**, 123 (1992).

269. F. Culot, Ph.D. Thesis, Université Libre de Bruxelles, 1993.

270. F. Culot and J. Liévin, in preparation (1998).

271. G. Fogarasi and P. Pulay, in *Vibrational Spectra and Structure* (J. R. Durig, ed.), p. 125. Elsevier, Amsterdam, 1985.

272. M. Abramowitz and I. A. Stegun, eds., *Handbook of Mathematical Functions*, Appl. Math. Ser., Vol. 55. Dover, New York, 1968.

273. J. Geratt and I. M. Mills, *J. Chem. Phys.* **49**, 1719 (1968).

274. P. Pulay, in *Ab initio Methods in Quantum Chemistry* (K. P. Lawley, ed.), p. 241. Wiley, New York, 1987.

275. T. Helgaker and P. Jørgensen, in *Methods in Computational Molecular Physics*, (S. Wilson and G. H. F. Diercksen, eds.), NATO ASI Ser., Ser. B: Phys., p. 353. Plenum, New York and London, 1992.

276. Y. Yamaguchi, Y. Osamura, J. D. Goddard, and H. F. Schaefer, III, *A New Dimension to Quantum Chemistry: Analytic Derivative Methods in ab initio Molecular Electronic Structure Theory*. Oxford University Press, New York, 1994.

277. M. J. Frisch, G. W. Trucks, H. B. Schlegel, P. M. W. Gill, B. G. Johnson, M. A. Robb, J. R. Cheeseman, T. A. Keith, G. A. Peterson, J. A. Montgomery, K. Raghavachari, M. A. Al-Laham, V. G. Zakrzewski, J. V. Ortiz, J. B. Foresman, J. Cioslowski, B. B. Stefanov, A. Nanayakkara, M. Challacombe, C. Y. Peng, P. Y. Ayala, W. Chen, M. W. Wong, J. L. Andres, E. S. Replogle, R. Gomperts, R. L. Martin, D. J. Fox, J. S. Binkley, D. J. Defrees, et al., *Gaussian 94*, Gaussian, Inc., Pittsburgh, PA., 1995.

278. M. Dupuis and S. A. Maluendes, in *Modern Techniques in Computational Chemistry: MOTECC-91*, E. Clementi, Ed., p. 469. ESCOM, Leiden, 1991.

279. R. D. Amos and J. E. Rice, *The Cambridge Analytic Derivatives Package*, CADPAC, Issue 4.0. Cambridge, 1987.

280. J. F. Gaw, Y. Yamaguchi, and H. F. Schaefer, III, *J. Chem. Phys.* **81**, 6395 (1984).

281. J. F. Gaw and N. C. Handy, in *Geometrical Derivatives of Energy Surfaces and Molecular Properties* (P. Jorgensen and J. Simons, eds.), NATO ASI Ser., Ser. C, p. 79. Reidel Publ., Dordrecht, The Netherlands, 1986.

282. W. H. Green, D. Jayatilaka, A. Willetts, R. D. Amos, and N. C. Handy, *J. Chem. Phys.* **93**, 4965 (1990).

283. H. B. Schlegel, in *New Theoretical Concepts for Understanding Organic Reactions* (J. Bertrán and I. G. Csizmadia, eds.), NATO ASI Ser., Ser. C, p. 79. Kluwer Academic Publishers, Dordrecht, The Netherlands, 1989.

284. F. Bernardi and M. A. Robb, *Adv. Chem. Phys.* **67**, 155 (1987).

284a. H. B. Schlegel, *Adv. Chem. Phys.* **67**, 249 (1987).

285. R. Fletcher and M. J. D. Powell, *Comput. J.* **6**, 163 (1963).

286. C. G. Broyden, *Math. Comput.* **21**, 368 (1967).

287. W. C. Davidon, *Comput. J.* **10**, 406 (1968).

288. R. Fletcher, *Comput. J.* **13**, 317 (1970).

289. B. A. Murtagh and R. W. H. Sargent, *Comput. J.* **13**, 185 (1970).

290. D. F. Shanno, *Math. Comput.* **24**, 647 (1970).

291. D. Goldfarb, *Math. Comput.* **24**, 23 (1970).

292. W. D. Allen and A. G. Császár, *J. Chem. Phys.* **98**, 2983 (1993).

293. P. Palmieri, R. Tarroni, M. M. Hühn, N. C. Handy, and A. Willetts, *Chem. Phys.* **190**, 327 (1995).

294. E. Kauppi, *J. Chem. Phys.* **105**, 7986 (1996).

295. J. Tennyson, S. Miller, and C. R. Le Sueur, *Comput. Phys. Commun.* **75**, 339 (1993).

296. J. A. Bentley, R. E. Wyatt, M. Menou, and C. Leforestier, *J. Chem. Phys.* **97**, 4255 (1992).

297. M. J. Bramley and N. C. Handy, *J. Chem. Phys.* **98**, 1378 (1993).

298. E. L. Sibert, III and R. C. Mayrhofer, *J. Chem. Phys.* **99**, 937 (1993).

299. M. J. Bramley and T. Carrington, Jr., *J. Chem. Phys.* **101**, 8494 (1994).

300. P. N. Day and D. G. Truhlar, *J. Chem. Phys.* **95**, 6615 (1991).

301. Z. Bačić and J. C. Light, *J. Chem. Phys.* **90**, 1774 (1987).

302. J. Antikainen, R. Friesner, and C. Leforestier, *J. Chem. Phys.* **102**, 1270 (1995).

303. G. Yan, D. Xie, and A. Tian, *J. Phys. Chem.* **98**, 8870 (1994).

304. C. L. Chen, B. Mässen, and M. Wolfsberg, *J. Chem. Phys.* **83**, 1795 (1985).

305. B. T. Sutcliffe and J. Tennyson, *Mol. Phys.* **58**, 1067 (1986).

306. S. E. Choi and J. C. Light, *J. Chem. Phys.* **92**, 2129 (1990).

307. J. M. Bowman, *J. Chem. Phys.* **68**, 608 (1978).

308. J. M. Bowman, K. Christoffel, and F. M. Tobin, *J. Phys. Chem.* **83**, 905 (1979).

309. M. Cohen, S. Greita, and R. P. McEachran, *Chem. Phys. Lett.* **60**, 445 (1979).

310. R. B. Gerber and M. A. Ratner, *Chem. Phys. Lett.* **68**, 195 (1979).

311. F. L. Tobin and J. M. Bowman, *Chem. Phys.* **47**, 151 (1980).

312. K. M. Christoffel and J. M. Bowman, *Chem. Phys. Lett.* **85**, 220 (1982).

313. J. M. Bowman, *Acc. Chem. Res.* **19**, 202 (1986).

314. J. M. Bowman, A. Wierzbicki, and J. Zúñiga, *Chem. Phys. Lett.* **150**, 269 (1988).

315. A. Hidalgo, J. Zúñiga, A. Bastida, and A. Requena, *Int. J. Quantum Chem.* **36**, 49 (1989).

316. F. Culot and J. Liévin, *Theor. Chim. Acta* **89**, 227 (1994).

317. G. Yan, D. Xie, J. Xie, and A. Tian, *J. Phys. Chem.* **97**, 1507 (1993).

318. D. W. Schwenke, *J. Chem. Phys.* **96**, 3426 (1992).

319. F. Grein and T. C. Chang, *Chem. Phys. Lett.* **91**, 149 (1971).

320. K. Ruedenberg, L. M. Cheung, and S. T. Elbert, *Int. J. Quantum Chem.* **16**, 1069 (1979).

321. B. Levy and G. Berthier, *Int. J. Quantum Chem.* **2**, 307 (1968).

322. M. Godefroid, J. Liévin, and J.-Y. Metz, *Int. J. Quantum Chem.* **40**, 243 (1991).

323. J. C. Light, I. P. Hamilton, and J. V. Lill, *J. Chem. Phys.* **82**, 1400 (1985).

324. Z. Bačić and J. C. Light, *J. Chem. Phys.* **85**, 4594 (1986).

325. R. M. Whitnell and J. C. Light, *J. Chem. Phys.* **90**, 1774 (1989).

326. W. Yang and A. C. Peet, *Chem. Phys. Lett.* **153**, 98 (1988).

327. H. Wei and T. Carrington, Jr., *J. Chem. Phys.* **97**, 3029 (1992).

328. M. J. Bramley, J. W. Tromp, T. Carrington, Jr., and G. C. Corey, *J. Chem. Phys.* **100**, 6175 (1994).

329. H. Wei and T. Carrington, Jr., *J. Chem. Phys.* **101**, 1343 (1994).

330. V. Szalay and L. Nemes, *Chem. Phys. Lett.* **231**, 225 (1994).

331. J. R. Henderson and J. Tennyson, *Comput. Phys. Commun.* **75**, 365 (1993).

332. J. R. Henderson, C. R. Le Sueur, and J. Tennyson, *Comput. Phys. Commun.* **75**, 379 (1993).

333. J. Tennyson, J. R. Henderson, and N. G. Fulton, *Comput. Phys. Commun.* **89**, 175 (1995).

334. N. M. Poulin, M. J. Bramley, T. Carrington, Jr., H. G. Kjaergaard, and B. R. Henry, *J. Chem. Phys.* **104**, 7807 (1996).

335. H. Wei, *J. Chem. Phys.* **106**, 6885 (1997).

336. R. E. Wyatt, Adv. *Chem. Phys.* **73**, 231 (1989).

337. A. Nauts and R. E. Wyatt, *Phys. Rev. Lett.* **51**, 2238 (1983).

338. A. Nauts and R. E. Wyatt, *Phys. Rev. A* **30**, 872 (1984).

339. A. McNichols and T. Carrington, Jr., *Chem. Phys. Lett.* **202**, 464 (1993).

340. J. Cullum and R. A. Willoughby, *Lancsoz Algorithms for Large Symmetric Eigenvalue Computations*, Birkhäuser, Boston, 1981.

341. J. Cullum and R. A. Willoughby, *J. Comput. Phys.* **44**, 329 (1981).

342. G. Charron and T. Carrington, Jr., *Mol. Phys.* **79**, 13 (1993).

343. P. N. Roy and T. Carrington, Jr., *Chem. Phys. Lett.* **257**, 98 (1996).

344. R. A. Friesner, J. A. Bentley, M. Menou, and C. Leforestier, *J. Chem. Phys.* **99**, 324 (1993).

345. S. E. Choi and J. C. Light, *J. Chem. Phys.* **97**, 7031 (1992).

346. B. R. Johnson and W. P. Reinhardt, *J. Chem. Phys.* **85**, 4538 (1986).

347. Z. Bačić, *J. Chem. Phys.* **95**, 3546 (1991).

348. N. M. Harvey and D. G. Truhlar, *Chem. Phys. Lett.* **74**, 252 (1980).

349. A. B. McCoy and E. L. Sibert, III, *J. Chem. Phys.* **105**, 459 (1996).

350. J. M. Bowman, *Chem. Phys. Lett.* **217**, 36 (1994).

351. P. Jensen, *Comput. Phys. Rep.* **1**, 1 (1983).

352. P. Jensen and P. R. Bunker, *J. Mol. Spectrosc.* **118**, 18 (1988).

353. J. H. Schryber, O. L. Polyansky, P. Jensen, and J. Tennyson, *J. Mol. Spectrosc.* **185**, 234 (1997).

354. S. A. Tashkun and P. Jensen, *J. Mol. Spectrosc.* **163**, 173 (1994).

355. P. C. Gomez, L. F. Pacios, and P. Jensen, *J. Mol. Spectrosc.* **186**, 99 (1997).

356. O. L. Polyansky, P. Jensen, and J. Tennyson, *J. Mol. Spectrosc.* **178**, 184 (1996).

357. P. C. Gomez and P. Jensen, *J. Mol. Spectrosc.* **185**, 282 (1997).

358. E. B. Wilson and J. B. Howard, *J. Chem. Phys.* **4**, 260 (1936).

359. H. H. Nielsen, *Rev. Mod. Phys.* **23**, 90 (1951).

360. H. H. Nielsen, *Handb. Physiol.* **37**/(1), 173 (1959).

361. S. Maes and G. Amat, *Can. J. Phys.* **43**, 321 (1965).

362. G. Amat and H. H. Nielsen, *J. Mol. Spectrosc.* **2**, 152 (1958).

363. G. Amat and H. H. Nielsen, *J. Mol. Spectrosc.* **2**, 163 (1958).

364. G. Amat and H. Henri, *Cah. Phys.* **12**, 273 (1958).

365. G. Amat and H. Henri, *Cah. Phys.* **14**, 230 (1960).

366. T. Oka and Y. Morino, *J. Mol. Spectrosc.* **6**, 472 (1961).

367. T. Oka, *J. Chem. Phys.* **47**, 5410 (1967).

368. D. Kivelson and E. B. Wilson, *J. Chem. Phys.* **20**, 575 (1952).

369. J. K. G. Watson, *J. Chem. Phys.* **45**, 1360 (1966).

370. J. K. G. Watson, *J. Chem. Phys.* **46**, 1935 (1967).

371. J. K. G. Watson, *Mol. Phys.* **15**, 479 (1968).

372. J. K. G. Watson, *Mol. Phys.* **19**, 465 (1970).

373. I. M. Mills, *Molecular Spectroscopy: Modern Research*, Vol. I, p. 1, K. Narahari Rao and C. W. Matthews, Eds., Academic Press, New York, 1972.

374. J. Pliva, *J. Mol. Spectrosc.* **139**, 278 (1990).

375. J. F. Gaw, A. Willets, W. H. Green, and N. C. Handy, in *Advances in Molecular Vibrations and Collision Dynamics* (J. M. Bowman, ed.), p. 169. JAI Press, Greenwich, CT, 1991.

376. J. M. L. Martin, T. J. Lee, P. R. Taylor, and J. P. François, *J. Chem. Phys.* **103**, 2589 (1995).

377. J. M. Martin and R. P. Taylor, *Spectrochim. Acta, Part A* **53**, 1039 (1997).

378. M. M. Law and J. L. Duncan, *Chem. Phys. Lett.* **212**, 172 (1993).

379. E. L. Sibert, III, *Chem. Phys. Lett.* **128**, 404 (1986).

380. E. L. Sibert, III, *J. Chem. Phys.* **88**, 4378 (1988).

381. E. L. Sibert, III, *Comput. Phys. Commun.* **51**, 149 (1988).

382. A. B. McCoy and E. L. Sibert, III, in *Advances in Molecular Vibrations and Collision Dynamics* (J. L. Bowman, ed.), 255, JAI Press, Greenwich, CT, 1991.

383. E. L. Sibert, III, *J. Chem. Phys.* **90**, 2672 (1989).

384. A. B. McCoy, D. C. Burleigh, and E. L. Sibert, III, *J. Chem. Phys.* **95**, 7449 (1991).

385. G. Herzberg, *Discuss. Faraday Soc.* **35**, 7 (1963).

386. R. Colin, M. Herman, and I. Kopp, *Mol. Phys.* **37**, 1397 (1979).

387. P. G. Wilkinson, *J. Mol. Spectrosc.* **6**, 1 (1961).

388. P. G. Wilkinson, *J. Mol. Spectrosc.* **2**, 387 (1958).

389. J. K. Lundberg, D. M. Jonas, Y. Chen, B. Rajaram, and R. W. Field, *J. Chem. Phys.* **97**, 7180 (1992).

390. M. N. R. Ashfold, B. Tutcher, B. Yang, Z. K. Jin, and S. L. Anderson, *J. Chem. Phys.* **87**, 5105 (1987).

391. J. H. Fillion, A. Campos, J. Pedersen, N. Shafizadeh, and D. Gauyacq, *J. Chem. Phys.* **105**, 22 (1996).

392. F. Laruelle, Ph.D. Thesis, Université Libre de Bruxelles, 1997.

393. J. T. Hougen and J. K. G. Watson, *Can. J. Phys.* **43**, 298 (1965).

394. J. K. G. Watson, M. Herman, J. C. Van Craen, and R. Colin, *J. Mol. Spectrosc.* **95**, 101 (1982).

395. R. T. Huet, M. Godefroid, and M. Herman, *J. Mol. Spectrosc.* **144**, 32 (1990).

396. P. Dupre, *Chem. Phys. Lett.* **212**, 555 (1993).

397. M. Drabbels, J. Heinze, and W. L. Meerts, *J. Chem. Phys.* **100**, 165 (1994).

398. S. Drucker, J. P. O'Brien, and R. W. Field, *J. Chem. Phys.* **106**, 3423 (1997).

399. Y.-C. Hsu, M.-S. Lin, and C. P. Hsu, *J. Chem. Phys.* **94**, 7832 (1991).

400. M. Fujii, A. Haijima, and M. Ito, *Chem. Phys. Lett.* **150**, 380 (1988).

401. M. Fujii, S. Tanabe, Y. Okuzawa, and M. Ito, *Appl. Phys. B* **14**, 161 (1994).

402. C. K. Ingold and G. W. King, *J. Chem. Soc.*, p. 2702 (1953).

403. K. K. Innes, *J. Chem. Phys.* **22**, 863 (1954).

404. A. L. Utz, J. D. Tobiason, L. J. Sanders, and F. F. Crim, *J. Chem. Phys.* **98**, 2742 (1993).

405. J. D. Tobiason, A. L. Utz, and F. F. Crim, *J. Chem. Phys.* **99**, 928 (1993).

406. J. D. Tobiason, A. L. Utz, E. L. Sibert, III, and F. F. Crim, *J. Chem. Phys.* **99**, 5762 (1993).

407. S. J. Humphrey, C. G. Morgan, A. M. Wodtke, K. L. Cunningham, S. Drucker, and R. W. Field, *J. Chem. Phys.* **107**, 49 (1997).

408. S. Trajmar, J. K. Rice, P. S. P. Wei, and A. Kupperman, *Chem. Phys. Lett.* **1**, 703 (1968).

409. H. R. Wendt, H. Hippler, and H. E. Hunziker, *J. Chem. Phys.* **70**, 4044 (1979).

410. J. M. Lisy and W. Klemperer, *J. Chem. Phys.* **72**, 3880 (1980).

411. C. D. Sherrill, G. Vacek, Y. Yamaguchi, H. F. Schaeffer, III, J. F. Stanton, and J. Gauss, *J. Chem. Phys.* **104**, 8507 (1996).

412. K. Tanigawa and H. Kanamori, *The Future of Spectroscopy, Conference, St Adèle, Québec*, poster T.44, A. R. W. McKellar ed., CNRC, (1994).

413. J. K. Lundberg, Y. Chen, J. P. Pique, and R. W. Field, *J. Mol. Spectrosc.* **156**, 104 (1992).

414. W. C. Price, *Phys. Rev.* **47**, 444 (1935).

415. M. Herman and R. Colin, *Phys. Scr.* **25**, 275 (1982).

416. M. Perić, S. D. Peyerimhoff, and R. J. Buenker, *Mol. Phys.* **62**, 1339 (1987).

417. M. Perić, R. J. Buenker, and S. D. Peyerimhoff, *Mol. Phys.* **53**, 1177 (1984).

418. M. Perić, S. D. Peyerimhoff, and R. J. Buenker, *Mol. Phys.* **55**, 1129 (1985).

419. P. G. Green, J. L. Kinsey, and R. W. Field, *J. Chem. Phys.* **91**, 5160 (1989).

420. J. Berkowitz, G. B. Ellison, and D. Gutman, *J. Phys. Chem.* **98**, 2744 (1994).

421. D. H. Mordaunt and M. N. R. Ashfold, *J. Chem. Phys.* **101**, 2630 (1994).

422. N. Hashimoto and T. Suzuki, *J. Chem. Phys.* **104**, 6070 (1996).

423. H. Hashimoto, N. Yonekura, and T. Suzuki, *Chem. Phys. Lett.* **264**, 554 (1997).

424. J. Zhang, C. W. Riehn, M. Dulligan, and C. Wittig, *J. Chem. Phys.* **103**, 6815 (1995).

425. J. H. Fillion, N. Shafizadeh, and D. Gauyacq, *AIP Conf. Proc.* **312**, 395 (1994).

426. M. Ukai, K. Kameta, K. Chiba, K. Nagano, N. Kouchi, K. Shinsaka, Y. Hatano, H. Umemoto, Y. Ito, and K. Tanaka, *J. Chem. Phys.* **95**, 4142 (1991).

427. F. A. Cotton, *Chemical Applications of Group Theory*. Wiley-Interscience, New York, 1971.

428. Y. Yamaguchi, G. Vacek, and H. F. Schaefer, III, *Theor. Chim. Acta* **86**, 1993 (1993).

429. P. Halvick, D. Liotard, and J. C. Rayez, *Chem. Phys.* **177**, 69 (1993).

430. N.-Y. Chang, M.-Y. Shen, and C.-H. Yu, *J. Chem. Phys.* **106**, 3237 (1997).

431. R. S. Urdahl, Y. Bao, and W. M. Jackson, *Chem. Phys. Lett.* **178**, 425 (1991).

432. K. M. Ervin, S. Gronert, S. E. Barlow, M. K. Gilles, A. G. Harrison, V. M. Bierbaum, C. H. DePuy, W. C. Lineberger, and G. B. Ellison, *J. Am. Chem. Soc.* **112**, 5750 (1990).

433. H. Partridge and C. W. Bauschlicher, Jr., *J. Chem. Phys.* **103**, 10589 (1995).

434. B. A. Balko, J. Zhang, and Y. T. Lee, *J. Chem. Phys.* **94**, 7958 (1991).

435. J. E. Reutt, L. S. Wang, J. E. Pollard, D. J. Trevor, Y. T. Lee, and D. A. Shirley, *J. Chem. Phys.* **84**, 3022 (1986).

436. M. Rerat, D. Liotard, and J. M. Robine, *Theor. Chim. Acta* **88**, 285 (1994).

437. S. Carter, I. M. Mills, and J. N. Murrell, *Mol. Phys.* **41**, 191 (1980).

438. T. A. Holme and R. D. Levine, *Chem. Phys.* **131**, 169 (1989).

439. P. Duffy, S. A. C. Clark, C. E. Brion, M. E. Casida, D. P. Chong, E. R. Davidson, and C. Maxwell, *Chem. Phys.* **165**, 183 (1992).

440. J. R. Thomas, B. J. DeLeeuw, G. Vacek, T. D. Crawford, Y. Yamaguchi, and H. F. Schaefer, III, *J. Chem. Phys.* **99**, 403 (1993).

441. C. W. Bauschlicher, Jr. and H. Partridge, *Chem. Phys. Lett.* **245**, 158 (1995).

442. R. Prosmiti and S. C. Farantos, *J. Chem. Phys.* **103**, 3299 (1995).

443. I. Suzuki and J. Overend, *Spectrochim. Acta, Part A* **25**, 977 (1969).

444. J. Liévin, M. Abbouti Temsamani, P. Gaspard, and M. Herman, *Chem. Phys.* **190**, 419 (1995).

445. K. K. Lehmann, G. J. Scherer, and W. Klemperer, *J. Chem. Phys.* **79**, 1369 (1983).

446. F. Iachello, S. Oss, and R. Lemus, *J. Mol. Spectrosc.* **146**, 56 (1991).

447. F. Iachello, S. Oss, and R. Lemus, *J. Mol. Spectrosc.* **149**, 132 (1991).

448. F. Iachello and S. Oss, *J. Chem. Phys.* **104**, 6956 (1996).

449. S. Carter and N. C. Handy, *Comput. Phys. Commun.* **51**, 49 (1988).

450. M. I. El Idrissi, J. Liévin, A. Campargue, and M. Herman, *J. Chem. Phys.*, **110**, in press (1999).

451. J. Vander Auwera, D. Hurtmans, M. Carleer, and M. Herman, *J. Mol. Spectrosc.* **157**, 337 (1993).

452. K. F. Palmer, M. E. Mickelson, and K. N. Rao, *J. Mol. Spectrosc.* **44**, 131 (1972).

453. Q. Kou, G. Guelachvili, M. Abbouti Temsamani, and M. Herman, *Can. J. Phys.* **72**, 1241 (1994).

454. D. Jayatikala, P. E. Maslen, R. D. Amos, and N. C. Handy, *Mol. Phys.* **75**, 271 (1992).

455. D. A. Clabo, Jr., W. D. Allen, R. B. Remington, Y. Yamaguchi, and H. F. Schaeffer, III, *Chem. Phys.* **123**, 187 (1988).

456. M. R. Aliev and J. K. G. Watson, in *Molecular spectroscopy: Modern research* (K. N. Rao and C. W. Mathews, eds.), p. 1. Academic Press, Orlando, FL, 1985.

457. E. D. Simandiras, J. E. Rice, T. J. Lee, R. D. Amos, and N. C. Handy, *J. Chem. Phys.* **88**, 3187 (1988).

458. A. Cantarella, F. Culot, and J. Liévin, *Phys. Scr.* **46**, 489 (1992).

459. P. Botschwina, *Chem. Phys.* **68**, 41 (1982).

460. H. Romanowski and J. M. Bowman, *QCPE Bull.* **5**, 64 (1985).

461. J.-L. Teffo, O. M. Lyulin, V. Perevalov, and E. I. Lobodenko, *J. Mol. Spectrosc.* **187**, 28 (1998).

462. V. I. Perevalov, E. I. Lobodenko, O. M. Lyulin, and J.-L. Teffo, *J. Mol. Spectrosc.* **171**, 435 (1995).

463. O. M. Lyulin, V. I. Perevalov, and J.-L. Teffo, *J. Mol. Spectrosc.* **174**, 566 (1995).

464. J. Almlöf and P. R. Taylor, *J. Chem. Phys.* **86**, 4070 (1987).

465. T. H. Dunning, *J. Chem. Phys.* **90**, 1007 (1989).

466. R. A. Kendall, T. H. Dunning, and R. J. Harrison, *J. Chem. Phys.* **96**, 6796 (1992).

467. J. E. Del Bene, *J. Phys. Chem.* **97**, 107 (1993).

468. M. Abbouti Temsamani and M. Herman, *J. Chem. Phys.* **102**, 6371 (1995).

469. M. Abbouti Temsamani, M. Herman, S. A. B. Solina, J. P. O'Brien, and R. W. Field, *J. Chem. Phys.* **105**, 11357 (1996).

470. A. F. Borro, I. M. Mills, and E. Venuti, *J. Chem. Phys.* **102**, 3938 (1995).

471. W. Kolos and L. Wolniewicz, *Rev. Mod. Phys.* **35**, 473 (1963).

472. J. Pliva, *J. Mol. Spectrosc.* **44**, 165 (1972).

473. D. M. Jonas, S. A. B. Solina, B. Rajaram, S. J. Cohen, R. J. Silbey, R. W. Field, K. Yamanouchi, and S. Tsuchiya, *J. Chem. Phys.* **99**, 7350 (1993).

474. S. A. B. Solina, J. P. O'Brien, R. W. Field, and W. F. Polik, *Ber. Bunsenges. Phys. Chem.* **99**, 555 (1995).

475. T. A. Holme and R. D. Levine, *J. Chem. Phys.* **89**, 3379 (1988).

476. E. Abramson, R. W. Field, D. Imre, K. K. Innes, and J. L. Kinsey, *J. Chem. Phys.* **83**, 453 (1985).

477. R. L. Sundberg, E. Abramson, J. L. Kinsey, and R. W. Field, *J. Chem. Phys.* **83**, 466 (1985).

478. J. P. Pique, Y. Chen, R. W. Field, and J. L. Kinsey, *Phys. Rev. Lett.* **58**, 475 (1987).

479. J. P. Pique, Y. M. Engel, R. D. Levine, Y. Chen, R. W. Field, and J. L. Kinsey, *J. Chem. Phys.* **88**, 5972 (1988).

480. J. P. Pique, M. Lombardi, Y. Chen, R. W. Field, and J. L. Kinsey, *Ber. Bunsenges. Phys. Chem.* **92**, 422 (1988).

481. P. Gaspard and P. van Ede van der Pals, unpublished results (1998).

482. P. van Ede van der Pals and P. Gaspard, unpublished results (1998).

483. J. P. Rose and M. E. Kellman, *J. Chem. Phys.* **105**, 10743 (1996).

484. M. E. Kellman, *J. Chem. Phys.* **83**, 3843 (1985).

485. M. E. Kellman and E. D. Lynch, *J. Chem. Phys.* **85**, 5855 (1986).

486. L. Xiao and M. E. Kellman, *J. Chem. Phys.* **90**, 6086 (1989).

487. L. Xiao and M. E. Kellman, *J. Chem. Phys.* **93**, 5805 (1990).

488. S. K. Gray and M. S. Child, *Mol. Phys.* **53**, 961 (1984).

489. J. Hornos and F. Iachello, *J. Chem. Phys.* **90**, 5284 (1989).

490. F. Iachello and R. D. Levine, *Algebraic Theory of Molecules*, Oxford University Press, Oxford, 1994.

491. A. Frank and P. Van Isacker, *Algebraic Methods in Molecular and Nuclear Structure Physics*, Wiley, New York, 1994.

492. S. Oss, *Adv. Chem. Phys.* **93**, 455 (1996).

493. M. E. Kellman, *Annu. Rev. Phys. Chem.* **46**, 395 (1995).

494. C. J. H. Schutte, J. E. Bertie, P. R. Bunker, J. T. Hougen, I. M. Mills, J. K. G. Watson, and B. P. Winnewisser, *Pure Appl. Chem.* **69**, 1633 (1997).

495. C. J. H. Schutte, J. E. Bertie, P. R. Bunker, J. T. Hougen, I. M. Mills, J. K. G. Watson, and B. P. Winnewisser, *Pure Appl. Chem.* **69**, 1641 (1997).

496. P. R. Bunker, C. J. H. Schutte, J. T. Hougen, I. M. Mills, J. K. G. Watson, and B. P. Winnewisser, *Pure Appl. Chem.* **69**, 1651 (1997).

497. L. Nemes, *Vib. Spectra Struct.* **10**, 395 (1981).

498. J. T. Hougen, *J. Chem. Phys.* **57**, 4207 (1962).

499. A. Sayvetz, *J. Chem. Phys.* **7**, 383 (1939).

500. G. Amat, H. H. Nielsen, and G. Tarrago, *Rotation-vibration of Polyatomic Molecules: Higher Order Energies and Frequencies of Spectral Transitions*, Dekker, New York, 1971.

501. M. Joyeux, *Chem. Phys.* **203**, 281 (1996).

502. M. E. Kellman, *J. Chem. Phys.* **82**, 3300 (1985).

503. W. H. Shaffer and B. J. Krohn, *J. Mol. Spectrosc.*, **63**, 323 (1976).

504. Y. Kabbadj, M. Herman, G. Di Lonardo, L. Fusina, and J. W. C. Johns, *J. Mol. Spectrosc.* **150**, 535 (1991).

505. T. R. Huet, M. Herman, and J. W. C. Johns, *J. Chem. Phys.* **94**, 3407 (1991).

506. G. V. Hartland, D. Qin, and H. L. Dai, *J. Chem. Phys.* **100**, 7832 (1994).

507. J. C. D. Brand, J. H. Callomon, K. K. Innes, J. Jortner, S. Leach, D. H. Levy, A. J. Merer, I. M. Mills, C. B. Moore, C. S. Parmenter, D. A. Ramsay, N. K. Rao, E. W. Schlag, J. K. G. Watson, and R. N. Zare, *J. Mol. Spectrosc.* **99**, 482 (1983).

508. E. Fermi, *Z. Phys.* **71**, 250 (1931).

509. B. T. Darling and D. M. Dennison, *Phys. Rev.* **57**, 128 (1940).

510. P. Botschwina, B. Schulz, H. Horn, and M. Matuschewski, *Chem. Phys.* **190**, 345 (1995).

511. F. Meguellati, G. Graner, K. Burczyk, and H. Bürger, *J. Mol. Spectrosc.* **185**, 392 (1997).

512. A. Amrein, H. R. Dübal, and M. Quack, *Mol. Phys.* **56**, 727 (1985).

513. A. L. Utz, E. Carrasquillo, J. D. Tobiason, and F. F. Crim, *Chem. Phys.* **190**, 311 (1995).

514. A. P. Milce and B. J. Orr, *J. Chem. Phys.* **104**, 6423 (1996).

515. G. Herzberg, *Infrared and Raman Spectra of Polyatomic Molecules*, Van Nostrand, New York, 1945.

516. S. F. Yang, L. Biennier, A. Campargue, M. Abbouti Temsamani, and M. Herman, *Mol. Phys.* **90**, 807 (1997).

517. O. Vaittinen, T. Lukka, L. Halonen, H. Bürger, and O. Polanz, *J. Mol. Spectrosc.* **172**, 503 (1995).

518. I. M. Mills and A. G. Robiette, *Mol. Phys.* **56**, 743 (1985).

519. K. K. Lehmann, *J. Chem. Phys.* **79**, 1098 (1983).

520. K. K. Lehmann, *J. Chem. Phys.* **84**, 6524 (1986).

521. R. G. Della Valle, *Mol. Phys.* **63**, 611 (1988).

522. M. M. Law and J. L. Duncan, *Mol. Phys.* **93**, 821 (1998).

523. M. M. Law and J. L. Duncan, *Mol. Phys.* **93**, 809 (1998).

524. M. E. Kellman, *J. Chem. Phys.* **76**, 4528 (1982).

525. M. Herman, M. I. El Idrissi, A. Pisarchik, A. Campargue, A.-C. Gaillot, L. Biennier, G. Di Lonardo, and L. Fusina, *J. Chem. Phys.* **108**, 1377 (1998).

526. J. L. Duncan and I. M. Mills, *Chem. Phys. Lett.* **145**, 347 (1988).

527. M. M. Law, Ph.D. Thesis, University of Aberdeen, 1997.

528. K. K. Lehmann, *J. Chem. Phys.* **96**, 8117 (1992).

529. L. Lubich, O. V. Boyarkin, R. D. F. Settle, D. S. Perry, and T. R. Rizzo, *Faraday Discuss. Chem. Soc.* **102**, 167 (1995).

530. K. K. Lehmann, *Mol. Phys.* **66**, 1129 (1989).

531. K. K. Lehmann, *Mol. Phys.* **75**, 739 (1992).

532. I. M. Mills, *Faraday Discuss. Chem. Soc.* **102**, 244 (1995).

533. M. Abbouti Temsamani, Ph.D. Thesis, Université Libre De Bruxelles, 1996.

534. M. Winnewisser and B. P. Winnewisser, *J. Mol. Spectrosc.* **41**, 143 (1972).

535. J. Pliva, *J. Mol. Spectrosc.* **44**, 145 (1972).

536. D. M. Jonas, Ph.D. Thesis, MIT, Cambridge, MA, 1992.

537. M. L. Grenier-Besson, *J. Phys. Radium* **25**, 757 (1964).

538. A. G. Maki, Jr. and D. R. Lide, *J. Chem. Phys.* **47**, 3206 (1967).

539. V. I. Perevalov and O. N. Sulakshina, *SPIE* **2205**, 182 (1993).

540. V. I. Perevalov, E. I. Lobodenko, and J. L. Teffo, *SPIE* **3090**, 143 (1997).

541. J. M. Brown, J. T. Hougen, K. P. Huber, J. W. C. Johns, I. Kopp, H. Lefebvre-Brion, A. J. Merer, D. A. Ramsay, J. Rostas, and R. N. Zare, *J. Mol. Spectrosc.* **55**, 500 (1975).

542. C. Jaffe, *J. Chem. Phys.* **89**, 3395 (1988).

543. L. E. Fried and G. S. Ezra, *J. Chem. Phys.* **86**, 6270 (1987).

544. M. Quack, *Annu. Rev. Phys. Chem.* **41**, 839 (1990).

545. G. Sitja, Ph.D. Thesis, Université J. Fourier, Grenoble (France), 1993.

546. M. E. Kellman, *J. Chem. Phys.* **93**, 6630 (1990).

547. M. E. Kellman and G. Chen, *J. Chem. Phys.* **95**, 8671 (1991).

548. D. A. Sadovskii and B. Zhilinskii, *Phys. Rev. A* **47**, 2653 (1993).

549. D. A. Sadovskii and B. Zhilinskii, *Phys. Rev. A* **48**, 1035 (1993).

550. J.-P. Champion, M. Loëte, and G. Pierre, in *Spectroscopy of the Earth's Atmosphere and Interstellar Medium*, (K. Narahari Rao and A. Weber eds.), pp. 339–422, Academic Press, Boston, 1992.

551. A. Nikitin, J. P. Champion, and V. G. Tyuterev, *J. Mol. Spectrosc.* **182**, 72 (1997).

552. D. J. Nesbitt and R. W. Field, *J. Phys. Chem.* **100**, 12735 (1996).

553. D. A. Sadovskii and B. I. Zhilinskii, *J. Chem. Phys.* **103**, 10520 (1995).

554. I. N. Kozin, P. Jensen, O. Polanz, S. Klee, L. Poteau, and J. Demaison, *J. Mol. Spectrosc.* **180**, 402 (1996).

555. S. Brodersen and B. I. Zhilinskii, *J. Mol. Spectrosc.* **172**, 303 (1995).

556. R. B. Wattson and L. S. Rothman, *J. Mol. Spectrosc.* **119**, 83 (1986).

557. J.-L. Teffo, O. N. Sulakshina, and V. I. Perevalov, *J. Mol. Spectrosc.* **156**, 48 (1992).

558. A. Campargue, A. Charvat, and D. Permogorov, *Chem. Phys. Lett.* **223**, 567 (1994).

559. J.-P. Pique, J. Manners, G. Sitja, and M. Joyeux, *J. Chem. Phys.* **96**, 6495 (1992).

560. M. Joyeux, *J. Mol. Spectrosc.* **175**, 262 (1996).

561. T. Platz, M. Matheis, C. Hornberger, and W. Demtröder, *J. Mol. Spectrosc.* **180**, 81 (1996).

562. A. M. Smith, S. L. Coy, W. Klemperer, and K. K. Lehmann, *J. Mol. Spectrosc.* **134**, 134 (1989).

563. S. Carter, I. M. Mills, and N. C. Handy, *J. Chem. Phys.* **99**, 4379 (1993).

564. F. Gatti, Y. Justum, M. Menon, A. Nauts, and X. Chapuisat, *J. Mol. Spectrosc.* **181**, 403 (1997).

565. A. Delon and R. Jost, *J. Chem. Phys.* **95**, 5686 (1991).

566. A. D. Bykov, O. V. Naumenko, M. A. Smirnov, L. Sinitsa, L. R. Brown, J. Crisp, and D. Crip, *Can. J. Phys.* **72**, 989 (1994).

567. O. N. Ulenikov, G. A. Onopenko, H. Lin, J.-H. Zhang, Z.-Y. Zhou, Q.-S. Zhu, and R. N. Tolchenov, *J. Mol. Spectrosc.* **189**, 29 (1998).

568. A. Fayt, R. Vandenhaute, and J. G. Lahaye, *J. Mol. Spectrosc.* **119**, 233 (1986).

569. E. Rbaihi, A. Belafhal, J. Vander Auwera, S. Naïm, and A. Fayt, *J. Mol. Spectrosc.* **191**, 32 (1998).

570. H. H. Hamann, A. Charvat, B. Abel, S. A. Kovalenko, and A. A. Kachanov, *J. Chem. Phys.* **106**, 3103 (1997).

571. E. Venuti, G. Di Lonardo, P. Ferracuti, L. Fusina, and I. M. Mills, *Chem. Phys.* **190**, 279 (1995).

572. E. S. Bernardes, Y. M. M. Hornos, and J. E. M. Hornos, *Chem. Phys.* **213**, 17 (1996).

573. A. F. Borro, I. M. Mills, and A. Mose, *Chem. Phys.* **190**, 363 (1995).

574. J. K. Holland, D. A. Newham, and I. M. Mills, *Mol. Phys.* **70**, 319 (1990).

575. J. K. Holland, D. A. Newnham, I. M. Mills, and M. Herman, *J. Mol. Spectrosc.* **151**, 346 (1992).

576. F. Iachello, S. Oss, and L. Viola, *Mol. Phys.* **78**, 545 (1993).

577. O. Vaittinen, M. Saarinen, L. Halonen, and I. M. Mills, *J. Chem. Phys.* **99**, 3277 (1993).

578. O. Vaittinen, M. Hämäläinen, P. Jungner, K. Pulkkinen, L. Halonen, H. Bürger, and O. Polanz, *J. Mol. Spectrosc.* **187**, 193 (1998).

579. J. M. L. Martin, *Chem. Phys. Lett.* **242**, 343 (1995).

580. S. Carter, N. C. Handy, and J. Demaison, *Mol. Phys.* **90**, 729 (1997).

581. D. C. Burleigh, A. B. McCoy, and E. L. Sibert, III, *J. Chem. Phys.* **104**, 480 (1996).

582. R. J. Bouwens, J. A. Hammerschmidt, M. M. Grzeskowiak, T. A. Stegink, P. M. Yorba, and W. F. Polik, *J. Chem. Phys.* **104**, 460 (1996).

583. J.-C. Hilico, J. P. Champion, S. Toumi, V. G. Tyuterev, and S. A. Tashkun, *J. Mol. Spectrosc.* **168**, 455 (1994).

584. J.-C. Hilico, S. Toumi, and L. R. Brown, 'Proc. XIVth Colloquim on High Resolution Molecular Spectroscopy, Dijon, (France), 11–15 September 1995, p. N21.

585. L. Wiesenfeld, *J. Mol. Spectrosc.* **184**, 277 (1997).

586. R. Georges, M. Herman, J.-C. Hillico, and D. Robert, *J. Mol. Spectrosc.* **187**, 13 (1998).

587. T. J. Lee, J. M. L. Martin, and P. R. Taylor, *J. Chem. Phys.* **102**, 254 (1995).

588. H. Hollenstein, R. R. Marquardt, M. Quack, and M. A. Suhm, *J. Chem. Phys.* **101**, 3588 (1994).

589. M. Bach, R. Georges, M. Herman, and A. Perrin, *Mol. Phys.*, accepted for publication (1999).

590. B. G. Sartakov, J. Oomens, J. Reuss, and A. Fayt, *J. Mol. Spectrosc.* **145**, 31 (1997).

591. J. L. Duncan and G. E. Robertson, *J. Mol. Spectrosc.* **145**, 251 (1991).

592. R. Georges, M. Bach, and M. Herman, *Mol. Phys.*, accepted for publication (1999).

593. M. Abbouti Temsamani and M. Herman, *Chem. Phys. Lett.* **260**, 253 (1996).

594. M. Abbouti Temsamani, J. Vander Auwera, and M. Herman, *Mol. Phys.* **79**, 359 (1993).

595. J.-L. Teffo, V. I. Perevalov, and O. M. Lyulin, *J. Mol. Spectrosc.* **168**, 390 (1994).

596. A. Campargue, D. Permogorov, M. Bach, M. Abbouti Temsamani, J. Vander Auwera, M. Herman, and M. Fujii, *J. Chem. Phys.* **102**, 5931 (1995).

597. L. Halonen, D. W. Noid, and M. S. Child, *J. Chem. Phys.* **78**, 2803 (1983).

598. L. Halonen, *J. Chem. Phys.* **86**, 588 (1987).

599. B. R. Henry, M. A. Mohammadi, and A. W. Tarr, *J. Chem. Phys.* **77**, 3295 (1982).

600. O. S. van Roosmalen, I. Benjamin, and R. D. Levine, *J. Chem. Phys.* **81**, 5986 (1984).

601. I. Benjamin, O. S. van Roosmalen, and R. D. Levine, *J. Chem. Phys.* **81**, 3352 (1984).

602. B. C. Smith and J. S. Winn, *J. Chem. Phys.* **89**, 4638 (1988).

603. B. C. Smith and J. S. Winn, *J. Chem. Phys.* **94**, 4120 (1990).

604. G. Di Lonardo and L. Fusina, private communication (1998).

605. M. Quack, *Faraday Discuss. Chem. Soc.* **71**, 359 (1981).

606. H. R. Dübal and M. Quack, *J. Chem. Phys.* **81**, 3779 (1984).

607. H. R. Dübal and M. Quack, *Mol. Phys.* **53**, 257 (1984).

608. H. Ishikawa, Y. T. Chen, Y. Ohsima, B. Rajaram, J. Wang, and R. W. Field, *J. Chem. Phys.* **105**, 7383 (1996).

609. S. A. B. Solina, J. P. O'Brien, R. W. Field, and W. F. Polik, *J. Phys. Chem.* **100**, 7797 (1996).

609a. J. P. O'Brien, M. P. Jacobson, J. J. Sokol, S. L. Coy, and R. W. Field, *J. Chem. Phys.* **108**, 7100 (1998).

609b. M. P. Jacobson, J. P. O'Brien, R. J. Silbey, and R. W. Field, *J. Chem. Phys.* **109**, 121 (1998).

610. M. Abbouti Temsamani and M. Herman, *J. Chem. Phys.* **105**, 1355 (1996).

611. M. Lewerenz and M. Quack, *J. Chem. Phys.* **88**, 5408 (1988).

612. A. Campargue, *Chem. Phys. Lett.* **259**, 563 (1996).

613. G. Di Lonardo, P. Ferracuti, L. Fusina, and E. Venuti, *J. Mol. Spectrosc.* **164**, 219 (1994).

614. A. Campargue, M. Abbouti Temsamani, and M. Herman, *Mol. Phys.* **90**, 793 (1997).

615. X. Zhan and L. Halonen, *J. Mol. Spectrosc.* **160**, 464 (1993).

616. J. Sakai and M. Katayama, *J. Mol. Spectrosc.* **154**, 277 (1992).

617. A. Campargue, M. Abbouti Temsamani, and M. Herman, *Chem. Phys. Lett.* **242**, 101 (1995).

618. J. J. Scherer, K. K. Lehmann, and W. Klemperer, *J. Chem. Phys.* **78**, 2817 (1983).

619. L. Biennier, A. Campargue, and M. Herman, *Mol. Phys.* **93**, 457 (1998).

620. J. M. L. Martin and P. R. Taylor, *Chem. Phys. Lett.* **248**, 336 (1996).

621. T. Feldman, J. Romanko, and H. L. Welsh, *Can. J. Phys.* **34**, 737 (1956).

622. M. E. Jacox, *J. Chem. Phys.* **36**, 140 (1962).

623. H. J. Becher and A. Adrian, *J. Mol. Struct.* **6**, 479 (1970).

624. D. Van Lerberghe, I. J. Wright, and J. L. Duncan, *J. Mol. Spectrosc.* **42**, 251 (1972).

625. Y. Osamura and H. F. Schaefer, III, *Chem. Phys. Lett.* **79**, 412 (1981).

626. T. Carrington, Jr., L. M. Hubbard, H. F. Schaefer, III, and W. H. Miller, *J. Chem. Phys.* **80**, 4347 (1984).

627. Y. Chen, D. M. Jonas, J. L. Kinsey, and R. W. Field, *J. Chem. Phys.* **91**, 3976 (1989).

628. K. M. Ervin, J. Ho, and W. C. Lineberger, *J. Chem. Phys.* **91**, 5974 (1989).

629. M. M. Gallo, T. P. Hamilton, and H. F. Schaeffer, III, *J. Am. Chem. Soc.* **112**, 8714 (1990).

630. M. S. Krishnan and T. Carrington, Jr., *Chem. Phys.* **219**, 31 (1997).

631. A. McIlroy and D. J. Nesbitt, *J. Chem. Phys.* **92**, 2229 (1990).

632. J. E. Gambogi, E. R. T. Kerstel, K. K. Lehmann, and G. Scoles, *J. Chem. Phys.* **100**, 2612 (1994).

633. J. E. Gambogi, R. Zachari Pearson, X. Yang, K. K. Lehmann, and G. Scoles, *Chem. Phys.* **190**, 191 (1995).

634. M. J. Davis, *J. Chem. Phys.* **98**, 2614 (1993).

635. K. K. Lehmann, B. H. Pate, and G. Scoles, *Annu. Rev. Phys. Chem.* **45**, 241 (1994).

636. W. G. Harter, *Principles of Symmetry, Dynamics and Spectroscopy*, Wiley, New York, 1993.

637. B. S. Ray, *Z. Phys.* **78**, 74 (1932).

638. E. U. Condon and G. H. Shortley, *The Theory of Atomic Spectra*, Cambridge University Press, Cambridge, UK, 1967.

639. H. Van Vleck, *Rev. Mod. Phys.* **23**, 213 (1951).

640. W. Gordy and R. L. Cook, *Microwave Molecular Spectra*, Interscience, New York, 1970.

641. H. W. Kroto, *Molecular Rotation Spectra*, Dover, New York, 1992.

642. J. K. G. Watson, *Vibrational Spectra and Structure. A Series of Advances*, Elsevier, Amsterdam, 1977.

643. J. K. G. Watson, *J. Mol. Spectrosc.* **101**, 83 (1983).

644. M. L. Grenier-Besson, *J. Phys. Radium* **21**, 555 (1960).

645. T. Oka, *J. Mol. Spectrosc.* **29**, 84 (1969).

646. S. Haas, K. M. T. Yamada, and G. Winnewisser, *J. Mol. Spectrosc.* **164**, 445 (1994).

647. K. M. T. Yamada, F. W. Birss, and M. R. Aliev, *J. Mol. Spectrosc.* **112**, 347 (1985).

648. H. Finsterhölzl, H. W. Schrötter, and G. Strey, *J. Raman Spectrosc.* **11**, 375 (1981).

649. L. Halonen and A. G. Robiette, *J. Chem. Phys.* **84**, 6861 (1986).

650. T. Lukka and L. Halonen, *J. Chem. Phys.* **101**, 8380 (1994).

651. O. N. Ulenikov, R. N. Tolchenov, and Z. Qing-Shi, *Spectrochim. Acta, Part A* **52**, 1829 (1996).

652. O. N. Ulenikov, S. N. Yurchenko, and R. N. Tolchenov, *Spectrochim. Acta, Part A* **53**, 329 (1997).

653. O. N. Ulenikov, R. N. Tolchenov, and Z. Qing-Shi, *Spectrochim. Acta, Part A* **53**, 845 (1997).

654. W. H. Shaffer and A. H. Nieldsen, *J. Chem. Phys.* **9**, 847 (1941).

655. M. Herman, T. R. Huet, Y. Kabbadj, and J. Vander Auwera, *Mol. Phys.* **72**, 75 (1991).

656. J. K. G. Watson, *J. Chem. Phys.* **48**, 181 (1968).

657. C. Di Lauro and I. M. Mills, *J. Mol. Spectrosc.* **21**, 386 (1966).

658. W. E. Blass, *J. Mol. Spectrosc.* **31**, 196 (1969).

659. A. G. Maki, W. B. Olson, and R. L. Sams, *J. Mol. Spectrosc.* **36**, 433 (1970).

660. T. Nakagawa and Y. Morino, *J. Mol. Spectrosc.* **38**, 84 (1971).

661. A. R. Hoy, *J. Mol. Spectrosc.* **86**, 55 (1981).

662. E. Willemot, *J. Mol. Spectrosc.* **128**, 246 (1986).

663. A. Perrin, J.-M. Flaud, C. Camy-Peyret, A. M. Vasserot, G. Guelachvili, A. Goldman, F. J. Murcray, and R. D. Blatherwick, *J. Mol. Spectrosc.* **154**, 391 (1992).

664. J. M. Flaud, R. Grossloss, S. B. Rai, R. Stuber, W. Demtröder, D. A. Tate, L. G. Wang, and T. F. Gallagher, *J. Mol. Spectrosc.* **172**, 275 (1995).

664a. C. Camy-Peyret and J.-M. Flaud in *Molecular Spectroscopy: Modern Research*, Vol. III, pp. 69–110, K. Narahari Rao ed., Academic Press, New York, 1985.

665. G. Wagner, B. P. Winnewisser, and M. Winnewisser, *J. Mol. Spectrosc.* **146**, 104 (1991).

666. H. A. Jahn, *Phys. Rev.* **56**, 680 (1939).

667. O. Vaittinen, L. Halonen, H. Bürger, and O. Polanz, *J. Mol. Spectrosc.* **165**, 55 (1994).

668. E. Kostyk and H. L. Welsh, *Can. J. Phys.* **58**, 534 (1980).

669. J. Hietanen and J. Kauppinen, *Mol. Phys.* **42**, 411 (1981).

670. A. Moravec, G. Winnewisser, K. M. T. Yamada, and C. E. Blom, *Z. Naturforsch., A* **45A**, 946 (1990).

671. J. J. Hillman, D. E. Jennings, G. W. Halsey, S. Nadler, and W. E. Blass, *J. Mol. Spectrosc.* **146**, 389 (1991).

672. M. Weber, W. E. Blass, and J. L. Salanave, *J. Quant. Spectrosc. Radiat. Transfer* **42**, 437 (1989).

673. M. Weber and W. E. Blass, *Spectrochim. Acta, Part A* **48**, 1203 (1992).

674. M. Weber and W. E. Blass, *Spectrochim. Acta, Part A* **49**, 1659 (1993).

675. M. Weber, W. E. Blass, G. W. Halsey, and J. J. Hillman, *J. Mol. Spectrosc.* **165**, 107 (1994).

676. T. R. Huet and M. Herman, *J. Mol. Spectrosc.* **132**, 361 (1988).

677. W. J. Lafferty, R. D. Suenram, and D. R. Johnson, *J. Mol. Spectrosc.* **64**, 147 (1977).

678. R. L. DeLeon and J. S. Muenter, *J. Mol. Spectrosc.* **126**, 13 (1987).

679. D. C. Mc Kean, *Spectrochim. Acta, Part A* **29**, 1559 (1973).

680. I. Mills, T. Cvitas, K. Homann, N. Kallay, and K. Kuchitsu, *Quantities, Units and Symbols in Physical Chemistry*. Blackwell, Oxford, 1993.

681. W. Demtröder, *Laser Spectroscopy*, Springer-Verlag, Berlin, 1996.

682. L. Pauling and E. W. Wilson, *Introduction to Quantum Mechanics*. McGraw-Hill International, Tokyo, 1935.

683. L. S. Rothman, C. P. Rinsland, A. Goldman, S. T. Massie, D. P. Edwards, J.-M. Flaud, A. Perrin, C. Camy-Peyret, V. Dana, J.-Y. Mandin, J. Schroeder, A. McCann, R. R. Gamache, R. B. Wattson, K. Yoshino, K. V. Chance, K. W. Jucks, L. R. Brown, V. Nemtchinov, and P. Varanasi, *J. Quant. Spectrosc. Radiat. Transfer*, in press (1998).

684. B. H. Armstrong, *J. Quant. Spectrosc. Radiat. Transfer* **7**, 61 (1967).

685. F. Schreier, *J. Quant. Spectrosc. Radiat. Transfer* **48**, 743 (1992).

686. R. H. Dicke, *Phys. Rev.* **89**, 472 (1953).

687. L. Galatry, *Phys. Rev.* **122**, 1218 (1961).

688. A. S. Pine and J. P. Looney, *J. Mol. Spectrosc.* **122**, 41 (1987).

689. J.-P. Bouanich, C. Boulet, G. Blanquet, J. Walrand, and D. Lambot, *J. Quant. Spectrosc. Radiat. Transfer* **46**, 317 (1991).

690. J. L. Domenech, D. Bermejo, J. Santos, J.-P. Bouanich, and B. Christian, *J. Mol. Spectrosc.* **169**, 211 (1995).

691. A. Lévy, N. Lacome, and C. J. Chackerian, in *Spectroscopy of the Earth's Atmosphere and Interstellar Medium*, (K.N. Rao and A. Weber, eds.), p. 261. Academic Press, San Diego, CA, 1992.

692. J. H. Van Vleck and V. F. Weisskopf, *Rev. Mod. Phys.* **17**, 227 (1945).

693. A. Bauer, M. Godon, J. Carlier, and R. R. Gamache, *J. Mol. Spectrosc.* **176**, 45 (1996).

694. R. J. Bell, *Introductory Fourier Transform Spectroscopy*. Academic press, New York and London, 1972.

695. J. W. Brault, Fourier Transform Spectrometry, *Saas Fe courses*, 1985.

696. P. R. Griffiths and J. A. de Haseth, *Fourier Transform Infrared Spectrometry*, Wiley, New York, 1986.

697. G. Herzberg, *Molecular Spectra and Molecular Structure. II. Infrared and Raman Spectra of Polyatomic Molecules*, Van Nostrand, Princeton, NJ, 1945.

698. R. S. McDowell, *J. Quant. Spectrosc. Radiat. Transfer* **38**, 337 (1987).

699. R. R. Gamache, R. L. Hawkins, and L. S. Rothman, *J. Mol. Spectrosc.* **142**, 205 (1990).

700. J. T. Hougen, I. Kleiner, and M. R. Godefroid, *J. Mol. Spectrosc.* **163**, 559 (1994).

701. B. P. Winnewisser, in *Molecular Spectroscopy: Modern Research* (K.N. Rao, ed.), p. 321. Academic Press, New York, 1985.

702. J. W. C. Johns and J. Vander Auwera, *J. Mol. Spectrosc.* **140**, 71 (1990).

703. H. G. Kjaergaard, B. R. Henry, H. Wei, S. Lefebvre, T. Carrington, Jr., O. S. Mortensen, and M. L. Sage, *J. Chem. Phys.* **100**, 6228 (1994).

704. H. G. Kjaergaard and B. R. Henry, *Mol. Phys.* **83**, 1099 (1994).

705. H. G. Kjaergaard, C. D. Daub, and B. R. Henry, *Mol. Phys.* **90**, 201 (1997).

706. K. K. Lehmann, *J. Chem. Phys.* **91**, 2759 (1989).

707. K. K. Lehmann and A. M. Smith, *J. Chem. Phys.* **93**, 6140 (1990).

708. S. Hassoon and D. L. Snavely, *J. Chem. Phys.* **99**, 2511 (1993).

709. D. L. Snavely, F. R. Balckburn, Y. Ranasinghe, V. A. Walters, and M. Gonzales del Riego, *J. Phys. Chem.* **96**, 3599 (1992).

710. H. G. Kjaergaard and B. R. Henry, *J. Phys. Chem.* **99**, 899 (1995).

711. H. G. Kjaergaard, D. M. Turnbull, and B. R. Henry, *J. Chem. Phys.* **99**, 9438 (1993).

712. H. G. Kjaergaard, D. M. Turnbull, and B. R. Henry, *J. Phys. Chem.* **101**, 2589 (1997).

713. B. I. Niefer, H. G. Kjaergaard, and B. R. Henry, *J. Chem. Phys.* **99**, 5682 (1993).

714. D. M. Turnbull, H. G. Kjaergaard, and B. R. Henry, *Chem. Phys.* **195**, 129 and references therein (1995).

715. J. Peng, N. Mina-Camilde, and C. I. Manzanarès, *Vib. Spectrosc.* **8**, 319 (1995).

716. O. M. Lyulin, V. I. Perevalov, and J.-L. Teffo, *J. Mol. Spectrosc.* **180**, 72 (1996).

717. B. J. Orr, *Chem. Phys.* **190**, 261 (1995).

718. A. F. Borro, I. M. Mills, and A. Mose, *Chem. Phys.* **190**, 365 (1995).

719. G. Graner, *J. Mol. Spectrosc.* **161**, 58 (1993).

720. H. C. J. Allen and P. C. Cross, *Molecular Vib-rotors*. New York, 1963.

721. A. Mellouki, R. Georges, M. Herman, D. L. Snavely, and S. Leytner, *Chem. Phys.* **220**, 311 (1997).

722. M. Herman, T. R. Huet, and M. Vervloet, *Mol. Phys.* **66**, 333 (1989).

723. G. Amat, H. H. Nielsen, and G. Tarrago, *Rotation-vibration of Polyatomic Molecules*. Dekker, New York, 1971.

724. R. Herman and R. F. Wallis, *J. Chem. Phys.* **23**, 637 (1955).

725. J. K. G. Watson, *J. Mol. Spectrosc.* **125**, 428 (1987).

726. J. W. C. Johns, *J. Mol. Spectrosc.* **48**, 567 (1992).

727. J. K. G. Watson, *J. Mol. Spectrosc.* **153**, 211 (1992).

728. R. A. Toth, *Appl. Opt.* **32**, 7326 (1993).

729. F. Rachet, M. Margottin-Maclou, M. El Azizi, A. Henry, and A. Valentin, *J. Mol. Spectrosc.* **166**, 79 (1994).

730. Q. Errera, J. Vander Auwera, A. Belafhal, and A. Fayt, *J. Mol. Spectrosc.* **173**, 347 (1995).

731. M. Dang-Nhu and G. Guelachvili, *Mol. Phys.* **58**, 535 (1986).

732. V. M. Devi, B. Fridovich, G. D. Jones, and D. G. S. Snyder, *J. Mol. Spectrosc.* **105**, 61 (1984).

733. J. W. C. Johns, Z. Lu, F. Thibault, R. Le Doucen, P. Arcas, and C. Boulet, *J. Mol. Spectrosc.* **159**, 259 (1993).

734. A. Maki, W. Quapp, S. Klee, G. C. Mellau, and S. Albert, *J. Mol. Spectrosc.* **185**, 356 (1997).

735. D. Bermejo, J. L. Domenech, J. Santos, J.-P. Bouanich, and G. Blanquet, *J. Mol. Spectrosc.* **185**, 26 (1997).

736. D. Romanini and K. K. Lehmann, *J. Chem. Phys.* **105**, 68 (1996).

737. P. A. Martin and G. Guelachvili, *Chem. Phys. Lett.* **180**, 344 (1991).

738. A. Benidar, R. Farrenq, G. Guelachvili, and C. J. Chackerian, *J. Mol. Spectrosc.* **147**, 383 (1991).

739. D. D. J. Nelson, A. Schiffman, and D. J. Nesbitt, *J. Chem. Phys.* **90**, 5455 (1989).

740. D. D. J. Nelson, A. Schiffman, D. J. Nesbitt, J. J. Orlando, and J. B. Burkholder, *J. Chem. Phys.* **93**, 7003 (1990).

741. J. K. G. Watson, *J. Mol. Spectrosc.* **132**, 483 (1988).

742. C. M. Deeley and J. W. C. Johns, *J. Mol. Spectrosc.* **129**, 151 (1988).

743. A. Belafhal, A. Fayt, and G. Guelachvili, *J. Mol. Spectrosc.* **174**, 1 (1995).

744. L. Halonen, I. M. Mills, and J. Kauppinen, *Mol. Phys.* **43**, 913 (1981).

745. A. Maki, W. Quapp, S. Klee, G. C. Mellau, and S. Albert, *J. Mol. Spectrosc.* **174**, 365 (1995).

746. J. K. G. Watson, *J. Mol. Spectrosc.* **188**, 78 (1998).

746a. J. Vander Auwera and Y. El Youssoufi, in preparation.

747. B. H. Pate, K. K. Lehmann, and G. Scoles, *Annu. Rev. Phys. Chem.* **45**, 241 (1994).

748. G. Herzberg and K. Narahari Rao, *J. Chem. Phys.* **17**, 1099 (1949).

749. G. Herzberg and L. Herzberg, *J. Chem. Phys.* **18**, 1551 (1950).

750. G. Herzberg and L. Herzberg, *J. Opt. Soc. Am.* **43**, 1037 (1953).

751. H. J. Bernstein and G. Herzberg, *J. Chem. Phys.* **16**, 30 (1948).

752. L. F. H. Bovey, *J. Chem. Phys.* **21**, 830 (1953).

753. J. K. Wilmshurst and H. J. Bernstein, *Can. J. Chem.* **35**, 226 (1957).

754. T. A. Wiggins, E. R. Shull, J. M. Bennett, and D. H. Rank, *J. Chem. Phys.* **21**, 1940 (1953).

755. M. Carlotti, G. Di Lonardo, G. Galloni, and A. Trombetti, *J. Chem. Soc., Faraday Trans. 2*, **68**, 1473 (1972).

756. K. Hedfeld and P. Lueg, *Z. Phys.* **77**, 446 (1932).

757. G. W. Funke and E. Lindholm, *Z. Phys.* **106**, 518 (1937).

758. S. Tranchart, I. Hadj Bachir, T. R. Huet, A. Olafsson, J.-L. Destombes, S. Naïm, and A. Fayt, to be published (1998).

759. J. Connes, *Rev. Opt., Theor. Instrum.* **40**, 45 (1961).

760. J. Connes, *Rev. Opt., Theor. Instrum.* **40**, 116 (1961).

761. J. Connes, *Rev. Opt., Theor. Instrum.* **40**, 171 (1961).

762. J. Connes, *Rev. Opt., Theor. Instrum.* **40**, 231 (1961).

763. J. Connes, H. Delouis, P. Connes, G. Guelachvili, J.-P. Maillard, and G. Michel, *Nouv. Rev. Opt. Appl.* **1**, 3 (1970).

764. P. Connes and G. Michel, *Appl. Opt.* **4**, 2067 (1975).

765. G. Guelachvili, *Appl. Opt.* **17**, 1322 (1978).

766. J. W. Cooley and J. W. Tuckey, *Math. Comput.* **19**, 297 (1965).

767. E. O. Brigham, *The Fast Fourier Transform.* Prentice-Hall, Englewood Cliffs, NJ, 1974.

768. P. Connes, *Annu. Rev. Astron. Astrophys.* **8**, 209 (1970).

769. S. T. Ridgway and J. W. Brault, *Annu. Rev. Astron. Astrophys.* **22**, 291 (1984).

770. S. Gerstenkorn and P. Luc, *Atlas du spectre d'absorption de la molécule d'iode.* Editions du CNRS, Paris, 1978.

771. S. Gerstenkorn and P. Luc, *Rev. Phys. Appl.* **8**, 791 (1979).

772. G. Guelachvili, M. Birk, C. J. Bordé, J. W. Brault, L. R. Brown, B. Carli, A. R. H. Cole, K. M. Evenson, A. Fayt, D. Hausamann, J. W. C. Johns, J. Kauppinen, Q. Kou, A. G. Maki, R. K. Narahari, R. A. Toth, W. Urban, A. Valentin, J. Vergès, G. Wagner, M. H. Wappelhorst, J. S. Wells, B. P. Winnewisser, and M. Winnewisser, *Pure Appl. Chem.* **68**, 193 (1996).

773. J. W. C. Johns, *J. Opt. Soc. Am. B* **2**, 1340 (1985).

774. J.-M. Flaud, C. Camy-Peyret, J. W. C. Johns, and B. Carli, *J. Chem. Phys.* **91**, 1504 (1989).

775. M. Carlotti, G. Di Lonardo, L. Fusina, A. Trombetti, and B. Carli, *J. Mol. Spectrosc.* **141**, 29 (1990).

776. J. Vander Auwera, *J. Mol. Spectrosc.* **155**, 136 (1992).

777. G. Wagner, B. P. Winnewisser, M. Winnewisser, and K. Sarka, *J. Mol. Spectrosc.* **162**, 82 (1993).

778. R. Rodrigues, P. De Natale, G. Di Lonardo, and J.-M. Hartmann, *J. Mol. Spectrosc.* **175**, 429 (1996).

779. I. Morino and K. Kawaguchi, *J. Mol. Spectrosc.* **182**, 428 (1997).

780. M. Niedenhoff, K. M. T. Yamada, and G. Winnewisser, *J. Mol. Spectrosc.* **183**, 176 (1997).

781. I. Mukhopadhyay, P. K. Gupta, G. Moruzzi, B. P. Winnewisser, and M. Winnewisser, *J. Mol. Spectrosc.* **186**, 15 (1997).

782. J. W. G. Seibert, B. P. Winnewisser, M. Winnewisser, F. C. DeLucia, P. Helminger, and G. Pawelke, *J. Mol. Spectrosc.* **176**, 258 (1996).

783. M. Birk, G. Wagner, and J.-M. Flaud, *J. Mol. Spectrosc.* **163**, 245 (1994).

784. F. Roux, F. Michaux, and M. Vervloet, *J. Mol. Spectrosc.* **158**, 270 (1993).

785. M. Vervloet, unpublished results.

786. K. Yoshino, J. R. Esmond, J. E. Murray, W. H. Parkinson, A. P. Thorne, R. C. M. Learner, and G. Cox, *J. Chem. Phys.* **103**, 1243 (1995).

787. A. C. Vandaele, C. Hermans, P. C. Simon, M. Carleer, R. Colin, S. Fally, M.-F. Merienne, A. Jenouvrier, and B. Coquart, *J. Quant. Spectrosc. Radiat. Transfer* **59**, 171 (1998).

788. R. Bacis, A. J. Bouvier, and J. M. Flaud, *Spectrochim. Acta, Part A* **54**, 17 (1998).

789. A. P. Thorne, *Phys. Scr.* **T65**, 31 (1997).

790. A. Barbe, J. J. Plateaux, S. Mikhailenko, and V. G. Tyuterev, *J. Mol. Spectrosc.* **185**, 408 (1997).

791. V. Dana, J.-Y. Mandin, M.-Y. Allout, A. Perrin, L. Régalia, A. Barbe, J.-J. Plateaux, and X. Thomas, *J. Quant. Spectrosc. Radiat. Transfer* **57**, 445 (1997).

792. R. D'Cunha, V. A. Job, G. Rajappan, V. M. Devi, W. J. Lafferty, and A. Weber, *J. Mol. Spectrosc.* **186**, 363 (1997).

793. J. Demaison, J. Cosléou, H. Bürger, and E. B. Mkadmi, *J. Mol. Spectrosc.* **185**, 384 (1997).

794. V. M. Devi, D. C. Benner, M. A. H. Smith, and C. P. Rinsland, *J. Mol. Spectrosc.* **182**, 221 (1997).

795. C. Di Lauro, P. R. Bunker, J. W. C. Johns, and A. R. W. McKellar, *J. Mol. Spectrosc.* **184**, 177 (1997).

796. J.-M. Flaud, P. Arcas, H. Bürger, O. Polanz, and L. Halonen, *J. Mol. Spectrosc.* **183**, 310 (1997).

797. M.-F. Le Moal, M. Margottin-Maclou, and A. Valentin, *J. Mol. Spectrosc.* **183**, 93 (1997).

798. J. Lummila, O. Vaittinen, P. Jungner, L. Halonen, and A.-M. Tolonen, *J. Mol. Spectrosc.* **185**, 296 (1997).

799. M. J. W. McPhail, G. Duxbury, and R. D. May, *J. Mol. Spectrosc.* **182**, 118 (1997).

800. A. Nikitin, J.-P. Champion, V. G. Tyuterev, and L. R. Brown, *J. Mol. Spectrosc.* **184**, 120 (1997).

801. P. Pracna, K. Sarka, J. Demaison, J. Cosléou, F. Herlemont, M. Khelkhal, H. Fichoux, D. Papousek, M. Paplewski, and H. Bürger, *J. Mol. Spectrosc.* **184**, 93 (1997).

802. L. R. Brown, J. A. Crisp, D. Crisp, O. V. Naumenko, M. A. Smirnov, L. N. Sinitsa, and A. Perrin, *J. Mol. Spectrosc.* **188**, 148 (1998).

803. H. Bürger and W. Jerzembeck, *J. Mol. Spectrosc.* **188**, 209 (1998).

804. G. Graner, O. Polanz, H. Bürger, H. Ruland, and P. Pracna, *J. Mol. Spectrosc.* **188**, 115 (1998).

805. A. E. Heathfield, C. Anastasi, J. Ballard, D. Newnham, and A. McCulloch, *J. Quant. Spectrosc. Radiat. Transfer* **59**, 91 (1998).

806. J.-Y. Mandin, V. Dana, L. Régalia, A. Barbe, and P. von der Heyden, *J. Mol. Spectrosc.* **187**, 200 (1998).

807. A. Perrin, J.-M. Flaud, A. Predoi-Cross, M. Winnewisser, B. P. Winnewisser, G. Mellau, and M. Lock, *J. Mol. Spectrosc.* **187**, 61 (1998).

808. R. A. Toth, L. R. Brown, and C. Plymate, *J. Quant. Spectrosc. Radiat. Transfer* **59**, 529 (1998).

809. Q.-S. Zhu and B. A. Thrush, *J. Chem. Phys.* **92**, 2691 (1990).

810. Q.-S. Zhu, H.-B. Qian, and B. A. Thrush, *Chem. Phys. Lett.* **186**, 436 (1993).

811. A. Dkhissi, P. Soulard, A. Perrin, and N. Lacome, *J. Mol. Spectrosc.* **183**, 12 (1997).

812. M. Mengel, B. P. Winnewisser, and M. Winnewisser, *J. Mol. Spectrosc.* **188**, 221 (1998).

813. J. Preusser, R. Schermaul, B. P. Winnewisser, and M. Winnewisser, *J. Mol. Spectrosc.* **176**, 99 (1996).

814. O. N. Ulenikov, G. A. Onopenko, N. E. Tyabaeva, S. Alanko, M. Koivusaari, and R. Antilla, *J. Mol. Spectrosc.* **186**, 293 (1997).

815. F. Winther and M. Schönhof, *J. Mol. Spectrosc.* **186**, 54 (1997).

816. J.-M. Flaud and R. Bacis, *Spectrochim. Acta, Part A* **54**, 3 (1998).

817. A. G. Maki and J. S. Wells, *Wavenumber Calibration Tables from Heterodyne Frequency Measurements*, NIST Spec. Publ. 821. National Institute of Standards and Technology, Washington, DC, 1991.

818. C. Chackerian, G. Guelachvili, and R. H. Tipping, *J. Quant. Spectrosc. Radiat. Transfer* **30**, 107 (1983).

819. J. J. Sloan and E. J. Kruus, in *Time Resolved Spectroscopy* (R. J. H. Clark and R. E. Hester, eds.) 219–253, Wiley, New York, 1989.

820. N. Balucani, L. Beneventi, P. Casavecchia, C. G. Volpi, E. J. Kruus, and J. J. Sloan, *Can. J. Phys.* **72**, 888 (1994).

821. G. Durry, Ph. D. Thesis, Université de Paris-Sud, 1994.

822. G. Durry and G. Guelachvili, *Vib. Spectrosc.* **8**, 255 (1995).

823. G. Durry and G. Guelachvili, *Appl. Opt.* **34**, 1971 (1995).

824. G. Durry and G. Guelachvili, *J. Opt. Soc. Am. B* **12**, 1555 (1995).

825. G. Hartland, D. Quin, and H. Dai, *J. Chem. Phys.* **98**, 6906 (1992).

826. J. K. Messer and F. C. De Lucia, *Phys. Rev. Lett.* **53**, 2555 (1984).

827. D. Newnham, J. Ballard, and M. Page, *Rev. Sci. Instrum.* **66**, 4475 (1995).

828. K. M. Smith, G. Duxbury, D. Newnham, and J. Ballard, *J. Chem. Soc., Faraday Trans.* **93**, 2735 (1997).

829. R. Schermaul, J. W. G. Seibert, G. C. Mellau, and M. Winnewisser, *Appl. Opt.* **35**, 2884 (1996).

830. J. Vergès, C. Amiot, R. Bacis, and A. J. Ross, *Spectrochim. Acta, Part A* **51**, 1191 (1995).

831. A. J. Ross, P. Crozet, R. Bacis, S. Churassy, B. Erba, S. H. Ashworth, N. M. Lakin, M. R. Wickham, L. R. Beattie, and J. M. Brown, *J. Mol. Spectrosc.* **177**, 134 (1996).

832. O. Launila, B. Schimmelpfennig, H. Fagerli, O. Gropen, A. G. Taklif, and U. Wahlgren, *J. Mol. Spectrosc.* **186**, 131 (1997).

833. A. J. Phillips, F. Peters, and P. A. Hamilton, *J. Mol. Spectrosc.* **184**, 162 (1997).

834. R. S. Ram and P. F. Bernath, *J. Mol. Spectrosc.* **186**, 113 (1997).

835. R. S. Ram and P. F. Bernath, *J. Mol. Spectrosc.* **186**, 335 (1997).

836. E. H. Fink and D. A. Ramsay, *J. Mol. Spectrosc.* **185**, 304 (1997).

837. D. Bailly and N. Legay, *J. Mol. Spectrosc.* **157**, 1 (1993).

838. D. Bailly, C. Camy-Peyret, and R. Lanquetin, *J. Mol. Spectrosc.* **182**, 10 (1997).

839. C. Camy-Peyret, J.-M. Flaud, J.-P. Maillard, and G. Guelachvili, *Mol. Phys.* **33**, 1641 (1977).

840. J.-M. Flaud, C. Camy-Peyret, and J.-P. Maillard, *Mol. Phys.* **32**, 499 (1976).

841. O. L. Polyansky, J. R. Busler, B. Guo, K. Zhang, and P. F. Bernath, *J. Mol. Spectrosc.* **176**, 305 (1996).

842. M. Dulick, K.-Q. Zhang, B. Guo, and P. F. Bernath, *J. Mol. Spectrosc.* **188**, 14 (1998).

843. P. Jensen, *Bruker Rep.* **1** (1989).

844. J. W. Brault and M. C. Abrams, *Opt. Soc. Am. Tech. Dig. Ser.* **62**, 110 (1989).

845. J. W. C. Johns, *J. Mol. Spectrosc.* **125**, 442 (1987).

846. J. W. C. Johns, *Mikrochim. Acta* **III**, 171 (1987).

847. M. Birk, D. Hausamann, G. Wagner, and J. W. C. Johns, Appl. Opt. **35**, 2971 (1996).

848. M. Carleer, in *Proceedings, 12th Colloquium of High Resolution Molecular Spectroscopy, Dijon, France, September 9–13, 1991* (J. Moret-Bailly, ed.). Presses de l'Université de Bourgogne, 1991.

849. D. C. Benner, C. P. Rinsland, V. M. Devi, M. A. H. Smith, and D. Atkins, *J. Quant. Spectrosc. Radiat. Transfer* **53**, 705 (1995).

850. J. W. Brault, *Mikrochim. Acta* **III**, 215 (1987).

851. D. L. Cohen, *Appl. Opt.* **36**, 4034 (1997).

852. P. Saarinen and J. Kauppinen, Appl. Opt. **31**, 2353 (1992).

853. J. Schreiber, T. Blumenstock, and H. Fischer, *Appl. Opt.* **35**, 6203 (1996).

854. J. Ballard, J. J. Remedios, and H. K. Roscoe, *J. Quant. Spectrosc. Radiat. Transfer* **43**, 733 (1992).

855. M. C. Abrams, G. C. Toons, and R. A. Schindler, *Appl. Opt.* **33**, 6307 (1994).

856. J. U. White, *J. Opt. Soc. Am.* **32**, 285 (1942).

857. A. R. W. McKellar, N. Rich, and V. Soots, *Appl. Opt.* **9**, 222 (1970).

858. D. Horn and G. C. Pimentel, *Appl. Opt.* **10**, 1892 (1971).

859. J. Ballard, K. Strong, J. J. Remedios, M. Page, and W. B. Johnston, *J. Quant. Spectrosc. Radiat. Transfer* **52**, 677 (1985).

860. T. Ahonen, S. Alanko, V.-M. Horneman, M. Koivusaari, A.-M. Tolonen, and R. Antilla, *J. Mol. Spectrosc.* **181**, 279 (1997).

861. P. Bernath, M. Carleer, R. Colin, A. Jennouvrier, M.-F. Merienne, O. L. Polyansky, J. Tennyson, and C. Van Daele, unpublished results.

862. M. Herman, M. Abbouti Temsamani, D. Lemaitre, and J. Vander Auwera, *Chem. Phys. Lett.* **185**, 220 (1991).

863. D. Bailly, *J. Mol. Spectrosc.* **157**, 1 (1993).

864. M. Herman, D. Hurtmans, J. Vander Auwera, and M. Vervloet, *J. Mol. Spectrosc.* **150**, 293 (1991).

865. P. Jungner and L. Halonen, *J. Chem. Phys.* **107**, 1680 (1997).

866. A. Mellouki, J. Vander Auwera, and M. Herman, *J. Mol. Spectrosc.*, **192**, in press (1998).

867. M. Bach, R. Georges, M. Hepp, and M. Herman, *Chem. Phys. Lett.*, **294**, 533 (1998).

868. J. Oomens and J. Reuss, *J. Mol. Spectrosc.* **173**, 14 (1995).

869. J. Oomens, L. Oudejans, J. Reuss, and A. Fayt, *Chem. Phys.* **187**, 57 (1994).

870. J. Oomens and J. Reuss, *J. Mol. Spectrosc.* **177**, 19 (1996).

871. J. Oomens, J. Reuss, G. C. Mellau, S. Klee, I. Gulaczyk, and A. Fayt, *J. Mol. Spectrosc.* **180**, 236 (1996).

872. D. L. Snavely, S. D. Colson, and K. B. Wiberg, *J. Chem. Phys.* **74**, 6975 (1981).

873. A. Amrein, M. Quack, and U. Schmitt, *Z. Phys. Chem* **154**, 59 (1987).

873a. A. Amrein, M. Quack, and U. Schmitt, *Mol. Phys.* **60**, 237 (1987).

874. A. Amrein, H. Hollenstein, P. Locher, M. Quack, U. Schmitt, and H. Bürger, *Chem. Phys. Lett.* **139**, 82 (1987).

875. A. Amrein, M. Quack, and U. Schmitt, *J. Phys. Chem.* **92**, 5455 (1988).

875a. A. Amrein, D. Luckhaus, F. Merkt, and M. Quack, *Chem. Phys. Lett.* **152**, 275 (1988).

876. A. Amrein, H. Hollenstein, M. Quack, and U. Schmitt, *Infrared Phys.* **29**, 561 (1989).

877. P. Asselin, P. Soulard, G. Tarrago, N. Lacome, and L. Manceron, *J. Chem. Phys.* **104**, 4427 (1996).

878. P. Asselin, B. Dupuis, J. P. Perchard, and P. Soulard, *Chem. Phys. Lett.* **268**, 265 (1997).

879. P. Asselin, P. Soulard, and N. Lacome, *J. Mol. Spectrosc.*, **190**, 274 (1998).

880. M. Bach, R. Georges, M. Herman, and A. Perrin, *Mol. Phys.*, accepted for publication (1999).

881. J. Ballard, D. Newnham, and M. Page, *Chem. Phys. Lett.* **208**, 295 (1993).

882. J. A. Barnes and T. E. Gough, *Chem. Phys. Lett.* **130**, 297 (1986).

883. J. A. Barnes and T. E. Gough, *J. Chem. Phys.* **86**, 6012 (1987).

884. J. A. Barnes, C. Douketis, J. G. Gibson, and T. E. Gough, *Chem. Phys. Lett.* **200**, 274 (1992).

884a. V. Boudon, M. Hepp, M. Herman, I. Pak, and G. Pierre, *J. Mol. Spectrosc.* in press (1998).

884b. H. Bürger, A. Rahner, A. Amrein, H. Hollenskein, and M. Quack, *Chem. Phys. Lett.* **156**, 557 (1989).

885. H. Bürger, U. Goergens, H. Ruland, M. Quack, and U. Schmitt, *Mol. Phys.* **87**, 469 (1996).

885a. H.-R. Dübal, M. Quack, and U. Schmitt, *Chimia* **38**, 438 (1984).

886. Y. El Youssoufi, R. Georges, J. Liévin, and M. Herman, *J. Mol. Spectrosc.* **186**, 239 (1997).

887. G. J. Frost, L. M. Goss, and V. Vaida, J. Geophys. Res. **101**, 3869 (1996).

888. M. Gauthier, *J. Chem. Phys.* **88**, 5439 (1988).

889. R. Georges, G. Durry, M. Bach, R. Pétrisse, R. Jost, and M. Herman, *Chem. Phys. Lett.* **246**, 601 (1995).

890. R. Georges, J. Liévin, M. Herman, and A. Perrin, *Chem. Phys. Lett.* **256**, 675 (1996).

891. R. Georges, M. Bach, and M. Herman, *Mol. Phys.* **90**, 381 (1997).

892. J. Han, Z. Wang, A. L. McIntosh, R. R. Lucchese, and J. W. Bevan, *J. Chem. Phys.* **100**, 7101 (1994).

893. M. Hepp, R. Georges, and M. Herman, *Chem. Phys. Lett.* **275**, 513 (1997).

893a. M. Hepp, R. Georges, M. Herman, W. J. Lafferty, and J.-M. Flaud, *J. Mol. Struct.*, submitted for publication (1998).

894. M. Hepp and M. Herman, *Mol. Phys.*, **94**, 829 (1998).

894a. M. Hepp and M. Herman, *J. Mol. Spectrosc.*, **194**, in press (1999).

895. M. Hepp, F. Herregodts, and M. Herman, *Chem. Phys. Lett.*, **294**, 528 (1998).

896. J. K. Holland, M. Carleer, R. Pétrisse, and M. Herman, *Chem. Phys. Lett.* **194**, 175 (1992).

897. H. Hollenstein, M. Quack, and E. Richard, *Chem. Phys. Lett.* **222**, 176 (1994).

897a. D. Luckhaus and M. Quack, *Chem. Phys. Lett.* **199**, 293 (1992).

897b. D. Luckhaus and M. Quack, *J. Mol. Struct.* **293**, 213 (1993).

898. D. Luckhaus, M. Quack, U. Schmitt, and M. A. Suhm, *Ber. Bunsenges. Phys. Chem.* **99**, 457 (1995).

899. D. McNaughton, D. McGilvery, and E. G. Robertson, *J. Chem. Soc., Faraday Trans.*, **90**, 1055 (1994).

900. D. McNaughton, C. Evans, and E. G. Robertson, *J. Chem. Soc., Faraday Trans.*, **91**, 1723 (1995).

901. D. McNaughton, D. McGilvery, and E. G. Robertson, *J. Mol. Struct.* **348**, 1 (1995).

902. D. McNaughton, C. Evans, and E. G. Robertson, *Mikrochim. Acta* (1996).

903. D. McNaughton, E. G. Robertson, and F. Shanks, *Chem. Phys.* **2304** (1996).

904. D. McNaughton and C. Evans, *J. Mol. Spectrosc.* **182**, 342 (1997).

905. D. McNaughton and C. Evans, *J. Phys. Chem.* **100**, 8660 (1996).

905a. R. F. Meads, A. L. McIntrosh, J. I. Arno, C. L. Hartz, R. R. Lucchese, and J. W. Bevan, *J. Chem. Phys.* **101**, 4593 (1994).

906. F. Mélen, M. Herman, G. Y. Matti, and D. M. McNaughton, *J. Mol. Spectrosc.* **160**, 601 (1993).

907. F. Mélen, M. Carleer, and M. Herman, *Chem. Phys. Lett.* **199**, 124 (1992).

907a. T. Nakanaga, F. Ito, and H. Takeo, *Bull. Inst. Chem. Res., Kyoto Univ.* **71**, 140 (1993).

908. M. Quack, U. Schmitt, and M. A. Suhm, *Chem. Phys. Lett.* **208**, 446 (1993).

909. E. C. Richard, D. J. Donaldson, and V. Vaida, *Chem. Phys. Lett.* **157**, 295 (1989).

910. E. C. Richard, C. T. Wickham-Jones, and V. Vaida, *J. Phys. Chem.* **93**, 6346 (1989).

911. E. C. Richard and V. Vaida, *J. Chem. Phys.* **94**, 163 (1991).

912. E. C. Richard and V. Vaida, *J. Chem. Phys.* **94**, 153 (1991).

912a. C. Rödig, D. Weidlich, and F. Siebert, in "Time resolved vibrational Spectrocopy," A. Lau, F. Siebert, and W. Werncke, eds., pp. 227–230, Springer, Berlin, 1994.

913. D. L. Snavely, K. B. Wiberg, and S. D. Colson, *Chem. Phys. Lett.* **96**, 319 (1983).

914. D. L. Snavely, V. A. Walters, S. D. Colson, and K. B. Wiberg, *Chem. Phys. Lett.* **103**, 423 (1984).

915. M. Snels, A. Beil, H. Hollenstein, M. Quack, U. Schmitt, and F. D'Amato, *J. Chem. Phys.* **103**, 8846 (1995).

916. A. D. Walters, M. Winnewisser, K. Lattner, and B. P. Winnewisser, *J. Mol. Spectrosc.* **149**, 542 (1991).

917. J. Arno and J. W. Bevan, *Infrared Spectroscopy in Supersonic Free Jets and Molecular Beams.*, in *Jet Spectroscopy and Molecular Dynamics*, pp. 27–73, J. M. Hollas and D. Phillips, eds., Blackie Academic Professioal, 1995.

918. D. S. Bomse, A. C. Stanton, and J. A. Silver, *Appl. Opt.* **31**, 718 (1992).

919. G. R. Janik, C. B. Carlisle, and T. F. Gallagher, *J. Opt. Soc. Am.* **3**, 1070 (1986).

920. G. C. Bjorklund, *Opt. Lett.* **5**, 15 (1980).

921. P. Werle, *Appl. Phys. B* **60**, 499 (1986).

922. G. C. Bjorklund, *Applied Opt. B* **32**, 145 (1983).

923. P. Pokrovsky, W. Zapka, F. Chu, and G. C. Bjorklund, *Opt. Commun.* **44**, 175 (1983).

924. N. Y. Chou and G. W. Sachse, *Appl. Opt.* **27**, 4438 (1988).

925. E. I. Moses and C. L. Tang, *Opt. Lett.* **1**, 115 (1977).

926. C. B. Carlisle, D. E. Cooper, and H. Preier, *Appl. Opt.* **28**, 2567 (1989).

927. L. G. Wang, H. Hiris, C. B. Carlisle, and T. F. Gallagher, *Appl. Opt.* **27**, 2071 (1988).

928. J. A. Silver and A. C. Stanton, *Appl. Opt.* **27**, 4438 (1988).

929. P. Cancio and F. S. Pavone, *Phys. Scr.* **T58**, 118 (1993).

930. E. D. Hinkley, *Topics in Applied Physics*, Springer-Verlag, Berlin, 1976.

931. P. Werle, F. Slehr, M. Gehrtz, and C. Brauchle, *Appl. Opt.* **49**, 99 (1989).

932. M. Gehrtz, G. C. Bjorklund, and E. A. Whittaker, *J. Opt. Soc. Am. B* **2**, 1510 (1985).

933. J. A. Silver, *Appl. Opt.* **31**, 707 (1992).

934. H. C. Sun, E. A. Whittaker, Y. W. Bae, C. K. Ng, V. Patel, W. H. Tam, S. Mc Guire, B. Singh, and B. Gallois, *Appl. Opt.* **32**, 885 (1993).

935. L. C. Philippe and R. K. Hanson, *Appl. Opt.* **32**, 6090 (1993).

936. P. C. D. Hobbs, *Appl. Opt.* **36**, 903 (1997).

937. P. C. D. Hobbs, *SPIE Laser Noise* **1376** (1991).

938. D. E. Cooper and R. E. Warren, *Appl. Opt.* **26**, 3726 (1987).

939. R. Grosskloss, H. Wenz, S. B. Rai, and W. Demtröder, *Mol. Phys.* **85**, 71 (1995).

940. R. Grosskloss, P. Kersten, and W. Demtröder, *Appl. Phys. B* **58**, 137 (1994).

941. N. Moriwaki, T. Tsuchida, Y. Takehisa, and N. Ohashi, *J. Mol. Spectrosc.* **137**, 230 (1989).

942. S. Kinugawa and H. Sasada, *Jpn. J. Appl. Phys.* **29**, 611 (1990).

943. J. Sakai and M. Katayama, *J. Mol. Spectrosc.* **160**, 217 (1993).

944. H. Sasada, T. Tsukamoto, Y. Kuba, N. Tanaka, and K. Uehara, *J. Opt. Soc. Am. B* **11**, 191 (1994).

945. T. Tsukamoto and H. Sasada, *J. Chem. Phys.* **102**, 5126 (1995).

946. H. Sasada, *J. Mol. Spectrosc.* **165**, 588 (1994).

947. T. Katayama, F. Matsushima, and H. Sasada, *J. Mol. Spectrosc.* **167**, 236 (1994).

948. D. Romanini and K. K. Lehmann, *J. Chem. Phys.* **99**, 6287 (1993).

949. D. Romanini, A. A. Kachanov, and F. Stoeckel, *Chem. Phys. Lett.* **270**, 546 (1997).

950. J. M. Herbelin, J. A. McKay, M. A. Kwok, R. H. Uenten, D. S. Urevig, D. J. Spencer, and D. J. Benard, *Appl. Opt.* **19**, 144 (1980).

951. A. O'Keefe and D. A. G. Deacon, *Rev. Sci. Instrum.* **59**, 2544 (1988).

952. J. J. Scherer, D. Voelkel, D. J. Rakestraw, J. B. Paul, C. P. Collier, R. J. Saykally, and A. O'keefe, *Chem. Phys. Lett.* **245**, 273 (1995).

953. P. Zalicki and R. N. Zare, *J. Chem. Phys.* **102**, 2708 (1995).

954. G. Meijer, M. G. H. Boogaarts, R. T. Jongma, D. H. Parker, and A. M. Wodtke, *Chem. Phys. Lett.* **217**, 112 (1994).

955. J. Martin, B. A. Paldus, P. Zalicki, E. H. Wahl, T. G. Owano, J. S. Harris, Jr., C. H. Kruger, and R. N. Zare, *Chem. Phys. Lett.* **258**, 63 (1996).

956. T. Hodges, J. Looney, and R. D. van Zee, *Cavity-ringdown Spectroscopy—A New Technique for Trace Absorption Measurements.* American Chemical Society, Washington, DC, 1998.

957. T. Hodges, J. Looney, and R. D. van Zee, *Appl. Opt.* **35**, 4112 (1996).

958. D. Romanini, A. A. Kachanov, and F. Stoeckel, *Chem. Phys. Lett.* **270**, 538 (1997).

959. D. Romanini, A. A. Kachanov, N. Sadeghi, and F. Stoeckel, *Chem. Phys. Lett.* **264**, 316 (1997).

960. B. A. Paldus, J. S. Harris, Jr., J. Martin, J. Xie, and R. N. Zare, *J. Appl. Phys.* **82**, 3199 (1997).

961. D. Z. Anderson, J. C. Frisch, and C. S. Masser, *Appl. Opt.* **23**, 1238 (1984).

962. R. Engeln, G. von Helden, G. Berden, and G. Meijer, *Chem. Phys. Lett.* **262**, 105 (1996).

963. R. T. Jongma, M. G. H. Boogaarts, and G. Meijer, *J. Mol. Spectrosc.* **165**, 303 (1994).

964. A. O'Keefe, J. J. Scherer, A. L. Cooksy, R. Sheeks, J. Heath, and R. J. Saykally, *Chem. Phys. Lett.* **172**, 214 (1990).

965. J. J. Scherer, J. B. Paul, C. P. Collier, and R. J. Saykally, *J. Chem. Phys.* **102**, 5190 (1995).

966. J. J. Scherer, J. B. Paul, C. P. Collier, and R. J. Saykally, *J. Chem. Phys.* **103**, 113 (1995).

967. J. J. Scherer, J. B. Paul, and R. J. Saykally, *Chem. Phys. Lett.* **242**, 395 (1995).

968. J. J. Scherer, J. B. Paul, A. O'Keefe, and R. J. Saykally, *Chem. Rev.* **97**, 25 (1997).

969. M. Kotterer and J. P. Maier, *Chem. Phys. Lett.* **266**, 342 (1997).

970. T. Yu and M. C. Lin, *J. Amer. Chem. Soc.* **115**, 4371 (1993).

971. T. Yu and M. C. Lin, *J. Phys. Chem.* **98**, 9697 (1994).

972. T. G. Slanger, D. L. Huestis, P. C. Cosby, H. Naus, and G. Meijer, *J. Chem. Phys.* **105**, 9393 (1996).

973. D. Huestis, R. A. Copeland, K. Knutsen, T. G. Slanger, R. T. Jongma, M. G. H. Boogaarts, and G. Meijer, *Can. J. Phys.* **72**, 1109 (1994).

974. M. D. Wheeler, A. J. Orr-Ewing, M. N. R. Ashfold, and T. Ishiwata, *Chem. Phys. Lett.* **268**, 421 (1997).

975. D. Romanini and K. K. Lehmann, *J. Chem. Phys.* **102**, 633 (1995).

976. D. Romanini and K. K. Lehmann, *J. Chem. Phys.* **105**, 68 (1996).

977. K. K. Lehmann and D. Romanini, *J. Chem. Phys.* **105**, 10263 (1996).

978. J. Pearson, A. J. Orr-Ewing, M. N. R. Ashfold, and R. N. Dixon, *J. Chem. Soc., Faraday Trans.* **92**, 1283 (1996).

979. P. Pearson, A. J. Orr-Ewing, M. N. R. Ashfold, and R. N. Dixon, *J. Chem. Phys.* **106**, 5850 (1997).

980. K. Nakagawa, T. Katsuda, A. S. Shelkovnikov, M. De Labachelerie, and M. Ohtsu, *Opt. Commun.* **107**, 369 (1994).

981. J. Xie, B. A. Paldus, E. W. Wahl, J. Martin, T. G. Owano, C. H. Kruger, J. S. Harris, and R. N. Zare, *Chem. Phys. Lett.* **284**, 387 (1998).

982. J. J. Scherer and D. J. Rakestraw, *Chem. Phys. Lett.* **265**, 342 (1997).

983. S. Cheskis, I. Derzy, V. A. Lozovsky, A. Kachanov, and D. Romanini, *Applied Phys. B.*, **66**, 377 (1998).

984. J. J. L. Spaanjaars, J. J. ter Meulen, and G. Meijer, *J. Chem. Phys.* **107**, 2242 (1997).

985. P. Zalicki, Y. Ma, R. N. Zare, E. H. Wahl, J. Dadamio, T. G. Owano, and C. H. Kruger, *Chem. Phys. Lett.* **243**, 269 (1995).

986. M. Kotterer, J. Conceicao, and J. P. Maier, *Chem. Phys. Lett.* **259**, 233 (1996).

987. A. Campargue, D. Romanini, and N. Sadeghi, *J. Phys. D.* **31**, 1168 (1998).

988. J. B. Paul, C. P. Collier, R. J. Saykally, J. J. Scherer, and A. O'Keefe, *J. Phys. Chem.* **101**, 5211 (1997).

989. M. G. H. Boogaarts and G. Meijer, *J. Chem. Phys.* **103**, 5269 (1995).

990. L. A. Pakhomycheva, E. A. Sviridenkov, A. F. Suchkov, L. V. Titova, and S. S. Churilov, *JETP Lett (Engl. Transl.)* **12**, 43 (1970).

991. V. M. Baev, T. P. Belikova, E. A. Sviridenkov, and A. F. Suchkov, *Sov. Phys.—JETP (Engl. Transl.)* **47**, 21 (1978).

992. E. A. Sviridenkov and M. P. Frolov, *Sov. J. Quantum Electron. (Engl. Transl.)* **7**, 576 (1977).

993. T. P. Belikova, E. A. Sviridenkov, A. F. Suchkov, L. V. Titova, and S. S. Churilov, *Sov. Phys.—JETP (Engl. Transl.)* **35**, 1076 (1972).

994. R. A. Keller, E. F. Zalewski, and N. C. Peterson, *J. Opt. Soc. Am.* **62**, 319 (1972).

995. N. C. Peterson, M. J. Kurylo, W. Braun, A. M. Bass, and R. A. Keller, *J. Opt. Soc. Am.* **61**, 746 (1971).

996. T. W. Hänsch, A. L. Schawlow, and P. E. Toschek, *IEEE J. Quantum Electron.* **8**, 802 (1972).

997. G. H. Atkinson, A. H. Laufer, and M. J. Kurylo, *J. Chem. Phys.* **59**, 350 (1973).

998. P. E. Toschek and V. M. Baev, *Laser Spectroscopy and New Ideas Opt. Ser.* Springer-Verlag, 1987.

999. A. Kachanov, S. A. Kovalenko, and I. K. Pashkovich, *Sov. J. Quantum. Electron. (Engl. Transl.)* **19**, 95 (1989).

1000. A. Kachanov, A. Charvat, and F. Stoeckel, *J. Opt. Soc. Am. B* **11**, 2412 (1994).

1001. A. Kachanov, A. Charvat, and F. Stoeckel, *J. Opt. Soc. Am. B* **12**, 970 (1995).

1002. D. Romanini, A. Kachanov, E. Lacot, and F. Stoeckel, *Phys. Rev. A* **54**, 920 (1996).

1003. J. Sierks, V. M. Baev, and P. E. Toschek, *Opt. Commun.* **96**, 436 (1993).

1004. W. Brunner and H. Paul, *Opt. Commun.* **18**, 252 (1974).

1005. K. Tohama, *Opt. Commun.* **15**, 17 (1975).

1006. V. M. Baev, J. Eschner, E. Paeth, R. Schüler, and P. E. Toschek, *Appl. Phys. B* **55**, 463 (1992).

1007. V. R. Mironenko and V. I. Yudson, *Opt. Commun.* **34**, 397 (1980).

1008. S. A. Kovalenko, *Sov. J. Quantum Electron. (Engl. Transl.)* **11**, 759 (1981).

1009. V. R. Mironenko and I. Pak, *Sov. J. Quantum Electron. (Engl. Transl.)* **8**, 1394 (1978).

1010. A. Campargue, F. Stoeckel, and M. Chenevier, *Spectrochim. Acta, A Rev.* **13**, 69 (1990).

1011. B. Abel, H. H. Hamann, A. A. Kachanov, and J. Troe, *J. Chem. Phys.* **104**, 3189 (1996).

1012. B. Abel, A. Charvat, and S. F. Deppe, *Chem. Phys. Lett.* **277**, 347 (1997).

1013. R. Georges, A. Delon, F. Bylicki, R. Jost, A. Campargue, A. Charvat, M. Chenevier, and F. Stoeckel, *Chem. Phys.* **190**, 207 (1995).

1014. K. Singh and J. O'Brien, *J. Mol. Spectrosc.* **167**, 99 (1994).

1015. F. Stoeckel, M. A. Melieres, and M. Chenevier, *J. Chem. Phys.* **76**, 2191 (1982).

1016. H. Atmanspacher, H. Scheingraber, and C. R. Vidal, *Phys. Rev. A* **33**, 1052 (1986).

1017. M. Chenevier, M. A. Melieres, and F. Stoeckel, *Opt. Commun.* **45**, 385 (1983).

1018. M. A. Melieres, M. Chenevier, and F. Stoeckel, *J. Quant. Spectrosc. Radiat. Transfer* **33**, 337 (1985).

1019. V. M. Baev, J. Eschner, J. Sierks, A. Weiler, and P. E. Toschek, *Opt. Commun.* **94**, 436, and references therein (1992).

1020. P. Vujkovic Cvijin, J. O'Brien, G. H. Atkinson, W. K. Wells, J. I. Lunine, and D. M. Hunten, *Chem. Phys. Lett.* **159**, 331 (1989).

1021. K. Singh and J. O'Brien, *Chem. Phys. Lett.* **229**, 29 (1994).

1022. K. Singh and J. O'Brien, *J. Quant. Spectrosc. Radiat. Transfer* **52**, 75 (1994).

1023. K. Singh and J. O'Brien, *J. Quant. Spectrosc. Radiat. Transfer* **54**, 607 (1995).

1024. K. Singh and J. O'Brien, *Astrophys. Space Sci.* **236**, 97 (1996).

1025. A. Campargue and D. Permogorov, *Chem. Phys. Lett.* **241**, 339 (1995).

1026. V. R. Mironenko and V. I. Yudson, *Sov. Phys.—JETP (Engl. Transl.)* **52**, 594 (1980).

1027. S. E. Vinogradov, A. A. Kachanov, S. A. Kovalenko, and E. A. Sviridenkov, *Sov. Lett.—JETP (Engl. Transl.)* **55**, 581 (1992).

1028. A. A. Kachanov and T. V. Plakhotnik, *Opt. Commun.* **47**, 257 (1983).

1029. A. Garnache, A. A. Kachanov, and F. Stoeckel, to be published.

1030. H. Atmanspacher, H. Scheingraber, and V. M. Baev, *Phys. Rev. A* **35**, 142 (1987).

1031. E. N. Antonov, A. A. Kachanov, V. R. Mironenko, and T. V. Plakhotnik, *Opt. Commun.* **46**, 126 (1983).

1032. Y. M. Aivazyan, V. V. Ivanov, S. A. Kovalenko, V. M. Baev, E. A. Sviridenkov, H. Atmanspacher, and H. Scheingraber, *Appl. Phys.* **46**, 175 (1988).

1033. E. Lacot and F. Stoeckel, *J. Opt. Soc. Am.* **9**, 2034, and references therein (1996).

1034. V. M. Baev, G. Gaida, H. Schroder, and P. E. Toschek, *Opt. Commun.* **38**, 309 (1981).

1035. A. A. Kachanov, R. Romanini, A. Charvat, and B. Abel, in preparation (1998).

1036. A. Del Olmo, C. Domingo, J. M. Orza, and D. Bermejo, *J. Mol. Spectrosc.* **145**, 325 (1991).

1037. C. Domingo, A. Del Olmo, R. Escribano, D. Bermejo, and J. M. Orza, *J. Chem. Phys.* **96**, 972 (1992).

1038. A. A. Kachanov, F. Stoeckel, A. Charvat, and J. O'Brien, *Appl. Opt.* **36**, 4062 (1997).

1039. W. R. Lambert, P. M. Felker, and A. H. Zewail, *J. Chem. Phys.* **74**, 4732 (1981).

1040. A. Campargue, M. Chenevier, and F. Stoeckel, *Chem. Phys. Lett.* **183**, 153 (1991).

1041. A. Campargue, M. Chenevier, A. Delon, R. Jost, and F. Stoeckel, *J. Phys. IV, Colloq.* **C7**, supp. *J. Phys. III* 471 (1991).

1042. A. Campargue, L. Biennier, A. Kachanov, R. Jost, R. Bacis, V. Veyret, and S. Churassy, *Chem. Phys. Lett.* **288**, 734 (1998).

1043. J. Vetterhöffer, A. Campargue, M. Chenevier, and F. Stoeckel, *Diamond Relat. Mater.* **2**, 481 (1993).

1044. M. Lipp and J. O'Brien, *Chem. Phys. Lett.* **227**, 1 (1994).

1045. M. Lipp and J. O'Brien, *Chem. Phys.* **192**, 355 (1995).

1046. R. Escribano and A. Campargue, *J. Chem. Phys.* **108**, 6249 (1998).

1047. K. Tachibana, T. Shirafuji, and Y. Matsui, *Jpn. J. Appl. Phys.* **31**, 2588 (1992).

1048. S. Cheskis, *J. Chem. Phys.* **102**, 1851 (1995).

1049. S. Cheskis, I. Derzy, V. A. Lozovsky, A. Kachanov, and F. Stoeckel, *Chem. Phys. Lett.* **277**, 423 (1997).

1050. V. A. Lozovsky, S. Cheskis, A. Kachanov, and F. Stoeckel, *J. Chem. Phys.* **106**, 8384 (1997).

1051. R. Krugel, C. Williams, M. Fred, J. G. Malm, W. T. Carnall, J. C. Hindmann, W. J. Childs, and L. S. Goodman, *J. Chem. Phys.* **65**, 3486 (1976).

1052. D. Permogorov, A. Campargue, M. Chenevier, and H. Ben Kraiem, *J. Mol. Spectrosc.* **170**, 10 (1995).

1053. D. Permogorov and A. Campargue, *Chem. Phys.* **182**, 281 (1994).

1054. A. Charvat, S. F. Deppe, H. H. Hamman, and B. Abel, *J. Mol. Spectrosc.* **185**, 336 (1997).

1055. A. Campargue, L. Biennier, and M. Herman, *Mol. Phys.* **93**, 457 (1998).

1056. V. M. Baev, V. P. Dubov, and E. A. Sviridenkov, *Sov. J. Quantum Electron. (Engl. Transl.)* **15**, 1648 (1985).

1057. L. N. Sinitsa, *J. Quant. Spectrosc. Radiat. Transfer* **48**, 721 (1992).

1058. L. N. Sinitsa, *Kvantovaya Electron. (Moscow)* **4**, 148 (1977).

1059. V. M. Baev, H. Schroeder, and P. E. Toschek, *Opt. Commun.* **36**, 57 (1981).

1060. V. P. Kochanov, V. I. Serdyukov, and L. N. Sinitsa, *Opt. Acta* **32**, 1273 (1985).

1061. A. Campargue, F. Stoeckel, and M. C. Terrile, *Chem. Phys.* **110**, 145 (1986).

1062. D. A. Gilmore, P. Vujkovic Cvijin, and G. H. Atkinson, *Opt. Commun.* **103**, 370 (1993).

1063. A. Kachanov, F. Stoeckel, J. Yu, E. Frijafon, and J. P. Wolf, submitted for publication (1998).

1064. A. D. Bykov, V. A. Kapitanov, O. V. Naumenko, T. M. Petrova, V. I. Serdyukov, and L. N. Sinitsa, *J. Mol. Spectrosc.* **153**, 197 (1992).

1065. A. D. Bykov, V. P. Lopasov, Y. S. Makushkin, L. N. Sinitsa, O. N. Ulenikov, and V. E. Zuev, *J. Mol. Spectrosc.* **94**, 1 (1982).

1066. A. D. Bykov, Y. S. Makushkin, V. I. Serdyukov, L. N. Sinitsa, O. N. Ulenikov, and V. E. Zuev, *J. Mol. Spectrosc.* **105**, 397 (1984).

1067. L. N. Sinitsa, *J. Mol. Spectrosc.* **84**, 57 (1980).

1068. V. E. Zuev, V. P. Lopasov, L. N. Sinitsa, and A. M. Solodov, *J. Mol. Spectrosc.* **94**, 208 (1982).

1069. V. P. Lopasov, L. N. Sinitsa, and A. M. Solodov, *Opt. Spectrosc.* **49**, 452 (1980).

1070. A. Garnache, A. Campargue, A. Kachanov, and F. Stoeckel, *Chem. Phys. Lett.*, **292**, 698 (1998).

1071. E. A. Sviridenkov and L. N. Sinitsa, *Intracavity Laser Spectroscopy.* SPIE, Bellingham, WA (USA), 1998.

1072. L. Biennier and A. Campargue, *J. Mol. Spectrosc.* **188**, 248 (1998).

1073. A. Campargue, D. Permogorov, and R. Jost, *J. Chem. Phys.* **102**, 5910 (1995).

1074. D. Romanini and A. Campargue, *Chem. Phys. Lett.* **254**, 52 (1996).

1075. S. Cheskis and S. A. Kovalenko, *Appl. Phys.* **59**, 547 (1994).

1076. S. Cheskis, A. Kachanov, M. Chenevier, and F. Stoeckel, *Appl. Phys.* **64**, 713 (1997).

1077. A. G. Bell, *Am. J. Soc.* **20**, 305 (1880).

1078. A. G. Bell, *Philos. Mag.* [5] **11**, 510 (1881).

1079. J. Tyndall, *Proc. R. Soc. London* **31**, 307 (1881).

1080. W. C. Röntgen, *Philos. Mag.* [5] **11**, 308 (1881).

1081. G. A. West, J. J. Barrett, D. R. Siebert, and K. V. Reddy, *Rev. Sci. Instrum.* **54**, 797 (1983).

1082. A. Rosencwaig, *Photoacoustics and Photoacoustic Spectroscopy.* Wiley, New York, 1980.

1083. P. Hess, *Photoacoustic, Photothermal and Photochemical Processes in Gases*, Springer-Verlag, Berlin, 1989.

1084. V. P. Zharov and V. S. Letokhov, *Laser Optoacoustic Spectroscopy*, Springer-Verlag, Berlin, 1986.

1085. B. R. Henry and M. G. Sowa, *Prog. Anal. Spectrosc.* **12**, 349 (1989).

1086. F. J. M. Harren, F. G. C. Bijnen, J. Reuss, L. A. C. J. Voesenek, and C. W. P. M. Blom, *Appl. Phys. B* **50**, 137 (1990).

1087. F. J. M. Harren, J. Reuss, E. J. Woltering, and D. D. Bicanic, *Appl. Spectrosc.* **44**, 1360 (1990).

1088. L. B. Kreuzer, *J. Appl. Phys.* **42**, 2934 (1971).

1089. E. G. Burkhart, C. A. Lambert, and C. K. N. Patel, *Science* **188**, 1111 (1975).

1090. M. W. Sigrist, *Infrared Phys. Technol.* **36**, 415 (1995).

1091. F. J. M. Harren and J. Reuss, in *SPIE* (A. Mandelis and P. Hess, eds.), p. 83. Bellingham, 1997.

1092. S. Schafer, A. Miklos, and P. Hess, in *SPIE* (A. Mandelis and P. Hess, eds.), p. 254. Bellingham, 1997.

1093. S. Schafer, A. Miklos, A. Pusel, and P. Hess, *Chem. Phys. Lett.* **285**, 235 (1998).

1094. R. D. Kamm, *J. Appl. Phys.* **47**, 3550 (1976).

1095. F. G. C. Bijnen, J. Reuss, and F. J. M. Harren, *Rev. Sci. Instrum.* **67**, 2914 (1996).

1096. C. Hornberger, M. Koening, S. B. Rai, and W. Demtröder, *Chem. Phys.* **190**, 171 (1995).

1097. X. Zhan, E. Kauppi, and L. Halonen, *Rev. Sci. Instrum.* **63**, 5546 (1992).

1098. X. Zhan, O. Vaittinen, E. Kauppi, and L. Halonen, *Chem. Phys. Lett.* **180**, 310 (1991).

1099. R. F. Menefee, R. R. Hall, and M. J. Berry, *Appl. Phys. B* **28**, 121 (1982).

1100. D. W. Chandler, W. E. Farneth, and R. N. Zare, *J. Chem. Phys.* **77**, 4447 (1982).

1101. J. J. Scherer, K. K. Lehmann, and W. Klemperer, *J. Chem. Phys.* **81**, 5319 (1984).

1102. G. Stella, J. Gelfand, and W. H. Smith, *Chem. Phys. Lett.* **39**, 146 (1976).

1103. J. Gelfand, W. Hermina, and W. H. Smith, *Chem. Phys. Lett.* **65**, 201 (1979).

1104. R. R. Hall, Ph.D. Thesis, Rice University, Houston, TX, 1984.

1105. K. K. Lehmann, G. J. Scherer, and W. Klemperer, *J. Chem. Phys.* **77**, 2853 (1982).

1106. S. L. Coy and K. K. Lehmann, *J. Chem. Phys.* **84**, 5239 (1986).

1107. X. Zhan, O. Vaittinen, and L. Halonen, *J. Mol. Spectrosc.* **160**, 172 (1993).

1108. J. Lummila, O. Vaittinen, P. Jungner, L. Halonen, and A. M. Tolonen, *J. Mol. Spectrosc.*, **185**, 296 (1998).

1109. M. Halonen and X. Zhan, *J. Chem. Phys.* **101**, 950 (1994).

1110. X. Zhan, M. Halonen, L. Halonen, H. Bürger, and O. Polanz, *J. Chem. Phys.* **102**, 3911 (1995).

1111. J. Lummila, T. Lukka, L. Halonen, H. Burger, and O. Polanz, *J. Chem. Phys.* **104**, 488 (1996).

1112. J. W. Perry, D. J. Moll, A. Kupperman, and A. H. Zewail, *J. Chem. Phys.* **82**, 1195 (1985).

1113. R. A. Bernheim, F. W. Lampe, J. F. O'Keefe, and J. R. Qualey, III, *J. Mol. Spectrosc.* **104**, 194 (1984).

1114. X. Yang, C. J. Petrillo, and C. Noda, *Chem. Phys. Lett.* **214**, 536 (1993).

1115. X. Yang and C. Noda, *J. Mol. Spectrosc.* **183**, 151 (1997).

1116. K. Boraas and J. P. Reilly, *Rev. Sci. Instrum.* **64**, 3108 (1993).

1117. K. Boraas, Z. Lin, and J. P. Reilly, *J. Chem. Phys.* **100**, 7916 (1994).

1118. K. Boraas and J. P. Reilly, *Chem. Phys.* **190**, 301 (1995).

1119. Z. Lin, K. Boraas, and J. P. Reilly, *J. Mol. Spectrosc.* **170**, 266 (1995).

1120. Z. Lin, K. Boraas, and J. P. Reilly, *J. Mol. Spectrosc.* **156**, 147 (1992).

1121. D. Luckhaus, M. J. Coffey, M. D. Fritz, and F. F. Crim, *J. Chem. Phys.* **104**, 3472 (1996).

1122. C. Douketis and J. P. Reilly, *J. Chem. Phys.* **96**, 3431 (1992).

1123. J. R. Fair, O. Votava, and D. J. Nesbitt, *J. Chem. Phys.* **108**, 72 (1998).

1124. C. Douketis and J. P. Reilly, *J. Chem. Phys.* **91**, 5239 (1989).

1125. J. L. Scott, D. Luckhaus, S. S. Brown, and F. F. Crim, *J. Chem. Phys.* **102**, 675 (1995).

1126. J. X. Han, O. N. Ulenikov, S. Yurchinko, L. Y. How, W. G. Wang, and Q. S. Zhu, *Spectrochim. Acta, Part A* **53**, 1705 (1997).

1127. C. Hornberger, B. Boor, R. Stuber, W. Demtröder, S. Naim, and A. Fayt, *J. Mol. Spectrosc.* **179**, 237 (1996).

1128. M. Takahashi, Y. Okuzawa, M. Fujii, and M. Ito, *Can. J. Chem.* **69**, 1656 (1991).

1129. K. V. Reddy and M. J. Berry, *Chem. Phys. Lett.* **52**, 111 (1977).

1130. K. V. Reddy and M. J. Berry, *Faraday Discuss. Chem. Soc.* **67**, 188 (1979).

1131. K. V. Reddy, D. F. Heller, and M. J. Berry, *J. Chem. Phys.* **76**, 2814 (1982).

1132. M. L. Sage, *J. Chem. Phys.* **80**, 2872 (1984).

1133. H. L. Fang and R. L. Swofford, *J. Chem. Phys.* **73**, 2607 (1980).

1134. H. L. Fang and R. L. Swofford, *Appl. Opt.* **21**, 55 (1982).

1135. L. A. Findsen, H. L. Fang, R. L. Swofford, and R. R. Birge, *J. Chem. Phys.* **84**, 16 (1986).

1136. M. W. Crofton, C. G. Stevens, D. Klenerman, J. H. Gutow, and R. N. Zare, *J. Chem. Phys.* **89**, 7100 (1988).

1137. J. L. Duncan and A. M. Ferguson, *J. Chem. Phys.* **89**, 4216 (1988).

1138. J. L. Duncan, C. A. New, and B. Leavitt, *J. Chem. Phys.* **102**, 4012 (1995).

1139. J. S. Wong and C. B. Moore, *J. Chem. Phys.* **77**, 603 (1982).

1140. J. S. Wong, W. H. Green, C. Cheng, and C. B. Moore, *J. Chem. Phys.* **86**, 5994 (1987).

1141. J. E. Baggott, H. J. Clase, and I. M. Mills, *J. Chem. Phys.* **84**, 4193 (1986).

1142. Z. Lin, K. Boraas, and J. P. Reilly, *Chem. Phys. Lett.* **217**, 239 (1994).

1143. C. Manzanares, N. L. S. Yamasaki, and E. Wertz, *Chem. Phys. Lett.* **144**, 43 (1988).

1144. J. Segall, R. N. Zare, H. R. Dübal, M. Lewerenz, and M. Quack, *J. Chem. Phys.* **86**, 634 (1987).

1145. J. Davidson, J. H. Gutow, R. N. Zare, H. A. Hollenstein, R. R. Marquadt, and M. Quack, *J. Phys. Chem.* **95**, 1201, and references therein (1991).

1146. M. F. Hineman, R. G. Rodriguez, and J. W. Nibler, *J. Chem. Phys.* **89**, 2630 (1988).

1147. T. E. Gough, R. E. Miller, and G. Scoles, Appl. Phys. *Lett.* **30**, 338 (1977).

1148. T. E. Gough, R. E. Miller, and G. Scoles, *J. Chem. Phys.* **69**, 1588 (1978).

1149. T. E. Gough, Gravel, and R. E. Miller, *Rev. Sci. Instrum.* **52**, 802 (1981).

1150. E. R. T. Kerstel, M. Becucci, G. Pietraperzia, and E. Castellucci, *Chem. Phys.* **199**, 263 (1995).

1151. D. Bassi, A. Boschetti, and M. Scotoni, *Applied Laser Spectroscopy*, Plenum, New York, 1990.

1152. C. Douketis, D. Anex, G. Ewing, and J. P. Reilly, *J. Phys. Chem.* **89**, 4173 (1985).

1153. C. Douketis, D. Anex, and J. P. Reilly, *SPIE* **669**, 137 (1986).

1154. K. Boraas, D. F. De Boer, Z. Lin, and J. P. Reilly, *J. Chem. Phys.* **96**, 3431 (1992).

1155. K. Boraas, D. F. De Boer, Z. Lin, and J. P. Reilly, *J. Chem. Phys.* **99**, 1429 (1993).

1156. M. Scotoni, A. Boschetti, N. Oberhoffer, and D. Bassi, *J. Chem. Phys.* **94**, 971 (1991).

1157. M. Scotoni, A. Boschetti, N. Oberhofer, and D. Bassi, *J. Chem. Phys.* **95**, 8655 (1991).

1158. D. Bassi, C. Corbo, L. Lubich, S. Oss, and M. Scotoni, *J. Chem. Phys.* **107**, 1106 (1997).

1159. R. H. Page, Y. R. Shen, and Y. T. Lee, *J. Chem. Phys.* **88**, 4621 (1988).

1160. R. H. Page, Y. R. Shen, and Y. T. Lee, *Phys. Rev. Lett.* **59**, 1293 (1987).

1161. E. R. T. Kerstel, M. Becucci, G. Pietraperzia, and E. Castellucci, *J. Chem. Phys.* **106**, 1318 (1997).

1162. A. S. Pine, G. T. Fraser, and J. M. Pliva, *J. Chem. Phys.* **89**, 2720 (1988).

1163. J. E. Gambogi, K. K. Lehmann, B. H. Pate, G. Scoles, and X. Yang, *J. Chem. Phys.* **98**, 1748 (1993).

1164. E. R. T. Kerstel, K. K. Lehmann, T. F. Mentel, B. H. Pate, and G. Scoles, *J. Chem. Phys.* **95**, 8282 (1991).

1165. A. McIlroy, D. J. Nesbitt, E. R. T. Kerstel, B. H. Pate, K. K. Lehmann, and G. Scoles, *J. Chem. Phys.* **100**, 2596 (1994).

1166. B. H. Pate, K. K. Lehmann, and G. Scoles, *J. Chem. Phys.* **98**, 3891 (1993).

1167. J. V. Dolce, A. Callegari, B. Meyer, K. K. Lehmann, and G. Scoles, *J. Chem. Phys.* **107**, 6549 (1997).

1168. D. Romanini, Cavity-ringdown Spectroscopy. *A New Technique for Trace Absorption Measurements.* American Chemical Society, Washington, DC, 1998.

1169. J. E. Baggott and D. W. Law, *J. Chem. Soc., Faraday Trans.* **284**, 1560 (1988).

1170. L. B. Kreuzer, *Optoacoustic Spectroscopy and Detection.* Academic Press, New York, 1977.

1171. A. M. Smith, W. Klemperer, and K. K. Lehmann, *J. Chem. Phys.* **90**, 4633, and references therein (1989).

1172. J. H. Gutow, J. Davidsson, and R. N. Zare, *Chem. Phys. Lett.* **185**, 120 (1991).

1173. R. Engeln and G. Meijer, *Rev. Sci. Instrum.* **67**, 2708 (1996).

1174. R. Engeln, E. Van den Berg, G. Meijer, L. Lin, G. M. H. Knippels, and A. F. G. Van der Meer, *Chem. Phys. Lett.* **269**, 293 (1997).

1175. O. Votava, J. R. Fair, D. F. Plusquellic, E. Riedle, and D. J. Nesbitt, *J. Chem. Phys.* **107**, 8854 (1997).

1176. F. F. Crim, *Science* **249**, 1387 (1990).

1177. F. F. Crim, *J. Phys. Chem.* **100**, 12725 (1996).

1178. M. Chevalier and A. De Martino, *J. Chem. Phys.* **90**, 2077 (1989).

1179. M. Chevalier and A. De Martino, *Chem. Phys. Lett.* **135**, 446 (1987).

1180. A. de Martino, R. Frey, and F. Pradere, *Mol. Phys.* **55**, 731 (1985).

1181. R. H. Page, Y. R. Shen, and Y. T. Lee, *J. Chem. Phys.* **88**, 5362 (1988).

1182. A. L. Utz, J. D. Tobiason, E. Carrasquillo, M. D. Fritz, and F. F. Crim, *J. Chem. Phys.* **97**, 389 (1992).

1183. A. P. Milce, H.-D. Barth, and B. J. Orr, *J. Chem. Phys.* **100**, 2398 (1994).

1184. M. A. Payne, A. L. Milce, M. J. Frost, and B. J. Orr, *Chem. Phys. Lett.* **265**, 244 (1997).

1185. T. R. Rizzo, C. C. Hayden, F. F. Crim, and X. Luo, *J. Chem. Phys.* **94**, 889 (1981).

1186. T. R. Rizzo, C. C. Hayden, F. F. Crim, and X. Luo, *Faraday Discuss. Chem. Soc.* **754**, 223 (1983).

1187. T. R. Rizzo, C. C. Hayden, F. F. Crim, and X. Luo, *J. Chem. Phys.* **81**, 4501 (1984).

1188. T. R. Rizzo, C. C. Hayden, F. F. Crim, and X. Luo, *J. Chem. Phys.* **96**, 5659 (1992).

1189. L. J. Butler, T. M. Ticich, M. D. Likar, and F. F. Crim, *J. Chem. Phys.* **85**, 2332 (1986).

1190. T. M. Ticich, M. D. Likar, H. R. Dübal, L. J. Butler, and F. F. Crim, *J. Chem. Phys.* **87**, 5820 (1987).

1191. X. Luo, P. R. Fleming, T. A. Seckel, and T. R. Rizzo, *J. Chem. Phys.* **93**, 9194 (1990).

1192. B. R. Foy, M. P. Casassa, J. C. Stephenson, and D. S. King, *J. Chem. Phys.* **92**, 2782 (1990).

1193. M. R. Wedlock, R. Jost, and T. R. Rizzo, *J. Chem. Phys.* **107**, 111 (1997).

1194. R. L. Van der Wal, J. C. Scott, F. F. Crim, K. Weide, and R. Schinke, *J. Chem. Phys.* **94**, 3548 (1991).

1195. R. L. Van der Wal, J. L. Scott, and F. F. Crim, *J. Chem. Phys.* **92**, 803 (1990).

1196. R. B. Metz, J. D. Thoemke, J. M. Pfeiffer, and F. F. Crim, *J. Chem. Phys.* **99**, 1744 (1993).

1197. F. F. Crim, *Annu. Rev. Phys. Chem.* **44**, 397 (1993).

1198. M. Brouard and S. R. Langford, *J. Chem. Phys.* **106**, 6354 (1997).

1199. M. Hippler and M. Quack, *Chem. Phys. Lett.* **231**, 65 (1994).

1200. M. Hippler and M. Quack, *J. Chem. Phys.* **104**, 7426 (1996).

1201. M. Hippler and M. Quack, *Ber. Bunsenges. Phys. Chem.* **99**, 417 (1995).

1202. H. M. Lambert and P. J. Dagdigian, *Chem. Phys. Lett.* **275**, 499 (1997).

1203. T. Arusi-Parpar, R. P. Schmid, R. J. Li, I. Bar, and S. Rosenwaks, *Chem. Phys. Lett.* **268**, 163 (1997).

1204. T. Arusi-Parpar, R. P. Schmid, Y. Ganot, I. Bar, and S. Rosenwaks, *Chem. Phys. Lett.* **287**, 347 (1998).

1205. R. P. Schmid, T. Arusi-Parpar, R.-J. Li, I. Bar, and S. Rosenwaks, *J. Chem. Phys.* **107**, 385 (1997).

1206. R. P. Schmid, Y. Ganot, T. Arusi-Parpar, R. J. Li, I. Bar, and S. Rosenwaks, *SPIE* **3271**, in press (1998).

1207. O. V. Boyarkin, R. D. F. Settle, and T. R. Rizzo, *Ber. Bunsenges. Phys. Chem.* **99**, 504 (1995).

1208. R. D. F. Settle and T. R. Rizzo, *J. Chem. Phys.* **97**, 2823 (1992).

1209. O. V. Boyarkin and T. R. Rizzo, *J. Chem. Phys.* **105**, 6285 (1996).

1210. C. E. Hamilton, J. L. Kinsey, and R. W. Field, *Annu. Rev. Phys. Chem.* **37**, 493 (1986).

1211. D. Hai-Lung, *J. Opt. Soc. Am. B* **7**, 1802 (1990).

1212. K. Yamanouchi, H. Yamada, and S. Tsuchiya, *J. Chem. Phys.* **88**, 4664 (1988).

1213. C. Kittrell, E. Abramson, J. L. Kinsey, S. A. McDonald, D. E. Reisner, R. W. Field, and D. H. Katayama, *J. Chem. Phys.* **75**, 2056 (1981).

1214. D. E. Reisner, R. W. Field, J. L. Kinsey, and H. L. Dai, *J. Chem. Phys.* **78**, 2817 (1983).

1215. Y. Chen, S. Halle, D. M. Jonas, J. L. Kinsey, and R. W. Field, *J. Opt. Soc. Am. B* **7**, 1805 (1990).

1216. K. Yamanouchi, N. Ikeda, S. Tsuchiya, D. M. Jonas, J. K. Lundberg, G. W. Adamson, and R. W. Field, *J. Chem. Phys.* **95**, 6330 (1991).

1217. K. Yamanouchi, S. Takeunchi, and S. Tsuchiya, *J. Chem. Phys.* **92**, 4044 (1990).

1218. Y. S. Choi and C. B. Moore, *J. Chem. Phys.* **90**, 3875 (1989).

1219. Y. S. Choi and C. B. Moore, *J. Chem. Phys.* **94**, 5414 (1991).

1220. J. D. Tobiason, J. R. Dunlop, and E. A. Rohlfing, *J. Chem. Phys.* **103**, 1448 (1995).

1221. D. W. Neyer, X. Luo, I. Burak, and P. L. Houston, *J. Chem. Phys.* **102**, 1645 (1995).

1222. C. Stöck, X. Li, H. M. Keller, R. Schinke, and F. Temps, *J. Chem. Phys.* **106**, 5333 (1997).

1223. A. Delon and R. Jost, *J. Chem. Phys.* (submitted for publication).

1224. R. E. Smalley, L. Wharton, and D. H. Levy, *J. Chem. Phys.* **63**, 4977 (1975).

1225. A. Delon, R. Georges, and R. Jost, *J. Chem. Phys.* **103**, 7740 (1995).

1226. G. Persch, E. Mehdizadeh, W. Demtröder, T. Zimmerman, H. Koppel, and L. S. Cederbaum, *Ber. Bunsenges. Phys. Chem.* **92**, 312 (1988).

1227. S. Hiraoka, K. Shibruja, and K. Obi, *J. Mol. Spectrosc.* **126**, 427 (1987).

1228. A. Delon, P. Dupré, and R. Jost, *J. Chem. Phys.* **99**, 9482 (1993).

1229. R. Jost, J. Nygard, A. Pasinski, and A. Delon, *J. Chem. Phys.* **105**, 1287 (1996).

1230. C. A. Biesheuvel, D. H. A. ter Steege, J. Bulthuis, M. H. M. Janssen, J. G. Snijders, and S. Stolte, *Chem. Phys. Lett.* **269**, 515 (1997).

1231. R. Titeica, *Actual. Sci. Ind.* **334**, 1 (1936).

APPENDIX A.   ABBREVIATIONS AND SYMBOLS

Abbreviations, acronyms, and symbols used in this chapter are listed here.

Abbreviations

AH	Abbouti Temsamani and Herman
ANO	atomic natural orbital
BCHM	Bramley, Carter, Handy, and Mills
BLYP	DFT with Becke long-range correction and Lee–Yang–Parr correlation function
CASPT2	complete active-space perturbation theory at second order
CASSCF	complete active-space self-consistent field
CASSCF-CI	complete active-space self-consistent field with configuration interaction
CCSD	coupled cluster with all single and double excitations
CCSD(T)	CCSD with perturbative treatment of triple excitations
CCSD(TQ)	CCSD with perturbative treatment of triple and quadruple excitations
CI	configuration interaction
CNPI	complete nuclear permutation inversion
CNPR	complete nuclear permutation rotation
CPHF	coupled Hartree–Fock equations
CRDS	cavity ringdown spectroscopy
CSF	configuration state functions
CVPT(n)	canonical Van Vleck perturbation theory at nth order
CW	continuous wavelength
DD	Darling–Dennison
DF	dispersed fluorescence
DFT	density function theory
DMS	dipole moment surface
DVR	discrete–variable representation
DZ	double zeta
EMS	extended molecular symmetry
F	Fermi
FBR	finite basis representation
FMDL	frequency modulation diode laser
FT	Fourier transform
FTS	Fourier transform spectroscopy
HCC	Halonen, Child, and Carter
HO	harmonic oscillator
HRFTS	high-resolution Fourier transform spectroscopy
ICLAS	intracavity laser absorption spectroscopy

IRLAPS	infrared laser–assisted photofragment spectroscopy
IVR	intramolecular vibrational energy redistribution
LAS	laboratory axis system
LCAO	linear combination of atomic orbitals
LIF	laser-induced fluorescence
MAS	molecular axis system
MCSCF	multiconfigurational self-consistent field
MIME	matrix image of the molecule
MORBID	Morse oscillator rigid bender internal dynamics
MP(n)	Møller–Plesset partition of the Hamiltonian for nth-order treatment
MRCI	multireference configuration interaction
MS	molecular symmetry
OA	optoacoustic
OPO	optoparametric oscillator
OSVADPI	overtone spectroscopy by vibrationally assisted dissociation and photofragment ionisation
OT	optothermal
PES	potential-energy surface
PO-DVR	potential optimized discrete-variable representation
REMPI	resonance-enhanced multiphoton ionization
RRGM	recursive residue generation method
R2PI	resonance two-photon ionization
SCF	self-consistent field
SDCI	single and double configuration interaction
SD(Q)CI	SDCI with Davidson's correction for unlinked quadruple excitations
SEP	stimulated emission pumping
SM	Strey and Mills
SM(M)	modified SM force field
SPF	Simon, Parr, and Finlan
SZ	single zeta
TZ	triple zeta
TZ2P	TZ with double polarisation
UAO	uncoupled anharmonic oscillator
UV	ultraviolet
VCASSCF	vibrational complete active-space self-consistent field
VCC	variable curvature coordinates
VCI	vibrational configuration interaction
VMCSCF	vibrational multiconfigurational self-consistent field
VMCSCF-CI	vibrational multiconfigurational self-consistent field with configuration interaction

VSCF	vibrational self-consistent field
VSCF-CI	vibrational SCF method with configuration interaction
ZPE	zero point energy

Symbols

$\otimes$	direct product of irreducible representations of a group of symmetry
a	principal axis of inertia
$a_i^{\alpha\beta}$	coefficient in development of reciprocal inertia tensor in function of normal-mode vibration coordinates (Q)
$a^{\pm}$	shift-up/shift-down operator
A	principal rotational constant along the a principal axis of inertia (cm^{-1})
$A_{1,2}^{RP,Q}$	Herman–Wallis factor
$A_e(\tilde{v})$	Naperian absorbance $[=-\ln(I(\tilde{v})/I_0(\tilde{v}))]$
$A(\tilde{v})$	absorption $[=1-\tau(\tilde{v})]$
A_{12}	Einstein coefficient for spontaneous emission
$\alpha(\tilde{v})$	absorption coefficient
α_{21}	integrated absorption coefficient (cm^{-2})
α_Ω	coefficient (cm^{-1}) of first term in vibrational dependence of principal rotation constant Ω_e
α,β	index running over the principal axis of rotation, a, b, and c (b and c for a linear top), defined as those diagonalizing the regular inertia tensor (and perpendicular to the molecular axis for a linear top)
b	principal axis of inertia; also, index running over the bending vibrations
b_i	coordinates associated with a two-degenerate bending vibration ($i=1,2$)
B	principal rotational constant along the b principal axis of inertia (cm^{-1})
B_{12}	Einstein coefficient for stimulated emission
B_{21}	Einstein coefficient for absorption
β	parameter related to anharmonicity in the Morse coordinate
β_l	coefficient,(cm^{-1}) of the first term in the vibrational dependence of the centrifugal distortion constant D
β_{ij}	coefficient, (cm^{-1}) of the second term in the vibrational dependence of the centrifugal distortion constant D
c	principal axis of inertia
c_0	speed of light in vacuum
C	principal rotational constant along the c principal axis of inertia, (cm^{-1})

χ	Euler angle related to z
d_i	degeneracy of the ith normal mode of vibration
D	centrifugal distortion constant (D_J, D_K, D_{JK}) (cm^{-1}); also bond dissociation energy
$D_{KM}^J(\Theta)$	Wigner functions
D_2	rotation group of symmetry
e	one of the two parities of the rotational levels in a linear top [level with parity $(-1)^J$ in even-electron molecules]; also, in subscript, reference to electronic properties; in sub- or superscript, reference (equilibrium) configuration; electron charge in Hartree atomic units; general electron charge, $e = 1.60217733(49) \times 10^{-19}$ C
E	eigensolution of Schrödinger equation, in Hartree units
E/hc	energy, (cm^{-1})
E^0/hc	unperturbed energy, (cm^{-1})
$\vec{E}_0$	amplitude of electromagnetic wave
E_i^0	coordinates of amplitude of the electromagnetic wave in LAS
$E_{en}^{(E)}$	total molecular energies including electronic and nuclear contributions, in Hartree atomic units
E^*	inversion symmetry operator defined in molecular symmetry group; defines $+$ and $-$ symmetry character of rotation–vibration wavefunction
E_{diss}	dissociation energy in Hartree atomic units
$\vec{\varepsilon}$	polarization vector of radiation wave
ε_0	permittivity of vacuum
E^m	Euclidian space (with $m = 3n$)
f	one of the two parities of rotational levels in a linear top [level with parity $(-1)^{J+1}$ in even-electron molecules]
f_{21}	oscillator strength of transition $|2\rangle \leftarrow |1\rangle$
$f_{ss...}$	force constant in internal coordinate
F	Fermi
$F(J)$	rotational energy (cm^{-1}) for spherical and linear tops
$F(J,K)$	rotational energy (cm^{-1}) for symmetric tops
$F(J,K_a,K_c)$	rotational energy (cm^{-1}) for asymmetric tops
ϕ	Euler angle
$\phi_{ll...}$	coefficients in development in series of potential $V^{(q)}$
Φ	radiant power (W)
g	line profile function (g_D, Doppler; g_L, Lorentzian; g_V, Voigt)
g_i	statistical weight
g_j	nuclear spin degeneracy for unpaired nuclei $(= \prod_j(2I_j + 1))$
g_n	total nuclear degeneracy
$g_1(g_2)$	total degeneracy of level 1 (2)

g_v	degeneracy of vibrational wavefunction
$g_{bb'}$	coefficient (cm^{-1}) of term in $l_b l_{b'}$ in Dunham expansion of linear tops
$G(v)$	vibrational energy (cm^{-1}) related to bottom of potential well
$G_{0(v)}$	vibrational energy (cm^{-1}) related to ground level
G_c	band center (cm^{-1}) (also v_0), with $G_c = G_0 - B'_v k'^2 + D'_v k'^4 + B'_v k''^2 - D'_v k''^4$
λ	(irreducible representation of a group of symmetry
γ_j	$2\pi c \tilde{\omega}/\hbar$
γ_{ij}	coefficient (cm^{-1}) of second term in vibrational dependence of principal rotation constant Ω_e
γ_{ijk}	coefficient (cm^{-1}) of third term in vibrational dependence of the principal rotation constant Ω_e
$\gamma^{bb'}$	coefficient (cm^{-1}) of the term $l_b l_{b'}$ in rotational energy of bending vibrations
$\gamma_i^{bb'}$	coefficient (cm^{-1}) of term $v_i l_b l_{b'}$ in rotational energy of bending vibrations
γ	half-width at half-maximum of a line
h	Planck's constant ($\hbar = h/2\pi$)
$\hat{h}_i$	part of the vibration–rotation Hamiltonian depending on a single vibration coordinate q_i
$\tilde{h}_{nm}$	part of transformed vibration–rotation Hamiltonian, including factor $1/hc$, with n and m representing the power in vibration and rotation operators, respectively
H	higher-order centrifugal distortion constant
H_{nm}	part of rovibrational Hamiltonian, with n and m denoting the power in vibrational and rotational operators, respectively
H_v	coefficient, (cm^{-1}) of first term in vibrational dependence of higher-order centrifugal distortion constant H
$\hat{H}_c$	vibrational Hamiltonian in curvilinear rectilinear coordinates
$\hat{H}_r$	rotation Hamiltonian
$\hat{H}_R$	vibrational Hamiltonian in rectilinear coordinates
$\hat{H}_{rovib}$	vibration–rotation coupling Hamiltonian
$\hat{H}_v$	vibrational Hamitonian
$\hat{H}_0$	time-independent Hamiltonian of a free molecule
$\hat{H}'$	term of molecular Hamiltonian
i	index running over $3n - 6$ ($3n - 5$ for a linear top) normal modes of vibration; also, index running over nuclei and/or electrons
I	moment of inertia; also, nuclear spin quantum number (I_j in definition of g_j); irradiance or conventional spectroscopic intensity (W/m^2)

$I(\tilde{v})$	spectral intensity or spectral irradiance
$I_0(\tilde{v})$	spectral intensity of a light beam incident on sample
j	index running over $3n - 6$ ($3n - 5$ for a linear top) normal modes of vibration; also, index running over nuclei and/or electrons
J	total quantum number, excluding electronic an nuclear spins [$J(J+1)$ is eigenvalue of $\hat{J}^2$]
$\hat{J}_\alpha$	αth component of the total angular momentum in units of $\hbar$, excluding electronic an nuclear spins
k	signed rovibrational symmetric top quantum number (k^2 is the eigenvalue of $\hat{J}_z^2$, at the same time $k = \sum_b l_b$ for singlet electronic states); also, Boltzmann's constant; force constant of spring in Hooke's law; index running over $3n - 6$ ($3n - 5$ for a linear top) normal modes of vibration; also refers to a given geometry of the nuclear skeleton; anharmonic strength
$\vec{k}$	propagation vector of radiation wave
K	rotation symmetric top quantum number ($K = |k|$)
K_a, K_c	pseudo-rotation asymmetric top quantum numbers
$K_{ij/kl}$	coefficient of the ij/kl anharmonic resonance
κ	Ray asymmetry parameter
l	signed symmetric top quantum number (l is the eigenvalue of $\hat{L}$ and, within a numerical coefficient, of $\hat{\pi}_a$); also, index running over $3n - 6$ ($3n - 5$ for a linear top) normal modes of vibration
ℓ	absorption pathlength
$\hat{L}$	vibrational angular momentum operator ($\hat{L} = q_{b_1}\hat{p}_{b_2} - q_{b_2}\hat{p}_{b_1}$)
L_J^a	line strength or Hönl–London factor
$\hat{L}_a$	electronic angular momentum operator
λ	direction cosine matrix; also, effective bond mode coupling
λ_i	$(2\pi c\tilde{\omega}_i)^2$
$\lambda_{\zeta\Xi}$	component of the direction cosine matrix
Λ	electronic angular momentum quantum number (Λ is the eigenvalue of $\hat{L}_a$)
m	mass of a particle; also the total number of nuclear degrees of freedom ($3n$)
m_e	electron rest mass; $m_e = 9.1093897(54) \times 10^{-23}$ g
m_{int}	number of vibration coordinates ($3n - 5$ or $3n - 6$)
m_0^ζ	value of $\mu_\zeta^{(E)}(q)$ at equilibrium geometry
$m_{ij...}^\zeta$	derivatives of $\mu_\zeta^{(E)}(q)$ with respect to $q_i, q_j \ldots$

M	magnetic quantum number (M^2 is eigenvalue of $\hat{J}_Z^2$); also $J(J+1)$
M_n	total nuclear mass
μ	reduced mass
$\mu_{\alpha\beta}$	$(\alpha\beta)$ component of a modified reciprocal inertia tensor, function of Q
$\hat{\mu}$	operator associated with the classical electric dipole moment vector of all the particles, defined in LAS
μ_0	permanent electric dipole moment of the molecule in its ground vibrational level
μ_e	permanent electric dipole moment of the molecule at equilibrium geometry of ground electronic state
μ_Ξ	components of the electric dipole moment in LAS
μ_ζ	components of the electric dipole moment in MAS
$\mu_\zeta^{(E)}(q)$	components of electric dipole moment in MAS
n	number of nuclei
n_e	number of electrons
n_L	Loschmidt's number: particle density under STP conditions ($2.687 \times 10^{19}\,\text{cm}^{-3}$)
n_{tot}	$n + n_e$
N	polyad or cluster quantum number; also, particle density (cm^{-3}); total number of particles in a gas
$N_1(N_2)$	population (in molecules per unit volume) of level 1 (2)
$\hat{N}$	product operator $a^+ a^-$
$\tilde{\nu}$	wavenumber (cm^{-1})
$\tilde{\nu}_0$	rotation–vibration band origin (cm^{-1}) (also G_c)
$\tilde{\nu}^0$	energy of fundamental band, (cm^{-1})
Ω	generic label for principal rotation constants (cm^{-1}); in subscript, principal axis of inertia
p	index running over energy states
$\hat{p}_i$	dimensionless conjugate momentum operator of the rth normal coordinate
$\hat{p}_\pm$	dimensionless conjugate momentum operator of the normal coordinate associated with a two-dimensional degenerate mode of vibration ($= \hat{p}_{b_1} \pm i\hat{p}_{b_2}$)
P_α^l	associated Legendre function
$\hat{P}_i$	conjugate momentum operator of the ith normal coordinate ($= -i\hbar\partial/\partial Q_r$)
$\hat{\pi}_\alpha$	αth component of the vibrational angular momentum operator in units of $\hbar$ along the rotating direction α
$\Psi(t)$	time-dependent wavefunction

Ψ_0	wavefunction of an unperturbed system
Ψ_e	electronic wavefunction
Ψ_v	vibrational wavefunction
Ψ_r	rotational wavefunction
Ψ_{es}	electron spin wavefunction
Ψ_{ns}	nuclear spin wavefunction
q_b	l-doubling constant in linear tops (with dependencies in J and v)
q_i	ith dimensionless normal coordinate and associated operator
$q_\pm$	dimensionless normal coordinate associated with a two-dimensional degenerate mode of vibration $(= q_{b_1} \pm i q_{b_2})$.
Q_i	ith normal coordinate and associated operator
$Q(T)$	total internal partition function [electronic, $Q_e(T)$, vibrational $Q_v(T)$ and rotational $Q_r(T)$]
θ	Euler angle
θ_{1x}	x component of first bending internal coordinate
Θ	general notation for Euler angles
r	in subscript, reference to rotation properties
$\bar{r}_i$	vector position for particle i in MAS
$r_{bb'}$	coefficient of the off-diagonal coupling term in $\Delta l_b = -\Delta l_{b'}$ $= 2$ for linear tops (with dependencies in J and v)
R_p	whole set of rotation quantum numbers of state p
$\bar{R}_i$	vector position for particle i in LAS
R_{21}	electric dipole transition moment between states 1 and 2
R_v	vibrational transition moment
R_i	vibrational transition moment associated with mode i
R_{21}^{Ξ}	components of R_{21} in LAS
R_{21}^{ζ}	components of R_{21} in MAS
$\mathscr{R}^{3n}$	$3n$-dimensional vector space associated with an Euclidean space E^{3n}
ρ	radiant energy density $[= W/V \ (\text{J/m}^3)]$
$\rho(\tilde{v})$	spectral radiant energy density
ρ_b	coefficient of off-diagonal coupling term in $\Delta l_b = 4$ for linear tops (with dependencies in J and v)
$\rho_{bb'}$	coefficient of off-diagonal coupling term in $\Delta l_b = \Delta l_{b'} = 2$ for linear tops
$\mathscr{R}^m$	m-dimensional vector space (with $m = 3n$)
s	internal coordinate; also, index running over stretching vibrations
S_v	band strength (conventionally provided in cm^{-2}/atm)

S_{21}	absolute intensity ($=\alpha_{21}/P$, conventionally provided in (cm^{-2}/atm)
σ	absorption cross section (cm^2)
σ_{21}	integrated absorption cross section (cm^2 molecule^{-1} cm^{-1})
σ_{xz}	plane symmetry operator defined in the point group
t	time; in subscript, reference to translation motion
$\{t\}$	translation-free coordinates ($m-3$)
T	kinetic energy; also, absolute temperature
T_0	273.15 K
$\hat{T}_e$	kinetic-energy operator of electrons
$\tau(\tilde{\nu})$	transmittance
τ	$K_a - K_c$
U	small, mass-dependent correction to vibrational potential energy, with $U = -(\hbar^2/8)\sum_a \mu_{aa}$ ($=0$ for a linear top)
U_W	Wang transformation matrix for asymmetric rotors
$\vec{u}$	direction of propagation of electromagnetic wave
v	vibrational quantum number; also, in subscript, reference to vibration properties
v_p	whole set of vibration quantum numbers of state p
$\hat{V}_e$	Coulomb potential-energy operator
$\hat{V}_n$	nuclear potential-energy operator
$\hat{V}_{rv}$	vibration–rotation potential-energy operator
$V^{(Q)}, V^{(q)}$	nuclear potential energy, function of $3n-6$ vibrational coordinates
$V^{(en)}(Q_r)[V^{(en)}(q_r)] = V^{(Q)}[V^{(q)}] + E_n^e(Q)[E_n^e(q)]$	total potential energy, including electronic contribution (E_n^e), within Born–Oppenheimer approximation
$\bar{\omega}$	harmonic angular vibration frequency (Hz)
$\tilde{\bar{\omega}}$	harmonic angular vibration frequency (cm^{-1})
ω	harmonic vibration frequency (Hz); also, experimental modulation frequency
$\tilde{\omega}$	harmonic vibration frequency, (cm^{-1})
$\tilde{\omega}^0$	harmonic vibration frequency, (cm^{-1}) defined with respect to ground state
W	radiant energy (J)
W_{21}	transition probability between states 1 and 2
x	axis in MAS
x^0	first anharmonic coefficients in Dunham expansion (cm^{-1}), defined with respect to ground state
x_{ij}	first anharmonic coefficients in Dunham expansion (cm^{-1})
X	axis in LAS

y	axis in MAS
y_{ijk}	second anharmonic coefficients in the Dunham expansion, (cm^{-1})
Y	axis in LAS
z_{ijk}	second anharmonic coefficients in the Dunham expansion, (cm^{-1})
$y_b^{b'b''}$	coefficient (cm^{-1}) of term in $v_b l_{b'} l_{b''}$ in Dunham expansion of linear tops
$\Psi_{en}^{(E)}(\bar{R}_e, \bar{R}_n) = \Psi_e^{(E)}(\bar{R}_e, \bar{R}_n)\Psi_n^{(E)}(\bar{R}_n)$	total wavefunction
z	axis in MAS
Z	axis in LAS
Z_i	charge of the ith particle
ζ_{lk}^{α}	Coriolis zeta constant

APPENDIX B

Bibliography on Acetylene in the Ground Electronic State

The following bibliography, arranged in alphabetic order, includes the high-resolution spectroscopic literature on acetylene in its ground electronic state through early 1998. Doctoral theses known to include unpublished results are mentioned.

Further information on early references is provided by Herzberg in Ref. [697] of the main list of references from the present review and by Titeica [R. Titeica, *Actualitiés Scientifiques et Industrielles* **334**, 1 (1936)]. The bibliography listed in this appendix is almost certainly not exhaustive, despite our efforts to achieve that aim.

M. Abbouti Temsamani and M. Herman, The vibrational energy levels in acetylene $^{12}C_2H_2$: Towards a regular pattern at higher energies. *J. Chem. Phys.* **102**, 6371–6385 (1995).

M. Abbouti Temsamani and M. Herman, Anharmonic resonances in monodeuteroacetylene ($^{12}C_2HD$). *Chem. Phys. Lett.* **260**, 253–256 (1996).

M. Abbouti Temsamani and M. Herman, Anharmonic resonances in monodeuteroacetylene ($^{12}C_2HD$). *Chem. Phys. Lett.* **264**, 556–557 (1996).

M. Abbouti Temsamani and M. Herman, The vibrational energy pattern in $^{12}C_2H_2$(II): Vibrational clustering and rotational structure. *J. Chem. Phys.* **105**, 1355–1362 (1996).

M. Abbouti Temsamani, J. Vander Auwera, and M. Herman, The absorption spectrum of C_2HD between 9000 and 13000 cm^{-1}. *Mol. Phys.* **79**, 359–371 (1993).

M. Abbouti Temsamani, M. Herman, S. A. B. Solina, J. P. O'Brien, and R. W. Field, Highly vibrationally excited $^{12}C_2H_2$ in the X $^1\Sigma_g^+$ state: Complementarity of absorption and dispersed fluorescence spectra. *J. Chem. Phys.* **105**, 11357–11359 (1996).

E. Abramson, R. W. Field, D. Imre, K. K. Innes, and J. L. Kinsey, Stimulated emission pumping of acetylene: Evidence for quantum chaotic behavior near $27\,900\,cm^{-1}$ of excitation? *J. Chem. Phys.* **80**, 2298–2300 (1984).

E. Abramson, R. W. Field, D. Imre, K. K. Innes, and J. L. Kinsey, Fluorescence and stimulated emission $S_1 - S_0$ spectra of acetylene: Regular and ergodic regions. *J. Chem. Phys.* **83**, 453 (1985).

J. E. Adams, Computer assisted processing of infrared spectra and bands of acetylene and acetylene-d_1 at 1.1 microns. *Diss. Abstr. Int. B* **38**, 2243 (1977).

J. E. Adams and K. N. Rao, $^{12}C_2H_2$ and $^{12}C_2HD$ bands at 1.1 µm. *J. Mol. Spectrosc.* **64**, 157–161 (1977).

K. E. Aiani and J. S. Hutchinson, An efficient method for optimal modes analysis of vibration: Application to HCCD. *J. Phys. Chem.* **97**, 8922–8928 (1993).

S. Alanko, R. Paso, and R. Anttila, A diode laser-grating study of the $v_2 + v_5$ band of C_2D_2, *Z. Naturforsch., A* **42A**, 1247–1252 (1987).

F. Alboni, G. Di Lonardo, P. Ferracuti, L. Fusina, E. Venuti, and K. A. Mohamed, Vibration-rotation spectra of ^{13}C-containing acetylene: The stretching fundamentals. *J. Mol. Spectrosc.* **169**, 148–153 (1995).

L. J. Allamandola and C. A. Norman, Infra-red emission lines from molecules in grain mantles. *Astron. Astrophys.* **63**, L23–L26 (1978).

L. J. Allamandola and C. A. Norman, Infra-red molecular line emission from grain surfaces in dense clouds. *Astron. Astrophys.* **66**, 129–135 (1978).

H. C. Allen Jr., L. R. Blaine, and E. K. Plyler, Vibrational constants of acetylene-d_2. *J. Res. Nat. Bur. Stand.* **56**, 279–283 (1956).

H. C. Allen Jr., E. D. Tidwell, and E. K. Plyler, The infrared spectrum of acetylene-d_1 *J. Am. Chem. Soc.* **78**, 3034–3040 (1956).

M. Allen, Y. L. Yung, and G. R. Gladstone, The relative abundance of ethane to acetylene in the Jovian stratosphere. *Icarus* **100**, 527–533 (1992).

W. D. Allen, Y. Yamaguchi, A. G. Császár, D. A. Clabo, Jr., R. B. Remington, and H. F. Schaefer, III, A systematic study of molecular vibrational anharmonicity and vibration-rotation interaction by self-consistent-field higher-derivative methods. Linear polyatomic molecules. *Chem. Phys.* **145**, 427–466 (1990).

R. D. Amos, SCF dipole moment derivatives, harmonic frequencies and infrared intensities for acetylene and ethylene. *Chem. Phys. Lett.* **114**, 10–14 (1985).

A. Amrein, M. Quack, and U. Schmitt, High-resolution interferometric Fourier transform infrared absorption spectroscopy in supersonic free jet expansions: Carbon monoxide, nitric oxide, methane, ethyne, propyne, and trifluoromethane. *J. Phys. Chem.* **92**, 5455–5466 (1988).

R. Anttila, J. Hietanen, and J. Kauppinen, The infra-red spectrum of C_2HD around $700\,cm^{-1}$. *Mol. Phys.* **37**, 925–935 (1979).

T. Arusi-Parpar, R. P. Schmid, R. J. Li, I. Bar, and S. Rosenwaks, Rotational-state dependent selectivity in bond fission of C_2HD. *Chem. Phys. Lett.* **268**, 163 (1997).

T. Arusi-Parpar, R. P. Schmid, Y. Ganot, I. Bar, and S. Rosenwaks, Enhanced action spectra of combination bands of acetylene via vibrationally mediated photodissociation and fragment ionization. *Chem. Phys. Lett.* **287**, 347 (1998).

M. N. R. Ashfold, B. Tutcher, B. Yang, Z. K. Jin, and S. L. Anderson, Gerade Rydberg states of acetylene by MPI and photoelectron spectroscopy. *J. Chem. Phys.* **87**, 5105–5115 (1987).

A. Babay, M. Ibrahimi, V. Lemaire, B. Lemoine, F. Rohart, and J.-P. Bouanich, Line frequency shifting in the v_5 band of C_2H_2 J. Quant. Spectrosc. Radiat. Transfer **59**, 195–202 (1998).

A. Baldacci and S. Ghersetti, Infrared combination bands of $^{12}C_2H_2$ involving $0000^0 1^1$ state. J. Mol. Spectrosc. **48**, 600–603 (1973).

A. Baldacci, S. Ghersetti, and K. Narahari Rao, Bands of $^{12}C_2HD$ in the region 6-10 μm. J. Mol. Spectrosc. **34**, 358–360 (1970).

A. Baldacci, S. Ghersetti, and K. Narahari Rao, Assignments of the $^{12}C_2H_2$ bands at 2.1 μm. J. Mol. Spectrosc. **41**, 222–225 (1972).

A. Baldacci, S. Ghersetti, S. C. Hurlock, and K. Narahari Rao, Spectrum of dideuteroacetylene near 18.6 microns. J. Mol. Spectrosc. **42**, 327–334 (1972).

A. Baldacci, S. Ghersetti, S. C. Hurlock, and K. Narahari Rao, Infrared bands of $^{12}C_2HD$. J. Mol. Spectrosc. **59**, 116–125 (1976).

A. Baldacci, S. Ghersetti, and K. Narahari Rao, Interpretation of the Acetylene spectrum at 1.5 μm J. Mol. Spectrosc. **68**, 183–194 (1977).

B. A. Balko, J. Zhang, and Y. T. Lee, 193 nm photodissociation of acetylene. J. Chem. Phys. **94**, 7958–7966 (1991).

J. A. Barnes, T. E. Gough, and M. Stoer, Stark field induced perturbations in the $v_1 + 3v_3$ vibrational overtone band of acetylene. Chem. Phys. Lett. **237**, 437–442 (1995).

A. Bar-Nun and M. Podolak, The contribution by thunderstorms to the abundances of CO, C_2H_2, and HCN on Jupiter. Icarus **64**, 112–124 (1985).

H. D. Barth, A. P. Milce, B. L. Chadwick, and B. J. Orr, Raman ultraviolet double-resonance spectroscopy of acetylene in a skimmed molecular beam, J. Chem. Soc., Faraday Trans. **88**, 2563–2564 (1992).

M. Becucci, E. Castellucci, L. Fusina, G. Di Lonardo, and H. W. Schrötter, Vibration-rotation Raman spectrum of ^{13}C-containing acetylene. J. Raman Spectrosc. **29**, 237–241 (1998).

I. Benjamin, O. S. van Roosmalen, and R. D. Levine, A model algebraic Hamiltonian for interacting nonequivalent local modes with application to HCCD and $H^{12}C^{13}CD$. J. Chem. Phys. **81**, 3352 (1984).

J. A. Bentley, R. E. Wyatt, M. Menou, and C. Leforestier, A finite basis-discrete variable representation calculation of vibrational levels of planar acetylene. J. Chem. Phys. **97**, 4255–4263 (1992).

D. Bermejo, P. Cancio, G. Di Lonardo, and L. Fusina, High resolution Raman spectroscopy from vibrationally excited states populated by a stimulated Raman process. $2v_2 - v_2$ of $^{12}C_2H_2$ and $^{13}C_2H_2$. J. Chem. Phys. **108**, 7224–7228 (1998).

J. Berryhill, S. Pramanick, M. Z. Zgierski, F. Zerbetto, and B. S. Hudson, Resonance Raman activity in odd quanta of the trans bending vibration of acetylene: Strong vibronic coupling in the X to A and X to B transitions. Chem. Phys. Lett. **205**, 39–45 (1993).

B. H. Besler, G. E. Scuseria, A. C. Scheiner, and H. F. Schaefer, III, A systematic theoretical study of harmonic vibrational frequencies: The single and double excitation coupled cluster (CCSD) method. J. Chem. Phys. **89**, 360–366 (1988).

L. Biennier, A. Campargue, and M. Herman, The visible absorption spectrum of $^{12}C_2H_2$ III. The region 14500–17000 cm^{-1}. Mol. Phys. **93**, 457–469 (1998).

W. E. Blass and W. L. Chin, Hydrogen and nitrogen broadening of the lines of C_2H_2 at 14 μm. J. Quant. Spectrosc. Radiat. Transfer **38**, 185–188 (1987).

W. E. Blass, S. J. Daunt, A. V. Peters, and M. C. Weber. Proc. Int. Con. Lab. Res. Planet. Atmos. 1st, pp. 54–63, (1990).

A. F. Borro, I. M. Mills, and A. Mose, Quartic anharmonic resonances in haloacetylenes. *Chem. Phys.* **190**, 363–372 (1995).

A. F. Borro, I. M. Mills, and E. Venuti, Quartic anharmonic resonances in acetylene and haloacetylene molecules. *J. Chem. Phys.* **102**, 3938–3944 (1995).

J. P. Bouanich, D. Lambot, G. Blanquet, and J. Walrand, N_2 and O_2-broadening coefficients of C_2H_2 IR lines. *J. Mol. Spectrosc.* **140**, 195–213 (1990).

J.-P. Bouanich, C. Boulet, G. Blanquet, J. Walrand, and D. Lambot, Lineshapes and broadening coefficients in the v_5 band of C_2H_2 in collision with Kr and He. *J. Quant. Spectrosc. Radiat. Transfer* **46**, 317–324 (1991).

J. P. Bouanich, G. Blanquet, J. C. Populaire, and J. Walrand, Nitrogen broadening of acetylene lines in the v_5 band at low temperature. *J. Mol. Spectrosc.* **190**, 7–14 (1998).

P. Bour, C. N. Tam, and T. A. Keiderling, Ab initio calculation of the vibrational magnetic dipole moment. *J. Phys. Chem.* **99**, 17810–17813 (1995).

P. Bour, C. N. Tam, and T. A. Keiderling, Vibrational magnetic dipole moment of acetylene in the v_5 mode. *J. Phys. Chem.* **100**, 2062 (1996).

R. Boyd, T.-S. Ho, and H. Rabitz, Inversion of absorption spectral data for relaxation matrix determination. II. Application to Q-branch line mixing in HCN, C_2H_2 and N_2O. *J. Chem. Phys.* **108**, 1780–1793 (1998).

M. J. Bramley and N. C. Handy, Efficient calculation of rovibrational eigenstates of sequentially bonded four-atom molecules. *J. Chem. Phys.* **98**, 1378–1397 (1993).

M. J. Bramley, S. Carter, N. C. Handy, and I. M. Mills, A refined quartic force field for acetylene: Accurate calculation of the vibrational spectrum. *J. Mol. Spectrosc.* **157**, 301–336 (1993).

D. B. Braund and A. R. H. Cole, The $v_5^1 - v_4^1$ far infrared bands of acetylene -d_1 and acetylene-d_2. *Austr. J. Chem.* **33**, 2053–2060 (1980).

D. B. Braund, A. R. H. Cole, and T. E. Grader, The difference band $v_5 - v_4$ of acetylene, $^{12}C_2H_2$, *J. Mol. Spectrosc.* **48**, 604–606 (1973).

T. Y. Brooke, A. T. Tokunaga, H. A. Weaver, J. Crovisier, D. Bockelee-Morvan, and D. Crisp, Detection of acetylene in the infrared spectrum of comet Hyakutake. *Nature (London)* **383**, 606–608 (1996).

M. I. Bruce, Organometallic chemistry of vinylidene and related unsaturated carbenes. *Chem. Rev.* **91**, 197–257 (1991).

L. E. Brus, Acetylene fluorescence. *J. Mol. Spectrosc.* **75**, 245–250 (1979).

P. Bundgen, A. J. Thakkar, F. Grein, M. Ernzerhof, C. M. Marian, and B. Nestmann, Moments of the quadrupole oscillator strength distribution for O_2, N_2, CO, HF, HCl, N_2O, CO_2, OCS, CS_2 and C_2H_2: Ab initio sum rule calculations. *Chem. Phys. Lett.* **261**, 625 (1996).

S. M. Burnett, A. E. Stevens, C. S. Feigerle, and W. C. Lineberger, Observation of X^1A_1 vinylidene by photoelectron spectroscopy of the C_2H_2- ion. *Chem. Phys. Lett.* **100**, 124–127 (1983).

J. H. Callomon and B. P. Stoicheff, High resolution Raman spectroscopy of gases. VIII. Rotational spectra of acetylene, diacetylene, diacetylene-d_2 and dimethylacetylene. *Can. J. Phys.* **35**, 373–382 (1957).

A. Campargue, M. Chenevier, and F. Stoeckel, Overtone spectroscopy of acetylene near $13\,500\,cm^{-1}$. *J. Mol. Spectrosc.* **151**, 275–281 (1992).

A. Campargue, M. Abbouti Temsamani, and M. Herman, The absorption spectrum of acetylene in the $2v_2 + 3v_3$ region. A test of the cluster model. *Chem. Phys. Lett.* **242**, 101–105 (1995).

A. Campargue, M. Abbouti Temsamani, and M. Herman, The absorption spectrum of $^{12}C_2H_2$ between 12800 and 18500 cm^{-1}. I. Vibrational assignments. *Mol. Phys.* **90**, 793–806 (1997).

A. Campargue, L. Biennier, and M. Herman, The visible absorption spectrum of $^{12}C_2H_2$. III. The region 14500–17000 cm^{-1}. *Mol. Phys.* **93**, 457–469 (1998).

M. Carrasquillo, A. L. Utz, and F. F. Crim, Collisional relaxation of single rotational states in highly vibrationally excited acetylene. *J. Chem. Phys.* **88**, 5976–5978 (1988).

T. Carrington Jr., L. M. Hubbard, H. F. Schaefer, III, and W. H. Miller, Vinylidene: Potential energy surface and unimolecular reaction dynamics. *J. Chem. Phys.* **80**, 4347–4354 (1984).

S. Carter, I. M. Mills, and J. N. Murrell, A potential energy surface for the ground state of acetylene, C_2H_2, *Mol. Phys.* **41**, 191–203 (1980).

B. L. Chadwick and B. J. Orr, Rotationally resolved V-V transfer in C_2D_2/Ar_2 collisions: Characterization of a vibrational bottleneck. *J. Chem. Phys.* **95**, 5476 (1991).

B. L. Chadwick and B. J. Orr, Raman-ultraviolet double resonance in acetylene: Rovibrational state preparation and spectroscopy. *J. Chem. Phys.* **97**, 3007–3020 (1992).

B. L. Chadwick, A. P. Milce, and B. J. Orr, State-to-state vibrational energy transfer in acetylene gas, measured by fluorescence-detected Raman-ultraviolet double resonance spectroscopy. *Chem. Phys.* **175**, 113–125 (1993).

B. L. Chadwick, A. P. Milce, and B. J. Orr, Fluorescence-detected infrared- and Raman-ultraviolet double resonance in acetylene gas: Studies of spectroscopy and rotational energy transfer. *Can. J. Phys.* **72**, 939–953 (1994).

N.-Y. Chang, M.-Y. Shen, and C.-H. Yu, Extended ab initio studies of the vinylidene-acetylene rearrangement. *J. Chem. Phys.* **106**, 3237–3242 (1997).

W.-C. Chen and C.-H. Yu, A study of the vinylidene-acetylene rearrangement using density functional theory. *Chem. Phys. Lett.* **277**, 245–251 (1997).

Y. Chen, D. M. Jonas, J. L. Kinsey, and R. W. Field, High resolution spectroscopic detection of acetylene-vinylidene isomerization by spectral cross correlation. *J. Chem. Phys.* **91**, 3976–3987 (1989).

Y. Chen, S. Halle, D. M. Jonas, J. L. Kinsey, and R. W. Field, Stimulated-emission pumping studies of acetylene $X^1\Sigma_g^+$ in the 11400–15700 cm^{-1} region: The onset of mixing. *J. Opt. Soc. Am. B* **7**, 1805–1815 (1990).

M. S. Child and L. Halonen, Overtone frequencies and intensities in the local mode picture. *Adv. Chem. Phys.* **57**, 1–58 (1984).

S. Civis, Z. Zelinger, and K. Tanaka, The infrared diode laser spectroscopy of the $\nu_2 + \nu_5 - \nu_2$ hot band of acetylene. *J. Mol. Spectrosc.* **187**, 82–88 (1998).

W. F. Colby, Isotope effects in acetylene. *Phys. Rev.* **47**, 388 (1935).

M. Combes, T. Encrenaz, L. Vapillon, Y. Zéau, and C. Lesqueren, Confirmation of the identification of C_2H_2 and C_2H_6 in the jovian atmosphere. *Astron. Astrophys.* **34**, 33 (1974).

M. P. Conrad and H. F. Schaefer, III, Absence of an energetically viable pathway for triplet 1,2 hydrogen shifts. A theoretical study of the vinylidene-acetylene isomerization. *J. Am. Chem. Soc.* **100**, 7820–7823 (1978).

S. Coriani, C. Hattig, P. Jørgensen, A. Halkier, and A. Rizzo, Coupled cluster calculations of Verdet constants. *Chem. Phys. Lett.* **281**, 445–451 (1997).

R. Courtin, D. Gautier, A. Marten, B. Bézard, and R. Hanel, The composition of Saturn's atmosphere at northern temperate latitudes from Voyager IRIS spectra-NH$_3$, PH$_3$, C_2H_2, C_2H_6, CH_3D, CH_4, and the Saturnian D/H isotopic ratio. *Astrophys. J.* **287**, 899–916 (1984).

A. Coustenis, T. Encrenaz, B. Bézard, G. Bjoraker, G. Graner, M. Dang-Nhu, and E. Arié, Modeling Titan's thermal infrared spectrum for high-resolution space observations. *Icarus* **102**, 240–260 (1993).

M. Couty, C. A. Bayse, and M. B. Hall, Extremely localized molecular orbitals (ELMO). A nonorthogonal Hartree-Fock method. *Theor. Acc.* **97**, 96–109 (1997).

J. S. Craw, M. A. C. Nascimento, and M. N. Ramos, Ab initio vibrational spectrum and infrared intensity parameters in hydrogen-bonded systems. Acetylene. *Spectrochim. Acta, Part A* **47**, 69 (1991).

D. Cronn and E. Robinson, Tropospheric and lower stratospheric vertical profiles of ethane and acetylene. *Geophys. Res. Lett.* **6**, 641–644 (1979).

F. Culot and J. Liévin, Stretch-bend interactions in $^{12}C_2H_2$: A vibrational CASSCF study. In preparation (1998).

R. D'Cunha, Current trends in high resolution spectroscopy of polyatomic molecules. *At., Mol. Cluster Phys.* 292–298, Ahmad S.A. Narosa ed., New Delhi, 1997.

R. D'Cunha, Y. A. Sarma, G. Guelachvili, R. Farrenq, Q. Kou, V. M. Devi, D. C. Benner, and K. N. Rao, Analysis of the high-resolution spectrum of acetylene in the 2.4 μm region. *J. Mol. Spectrosc.* **148**, 213–225 (1991).

R. D'Cunha, Y. A. Sarma, V. A. Job, G. Guelachvili, and K. N. Rao, Fermi coupling and *l*-type resonance effects in the hot bands of acetylene: The 2650 cm^{-1} region. *J. Mol. Spectrosc.* **157**, 358–368 (1993).

R. D'Cunha, Y. A. Sarma, G. Guelachvili, R. Farrenq, and K. N. Rao, Analysis of the $^{13}C^{12}CH_2$ bands in the FTIR spectrum of acetylene. *J. Mol. Spectrosc.* **160**, 181–185 (1993).

H. L. Dai, R. W. Field, and J. L. Kinsey, Intramolecular vibrational dynamics including rotational degrees of freedom: Chaos and quantum spectra. *J. Chem. Phys.* **82**, 2161–2163 (1985).

N. Dam, C. Liedenbaum, S. Stolte, and J. Reuss, Rotational distributions in seeded molecular beams. *Chem. Phys. Lett.* **136**, 73–80 (1987).

P. P. Das, V. M. Devi, and K. N. Rao, Diode laser spectra of acetylene: v_5 region at 15 μm. *J. Mol. Spectrosc.* **84**, 313–317 (1980).

M. de Labachelerie, K. Nakagawa, and M. Ohtsu, Ultranarrow $^{13}C_2H_2$ saturated-absorption lines at 1.5 μm. *Opt. Lett.* **19**, 840–842 (1994).

Y. Delaval, J. Cartigny, and J. Lecomte, Structure fine de bandes d'émission observées vers 2,4 μm (4.100 cm^{-1}) avec l'acétylène chauffé vers 700 K. *Mem. Soc. R. Sci. Liège* **6**, 65–74 (1971).

R. L. DeLeon and J. S. Muenter, Radiofrequency spectroscopy of DCCD: Deuterium quadrupole coupling in acetylene. *J. Mol. Spectrosc.* **126**, 13–18 (1987).

M. De Rosa, A. Ciucci, D. Pelliccia, C. Gabbanini, S. gozzini, and A. Lucchesini, On the measurement of pressure induced shift by diode lasers and harmonic detection. *Opt. Commun.* **147**, 55–60 (1998).

V. M. Devi, D. C. Benner, C. P. Rinsland, M. A. H. Smith, and B. D. Sidney, Tunable diode laser measurements of N_2- and air-broadened halfwidths: Lines in the $(v_4 + v_5)^0$ band of $^{12}C_2H_2$ near 7.4 μm. *J. Mol. Spectrosc.* **114**, 49–53 (1985).

G. Di Lonardo, P. Ferracuti, L. Fusina, E. Venuti, and J. W. C. Johns, Vibration-rotation spectra of ^{13}C containing acetylenes: I. The bending states up to $V_4 + V_5 = 2$. *J. Mol. Spectrosc.* **161**, 466–486 (1993).

G. Di Lonardo, P. Ferracuti, L. Fusina, and E. Venuti, Vibration-rotation spectra of ^{13}C containing acetylenes: The $v_1/v_2 + 2v_5$ Fermi diad. *J. Mol. Spectrosc.* **164**, 219–232 (1994).

G. Di Lonardo, P. Ferracuti, L. Fusina, E. Venuti, and K. A. Mohamed, Vibration-rotation spectra of ^{13}C containing acetylenes: The stretching fundamentals. *J. Mol. Spectrosc.* **169**, 148–153 (1995).

C. Domingo, R. Escribano, W. F. Murphy, and S. Montero, Raman intensities of overtones and combination bands of C_2H_2, C_2HD and C_2D_2. *J. Chem. Phys.* **77**, 4353–4359 (1982).

R. Dopheide and H. Zacharias, Rotational alignment by stimulated Raman pumping: C_2H_2 $(v_2'' = 1, J'')$. *J. Chem. Phys.* **99**, 4864–4866 (1993).

R. Dopheide, W. B. Gao, and H. Zacharias, Direct measurements of collision-induced state-to-state rotational energy transfer rates in $C_2H_2(v_2'' = 1)$. *Chem. Phys. Lett.* **182**, 21–26 (1991).

R. Dopheide, W. Conrath, and H. Zacharias, Rotational energy transfer in vibrationally excited acetylene X $^1\sum_g$ $(v_2'' = 1, J'')$: ΔJ propensities. *J. Chem. Phys.* **101**, 5804 (1994).

P. Drossart, J. Lacy, E. Serabyn, A. Tokunaya, B. Bézard, and T. Encrenaz, Detection of $^{13}C^{12}CH_2$ on Jupiter at 13 μm. *Astron. Astrophys.* **149**, L10–L12 (1985).

P. Drossart, B. Bézard, S. Atreya, J. Lacy, E. Serabyn, A. Tokunaga, and T. Encrenaz, Enhanced acetylene emission near the north pole of Jupiter. *Icarus* **66**, 610–618 (1986).

P. Drossart, B. Bezard, T. Encrenaz, S. Atreya, J. Lacy, E. Serabyn, and A. Tokunaga. *Proc. NASA Goddard Inst. Space Stud., Jovian Atmosp. 1986* p. 94.

P. Dube, L. S. Ma, J. Ye, P. Jungner, and J. L. Hall, Thermally induced self-locking of an optical cavity by overtone absorption in acetylene gas. *J. Opt. Soc. Am.* **13**, 2041 (1996).

T. Dudev and B. Galabov, *Ab initio* calculations of Raman intensity parameters and geometry of polyynes and polyynenitriles. *Spectrochim. Acta Part A* **53**, 2053–2059 (1997).

P. Duffy, S. A. C. Clark, C. E. Brion, M. E. Casida, D. P. Chong, E. R. Davidson, and C. Maxwell, Electron momentum spectroscopy of the valence orbitals of acetylene: Quantitative comparisons using near Hartree-Fock limit and correlated wavefunctions. *Chem. Phys.* **165**, 183–199 (1992).

M. D. Duncan, P. Oesterlin, F. König, and R. L. Byer, Observation of saturation broadening of the coherent anti-stokes Raman spectrum (CARS) of acetylene in a pulsed molecular beam. *Chem. Phys. Lett.* **80**, 253–256 (1981).

P. Dupre, Anomalous behavior of the anticrossing density as a function of excitation energy in the C_2H_2 molecule. *Chem. Phys.* **152**, 293 (1991).

P. Dupre, Characterization of a large singlet-triplet coupling in the A state of the acetylene molecule. *Chem. Phys. Lett.* **212**, 555 (1993).

P. Dupre, Quantum beat spectroscopy studies of Zeeman anticrossings in the A ^1A_u state of the acetylene molecule (C_2H_2). *Chem. Phys.* **196**, 211 (1995).

P. Dupre, Study of Zeeman anticrossing spectra of the A ^1A_u state of the acetylene molecule (C_2H_2) by Fourier transform: Product ρ_{vib} <V> and isomerization barrier. *Chem. Phys.* **196**, 239 (1996).

M. Duran, Y. Yamaguchi, R. B. Remington, and H. F. Schaefer, III, Analytic energy second derivatives for paired-excited multi-configuration self-consistent-field wavefunctions. Application of the MCSCF model to H_2O, CH_2, HCN, HCCH, H_2CO, NH_3, CH_4 and C_2H_4. *Chem. Phys.* **122**, 201–231 (1988).

C. E. Dykstra and H. F. Schaefer, III, The vinylidene-acetylene rearrangement. A self-consistent electron pairs study of a model unimolecular reaction. *J. Am. Chem. Soc.* **100**, 1378–1382 (1978).

S. G. Edgington, S. K. Atreya, L. M. Trafton, J. J. Caldwell, R. F. Beebe, A. A. Simon, and R. A. West, Latitudinal distribution of ammonia and upper limits of acetylene and phosphine in Jupiter's upper troposphere and lower stratosphere. *Bull. Am. Astron. Soc.* (1996).

M. I. El Idrissi, J. Liévin, A. Campargue, and M. Herman, The vibrational energy pattern in acetylene. IV. Updated vibrational constants in $^{12}C_2H_2$. *J. Chem. Phys.*, **110**, in press (1999).

K. M. Ervin, J. Ho, and W. C. Lineberger, A study of the singlet and triplet states of vinylidene by photoelectron spectroscopy of $H_2C=C^-$, $D_2C=C^-$, and $HDC=C^-$. Vinylidene-acetylene isomerization. *J. Chem. Phys.* **91**, 5974–5992 (1989).

K. M. Ervin, S. Gronert, S. E. Barlow, M. K. Gilles, A. G. Harrison, V. M. Bierbaum, C. H. DePuy, W. C. Lineberger, and G. B. Ellison, Bond strengths of ethylene and acetylene. *J. Am. Chem. Soc.* **112**, 5750 (1990).

N. J. Evans, J. H. Lacy, and J. S. Carr, Infrared molecular spectroscopy toward the Orion IRc2 and IRc7 sources: A new probe of physical conditions and abundances in molecular clouds. *Astrophys. J.* **383**, 674–692 (1991).

S. C. Farantos, Chaotic structure in the phase space of acetylene. *J. Chem. Phys.* **85**, 641–642 (1986).

H. Fast and H. L. Welsh, High-resolution Raman spectra of acetylene, acetylene-d_1, and acetylene-d_2. *J. Mol. Spectrosc.* **41**, 203–221 (1972).

T. Feldman, G. G. Shepherd, and H. L. Welsh, The Raman spectrum of acetylene. *Can. J. Phys.* **34**, 1425–1431 (1956).

R. W. Field, S. L. Coy, and S. A. B. Solina, Pure sequence vibrational spectra of small polyatomic molecules. *Prog. Theor. Phys.* **116**, 143–166 (1994).

R. W. Field, J. P. O'Brien, M. P. Jacobson, S. A. B. Solina, W. F. Polik, and H. Ishikawa, Intramolecular dynamics in the frequency domain. *Adv. Chem. Phys.* **CI**, 463–490 (1997).

J. H. Fillion, N. Shafizadeh, and D. Gauyacq, Photodynamics of acetylene. *AIP Conf. Proce.* **312**, 395–400 (1994).

H. Finsterhölzl, J. G. Hochenbleicher, and G. Strey, Intensity distribution in pure rotational Raman spectra of linear molecules in the ground and vibrational Π states application to acetylene. *J. Raman Spectrosc.* **6**, 13–19 (1977).

H. Finsterhölzl, H. W. Schrötter, and G. Strey, Determination of anharmonicity constants from the Raman spectrum of gaseous acetylene. *J. Raman Spectrosc.* **11**, 375–383 (1981).

S. F. Fischer and A. Irgens-Defregger, Vibrational relaxation pathways in C_2H_2 and C_2D_2. *J. Phys. Chem.* **87**, 2054–2059 (1983).

L. W. Flynn and G. Thodos, Lennard-Jones force constants from viscosity data: Their relationship to critical properties. *AIChE J.* **8**, 362–365 (1962).

J. Forster, T. Hagen, and J. Uhlenbusch, Infrared absorption of silane, ammonia, acetylene and diborane in the range of the CO_2 laser emission line: Measurements and modelling. *Appl. Phys. B* **62**, 263 (1996).

M. J. Frost and I. W. M. Smith, Infrared-ultraviolet double resonance measurements on the relaxation of rotational energy in the $(3_12_14_15_1)$ Fermi resonance states of C_2H_2. *Chem. Phys. Lett.* **191**, 574 (1992).

M. J. Frost and I. W. M. Smith, Energy transfer in the $3_12_14_15_1$ Fermi resonant states of acetylene. 2. Vibrational energy transfer. *J. Phys. Chem.* **99**, 1094 (1995).

A. Fuente, J. Cernicharo, and A. Omont, Inferring acetylene abundances from C_2H: the C_2H_2/HCN abundance ratio. *Astron. Astrophys.* **330**, 232–242 (1998).

M. Fujii, A. Haijima, and M. Ito, Predissociation of acetylene in Ã ^1A_u state, *Chem. Phys. Lett.* **150**, 380–385 (1988).

M. Fujii, S. Tanabe, Y. Okuzawa, and M. Ito, IR-UV double resonance spectrum of acetylene below and above the predissociation threshold. *Appl. Phys. B* **14**, 161–182 (1994).

G. W. Funke, Feinstruktur und störungen im rotationsschwingungsspektrum von acetylen. *Z. Phys.* **104**, 169–187 (1937).

G. W. Funke and E. Lindholm, Uber das spektrum des acetylens im photographischen ultrarot. *Z. Phys.* **106**, 518–531 (1937).

M. M. Gallo, T. P. Hamilton, and H. F. Schaeffer, III, Vinylidene: The final chapter? *J. Am. Chem. Soc.* **112**, 8714–8719 (1990).

P. Gaspard and P. van Ede van der Pals, The semi classical regime of intramolecular vibrational relaxation: Theory unpublished results.

R. Georges, D. Van Der Vorst, M. Herman, and D. Hurtmans, Ar and self pressure broadening coefficient of the R(11), $5v_3$ line of $^{12}C_2H_2$. *J. Mol. Spectrosc.* **185**, 187–188 (1997).

S. Ghersetti and K. N. Rao, High resolution spectra of dideuteroacetylene. Analysis of the bands at 4.1 µm. *J. Mol. Spectrosc.* **28**, 27–43 (1968).

S. Ghersetti and K. N. Rao, High resolution infrared spectra of dideuteroacetylene. Analysis of bands in the InSb region. *J. Mol. Spectrosc.* **28**, 373–393 (1968).

S. Ghersetti, J. Pliva, and K. N. Rao, Dideuteroacetylene bands in the 2–2.5 and 5–10 micron regions. *J. Mol. Spectrosc.* **38**, 53–69 (1971).

S. Ghersetti, A. Baldacci, S. Giorgianni, R. H. Barnes, and K. N. Rao, Infrared spectra of the carbon-13 isotopic varieties of acetylene, monodeuteroacetylene and dideuteroacetylene. *Gazz. Chim. Ital.* **105**, 875–900 (1975).

S. Ghersetti, J. E. Adams, and K. N. Rao, $^{12}C_2H_2$ and $^{12}C_2HD$ bands at 1.1 µm. *J. Mol. Spectrosc.* **64**, 157–161 (1977).

G. Glockler and C. E. Morrell, Raman effect of acetylenes. II. Diiodoacetylene, liquid acetylene and deuteroacetylenes. *J. Chem. Phys.* **4**, 15–22 (1936).

J. H. Goebel, J. D. Bregman, F. C. Witteborn, B. J. Taylor, and S. P. Willner, Identification of new infrared bands in a carbon-rich mira variable. *Astrophys. J.* **246**, 455–463 (1981).

A. Goldman, F. J. Murcray, R. D. Blatherwick, J. R. Gillis, F. S. Bonomo, F. H. Murcray, D. G. Murcray, and R. J. Cicerone, Identification of acetylene (C_2H_2) in infrared atmospheric absorption spectra. *J. Geophys. Res.* **86**, 143–146 (1981).

P. G. Green, J. L. Kinsey, and R. W. Field, A new determination of the dissociation energy of acetylene. *J. Chem. Phys.* **91**, 5160–5163 (1989).

R. Grosskloss, H. Wenz, S. B. Rai, and W. Demtröder, Near infrared overtone spectroscopy of C_2D_2. *Mol. Phys.* **85**, 71–80 (1995).

G. Guelachvili and K. N. Rao, *Handbook of Infrared Standards II.* Academic Press, San Diego, CA, 1993.

J. Häger, W. Krieger, and T. Rüegg, Vibrational relaxation of acetylene and acetylene-rare-gas mixtures. *J. Chem. Phys.* **72**, 4286–4290 (1980).

J. Häger, W. Krieger, and J. Pfab, Collisional deactivation of laser-excited acetylene by H_2, HBr, N_2 and CO. *J. Chem. Soc. Faraday Trans.* 2 **77**, 469–476 (1981).

R. R. Hall, Laser photoacoustic spectroscopy of forbidden transitions: Acetylene and alkyne high energy vibrational states and their interactions. Ph.D. Thesis, Rice University, Houston, TX, 1984.

L. Halonen, Rotational energy level structure of stretching vibrational states in some small symmetrical molecules. *J. Chem. Phys.* **86**, 588–596 (1987).

L. Halonen, M. S. Child, and S. Carter, Potential models and local mode vibrational eigenvalue calculations for acetylene. *Mol. Phys.* **47**, 1097–1112 (1982).

L. Halonen, D. W. Noid, and M. S. Child, Local mode predictions for excited stretching vibrational states of HCCD and $H^{12}C^{13}CH$. *J. Chem. Phys.* **78**, 2803–2804 (1983).

J. B. Halpern, R. Dopheide, and H. Zacharias, How a collision causes misalignment: Alignment decay in acetylene. *J. Phys. Chem.* **99**, 13611–13619 (1995).

P. Halvick, D. Liotard, and J. C. Rayez, A theoretical study of acetylene: Toward the complete characterization of the singlet ground state potential energy surface. *Chem. Phys.* **177**, 69–78 (1993).

N. C. Handy, The derivation of vibration-rotation kinetic energy operators, in internal coordinates. *Mol. Phys.* **61**, 207–223 (1987).

N. C. Handy, C. W. Murray, and R. D. Amos, Study of CH_4, C_2H_2, C_2H_4 and C_6H_6 using Kohn-Sham theory. *J. Phys. Chem.* **97**, 4392–4396 (1993).

R. A. Hanel, B. Conrath, F. M. Flasar, V. Kunde, W. Maguire, J. Pearl, J. Pirraglia, R. Samuelson, L. Herath, M. Allison, D. Cruishank, D. Gauthier, P. Gierasch, L. Horn, R. Koppany, and C. Ponnamperuma, Infrared observations of the Saturnian system from Voyager 1. *Science* **212**, 192–200 (1981).

R. A. Hanel, B. Conrath, F. M. Flasar, V. Kunde, W. Maguire, J. Pearl, J. Pirraglia, R. Samuelson, L. Herath, M. Allison, D. Cruishank, D. Gautier, R. Gierasch, L. Horn, and C. Ponnamperuma, Infrared observations of the Saturnian system from Voyager 2. *Science* **215**, 544–548 (1982).

H. Hashimoto, N. Yonekura, and T. Suzuki, Pump-probe measurements of the predissociation reaction time of C_2H_2 from A^1A_u state. *Chem. Phys. Lett.* **264**, 554–550 (1997).

N. Hashimoto and T. Suzuki, Energy and state dependence in the quenching and predissociation of C_2H_2. *J. Chem. Phys.* **104**, 6070–6073 (1996).

N. Hashimoto and K. Takatsuka, Expected significance of weakly chaotic vibrational motions in single molecule spectroscopy. *J. Chem. Phys.* **103**, 6914–1929 (1995).

J. G. Haub, M. J. Johnson, and B. J. Orr, Continuously tunable, injection-seeded β-barium borate optical parametric oscillator: Spectroscopic applications. *Appl. Phys. Lett.* **58**, 1718–1721 (1991).

J. G. Haub, M. J. Johnson, and B. J. Orr, Spectroscopic and nonlinear-optical applications of a tunable beta-Barium Borate optical parametric oscillator. *J. Opt. Soc. Am. B* **10**, 1765–1777 (1993).

R. J. Hayward and B. R. Henry, Experimental manifestations of the local-mode description of high energy polyatomic overtone spectra. *Chem. Phys.* **12**, 387–396 (1976).

K. Hedfeld and P. Lueg, Das rotationsschwingungsspektrum des acetylens. *Z. Phys.* **77**, 446–458 (1932).

K. Hedfeld and R. Mecke, Das rotationsschwingungsspektrum des acetylens. I. Bandenanalyse. *Z. Phys.* **64**, 151–161 (1930).

B. R. Henry, M. A. Mohammadi, and A. W. Tarr, The overtone spectrum of acetylene: A rotational analysis based on a local model description. *J. Chem. Phys.* **77**, 3295–3300 (1982).

S. Henton, M. Islam, S. Gatenby, and I. W. M. Smith, Rotational energy transfer and rotationally specific vibration-vibration intaradyad transfer in collisions of $C_2H_2\tilde{X}^1\Sigma_g^+$ ($3_1/2_14_15_1$, $J = 10$) with C_2H_2, Ar, He and H_2. *J. Chem. Soc. Faraday Trans.* **94**, 3219–3228 (1998).

S. Henton, M. Islam, and I. W. M. Smith. Intramolecular V–V transfer between the components of the $(3_1/2_14_15_1)$ and $(3_14_1/2_14_25_1)$ Fermi dyads in acetylene in C_2H_2–C_2H_2 collisions. *Chem. Phys. Letters* **291**, 223–230 (1998).

S. Henton, M. Islam, and I. W. M. Smith. Relaxation within and from the $(3_1/2_14_15_1)$ and $(3_14_1/2_14_25_1)$ Fermi dyads in acetylene: Vibrational energy transfer in collisions with C_2H_2, N_2 and H_2, *J. Chem. Soc. Faraday Trans.*, **94**, 3207–3217 (1998).

E. Herbst and C. M. Leung, Gas-phase production of complex hydrocarbons, cyanopolyynes, and related compounds in dense interstellar clouds. *Astrophys. J., Suppl. S* **69**, 271–300 (1989).

M. Herman and R. Colin, The geometrical structure of the acetylene molecule in the X, G and I states. *Bull. Soc. Chim. Belg.* **89**, 335–342 (1980).

M. Herman and T. R. Huet, Corrigendum—Vibrational couplings in acetylene. *Mol. Phys.* **70**, 545–546 (1990).

M. Herman and J. Lievin, Acetylene: From intensity alternation in spectra to ortho and para molecules. *J. Chem. Educ.* **59**, 17–21 (1982).

M. Herman and A. Pisarchik, Spectroscopy of C_2D_2 in the $3v_3$ region. *J. Mol. Spectrosc.* **164**, 210–218 (1994).

M. Herman, T. R. Huet, and M. Vervloet, Spectroscopy and vibrational couplings in the $3v_3$ region of acetylene. *Mol. Phys.* **66**, 333–353 (1989).

M. Herman, T. R. Huet, Y. Kabbadj, and J. Vander Auwera, *l*-type resonances in C_2H_2. *Mol. Phys.* **72**, 75–88 (1991).

M. Herman, M. Abbouti Temsamani, J. Vander Auwera, and R. Ottinger, Near-infrared absorption spectroscopy: The n = 4 Polyad of C_2HD. *Chem. Phys. Lett.* **185**, 215–219 (1991).

M. Herman, M. Abbouti Temsamani, D. Lemaitre, and J. Vander Auwera, The Fourier-transform vibrational spectrum of acetylene in the visible range. *Chem. Phys. Lett.* **185**, 220–224 (1991).

M. Herman, D. Hurtmans, J. Vander Auwera, and M. Vervloet, Lack of intensity alternation in C_2H_2. *J. Mol. Spectrosc.* **150**, 293–295 (1991).

M. Herman, M. I. El Idrissi, A. Pisarchik, A. Campargue, A.-C. Gaillot, L. Biennier, G. Di Lonardo, and L. Fusina, The vibrational energy levels in acetylene. III. $^{12}C_2D_2$. *J. Chem. Phys.* **108**, 1377–1389 (1998).

J. Hietanen, l-Resonance effects in the hot bands $3v_5 - 2v_5(v_4 + 2v_5) - (v_4 + v_5)$ and $(2v_4 + v_5) - 2v_4$ of acetylene. *Mol. Phys.* **49**, 1029–1038 (1983).

J. Hietanen and J. Kauppinen, High-resolution infrared spectrum of acetylene in the region of the bending fundamental v_5. *Mol. Phys.* **42**, 411–423 (1981).

J. Hietanen, R. Anttila, and J. Kauppinen, The infra-red spectrum of C_2HD in the region of the bending fundamental. *Mol. Phys.* **38**, 1367–1377 (1979).

J. Hietanen, V. M. Horneman, and J. Kauppinen, Vibration-rotation infrared spectra of the carbon-13 isotopic varieties of acetylene at 13.7 microns. *Mol. Phys.* **59**, 587–593 (1986).

J. J. Hillman, D. E. Jennings, G. W. Halsey, S. Nadler, and W. E. Blass, An infrared study of the bending region of acetylene. *J. Mol. Spectrosc.* **146**, 389–401 (1991).

M. F. Hineman, R. G. Rodriguez, and J. W. Nibler, Photothermal lensing spectroscopy of supersonic jet expansions of acetylene. *J. Chem. Phys.* **89**, 2630–2634 (1988).

R. Höller and H. Lischka, Coupled-Hartree-Fock calculations of susceptibilities and magnetic shielding constants. I. The first now hydrides LiH, BeH_2, BH_3, CH_4, NH_3, H_2O and HF, and the hydrocarbons C_2H_2, C_2H_4 and C_2H_6. *Mol. Phys.* **41**, 1017–1040 (1980).

T. A. Holme and R. D. Levine, Short-time vibrational dynamics of acetylene versus its isotopic variants. *Chem. Phys. Lett.* **150**, 393–398 (1988).

T. A. Holme and R. D. Levine, Energy flow pathways and their spectral signatures in vibrationally excited acetylene *J. Chem. Phys.* **89**, 3379–3380 (1988).

T. A. Holme and R. D. Levine, Theoretical and computational studies of highly vibrationally excited acetylene. *Chem. Phys.* **131**, 169–190 (1989).

C. Hornberger, M. Koening, S. B. Rai, and W. Demtröder, Sensitive photoacoustic overtone spectroscopy of acetylene with a multipass photoacoustic cell and a colour centre laser at 1.5 μm. *Chem. Phys.* **190**, 171–177 (1995).

V. M. Horneman, S. Alanko, and J. Hietanen, Difference band $\nu_5 - \nu_4$ of acetylene C_2H_2. *J. Mol. Spectrosc.* **135**, 191–193 (1989).

J. Hornos and F. Iachello, The overtone spectrum of acetylene in the vibron model. *J. Chem. Phys.* **90**, 5284–5291 (1989).

T. R. Huet and M. Herman, Levels of the transbending normal mode of vibration in C_2D_2. *J. Mol. Spectrosc.* **132**, 361–368 (1988).

T. R. Huet, M. Herman, and J. W. C. Johns, The bending vibrational levels in C_2D_2 X $^1\Sigma_g^+$). *J. Chem. Phys.* **94**, 3407–3414 (1991).

M. Huhanantti, J. Hietanen, R. Anttila, and J. Kauppinen, The infra-red spectrum of C_2D_2 in the region of the bending fundamental ν_5. *Mol. Phys.* **37**, 905–923 (1979).

S. C. Hurlock, S. Ghersetti, and K. Narahari Rao, Energy levels and molecular constants of $^{12}C_2H_2$, $^{12}C_2D_2$ and $^{12}C_2HD$ determined from infrared spectra. *Mem. Soc. R. Sci. Liège*, 87–97 (1971).

N. Husson, A. Chedin, N. A. Scott, T. Cohen-Hallaleh, and I. Berroir, La banque de données GEISA. *Collog. Int. CNRS* **116** (1982).

F. Iachello and S. Oss, Algebraic approach to molecular spectra: Two-dimensional problems. *J. Chem. Phys.* **104**, 6956 (1996).

F. Iachello, S. Oss, and R. Lemus, Linear four-atomic molecules in the Vibron model. *J. Mol. Spectrosc.* **149**, 132 (1991).

M. P. Jacobson, J. P. O'Brien, and R. W. Field, Field, Anomalously slow intramolecular vibrational redistribution in the acetylene $\tilde{X}^1\Sigma_g^+$ state above $10\,000\,\mathrm{cm}^{-1}$ of internal energy. *J. Chem. Phys.* **109**, 3831–3840 (1998).

M. P. Jacobson, J. P. O'Brien, R. J. Silbey, and R. W. Field, Pure bending dynamics in the acetylene $X^1\Sigma_g^+$ state to $15{,}000\,\mathrm{cm}^{-1}$ of internal energy. *J. Chem. Phys.* **109**, 121–133 (1998).

M. Jaszunski, P. Jorgensen, A. Rizzo, K. Ruud, and T. Helgaker, MCSCF calculations of Verdet constants. *Chem. Phys. Lett.* **222**, 263–266 (1997).

P. Jona, M. Gussoni, and G. Zerbi, Interpretation of infrared intensities of acetylene, propyne, and 2-butyne. A common set of electrooptical parameters. *J. Phys. Chem.* **85**, 2210–2218 (1981).

D. M. Jonas, S. A. B. Solina, B. Rajaram, R. J. Silbey, R. W. Field, K. Yamanouchi, and S. Tsuchiya, Intramolecular vibrational relaxation and forbidden transitions in the SEP spectrum of acetylene. *J. Chem. Phys.* **97**, 2813–2816 (1992).

D. M. Jonas, S. A. B. Solina, B. Rajaram, S. J. Cohen, R. J. Silbey, R. W. Field, K. Yamanouchi, and S. Tsuchiya, Intramolecular vibrational relaxation in the SEP spectrum of acetylene. *J. Chem. Phys.* **99**, 7350–7370 (1993).

L. H. Jones, M. Goldblatt, R. S. McDowell, and D. E. Armstrong, Vibration rotation bands of C_2T_2. Part I. ν_3. *J. Mol. Spectrosc.* **23**, 9–14 (1967).

P. Jungner and L. Halonen, Laser induced vibration-rotation fluorescence and infrared forbidden transitions in acetylene. *J. Chem. Phys.* **107**, 1680–1682 (1997).

Y. Kabbadj, M. Herman, G. Di Lonardo, L. Fusina, and J. W. C. Johns, The bending energy levels of C_2H_2. *J. Mol. Spectrosc.* **150**, 535–565 (1991).

M. Kaluza and J. T. Muckerman, Mode-selective infrared excitation of linear acetylene. *J. Chem. Phys.* **102**, 3897 (1995).

J. Kaski, P. Lantto, J. Vaara, and J. Jokisaari, Experimental and theoretical ab initio study of the $^{13}C-^{13}C$ spin–spin coupling and ^1H and ^{13}C shielding tensors in ethane, ethene, and ethyne. *J. Am. Chem. Soc.* **120**, 3993–4005 (1998).

M. E. Kellman, Approximate constants of motion for vibrational spectra of many-oscillator systems with multiple anharmonic resonances. *J. Chem. Phys.* **93**, 6630–6635 (1990).

M. E. Kellman, Algebraic methods in spectroscopy. *Annu. Rev. Phys. Chem.* **46**, 395–421 (1995).

M. E. Kellman, *Molecular dynamics and spectroscopy by Stimulated Emission Pumping.* World Scientific Publ. Co., Singapore, 1995.

M. E. Kellman and G. Chen, Approximate constants of motion and energy transfer pathways in highly excited acetylene. *J. Chem. Phys.* **95**, 8671–8672 (1991).

R. L. Kelly, R. Rollefson, and B. S. Schurin, The infrared dispersion of acetylene and the dipole moment of the C-H bond. *J. Chem. Phys.* **19**, 1595–1599 (1951).

K. A. Keppler, G. C. Mellau, S. Klee, B. P. Winnewisser, M. Winnewisser, J. Pliva, and K. N. Rao, Precision measurements of acetylene spectra at 1.4–1.7 microns recorded with 352.5 m pathlength. *J. Mol. Spectrosc.* **175**, 411–420 (1996).

K. Kim, Prediction of absolute infrared intensities of isotope-substituted acetylenes. *Chayon Kwahak Taehak Nomunjip (Soul Taehakkyo)* **9**, 45–53 (1984).

K. Kim and W. T. King, Integrated infrared intensities in HCCH, DCCD and HCCD. *J. Mol. Struct.* **57**, 201–207 (1979).

S. Kinugawa and H. Sasada, Wavenumber measurement of the 1.5-μm band of acetylene by semiconductor laser spectrometer. *Jpn. J. Appl. Phys.* **29**, 611–612 (1990).

R. F. Knacke, Y. H. Kim, and W. M. Irvine, An upper limit to the acetylene abundance toward BN in the Orion molecular cloud. *Astrophys. J.* **345**, 265–267 (1989).

T. Koike and W. C. Gardiner, Jr., High-temperature absorption of the 3.39 μm helium-neon laser line by acetylene, ethylene and propylene. *Appl. Spectrosc.* **34**, 81–84 (1980).

T. Koops, W. M. A. Smit, and T. Visser, Measurement and interpretation of the absolute infrared intensities of acetylene: Fundamentals and combination bands. *J. Mol. Struct.* **112**, 285–299 (1984).

T. Kostiuk, F. Espenak, D. Deming, M. J. Mumma, D. Zipoy, and J. Keady, Study of velocity structure in IRC + 10216 using acetylene line profiles. *Bull. Am. Astron. Soc.* **17**, 570 (1985).

E. Kostyk and H. L. Welsh, High resolution rotation-vibration Raman spectra of acetylene. I. The spectrum of C_2H_2. *Can. J. Phys.* **58**, 534–543 (1980).

E. Kostyk and H. L. Welsh, High resolution rotation-vibration Raman spectra of acetylene. II. The spectra of C_2D_2 and C_2HD. *Can. J. Phys.* **58**, 912–920 (1980).

Q. Kou, G. Guelachvili, M. Abbouti Temsamani, and M. Herman, The absorption spectrum of C_2H_2 around $v_1 + v_3$: Energy standards in the 1.5 μm region and vibrational clustering. *Can. J. Phys.* **72**, 1241–1250 (1994).

L. Krause and H. L. Welsh, The Raman spectrum of dideuteroacetylene. *Can. J. Phys.* **34**, 1431–1435 (1956).

J. H. Lacy, N. J. I. Evans, J. M. Achtermann, D. E. Bruce, J. F. Arens, and J. S. Carr, Discovery of interstellar acetylene. *Astrophys. J.* **342**, L43–L46 (1989).

W. J. Lafferty and A. S. Pine, Spectroscopic constants for the 2.5 and 3.0 μm bands of acetylene. *J. Mol. Spectrosc.* **141**, 223–230 (1990).

W. J. Lafferty and R. J. Thibault, High resolution infrared spectra of $^{12}C_2H_2$, $^{12}C^{13}CH_2$, and $^{13}C_2H_2$. *J. Mol. Spectrosc.* **14**, 79–96 (1964).

W. J. Lafferty, E. K. Plyler, and E. D. Tidwell, Infrared spectrum of acetylene-d_1. *J. Chem. Phys.* **37**, 1981–1988 (1962).

W. J. Lafferty, R. D. Suenram, and D. R. Johnson, Microwave spectrum of acetylene-d_2. *J. Mol. Spectrosc.* **64**, 147–156 (1977).

D. Lambot, G. Blanquet, and J. P. Bouanich, Diode laser measurements of collisional broadening in the v_5-band of C_2H_2 perturbed by O_2 and N_2. *J. Mol. Spectrosc.* **136**, 86–92 (1989).

D. Lambot, A. Olivier, G. Blanquet, J. Walrand, and J. P. Bouanich, Diode-laser measurements of collisional line broadening in the v_5-band of C_2H_2. *J. Quant. Spectrosc. Radiat. Transfer* **45**, 145–155 (1991).

D. Lambot, G. Blanquet, J. Walrand, and J. P. Bouanich, Diode-laser measurements of H_2-broadening coefficients in the v_5 band of C_2H_2. *J. Mol. Spectrosc.* **150**, 164–172 (1991).

B. Lance, G. Blanquet, J. Walrand, and J. P. Bouanich, On the speed-dependent hard collision lineshape models: Application to C_2H_2 perturbed by Xe. *J. Mol. Spectrosc.* **185**, 262–275 (1997).

C. Latrasse and M. Delabachelerie, Frequency stabilisation of a 1.5 μm distributed feedback laser using a heterodyne spectroscopy method for a space application, *Opt. Eng.* **33**, 1638–1641 (1994).

K. K. Lehmann, The absolute intensity of visible overtone bands of acetylene. *J. Chem. Phys.* **91**, 2759–2760 (1989).

K. K. Lehmann, Harmonically coupled, anharmonic oscillator model for the bending modes of acetylene. *J. Chem. Phys.* **96**, 8117–8119 (1992).

K. K. Lehmann, G. J. Scherer, and W. Klemperer, Comment on The overtone spectrum of acetylene: a rotational analysis based on a local model description. *J. Chem. Phys.* **79**, 530–532 (1983).

K. K. Lehmann, G. J. Scherer, and W. Klemperer, Variational calculation of the rotational constants for acetylene and its isotopic derivatives. *J. Chem. Phys.* **79**, 1369–1376 (1983).

A. Levin and C. F. Meyer, The infrared absorption spectra of acetylene, ethylene and ethane. *J. Opt. Soc. Am.* **16**, 137–164 (1928).

F. Liebrecht, H. Finsterhölzl, H. W. Schrötter, and S. Montero, Intensity distribution in the $2v_4$ Raman band of acetylene. *Indian J. Pure Ap. Phys.* **26**, 51–59 (1988).

J. Liévin, Ab initio characterization of the C' ^1A_g state of the acetylene molecule. *J. Mol. Spectrosc.* **156**, 123–146 (1992).

J. Liévin, M. Abbouti Temsamani, P. Gaspard, and M. Herman, Overtone spectroscopy and dynamics in monodeuteroacetylene (C_2HD). *Chem. Phys.* **190**, 419–445 (1995).

L. Liu and J. T. Muckerman, Vibrational eigenvalues and eigenfunctions for planar acetylene by wave-packet propagation, and its mode-selective infrared excitation. *J. Chem. Phys.* **107**, 3402 (1997).

R. Locht and M. Davister, The dissociative ionization of C_2H_2. The C^+, C_2^+ and CH_2^+ dissociation channels. The vinylidene ion as a transient? *Chem. Phys.* **195**, 443–456 (1995).

M. Loewenstein, J. R. Podolske, and P. Varanasi, High-resolution line-intensity measurements of the $v_4 + v_5$ band of acetylene. *Proc. SPIE—Int. Soc. Opt. Eng.* **438**, 189–192 (1983).

V. P. Lopasov, L. N. Sinitsa, and A. M. Solodov, Rotational structure of the absorption spectrum of C_2H_2 in the neodynium-laser emission band, *Opt. Spectrosc.* **49**, 452 (1980).

J. C. Lorquet, Y. M. Engel, and R. D. Levine, On the separation of time scales in the exploration of phase space of an isolated molecule. *Chem. Phys. Lett.* **175**, 461–466 (1990).

A. Lucchesini, M. DeRosa, D. Pelliccia, A. Ciucci, C. Galbanini, and S. Gozzini, Diode laser spectroscopy of overtone bands of acetylene. *Appl. Phys. B* **63**, 277–282 (1996).

J. K. Lundberg, R. W. Field, C. D. Sherrill, E. T. Seidl, Y. Xie, and H. F. Schaefer III, Acetylene: Synergy between theory and experiment. *J. Chem. Phys.* **98**, 8384–8391 (1993).

W. Macy, Mixing ratios of methane, ethane, and acetylene in Neptune's stratosphere. *Icarus* **41**, 153–158 (1980).

W. C. Maguire, A review of acetylene, ethylene and ethane molecular spectroscopy for planetary applications. *Vib. Spectrosc. Planet. Atmos.* **2**, 473–496 (1982).

W. C. Maguire, R. E. Samuelson, R. A. Hanel, and V. G. Kunde, Latitudinal variation of acetylene and ethane in the Jovian atmosphere from Voyager Iris observations. *Bull. Am. Astron. Soc.* **16**, 647 (1984).

J. Marquez, S. N. Orlov, Y. N. Polivanov, V. V. Smirnov, D. Voelkel, F. Huisken, and Y. L. Chuzakov, Narrow-band source of coherent radiation, tunable in the mid-IR range, for spectroscopic applications. *Kvantovaya Elektron.* **25**, 165–169 (1998).

M. D. Marshall and W. Klemperer, Deuterium nuclear quadrupole coupling constants in vibrationally excited C_2HD: Evidence for electron reorganization. *J. Chem. Phys.* **81**, 2928–2932 (1984).

J. M. L. Martin, T. J. Lee, and P. R. Taylor, A purely ab initio spectroscopic quality quartic force field for acetylene. *J. Chem. Phys.* **108**, 676–691 (1998).

K. Matsumura and T. Tanaka, The effect of bending vibrations on the dipole moment of a linear polyatomic molecule: Analysis of the vibrational changes and transition moments for acetylene and diacetylene. *J. Mol. Spectrosc.* **116**, 334–350 (1986).

K. Matsumura, T. Tanaka, Y. Endo, S. Salto, and E. Hirota, Microwave spectrum of acetylene-d in excited vibrational states. *J. Phys. Chem.* **84**, 1793–1797 (1980).

D. S. Matteson, J. I. Lunine, D. J. Stevenson, and Y. L. Yung, Acetylene on Titan. *Science* **223**, 1127 (1984).

A. B. McCoy and E. L. Sibert, III, Perturbative calculations of vibrational ($J = 0$) energy levels of linear molecules in normal coordinate representations. *J. Chem. Phys.* **95**, 3476–3487 (1991).

A. B. McCoy and E. L. Sibert, III, The bending dynamics of acetylene, *J. Chem. Phys.* **105**, 459–467 (1996).

R. Mecke and R. Ziegler, Das rotationsschwingungsspektrum des acetylens (C_2H_2). *Z. Phys.* **101**, 405–417 (1936).

J. V. Michael, D. F. Nava, R. P. Borkowski, W. A. Payne, and L. J. Stief, Pressure dependence of the absolute rate constant for the reaction OH + C_2H_2 from 228 to 413K. *J. Chem. Phys.* **73**, 6108–6116 (1980).

A. P. Milce and B. J. Orr, Symmetry-breaking perturbations in the $v_2 + 3v_3$ rovibrational manifold of acetylene: Spectroscopic and energy-transfer effects. *J. Chem. Phys.* **104**, 6423–6434 (1996).

A. P. Milce and B. J. Orr, The $v_{CC} + 3v_{CH}$ rovibrational manifold of acetylene. I. Collision-induced state-to-state transfer kinetics. *J. Chem. Phys.* **106**, 3592–3606 (1997).

A. P. Milce, H.-D. Barth, and B. J. Orr, Collision-induced state-to-state energy transfer in the $v_2 + 3v_3$ rovibrational manifold of gas-phase acetylene. *J. Chem. Phys.* **100**, 2398–2401 (1994).

I. M. Mills and A. G. Robiette, On the relationship of normal modes to local modes in molecular vibrations. *Mol. Phys.* **56**, 743–765 (1985).

K. A. Mohamed, High resolution Fourier transform infrared spectroscopy of $^{12}C^{13}CH_2$ in the three micron region. *Indian J. Pure Appl. Phys.* **35**, 402–407 (1997).

A. Moravec, G. Winnewisser, K. M. T. Yamada, and C. E. Blom, Improved molecular constants of acetylene obtained from the v_5 band system. *Z. Naturforsch.*, **45A**, 946–952 (1990).

D. H. Mordaunt and M. N. R. Ashfold, Near ultraviolte photolysis of C_2H_2: A precise determination of $D_0(HCC-H)$. *J. Chem. Phys.* **101**, 2630–2631 (1994).

D. H. Mordaunt, M. N. R. Ashfold, R. N. Dixon, P. Loffler, L. Schnieder, and K. H. Welge, Near threshold photodissociation of acetylene. *J. Chem. Phys.* **108**, 519–526 (1998).

N. Moriwaki, T. Tsuchida, Y. Takehisa, and N. Ohashi, 1.3 μm DFB diode laser spectroscopy of acetylene. *J. Mol. Spectrosc.* **137**, 230–234 (1989).

J. S. Muenter and V. W. Laurie, The microwave spectrum, dipole moment and polarizability of acetylene-d_1. *J. Am. Chem. Soc.* **86**, 3901–3902 (1964).

W. F. Murphy and S. Montero, The determination of the harmonic force field of acetylene using Raman intensities. *Mol. Phys.* **44**, 187–196 (1981).

K. K. Murray, C. L. Morter, and R. F. Curl, Vibrational relaxation of acetylene produced by the photolysis of vinyl bromide. *J. Chem. Phys.* **96**, 5047–5053 (1992).

K. Nakagawa, M. de Labachelerie, Y. Awaji, M. Kourogi, T. Enami, and M. Oktsu, Highly precise 1-THz optical frequency-difference measurements of 1.5-μm molecular absorption lines. *Opt. Lett.* **20**, 410–412 (1995).

K. Nakagawa, M. de Labachelerie, Y. Awaji, and M. Kourogi, Accurate optical frequency atlas of 1.5 microns band of acetylene. *J. Opt. Soc. Am. B* **13**, 2708–2714 (1996).

D. J. Nesbitt and R. W. Field, Vibrational energy flow in highly excited molecules: Role of intramolecular vibrational redistribution. *J. Phys. Chem.* **100**, 12735–12756 (1996).

B. Nikolova, B. Galabov, and W. J. Orville-Thomas, Transferability of bond polar parameters: Interpretation of infrared intensities in acetylene, propyne and 2-butyne. *J. Mol. Struct.* **125**, 197–209 (1984).

K. S. Noll, R. F. Knacke, A. T. Tokunaga, J. H. Lacy, S. Beck, and E. Serabyn, The abundances of ethane and acetylene in the atmospheres of Jupiter and Saturn. *Icarus* **65**, 257–263 (1986).

K. S. Noll, R. F. Knacke, A. T. Tokunaga, J. H. Lacy, S. Beck, and E. Serabyn, *Proce. NASA Goddard Int. Space Stud., Jt. Atoms., 1986*.

J. P. O'Brien, M. P. Jacobson, J. J. Sokol, S. L. Coy, and R. W. Field, Numerical pattern recognition in acetylene dispersed fluorescence spectra. *J. Chem. Phys.* **108**, 7100–7113 (1998).

Y. Ohsugi and N. Ohashi, 0.85-μm diode laser spectroscopy of $^{12}C_2H_2$. Rotational analysis and intensity measurements. *J. Mol. Spectrosc.* **131**, 215–222 (1988).

Y. Ohsugi, A. Katamura, and N. Ohashi, Near-infrared diode laser spectroscopy of the $3\nu_1 + \nu_3$ band of acetylene-d$_1$. *J. Mol. Spectrosc.* **128**, 592–593 (1988).

A. Onae, K. Okumura, Y. Miki, T. Kurosawa, E. Sakuma, J. Yoda, and K. Nakagawa, Saturation spectroscopy of an acetylene molecule in the 1550 nm region using an erbium doped fiber amplifier. *Opt. Commun.* **142**, 41–44 (1997).

J. Omens and J. Reuss. Hot band spectroscopy of acetylene after intermolecular vibrational energy transfer from ethylene. *J. Mol. Spectrosc.* **173**, 14–24 (1995).

F. Orduna, C. Domingo, and S. Montero, Gas phase Raman intensities of C_2H_2, C_2HD and C_2D_2. *Mol. Phys.* **45**, 65–75 (1982).

B. J. Orr, Collision-induced state-to-state energy transfer in perturbed rovibrational manifolds of small polyatomic molecules: mechanistic insights and observations. *Chem. Phys.* **190**, 261–278 (1995).

G. S. Orton, A. T. Tokunaga, and J. Caldwell, Observational constraints on the atmospheres of Uranus and Neptune from new measurements near 10 μm. *Icarus* **56**, 147–164 (1983).

G. S. Orton, D. K. Aitken, C. Smith, P. F. Rouche, J. Caldwell, and R. Snyder, The spectra of Uranus and Neptune at 8–14 and 17–23 μm. *Icarus* **70**, 1–12 (1987).

Y. Osamura and H. F. Schaefer III, Toward the spectroscopic identification of vinylidene, $H_2C=$ C:. *Chem. Phys. Lett.* **79**, 412–415 (1981).

S. Oss, Algebraic models in molecular spectroscopy. *Adv. Chem. Phys.* **93**, 455–649 (1996).

J. Overend, The equilibrium bond lengths in acetylene and HCN. *Trans. Faraday Soc.* **56**, 310 (1960).

J. Overend and H. W. Thompson, Vibration-rotation bands and rotational constants of dideuteroacetylene. *Proc. Roy. Soc. London Ser. A* **232**, 291–309 (1955).

M. J. Packer, S. P. A. Sauer, and J. Oddershede, Correlated dipole oscillator sum rules. *J. Chem. Phys.* **100**, 8969 (1994).

K. F. Palmer, S. Ghersetti, and K. N. Rao, $(\nu_4 + \nu_5)^0$ Bands of $^{12}C_2D_2$, $^{13}C_2D_2$ and $^{12}C^{13}CD_2$. *J. Mol. Spectrosc.* **30**, 146 (1969).

K. F. Palmer, M. E. Mickelson, and K. N. Rao, Investigations of several infrared bands of $^{12}C_2H_2$ and studies of the effects of vibrational rotational interactions. *J. Mol. Spectrosc.* **44**, 131–144 (1972).

S. T. Park, J. H. Moon, and M. S. Kim, Rotational energy analysis for rotating-vibrating linear molecules in classical trajectory simulation. *J. Chem. Phys.* **107**, 9899–9906 (1997).

H. Partridge and C. W. Bauschlicher, Jr., The dissociation energies of CH_4 and C_2H_2 revisited. *J. Chem. Phys.* **103**, 10589–10596 (1995).

F. S. Pavone, F. Marin, M. Inguscio, K. Ernst, and G. Di Lonardo, Sensitive detection of acetylene absorption in the visible by using a stabilized AlGaAs diode laser. *Appl. Opt.* **32**, 259–262 (1993).

M. A. Payne, A. P. Milce, M. J. Frost, and B. J. Orr, Dynamically symmetry breaking in the 4 ν(CH) rovibrational manifold of acetylene. *Chem. Phys. Lett.* **265**, 244–252 (1997).

V. I. Perevalov and O. N. Sulakshina, Reduced effective vibrational-rotational Hamiltonian for bending vibrational levels of acetylene molecule. *SPIE* **2205**, 182–187 (1993).

V. I. Perevalov, E. I. Lobodenko, and J. L. Teffo, Reduced effective Hamiltonian for global fitting of C_2H_2 rovibrational lines. *SPIE* **3090**, 143–149 (1997).

M. Perić, R. J. Buenker, and S. D. Peyerimhoff, Theoretical study of the U.V. spectrum of acetylene I. Ab initio calculation of singlet electronic states of acetylene by a large-scale CI method. *Mol. Phys.* **53**, 1177–1193 (1984).

M. Perić, S. D. Peyerimhoff, and R. J. Buenker, Theoretical study of the U.V. spectrum of acetylene III. Ab initio investigation of the valence-type singlet electronic states. *Mol. Phys.* **62**, 1339–1356 (1987).

K. A. Peterson and T. H. Dunning, Jr., Benchmark calculations with correlated molecular wave functions. VIII. Bond energies and equilibrium geometries of the CH_n and C_2H_n (n = 1–4) series. *J. Chem. Phys.* **106**, 4119–4140 (1997).

A. S. Pine, Self-, N_2 and Ar-broadening and line mixing in HCN and C_2H_2. *J. Quant. Spectrosc. Radiat. Transfer* **50**, 149 (1993).

J. P. Pique, Y. Chen, R. W. Field, and J. L. Kinsey, Chaos and dynamics on 0.5–300 ps time scales in vibrationally excited acetylene: Fourier transform of stimulated-emission pumping spectrum. *Phys. Rev. Lett.* **58**, 475 (1987).

J.-P. Pique, M. Lombardi, Y. Chen, R. W. Field, and J. L. Kinsey, New order out of the chaotic bath of highly vibrational states of acetylene. *Ber. Bunsenges. Phys. Chem.* **92**, 422–424 (1988).

J.-P. Pique, Y. M. Engel, R. D. Levine, Y. Chen, R. W. Field, and J. L. Kinsey, Broad spectral features in the stimulated emission pumping spectrum of acetylene. *J. Chem. Phys.* **88**, 5972 (1988).

A. Pisarchik, M. Abbouti Temsamani, J. Vander Auwera, and M. Herman, The Fourier transform spectrum of the coloured vibrations in mono- and dideuteroacetylene. *Chem. Phys. Lett.* **206**, 343–348 (1993).

J. Plivà, Spectrum of acetylene in the 5-micron region. *J. Mol. Spectrosc.* **44**, 145–164 (1972).

J. Plivà, Molecular constants for the bending modes of acetylene $^{12}C_2H_2$. *J. Mol. Spectrosc.* **44**, 165–182 (1972).

E. K. Plyler and N. Gailar, Near infrared absorption spectra of deuterated acetylene. *J. Res. Natl. Bur. Stand.* **47**, 248–251 (1951).

E. K. Plyler, E. D. Tidwell, and T. A. Wiggins, Rotation-vibration constants of acetylene. *J. Opt. Soc. Am.* **53**, 589–593 (1963).

J. R. Podolske, M. Loewenstein, and P. Varanasi, Diode laser line strength measurements of the $(v_4 + v_5)^0$ band of $^{12}C_2H_2$. *J. Mol. Spectrosc.* **107**, 241–249 (1984).

P. L. Prasad, Infrared intensities. Acetylene. *J. Phys. Chem.* **82**, 312–314 (1978).

R. Prosmiti and S. C. Farantos, Periodic orbits, bifurcation diagrams and the spectroscopy of C_2H_2 system. *J. Chem. Phys.* **103**, 3299–3314 (1995).

H. M. Randall and E. F. Barker, Infrared spectra of acetylene containing H_2. *Phys. Rev.* **45**, 124 (1934).

R. A. Rasmussen, M. A. K. Khalil, and R. J. Fox, Altitudinal and temporal variation of hydrocarbons and other gaseous tracers of arctic haze. *Geophys. Res. Lett.* **10**, 144–147 (1983).

S. P. Reddy, V. M. Devi, A. Baldacci, W. Ivancic, and K. N. Rao, Acetylene spectra with a tunable diode laser: $(v_4 + v_5)^{0+} - v_4^1$ Q branches of $^{12}C_2H_2$ and $^{12}C^{13}CH_2$. *J. Mol. Spectrosc.* **74**, 217–223 (1979).

F. Remacle and R. D. Levine, Time-domain information from frequency-resolved or time-resolved experiments using maximum entropy. *J. Phys. Chem.* **97**, 12553–12560 (1993).

M. Rerat, D. Liotard, and J. M. Robine, Topological complexity of potential surfaces and application to C_2H_2 molecule. *Theor. Chim. Acta* **88**, 285–298 (1994).

N. H. Rich and H. L. Welsh, High-resolution rotation-vibrational Raman spectroscopy of gases with laser excitation. *Indian J. Pure Appl. Phys.* **9**, 944–949 (1971).

R. L. Richardson Jr., and P. R. Griffiths, Evaluation of a system for generating quantatively accurate vapor-phase infrared reference spectra. *Appl. Spectrosc.* **52**, 143–153 (1998).

S. T. Ridgway, Jupiter: Identification of ethane and acetylene, *Astrophys. J.* **187**, L41–L43 (1974).

S. T. Ridgway, Jupiter: Identification of ethane and acetylene. *Astrophys. J.* **192**, L51 (1974).

S. T. Ridgway, D. N. B. Hall, S. G. Kleinmann, D. A. Weinberger, and R. S. Wojslam, Circumstellar acetylene in the infrared spectrum of IRC $+10°216$. *Nature (London)* **264**, 345–346 (1976).

S. T. Ridgway, D. F. Carbon, and D. N. B. Hall, Polyatomic species contributing to the carbon star 3 μm band. *Astrophys. J.* **225**, 138–147 (1978).

C. P. Rinsland, A. Baldacci, and K. N. Rao, Acetylene bands observed in carbon stars: A laboratory study and an illustrative example of its application to IRC + 10216. *Astrophys. J., Suppl. S49*, 487–513 (1982).

C. P. Rinsland, A. Goldman, and G. M. Stokes, Identification of atmospheric C_2H_2 lines in the $3230–3340\,cm^{-1}$ region of high resolution solar absorption spectra recorded at the National Solar Observatory. *Appl. Opt.* **24**, 2044–2046 (1985).

C. P. Rinsland, R. Zander, C. B. Farmer, R. H. Norton, and J. M. Russell, Concentrations of ethane (C_2H_6) in the lower stratosphere and upper troposphere and acetylene (C_2H_2) in the upper troposphere deduced from atmospheric trace molecule spectroscopy/Spacelab 3 spectra. *J. Geophys. Res.* **92**, 11951–11964 (1987).

E. Robinson, Hydrocarbons in the atmosphere. *Pure Appl. Geophys.* **116**, 372–384 (1978).

P. N. Romani and S. K. Atreya, Methane photochemistry and haze production on Neptune. *Icarus* **74**, 424–445 (1988).

J. P. Rose and M. E. Kellman, The 2345 multimode resonance in acetylene: a bifurcation analysis. *J. Chem. Phys.* **103**, 7255 (1995).

J. P. Rose and M. E. Kellman, Bending dynamics from acetylene spectra: Normal, local, and precessional modes. *J. Chem. Phys.* **105**, 10743–10754 (1996).

L. S. Rothman, R. R. Gamache, R. H. Tipping, C. P. Rinsland, M. A. H. Smith, D. C. Benner, V. M. Devi, J.-M. Flaud, C. Camy-Peyret, A. Perrin, A. Goldman, S. T. Massie, L. R. Brown, and R. A. Toth, The HITRAN molecular database: Editions of 1991 and 1992. *J. Quant. Spectrosc. Radiat. Transfer* **48**, 469–508 (1992).

M. C. Rovira, J. J. Novoa, M. H. Whangbo, and J. M. Williams, Ab initio computation of the potential energy surfaces of the water hydrocarbon complexes H_2O, C_2H_2, H_2O, C_2H_4 and H_2O, CH_4: Minimum energy structures, vibrational frequencies and hydrogen bond energies. *Chem. Phys.* **200**, 319 (1995).

A. J. Russell and M. A. Spackman, Accurate ab initio study of acetylene. Vibrational and rotational corrections to electrical properties. *Mol. Phys.* **88**, 1109–1136 (1996).

K. Ruud, T. Helgaker, and P. Jørgensen, The effect of correlation on molecular magnetizabilities and rotational g tensors. *J. Chem. Phys.* **107**, 10599–10606 (1998).

M. Saarinen, D. Permogorov and L. Halonen, *HCCH vibrational overtone states by laser induced dispersed fluorescence*, the 15th international conference on high resolution molecular spectroscopy, Prague, Czech Republic, August 30–September 3, 1998, p 43.

J. Sakai and M. Katayama, Diode laser spectroscopy of acetylene: $3v_1 + v_3$ region at 0.77 μm. *J. Mol. Spectrosc.* **154**, 277–287 (1992).

J. Sakai and M. Katayama, Diode laser spectroscopy of the $2v_1 + 2v_3 + v_5 - v_4$ band of acetylene. *J. Mol. Spectrosc.* **155**, 424–426 (1992).

J. Sakai and M. Katayama, Diode laser spectroscopy of acetylene: The $2\nu_1 + 2\nu_3 + \nu_4^1 - \nu_5^1$ and $4\nu_1 - \nu_5^1$ interacting band system. *J. Mol. Spectrosc.* **157**, 532–535 (1993).

J. Sakai and M. Katayama, Diode laser spectroscopy of acetylene: $2\nu_1 + \nu_2 + \nu_3$ region at 0.85 μm. *J. Mol. Spectrosc.* **160**, 217–224 (1993).

Y. Sakai, S. Sudo, and T. Ikegami, Frequency stabilisation of laser diodes using 1.51–1.55 μm absorption lines of $^{12}C_2H_2$ and $^{13}C_2H_2$. *IEEE J. Quantum Electron.* **28**, 75–81 (1982).

Y. Sakai, I. Yokohama, T. Kominato, and S. Sudo, Frequency stabilization of laser using a frequency-locked ring resonater to acetylene gas absorption lines. *IEEE Photon. Technol. Lett.* **3**, 568 (1991).

Y. Sakai, S. Sudo, and T. Ikegami, Frequency stabilization of laser diodes using 1.51–1.55 μm absorption lines of $^{12}C_2H_2$ and $^{13}C_2H_2$. *IEEE J. Quantum Elec.* **28**, 75–81 (1992).

J. Sakai, H. Segawa, and M. Katayama, Diode laser spectroscopy of the $2\nu_1 + 2\nu_2 + \nu_3$ band of acetylene. *J. Mol. Spectrosc.* **164**, 580–582 (1994).

B. D. Saksena, Rotation-vibration spectra of diatomic and simple polyatomic molecules with long absorbing paths. VII. The spectrum of dideuteroacetylene (C_2D_2) in the photographic infrared. *J. Chem. Phys.* **20**, 95–100 (1952).

Y. A. Sarma, R. D'Cunha, G. Guelachvili, R. Farrenq, and K. Narahari Rao, Stretch-bend levels of acetylene: Analysis of the hot bands in the 3800 cm^{-1} region. *J. Mol. Spectrosc.* **173**, 561–573 (1995).

H. Sasada, S. Takeuchi, M. Iritani, and K. Nakatani, Semiconductor laser heterodyne frequency measurements on 1.52 μm molecular transitions. *J. Opt. Soc. Am. B* **8**, 713–718 (1991).

H. Sasada, T. Tsukamoto, Y. Kuba, N. Tanaka, and K. Uehara, Ti:Sapphireaser spectrometer for Doppler-limited molecular spectroscopy. *J. Opt. Soc. Am. B* **11**, 191–197 (1994).

J. J. Scherer, K. K. Lehmann, and W. Klemperer, The high-resolution visible overtone spectrum of acetylene. *J. Chem. Phys.* **78**, 2817–2832 (1983).

R. P. Schmid, T. Arusi-Parpar, R.-J. Li, I. Bar, and S. Rosenwaks, Photodissociation of rovibrationally excited C_2H_2: Observation of two pathways. *J. Chem. Phys.* **107**, 385–391 (1997).

R. P. Schmid, Y. Ganot, T. Arusi-Parpar, R. J. Li, I. Bar, and S. Rosenwaks, State selective dissociation of acetylene isotopomers. *SPIE* **3271**, in press (1998).

D. W. Schwenke, Variational calculations of rovibrational energy levels and transition intensities for tetratomic molecules. *J. Phys. Chem.* **100**, 2867–2883 (1996).

J. F. Scott and K. N. Rao, Infrared absorption bands of acetylene. Part I. analysis of the bands at 13.7 μm. *J. Mol. Spectrosc.* **16**, 15–23 (1965).

J. F. Scott and K. N. Rao, Infrared absorption bands of acetylene. Part II. Effect of vibrational l-type doubling. *J. Mol. Spectrosc.* **18**, 152–157 (1965).

J. F. Scott and K. N. Rao, Reconsideration of vibrational l-type doubling in acetylene. *J. Mol. Spectrosc.* **18**, 451–452 (1965).

J. F. Scott and K. N. Rao, Infrared absorption bands of acetylene. *J. Mol. Spectrosc.* **20**, 438–460 (1966).

Y. Segui, and P. Raynaud, Plasma diagnostic by infrared absorption spectroscopy. *NATO ASI Ser., Ser. E* **346**, 81–100 (1997).

W. H. Shaffer and A. H. Nieldsen, The near infra-red spectra of linear Y_2X_2 molecules. *J. Chem. Phys.* **9**, 847–852 (1941).

Q. Shi, S. Hu, J. Chen, L. Hao, J. Han, and Q. Zhu, New type of high quality spherical photoacoustic cell for laser photoacoustic spectroscopy. *Huaxue Wuli Xuebao* **11**, 20–25 (1998).

E. L. Sibert, III and R. C. Mayrhofer, Highly excited vibrational states of acetylene: A variational calculation. *J. Chem. Phys.* **99**, 937–944 (1993).

E. L. Sibert, III and A. B. McCoy, Quantum, semiclassical and classical dynamics of the bending modes of acetylene. *J. Chem. Phys.* **105**, 469–478 (1996).

E. D. Simandiras, J. E. Rice, T. J. Lee, R. D. Amos, and N. C. Handy, On the necessity of f basis functions for bending frequencies. *J. Chem. Phys.* **88**, 3187–3195 (1988).

L. N. Sinitsa, The C_2H_2 absorption spectrum in the Nd laser range. *J. Mol. Spectrosc.* **84**, 57–59 (1980).

L. N. Sinitsa, High sensitive laser spectroscopy of highly excited molecular states. *J. Quant. Spectrosc. Radiat. Transfer* **48**, 721–723 (1992).

W. M. A. Smit and T. Van Dam, Ab initio study of the infrared intensities of acetylene. *J. Mol. Struct.* **88**, 273–281 (1982).

W. M. A. Smit, A. J. Van Straten, and T. Visser, Measurement and interpretation of the integrated infrared intensities of acetylene and perdeuterioacetylene. *J. Mol. Struct.* **48**, 177–189 (1978).

B. C. Smith and J. S. Winn, The C-H overtone spectra of acetylene: Bend/stretch interactions below $10\,000\,\text{cm}^{-1}$. *J. Chem. Phys.* **89**, 4638–4645 (1988).

B. C. Smith and J. S. Winn, The overtone dynamics of acetylene above $10\,000\,\text{cm}^{-1}$. *J. Chem. Phys.* **94**, 4120–4130 (1990).

J. Sneider, Z. Bozoki, A. Miklos, Z. Bor, and G. Szabo, On the possibility of combining external cavity diode laser with photoacoustic detector for high sensitivity gas monitoring. *Int. J. Environ. Anal. Chem.* **67**, 253–260 (1997).

S. A. B. Solina, J. P. O'Brien, R. W. Field, and W. F. Polik, The acetylene S_0 surface: From dispersed fluorescence spectra to polyads to dynamics. *Ber. Bunsenges. Phys. Chem.* **99**, 555–560 (1995).

S. A. B. Solina, J. P. O'Brien, R. W. Field, and W. F. Polik, Dispersed fluorescence spectrum of acetylene from the A ^1A_u origin: Recognition of polyads and test of multiresonant effective Hamiltonian model for the X state. *J. Phys. Chem.* **100**, 7797–7809 (1996).

F. Stitt, Infra-red and Raman spectra of polyatomic molecules. X. C_2D_2, C_2HD, and C_2H_2. *J. Chem. Phys.* **8**, 56–59 (1940).

G. Strey and I. M. Mills, Anharmonic force field of acetylene. *J. Mol. Spectrosc.* **59**, 103–115 (1976).

R. L. Sundberg, E. Abramson, J. L. Kinsey, and R. W. Field, Evidence of quantum ergodicity in stimulated emission pumping spectra of acetylene. *J. Chem. Phys.* **83**, 466 (1985).

I. Suzuki and J. Overend, Anharmonic force constants of acetylene. Cubic and quartic constants of acetylene; calculations with a general quartic force field and an empirical anharmonic potential function. *Spectrochim. Acta, Part A* **25**, 977–987 (1969).

E. A. Sviridenkov, Intracavity laser spectroscopy. *Proc. SPIE-Int. Soc. Opt. Eng.* **3342**, 1–21 (1998).

M. Takahashi, Y. Okuzawa, M. Fujii, and M. Ito, Thermal lensing and photoacoustic spectra of gaseous acetylene by pulsed tunable infrared laser. *Can. J. Chem.* **69**, 1656–1658 (1991).

Y. Takehisa, M. Oda, and N. Ohashi, Deperturbation and band strength for the (015^u00)–(00000) band of acetylene. *J. Mol. Spectrosc.* **128**, 590–591 (1988).

R. M. Talley and A. H. Nielsen, Infrared spectrum and molecular constants of C_2D_2. *J. Chem. Phys.* **22**, 2030–2037 (1954).

C. N. Tam and T. A. Keiderling, Direct measurement of the rotational g-value in the ground vibrational state of acetylene by magnetic vibrational circular dichroism. *Chem. Phys. Lett.* **243**, 55 (1995).

C. N. Tam, P. Bour, and T. A. Keiderling, Observations of rotational magnetic moments in the ground and some excited vibrational S states of C_2H_2, C_2HD, and C_2D_2 by magnetic vibrational circular dichroism. *J. Chem. Phys.* **104**, 1813 (1996).

Y. Tang and S. A. Reid, Infrared degenerate four wave mixing spectroscopy of jet-cooled C_2H_2. *Chem. Phys. Lett.* **248**, 476 (1996).

J. R. Thomas, B. J. DeLeeuw, G. Vacek, and H. F. Schaefer, III, A systematic theoretical study of the harmonic vibrational frequencies for polyatomic molecules: The single, double, and perturbative triple excitation coupled-cluster [CCSD(T)] method. *J. Chem. Phys.* **98**, 1336 (1993).

J. R. Thomas, B. J. DeLeeuw, G. Vacek, T. D. Crawford, Y. Yamaguchi, and H. F. Schaefer, III, The balance between theoretical method and basis set quality: A systematic study of equilibrium geometries, dipole moments, harmonic vibrational frequencies, and infrared intensities. *J. Chem. Phys.* **99**, 403–416 (1993).

E. D. Tidwell and E. K. Plyler, Infrared spectrum of dideuteroacetylene (C_2D_2). *J. Opt. Soc. Am.* **52**, 656–664 (1962).

R. Titeica, Spectres de vibration et structure des molécules polyatomiques. *Actual. Sci. Ind.* **334**, 1–68, and references therein (1936).

J. D. Tobiason, A. L. Utz, and F. F. Crim, State-to-state rotational energy transfer in highly vibrationally excited acetylene. *J. Chem. Phys.* **97**, 7437–7447 (1992).

J. D. Tobiason, A. L. Utz, and F. F. Crim, Direct measurements of state-to-state rotational and vibrational energy transfer in highly vibrationally excited acetylene: Vibrational overtone excitation-LIF detection. *SPIE* **1858**, 317 (1993).

J. D. Tobiason, A. L. Utz, and F. F. Crim, Direct measurements of rotation specific, state-to-state vibrational energy transfer in highly vibrationally excited acetylene. *J. Chem. Phys.* **101**, 1108 (1994).

J. D. Tobiason, M. D. Fritz, and F. F. Crim, State-to-state relaxation of highly vibrationally excited acetylene by argon. *J. Chem. Phys.* **101**, 9642 (1994).

A. Tokunaga, R. F. Knacke, and T. Owen, Ethane and acetylene abundances in the jovian atmosphere. *Astrophys. J.* **209**, 294–301 (1976).

A. Tokunaga, R. F. Knacke, S. T. Ridgway, and L. Wallace, High resolution spectra of Jupiter in the 744–980 inverse centimeter spectral range. *Astrophys. J.* **232**, 603–615 (1979).

A. M. Tolonen and S. Alanko, Investigations on the infrared spectrum of acetylene between $2500\,cm^{-1}$ and $2800\,cm^{-1}$ and studies of the resonance effects. *Mol. Phys.* **75**, 1155–1165 (1992).

M. Ukai, K. Kameta, K. Chiba, K. Nagano, N. Kouchi, K. Shinsaka, Y. Hatano, H. Umemoto, Y. Ito, and K. Tanaka, Ionizing and nonionizing decays of superexcited acetylene molecules in the extreme-ultraviolet region. *J. Chem. Phys.* **95**, 4142–4153 (1991).

A. L. Utz, J. D. Tobiason, E. Carrasquillo, M. D. Fritz, and F. F. Crim, Energy transfer in highly vibrationally excited acetylene: Relaxation for vibrational energies from 6500 to $13\,000\,cm^{-1}$. *J. Chem. Phys.* **97**, 389–396 (1992).

A. L. Utz, E. Carrasquillo, J. D. Tobiason, and F. F. Crim, Direct observation of weak state mixing in highly vibrationally excited acetylene. *Chem. Phys.* **190**, 311–326 (1995).

J. C. Van Craen, M. Herman, and R. Colin, The A-X band system of acetylene: Analysis of medium-wavelength bands, and vibration-rotation constants for the levels $nv'_3(n = 4 - 6)$, $v'_2 + nv'_3(n = 3 - 5)$, and $v'_1 + nv'_3(n = 2,3)$. *J. Mol. Spectrosc.* **111**, 185–197 (1985).

J. C. Van Craen, M. Herman, R. Colin, and J. K. G. Watson, The A-X band system of acetylene: Bands of the short-wavelength region. *J. Mol. Spectrosc.* **119**, 137–143 (1986).

J. Vander Auwera, D. Hurtmans, M. Carleer, and M. Herman, The v_3 fundamental in C_2H_2. *J. Mol. Spectrosc.* **157**, 337–357 (1993).

P. van Ede van der Pals and P. Gaspard, The vibrational dynamics of acetylene $^{12}C_2H_2$: The semiclassical regime, unpublished results.

O. S. van Roosmalen, I. Benjamin, and R. D. Levine, A unified model description for interacting vibrational modes in ABA molecules. *J. Chem. Phys.* **81**, 5986 (1984).

O. S. van Roosmalen, F. Iachello, R. D. Levine, and A. E. L. Dieperink, Algebraic approach to molecular rotation-vibration spectra. II. Triatomic molecules. *J. Chem. Phys.* **79**, 2515 (1983).

P. Varanasi, Intensity and linewidth measurements in the 13.7 μm fundamental bands of $^{12}C_2H_2$ and $^{12}C^{13}CH_2$ and planetary atmospheric temperatures. *J. Quant. Spectrosc. Radiat. Transfer* **47**, 263 (1992).

P. Varanasi and B. R. P. Bangaru, Measurement of integrated intensities of acetylene bands at 3.04, 7.53 and 13.7 μm. *J. Quant. Spectrosc. Radiat. Transfer* **14**, 839–844 (1974).

P. Varanasi, L. P. Giver, and F. P. J. Valero, Measurements of nitrogen-broadened line widths of acetylene at low temperatures. *J. Quant. Spectrosc. Radiat. Transfer* **30**, 505–509 (1983).

E. Venuti, G. Di Lonardo, P. Ferracuti, L. Fusina, and I. M. Mills, Vibration-rotation spectra of ^{13}C containing acetylenes: Anharmonic resonances. *Chem. Phys.* **190**, 279–290 (1995).

S. P. Walch and P. R. Taylor, Characterisation of the minimum energy paths and energetics for the reaction of vinylidene with acetylene. *J. Chem. Phys.* **103**, 4975–4979 (1995).

B. Wang, J. Guo, Y. Gu, W. Mao, and F. Kong, Vibrational energy transfer from highly excited state CO to C_2H_2. *Wuli Huaxue Xuebao* **14**, 327–331 (1998).

I. Y. Wang and A. Weber, High resolution Raman spectroscopy of gases with laser sources: the spectrum of acetylene. *Indian J. Pure Appl. Phys.* **16**, 358–369 (1978).

J. K. G. Watson, Intensities of linear molecule vibration–rotation transitions with $|\Delta k| = 2$, with applications to HCN, DCN and HCCH. *J. Mol. Spectrosc.* **188**, 78–84 (1998).

J. K. G. Watson, M. Herman, J. C. Van Craen, and R. Colin, The A-X band system of acetylene. Analysis of long wavelengths bands and vibration rotation constants for the levels $nv''_4(n = 0 - 4), nv'_3(n = 0 - 3)$ and $v'_2 + nv'_3(n = 0 - 2)$. *J. Mol. Spectrosc.* **95**, 101–132 (1982).

M. Weber, W. E. Blass, and J. L. Salanave, Tunable diode-laser measurements of the 14 μm line strengths in $^{12}C_2H_2$. *J. Quant. Spectrosc. Radiat. Transfer* **42**, 437 (1989).

M. Weber and W. E. Blass, l-Resonance effects in the v_5, $2v_5 - v_5$, and $v_4 + v_5 - v_4$ bands of C_2H_2 and $^{13}C^{12}CH_2$ near 13.7 μm. *Spectrochim. Acta, Part A* **48**, 1203–1226 (1992).

M. Weber and W. E. Blass, l-resonance effects in acetylene near 13.7 μm. Part II: The two quantum hot bands. *Spectrochim. Acta, Part A* **49**, 1659–1681 (1993).

M. Weber, W. E. Blass, G. W. Halsey, and J. J. Hillman, l-resonance perturbation of IR intensities in $^{12}C_2H_2$ near 13.7 μm, *J. Mol. Spectrosc.* **165**, 107–123 (1994).

K. A. White and G. C. Schatz, Analytical potential energy surfaces for ethynyl, acetylene and vinyl. *J. Phys. Chem.* **88**, 2049 (1984).

K. B. Wiberg and J. J. Wendoloski, The electrical nature of C–H bond and its relationship to infrared intensities. *J. Comput. Chem.* **2**, 53–57 (1981).

T. A. Wiggins, E. K. Plyler, and E. D. Tidwell, Infrared spectrum of acetylene, *J. Opt. Soc. Am.* **51**, 1219–1225 (1961).

G. Winnewisser and E. Herbst, *Organic Molecules in Space*. Springer-Verlag, Berlin, 1987.

G. Wlodarczak, J. Demaison, J. Burie, and M. C. Lasne, The rotational constants of acetylene-d. *Mol. Phys.* **66**, 669–674 (1989).

J. S. Wong, Pressure broadening of single vibrational-rotational transitions of acetylene at $v = 5$. *J. Mol. Spectrosc.* **82**, 449 (1980).

C. M. Wright, E. F. Van Dishoeck, F. P. Helmich, F. Lahuis, A. C. A. Boogert, and Th. De Graauw, Gas phase infrared molecular spectroscopy of young stellar objects. *ESA SP-419*, 37–41, First ISO Workshop on Analytical Spectroscopy, 1997.

K. Yamanouchi and R. W. Field, Short time dynamics of highly excited small polyatomic molecules extracted from laser spectra. *Laser Chem.* **16**, 31–41 (1995).

K. Yamanouchi, N. Ikeda, S. Tsuchiya, D. M. Jonas, J. K. Lundberg, G. W. Adamson, and R. W. Field, Vibrationally highly excited acetylene as studied by dispersed fluorescence and stimulated emission pumping spectroscopy: Vibrational assignment of the feature states. *J. Chem. Phys.* **95**, 6330 (1991).

K. Yamanouchi, J. Miyawaki, S. Tsuchiya, D. M. Jonas, and R. W. Field, New scheme for extracting molecular dynamics from spectra: Case study on vibrationally highly excited acetylene. *Laser Chem.* **14**, 183–190 (1994).

T. Yamasaki, W. And A. Goddard III, Correlation analysis of chemical bonds. *J. Phys. Chem.* **102A**, 2919–2933 (1998).

S. F. Yang, L. Biennier, A. Campargue, M. Abbouti Temsamani, and M. Herman, The absorption spectrum of $^{12}C_2H_2$ between 12800 and 18500 cm^{-1}. II. Rotational analysis. *Mol. Phys.* **90**, 807–816 (1997).

J. Ye, L.-S. Ma, and J. L. Hall, Sub-Doppler optical frequency reference at 1.064 μm by means of ultrasensitive cavity-enhanced frequency modulation spectroscopy of a C_2HD overtone transition. *Opt. Lett.* **21**, 1000 (1996).

J. Ye, L.-S. Ma, and J. L. Hall, Ultrasensitive detections in atomic and molecular physics: Demonstration in molecular overtone spectroscopy, *J. Opt. Soc. Am. B* **15**, 6–15 (1998).

T. Yoshida and H. Sasada, Near-infrared spectroscopy with a wavemeter. *J. Mol. Spectrosc.* **153**, 208–210 (1992).

M. Yu, M. Dolg, P. Fulde, and H. Stoll, Charge fluctuations and correlation strength in chemical bonds: First-row homonuclear diatomic molecules. *Int. J. Quantum Chem.* **67**, 157–173 (1998).

R. Zander, G. Stokes, and J. Brault, Physique de l'atmosphère - Détection par voie spectro-scopique, de l'acétylène et de l'éthane dans l'atmosphère terrestre, à partir d'observations solaires infrarouges au sol. *C. R. Seances Acad. Sci.* **295**, 583–586 (1982).

X. Zhan and L. Halonen, High-resolution photoacoustic study of the $v_1 + 3v_3$ band system of acetylene with a titanium: Sapphire ring laser. *J. Mol. Spectrosc.* **160**, 464–470 (1993).

X. Zhan, O. Vaittinen, E. Kauppi, and L. Halonen, High-resolution photoacoustic overtone spectrum of acetylene near 570 nm using a ring-dye-laser spectrometer. *Chem. Phys. Lett.* **180**, 310–316 (1991).

X. Zhan, O. Vaittinen, and L. Halonen, High-resolution photoacoustic study of acetylene between 11 500 and 11 900 cm $^{-1}$ using a titanium: Sapphire ring laser. *J. Mol. Spectrosc.* **160**, 172–180 (1993).

J. Zhang, C. W. Riehn, M. Dulligan, and C. Wittig, Propensities toward C_2H ($A\,^2\Pi$) in acetylene photodissociation. *J. Chem. Phys.* **103**, 6815–6818 (1995).

V. E. Zuev, V. P. Lopasov, L. N. Sinitsa, and A. M. Solodov, High resolution Q-branch spectrum of acetylene at 9366 and 9407 cm $^{-1}$. *J. Mol. Spectrosc.* **94**, 208–210 (1982).

APPENDIX C

Vibrational Energy Levels of $^{12}C_2H_2$ ($\tilde{X}\,^1\Sigma_g^+$)

Calculated (cm $^{-1}$) vibrational energy levels in $^{12}C_2H_2$ ($\tilde{X}\,^1\Sigma_g^+$) with $k < 3$, up to 10000 cm $^{-1}$ above zero point energy. Modes 1–5 are, respectively, the symmetric CH stretch, the CC stretch, the asymmetric CH stretch, the *trans* bend, and the *cis* bend. The calculation was performed using the cluster model as detailed in Section III.C. The vibrational constants are those from the work of El Idrissi et al., to be published in Ref. 450 of the present review. The levels are arranged by increasing order of energy. The two first columns, containing the energy and the k value with the u/g symmetry, respectively, are intended to provide firsthand information. Predicted B_v values are then provided, calculated from the eigenvectors in the cluster matrices diagonalisation, whose dominant components are listed in the remaining part of each line, provided they are larger than 25%.

The listed energies are expected to be accurate within a few cm $^{-1}$ for all levels (see Ref. [450]), except those involving higher excited *cis*-bend states (v_5) which are expected to be less accurate.

$G(v)$	B_v	$k,u/g$	v_1	v_2	v_3	v_4	l_4	v_5	l_5	%
612.7	1.178	1,g	0	0	0	1	1	0	0	100
730.2	1.179	1,u	0	0	0	0	0	1	1	100
1230.8	1.179	0,g	0	0	0	2	0	0	0	100
1233.0	1.179	2,g	0	0	0	2	2	0	0	100
1328.1	1.180	0,u	0	0	0	1	1	1	-1	100
1339.8	1.180	0,u	0	0	0	1	-1	1	1	100
1347.6	1.180	2,u	0	0	0	1	1	1	1	100
1449.0	1.181	0,g	0	0	0	0	0	2	0	100
1462.8	1.181	2,g	0	0	0	0	0	2	2	100
1856.4	1.181	1,g	0	0	0	3	1	0	0	100
1941.1	1.182	1,u	0	0	0	2	2	1	-1	78
1960.6	1.182	1,u	0	0	0	2	0	1	1	78
1974.3	1.170	0,g	0	1	0	0	0	0	0	100
2049.2	1.182	1,g	0	0	0	1	1	2	0	49
		1,g	0	0	0	1	-1	2	2	51
2066.5	1.182	1,g	0	0	0	1	1	2	0	50
		1,g	0	0	0	1	-1	2	2	49
2170.1	1.183	1,u	0	0	0	0	0	3	1	100
2487.0	1.182	0,g	0	0	0	4	0	0	0	99
2489.3	1.182	2,g	0	0	0	4	2	0	0	99
2560.6	1.183	2,u	0	0	0	3	3	1	-1	88
2561.6	1.183	0,u	0	0	0	3	1	1	-1	100
2574.5	1.172	1,g	0	1	0	1	1	0	0	100
2583.7	1.183	0,u	0	0	0	3	-1	1	1	99
2589.8	1.183	2,u	0	0	0	3	1	1	1	87
2648.2	1.184	0,g	0	0	0	2	2	2	-2	66
		0,g	0	0	0	2	0	2	0	34
2660.3	1.184	0,g	0	0	0	2	-2	2	2	100
2666.6	1.184	2,g	0	0	0	2	2	2	0	70
		2,g	0	0	0	2	0	2	2	30
2683.8	1.184	0,g	0	0	0	2	2	2	-2	34
		0,g	0	0	0	2	0	2	0	65
2692.7	1.184	2,g	0	0	0	2	2	2	0	30
		2,g	0	0	0	2	0	2	2	69
2703.0	1.173	1,u	0	1	0	0	0	1	1	100
2758.1	1.185	0,u	0	0	0	1	1	3	-1	100
2773.3	1.185	2,u	0	0	0	1	1	3	1	48
		2,u	0	0	0	1	-1	3	3	52
2783.7	1.185	0,u	0	0	0	1	-1	3	1	99
2795.4	1.185	2,u	0	0	0	1	1	3	1	51
		2,u	0	0	0	1	-1	3	3	48
2879.9	1.186	0,g	0	0	0	0	0	4	0	99
2893.7	1.186	2,g	0	0	0	0	0	4	2	99
3124.8	1.183	1,g	0	0	0	5	1	0	0	99
3180.8	1.173	0,g	0	1	0	2	0	0	0	100
3183.3	1.173	2,g	0	1	0	2	2	0	0	100
3187.4	1.184	1,u	0	0	0	4	2	1	-1	68
		1,u	0	0	0	4	0	1	1	31
3215.8	1.184	1,u	0	0	0	4	2	1	-1	30
		1,u	0	0	0	4	0	1	1	68
3261.5	1.185	1,g	0	0	0	3	3	2	-2	66
		1,g	0	0	0	3	1	2	0	26
3282.2	1.185	1,g	0	0	0	3	3	2	-2	27
		1,g	0	0	0	3	-1	2	2	53
3282.4	1.173	0,u	0	0	1	0	0	0	0	46
		0,u	0	1	0	1	1	1	-1	54
3295.1	1.172	0,u	0	0	1	0	0	0	0	54

$G(v)$	B_v	$k,u/g$	v_1	v_2	v_3	v_4	l_4	v_5	l_5	%
		0,u	0	1	0	1	1	1	-1	46
3300.0	1.174	0,u	0	1	0	1	-1	1	1	100
3307.8	1.174	2,u	0	1	0	1	1	1	1	100
3311.4	1.185	1,g	0	0	0	3	1	2	0	53
		1,g	0	0	0	3	-1	2	2	38
3360.3	1.186	1,u	0	0	0	2	2	3	-1	33
		1,u	0	0	0	2	0	3	1	38
		1,u	0	0	0	2	-2	3	3	29
3372.4	1.170	0,g	1	0	0	0	0	0	0	99
3378.6	1.186	1,u	0	0	0	2	2	3	-1	43
		1,u	0	0	0	2	-2	3	3	56
3408.2	1.186	1,u	0	0	0	2	0	3	1	59
3420.5	1.175	0,g	0	1	0	0	0	2	0	99
3434.1	1.175	2,g	0	1	0	0	0	2	2	100
3470.3	1.187	1,g	0	0	0	1	1	4	0	49
		1,g	0	0	0	1	-1	4	2	50
3503.0	1.187	1,g	0	0	0	1	1	4	0	50
		1,g	0	0	0	1	-1	4	2	48
3592.1	1.188	1,u	0	0	0	0	0	5	1	99
3767.0	1.185	0,g	0	0	0	6	0	0	0	97
3769.4	1.185	2,g	0	0	0	6	2	0	0	98
3795.4	1.175	1,g	0	1	0	3	1	0	0	99
3819.3	1.186	0,u	0	0	0	5	1	1	-1	98
3819.4	1.186	2,u	0	0	0	5	3	1	-1	78
3850.3	1.186	0,u	0	0	0	5	-1	1	1	96
3855.6	1.186	2,u	0	0	0	5	1	1	1	77
3880.4	1.187	2,g	0	0	0	4	4	2	-2	79
3882.6	1.174	1,u	0	0	1	1	1	0	0	44
		1,u	0	1	0	2	2	1	-1	42
3884.0	1.187	0,g	0	0	0	4	2	2	-2	60
		0,g	0	0	0	4	0	2	0	38
3898.4	1.174	1,u	0	0	1	1	1	0	0	55
		1,u	0	1	0	2	2	1	-1	39
3906.2	1.187	0,g	0	0	0	4	-2	2	2	98
3909.1	1.175	1,u	0	1	0	2	0	1	1	80
3911.5	1.187	2,g	0	0	0	4	2	2	0	38
		2,g	0	0	0	4	0	2	2	42
3933.8	1.164	0,g	0	2	0	0	0	0	0	100
3940.3	1.187	0,g	0	0	0	4	2	2	-2	37
		0,g	0	0	0	4	0	2	0	56
3947.4	1.187	2,g	0	0	0	4	2	2	0	39
		2,g	0	0	0	4	0	2	2	52
3960.8	1.187	0,u	0	0	0	3	3	3	-3	49
		0,u	0	0	0	3	1	3	-1	50
3969.8	1.171	1,g	1	0	0	1	1	0	0	94
3973.5	1.187	0,u	0	0	0	3	-3	3	3	90
3978.1	1.187	2,u	0	0	0	3	3	3	-1	53
		2,u	0	0	0	3	1	3	1	34
3996.9	1.187	0,u	0	0	0	3	3	3	-3	51
		0,u	0	0	0	3	1	3	-1	46
4002.5	1.175	1,g	0	0	1	0	0	1	1	35
		1,g	0	1	0	1	1	2	0	26
		1,g	0	1	0	1	-1	2	2	35
4005.7	1.187	2,u	0	0	0	3	3	3	-1	34
		2,u	0	0	0	3	-1	3	3	55
4017.0	1.174	1,g	0	0	1	0	0	1	1	59
		1,g	0	1	0	1	1	2	0	29

cont.

G(v)	B_v	k,u/g	v_1	v_2	v_3	v_4	l_4	v_5	l_5	%
4025.3	1.176	1,g	0	1	0	1	1	2	0	44
		1,g	0	1	0	1	-1	2	2	54
4032.7	1.187	0,u	0	0	0	3	-1	3	1	82
4041.5	1.187	2,u	0	0	0	3	1	3	1	52
		2,u	0	0	0	3	-1	3	3	31
4060.0	1.188	0,g	0	0	0	2	2	4	-2	59
		0,g	0	0	0	2	0	4	0	40
4076.0	1.188	2,g	0	0	0	2	2	4	0	32
		2,g	0	0	0	2	0	4	2	40
		2,g	0	0	0	2	-2	4	4	27
4086.9	1.188	0,g	0	0	0	2	-2	4	2	98
4092.2	1.172	1,u	1	0	0	0	0	1	1	99
4099.7	1.188	2,g	0	0	0	2	2	4	0	40
		2,g	0	0	0	2	-2	4	4	58
4124.6	1.188	0,g	0	0	0	2	2	4	-2	39
		0,g	0	0	0	2	0	4	0	54
4135.1	1.188	2,g	0	0	0	2	0	4	2	55
4140.3	1.177	1,u	0	1	0	0	0	3	1	99
4170.8	1.189	0,u	0	0	0	1	1	5	-1	98
4185.4	1.189	2,u	0	0	0	1	1	5	1	47
		2,u	0	0	0	1	-1	5	3	51
4212.8	1.189	0,u	0	0	0	1	-1	5	1	96
4224.9	1.189	2,u	0	0	0	1	1	5	1	50
		2,u	0	0	0	1	-1	5	3	47
4293.1	1.190	0,g	0	0	0	0	0	6	0	98
4306.8	1.190	2,g	0	0	0	0	0	6	2	98
4415.6	1.176	0,g	0	1	0	4	0	0	0	98
4415.9	1.186	1,g	0	0	0	7	1	0	0	96
4418.2	1.176	2,g	0	1	0	4	2	0	0	99
4456.7	1.187	1,u	0	0	0	6	2	1	-1	62
		1,u	0	0	0	6	0	1	1	34
4489.3	1.175	0,u	0	0	1	2	0	0	0	48
		0,u	0	1	0	3	1	1	-1	51
4490.7	1.175	2,u	0	0	1	2	2	0	0	41
		2,u	0	1	0	3	3	1	-1	49
4492.7	1.187	1,u	0	0	0	6	2	1	-1	33
		1,u	0	0	0	6	0	1	1	60
4508.7	1.175	0,u	0	0	1	2	0	0	0	51
		0,u	0	1	0	3	1	1	-1	49
4509.1	1.175	2,u	0	0	1	2	2	0	0	58
		2,u	0	1	0	3	3	1	-1	40
4509.9	1.188	1,g	0	0	0	5	3	2	-2	52
		1,g	0	0	0	5	1	2	0	32
4521.0	1.177	0,u	0	1	0	3	-1	1	1	99
4521.0	1.166	1,g	0	2	0	1	1	0	0	100
4527.3	1.177	2,u	0	1	0	3	1	1	1	87
4538.5	1.188	1,g	0	0	0	5	3	2	-2	35
		1,g	0	0	0	5	-1	2	2	52
4570.0	1.173	0,g	1	0	0	2	0	0	0	70
4574.5	1.189	1,u	0	0	0	4	4	3	-3	58
		1,u	0	0	0	4	2	3	-1	26
4575.3	1.173	2,g	1	0	0	2	2	0	0	90
4577.7	1.188	1,g	0	0	0	5	1	2	0	48
		1,g	0	0	0	5	-1	2	2	31
4589.8	1.176	0,g	0	1	0	2	2	2	-2	40
4595.9	1.189	1,u	0	0	0	4	4	3	-3	29
		1,u	0	0	0	4	0	3	1	29
		1,u	0	0	0	4	-2	3	3	35
4599.3	1.176	0,g	0	0	1	1	-1	1	1	43
		0,g	0	1	0	2	-2	2	2	57
4608.1	1.176	2,g	0	0	1	1	1	1	1	28
		2,g	0	1	0	2	2	2	0	42
4609.3	1.175	0,g	0	0	1	1	1	1	-1	60
4617.5	1.176	0,g	0	0	1	1	-1	1	1	57
		0,g	0	1	0	2	-2	2	2	43
4625.1	1.175	2,g	0	0	1	1	1	1	1	62
		2,g	0	1	0	2	2	2	0	31
4625.4	1.189	1,u	0	0	0	4	2	3	-1	36
		1,u	0	0	0	4	-2	3	3	45
4630.4	1.178	0,g	0	1	0	2	2	2	-2	34
		0,g	0	1	0	2	0	2	0	64
4639.6	1.178	2,g	0	1	0	2	2	2	0	26
		2,g	0	1	0	2	0	2	2	71
4660.9	1.167	1,u	0	2	0	0	0	1	1	100
4664.2	1.190	1,g	0	0	0	3	1	4	0	30
		1,g	0	0	0	3	-1	4	2	26
4666.6	1.189	1,u	0	0	0	4	0	3	1	46
4673.8	1.174	0,u	1	0	0	1	1	1	-1	90
4683.9	1.190	1,g	0	0	0	3	3	4	-2	35
		1,g	0	0	0	3	-3	4	4	52
4688.1	1.173	0,u	1	0	0	1	-1	1	1	99
4692.7	1.174	2,u	1	0	0	1	1	1	1	89
4710.8	1.177	0,u	0	0	1	0	0	2	0	26
		0,u	0	1	0	1	1	3	-1	66
4713.4	1.190	1,g	0	0	0	3	3	4	-2	29
		1,g	0	0	0	3	-1	4	2	25
		1,g	0	0	0	3	-3	4	4	26
4725.7	1.177	2,u	0	0	1	0	0	2	2	27
		2,u	0	1	0	1	1	3	1	26
		2,u	0	1	0	1	-1	3	3	39
4727.3	1.176	0,u	0	0	1	0	0	2	0	67
		0,u	0	1	0	1	1	3	-1	31
4740.8	1.178	0,u	0	1	0	1	-1	3	1	99
4741.3	1.176	2,u	0	0	1	0	0	2	2	61
		2,u	0	1	0	1	1	3	1	29
4752.9	1.178	2,u	0	1	0	1	1	3	1	43
		2,u	0	1	0	1	-1	3	3	52
4756.1	1.190	1,g	0	0	0	3	1	4	0	40
		1,g	0	0	0	3	-1	4	2	34
4763.6	1.190	1,u	0	0	0	2	2	5	-1	28
		1,u	0	0	0	2	0	5	1	41
		1,u	0	0	0	2	-2	5	3	28
4798.2	1.190	1,u	0	0	0	2	2	5	-1	45
		1,u	0	0	0	2	-2	5	3	51
4800.2	1.174	0,g	1	0	0	0	0	2	0	98
4814.3	1.174	2,g	1	0	0	0	0	2	2	98
4843.4	1.190	1,u	0	0	0	2	0	5	1	50
4848.9	1.179	0,g	0	1	0	0	0	4	0	98
4862.5	1.179	2,g	0	1	0	0	0	4	2	98
4874.4	1.191	1,g	0	0	0	1	1	6	0	48
		1,g	0	0	0	1	-1	6	2	50
4925.3	1.191	1,g	0	0	0	1	1	6	0	47
		1,g	0	0	0	1	-1	6	2	46
4996.7	1.192	1,u	0	0	0	0	0	7	1	96

G(v)	B_v	k,u/g	v_1	v_2	v_3	v_4	l_4	v_5	l_5	%
5043.6	1.177	1,g	0	1	0	5	1	0	0	97
5068.5	1.188	0,g	0	0	0	8	0	0	0	93
5071.0	1.188	2,g	0	0	0	8	2	0	0	93
5099.0	1.188	0,u	0	0	0	7	1	1	-1	94
5099.7	1.188	2,u	0	0	0	7	3	1	-1	70
5103.2	1.176	1,u	0	0	1	3	1	0	0	49
		1,u	0	1	0	4	2	1	-1	32
5115.0	1.167	0,g	0	2	0	2	0	0	0	100
5117.8	1.167	2,g	0	2	0	2	2	0	0	100
5125.5	1.176	1,u	0	0	1	3	1	0	0	50
		1,u	0	1	0	4	2	1	-1	38
5137.0	1.189	0,g	0	0	0	7	-1	1	1	89
5141.0	1.189	2,g	0	0	0	6	4	2	-2	66
5142.0	1.189	2,u	0	0	0	7	1	1	1	67
5142.7	1.178	1,u	0	1	0	4	2	1	-1	28
		1,u	0	1	0	4	0	1	1	68
5143.0	1.189	0,g	0	0	0	6	2	2	-2	58
		0,g	0	0	0	6	0	2	0	35
5172.9	1.189	0,g	0	0	0	6	-2	2	2	93
5176.4	1.175	1,g	1	0	0	3	1	0	0	43
5178.0	1.189	2,g	0	0	0	6	0	2	2	45
5192.9	1.190	2,u	0	0	0	5	5	3	-3	74
5194.2	1.177	1,g	1	0	0	3	1	0	0	39
		1,g	0	1	0	3	3	2	-2	39
5198.7	1.190	0,u	0	0	0	5	3	3	-3	42
		0,u	0	0	0	5	1	3	-1	53
5209.1	1.177	1,g	0	1	0	3	-1	2	2	35
5214.7	1.177	1,g	0	0	1	2	2	1	-1	45
		1,g	0	1	0	3	3	2	-2	27
5216.5	1.189	0,g	0	0	0	6	2	2	-2	35
		0,g	0	0	0	6	0	2	0	47
5220.6	1.190	0,u	0	0	0	5	-3	3	3	84
5222.7	1.189	2,g	0	0	0	6	2	2	0	38
		2,g	0	0	0	6	0	2	2	40
5225.3	1.190	2,u	0	0	0	5	1	3	1	34
5229.9	1.177	1,g	0	0	1	2	0	1	1	48
5230.1	1.168	0,u	0	2	0	1	1	1	-1	90
5244.9	1.168	0,u	0	2	0	1	-1	1	1	100
5247.0	1.179	1,g	0	1	0	3	1	2	0	51
		1,g	0	1	0	3	-1	2	2	39
5252.6	1.168	2,u	0	2	0	1	1	1	1	100
5254.6	1.190	0,u	0	0	0	5	3	3	-3	54
		0,u	0	0	0	5	1	3	-1	35
5260.4	1.165	0,u	0	1	1	0	0	0	0	90
5261.6	1.190	2,u	0	0	0	5	3	3	-1	35
		2,u	0	0	0	5	-1	3	3	49
5266.3	1.191	0,u	0	0	0	4	4	4	-4	38
		0,g	0	0	0	4	2	4	-2	39
5270.2	1.175	1,u	1	0	0	2	2	1	-1	53
5279.7	1.191	0,g	0	0	0	4	-4	4	4	80
5282.6	1.191	2,g	0	0	0	4	4	4	-2	41
		2,g	0	0	0	4	2	4	0	32
5290.0	1.176	1,u	1	0	0	2	2	1	-1	41
		1,u	1	0	0	2	0	1	1	30
5300.9	1.190	0,u	0	0	0	5	-1	3	1	66
5302.4	1.178	1,u	1	0	0	2	0	1	1	33
5303.0	1.191	0,g	0	0	0	4	4	4	-4	59
5308.3	1.190	2,u	0	0	0	5	1	3	1	42
5311.5	1.191	2,g	0	0	0	4	4	4	-2	34
		2,g	0	0	0	4	-2	4	4	39
5316.5	1.178	1,u	0	1	0	2	2	3	-1	28
		1,u	0	1	0	2	-2	3	3	31
5323.0	1.177	1,u	0	0	1	1	1	2	0	43
5335.0	1.164	0,g	1	1	0	0	0	0	0	98
5336.0	1.178	1,u	0	0	1	1	1	2	0	25
		1,u	0	0	1	1	-1	2	2	34
5338.8	1.191	0,g	0	0	0	4	-2	4	2	68
5347.6	1.191	2,g	0	0	0	4	2	4	0	25
		2,g	0	0	0	4	-2	4	4	41
5353.2	1.180	1,u	0	1	0	2	0	3	1	59
5355.3	1.192	0,u	0	0	0	3	3	5	-3	38
		0,u	0	0	0	3	1	5	-1	58
5371.7	1.192	1,u	0	0	0	3	1	5	1	31
		2,u	0	0	0	3	-1	5	3	27
5377.2	1.169	0,g	0	2	0	0	0	2	0	99
5383.3	1.176	1,g	1	0	0	1	1	2	0	46
		1,g	1	0	0	1	-1	2	2	40
5383.9	1.192	1,u	0	0	0	3	-3	5	3	83
5387.1	1.191	0,g	0	0	0	4	2	4	-2	35
		0,g	0	0	0	4	0	4	0	37
5390.4	1.169	2,g	0	2	0	0	0	2	2	100
5395.6	1.191	2,g	0	0	0	4	2	4	0	26
		2,g	0	0	0	4	0	4	2	38
5397.8	1.192	2,u	0	0	0	3	3	5	-1	30
		2,u	0	0	0	3	-3	5	5	55
5404.3	1.176	1,g	1	0	0	1	1	2	0	42
		1,g	1	0	0	1	-1	2	2	55
5421.9	1.192	0,u	0	0	0	3	3	5	-3	57
		0,u	0	0	0	3	1	5	-1	32
5422.3	1.180	1,g	0	1	0	1	1	4	0	32
		1,g	0	1	0	1	-1	4	2	37
5432.7	1.192	2,u	0	0	0	3	3	5	-1	30
		2,u	0	0	0	3	-1	5	3	27
		2,u	0	0	0	3	-3	5	5	26
5439.9	1.178	1,g	0	0	1	0	0	3	1	70
5455.4	1.193	0,g	0	0	0	2	2	6	-2	55
		0,g	0	0	0	2	0	6	0	41
5458.6	1.181	1,g	0	1	0	1	1	4	0	48
		1,g	0	1	0	1	-1	4	2	49
5470.6	1.193	2,g	0	0	0	2	2	6	0	27
		2,g	0	0	0	2	0	6	2	41
		2,g	0	0	0	2	-2	6	4	28
5472.4	1.192	0,u	0	0	0	3	-1	5	1	64
5482.1	1.192	2,u	0	0	0	3	1	5	1	37
		2,u	0	0	0	3	-1	5	3	29
5499.9	1.193	0,g	0	0	0	2	-2	6	2	93
5510.7	1.177	1,u	1	0	0	0	0	3	1	96
5512.7	1.193	2,g	0	0	0	2	2	6	0	42
		2,g	0	0	0	2	-2	6	4	52
5554.0	1.193	0,g	0	0	0	2	2	6	-2	37
		0,g	0	0	0	2	0	6	0	44
5560.0	1.181	1,u	0	1	0	0	0	5	1	96
5564.9	1.193	2,g	0	0	0	2	0	6	2	45
5566.8	1.194	0,u	0	0	0	1	1	7	-1	96

cont.

G(v)	Bv	k,u/g	v1	v2	v3	v4	l4	v5	l5	%
5581.1	1.194	2,u	0	0	0	1	1	7	1	46
		2,u	0	0	0	1	-1	7	3	50
5628.6	1.193	0,u	0	0	0	1	-1	7	1	90
5640.7	1.193	2,u	0	0	0	1	1	7	1	46
		2,u	0	0	0	1	-1	7	3	44
5676.4	1.179	0,g	0	1	0	6	0	0	0	94
5679.2	1.179	2,g	0	1	0	6	2	0	0	95
5689.5	1.194	0,g	0	0	0	0	0	8	0	94
5703.1	1.194	2,g	0	0	0	0	0	8	2	94
5718.3	1.168	1,g	0	2	0	3	1	0	0	99
5722.3	1.178	0,u	0	0	1	4	0	0	0	52
		0,u	0	1	0	5	1	1	-1	44
5724.4	1.178	2,u	0	0	1	4	2	0	0	49
		2,u	0	1	0	5	3	1	-1	36
5726.8	1.189	1,g	0	0	0	9	1	0	0	87
5746.1	1.190	1,u	0	0	0	8	2	1	-1	55
		1,u	0	0	0	8	0	1	1	34
5748.5	1.178	0,u	0	0	1	4	0	0	0	46
		0,u	0	1	0	5	1	1	-1	52
5749.2	1.178	2,u	0	0	1	4	2	0	0	49
		2,u	0	1	0	5	3	1	-1	43
5767.3	1.179	0,u	0	1	0	5	-1	1	1	94
5772.9	1.179	2,u	0	1	0	5	1	1	1	77
5780.1	1.191	1,g	0	0	0	7	3	2	-2	45
		1,g	0	0	0	7	1	2	0	28
5785.7	1.177	0,g	1	0	0	4	0	0	0	26
		0,g	0	0	1	3	1	1	-1	47
5787.9	1.178	2,g	0	0	1	3	3	1	-1	34
5788.0	1.190	1,u	0	0	0	8	2	1	-1	32
		1,u	0	0	0	8	0	1	1	52
5807.3	1.178	2,g	1	0	0	4	2	0	0	48
		2,g	0	1	0	4	4	2	-2	39
5809.9	1.178	0,g	1	0	0	4	0	0	0	47
		0,g	0	1	0	4	2	2	-2	31
5814.4	1.191	1,g	0	0	0	7	3	2	-2	35
		1,g	0	0	0	7	-1	2	2	48
5818.0	1.169	1,u	0	2	0	2	2	1	-1	66
5819.9	1.179	0,g	0	0	1	3	-1	1	1	46
		0,g	0	1	0	4	-2	2	2	51
5824.9	1.191	1,u	0	0	0	6	4	3	-3	43
		1,u	0	0	0	6	2	3	-1	27
5825.3	1.178	2,g	0	0	1	3	3	1	-1	55
5828.9	1.178	2,g								<25
5830.9	1.177	0,g	0	0	1	3	1	1	-1	52
5841.3	1.169	1,u	0	2	0	2	2	1	-1	29
		1,u	0	2	0	2	0	1	1	69
5844.1	1.179	0,g	0	0	1	3	-1	1	1	52
		0,g	0	1	0	4	-2	2	2	46
5850.6	1.178	2,g	0	0	1	3	1	1	1	51
5851.8	1.166	1,u	0	1	1	1	1	0	0	85
5853.1	1.192	1,u	0	0	0	6	4	3	-3	33
		1,u	0	0	0	6	-2	3	3	36
5862.6	1.191	1,g	0	0	0	7	1	2	0	39
5865.2	1.180	0,g	0	1	0	4	2	2	-2	37
		0,g	0	1	0	4	0	2	0	56
5866.4	1.178	0,u	1	0	0	3	1	1	-1	42
5872.6	1.180	2,g	0	1	0	4	2	2	0	37
		2,g	0	1	0	4	0	2	2	54
5875.0	1.177	2,u	1	0	0	3	3	1	-1	58
5878.3	1.158	0,g	0	3	0	0	0	0	0	100
5880.7	1.192	1,g	0	0	0	5	5	4	-4	52
5889.3	1.179	0,u	0	0	1	2	-2	2	2	44
		0,u	0	1	0	3	-3	3	3	28
5891.8	1.192	1,u	0	0	0	6	2	3	-1	26
		1,u	0	0	0	6	-2	3	3	37
5893.3	1.179	0,u	1	0	0	3	1	1	-1	39
		0,u	0	1	0	3	3	3	-3	31
		0,u	0	1	0	3	1	3	-1	26
5899.2	1.178	2,u	1	0	0	3	3	1	-1	36
5902.5	1.192	1,g	0	0	0	5	5	4	-4	29
5908.2	1.178	0,u	1	0	0	3	-1	1	1	57
		0,u	0	1	0	3	-3	3	3	38
5911.7	1.179	2,u	1	0	0	3	1	1	1	42
		2,u	0	1	0	3	3	3	-1	33
5916.8	1.179	0,u	0	0	1	2	2	2	-2	54
5920.4	1.165	1,g	1	1	0	1	1	0	0	94
5921.9	1.180	0,u	0	0	1	2	0	2	0	37
		0,u	0	1	0	3	3	3	-3	37
5927.0	1.179	0,u	0	0	1	2	-2	2	2	56
5932.0	1.179	2,u	0	0	1	2	2	2	0	43
		2,u	0	1	0	3	-1	3	3	35
5932.1	1.192	1,g	0	0	0	5	-3	4	4	42
5933.5	1.179	2,u	0	1	0	3	3	3	-1	30
5943.2	1.192	1,u	0	0	0	6	0	3	1	32
5944.0	1.179	0,u	0	0	1	2	0	2	0	39
5948.1	1.170	1,g	0	2	0	1	1	2	0	38
		1,g	0	2	0	1	-1	2	2	49
5953.9	1.179	2,u	0	0	1	2	0	2	2	47
5961.6	1.193	1,u								<25
5964.4	1.178	0,g	1	0	0	2	2	2	-2	43
		0,g	1	0	0	2	0	2	0	29
5966.3	1.181	0,u	0	1	0	3	-1	3	1	83
5968.1	1.170	1,g	0	2	0	1	1	2	0	58
		1,g	0	2	0	1	-1	2	2	39
5973.0	1.192	1,g	0	0	0	5	3	4	-2	27
5975.3	1.181	2,u	0	1	0	3	1	3	1	50
		2,u	0	1	0	3	-1	3	3	33
5981.7	1.167	1,g	0	1	1	0	0	1	1	86
5982.0	1.178	2,g	1	0	0	2	2	2	0	44
		2,g	1	0	0	2	0	2	2	25
5983.0	1.193	1,u	0	0	0	4	4	5	-3	28
		1,u	0	0	0	4	-4	5	5	46
5983.1	1.177	0,g	1	0	0	2	-2	2	2	92
5997.5	1.180	0,g	1	0	0	2	2	2	-2	33
		0,g	0	1	0	2	2	4	-2	27
6008.5	1.178	0,g	1	0	0	2	0	2	0	52
6009.1	1.178	2,g	1	0	0	2	2	2	0	43
6012.1	1.193	1,u	0	0	0	4	4	5	-3	30
		1,u	0	0	0	4	-4	5	5	32
6018.6	1.180	2,g	1	0	0	2	0	2	2	40
6023.5	1.181	0,g	0	0	1	1	-1	3	1	31
		0,g	0	1	0	2	-2	4	2	62
6024.6	1.179	0,g	0	0	1	1	1	3	-1	62
6026.5	1.192	1,g	0	0	0	5	1	4	0	27

$G(v)$	B_v	$k,u/g$	v_1	v_2	v_3	v_4	l_4	v_5	l_5	%	$G(v)$	B_v	$k,u/g$	v_1	v_2	v_3	v_4	l_4	v_5	l_5	%
6036.5	1.180	2,g	0	0	1	1	-1	3	3	25	6386.7	1.191	0,g	0	0	0	10	0	0	0	74
		2,g	0	1	0	2	2	4	0	30	6389.4	1.191	2,g	0	0	0	10	2	0	0	75
6040.1	1.180	2,g	0	0	1	1	1	3	1	53	6396.1	1.191	0,u	0	0	0	9	1	1	-1	80
6044.1	1.180	0,g	0	0	1	1	-1	3	1	63	6397.6	1.191	2,u	0	0	0	9	3	1	-1	58
		0,g	0	1	0	2	-2	4	2	35	6400.2	1.181	1,u	0	1	0	6	2	1	-1	31
6050.8	1.194	1,g	0	0	0	3	1	6	0	29			1,u	0	1	0	6	0	1	1	60
		1,g	0	0	0	3	-1	6	2	29	6401.4	1.179	1,g	0	0	1	4	2	1	-1	32
6052.9	1.166	1,u	1	1	0	0	0	1	1	97	6413.6	1.170	0,u	0	2	0	3	1	1	-1	81
6054.6	1.193	1,u	0	0	0	4	-2	5	3	32	6414.5	1.170	2,u	0	2	0	3	3	1	-1	72
6057.0	1.180	2,g	0	0	1	1	-1	3	3	36	6422.5	1.192	2,g	0	0	0	8	4	2	-2	54
6067.7	1.182	2,g	0	1	0	2	2	4	-2	39	6423.4	1.192	0,g	0	0	0	8	2	2	-2	52
		0,g	0	1	0	2	0	4	0	54	6429.7	1.179	1,g	1	0	0	5	1	0	0	41
6078.4	1.182	2,g	0	1	0	2	0	4	2	55			1,g	0	1	0	5	3	2	-2	33
6080.5	1.178	0,u	1	0	0	1	1	3	-1	83	6439.9	1.192	0,u	0	0	0	9	-1	1	1	76
6087.9	1.194	1,g	0	0	0	3	3	6	-2	36	6440.8	1.180	1,g	0	1	0	5	-1	2	2	31
		1,g	0	0	0	3	-3	6	4	44	6442.1	1.171	0,u	0	2	0	3	-1	1	1	99
6095.7	1.171	1,u	0	2	0	0	0	3	1	97	6444.7	1.192	2,u	0	0	0	9	1	1	1	54
6096.0	1.178	2,u	1	0	0	1	1	3	1	45	6446.8	1.170	2,u	0	2	0	3	3	1	-1	27
		2,u	1	0	0	1	-1	3	3	38			2,u	0	2	0	3	1	1	1	51
6110.1	1.193	1,u	0	0	0	4	0	5	1	29	6449.2	1.168	0,u	0	1	1	2	0	0	0	82
6111.0	1.178	0,u	1	0	0	1	-1	3	1	97	6451.8	1.178	1,g	1	0	0	5	1	0	0	36
6122.1	1.182	0,u	0	1	0	1	1	5	-1	69			1,g	0	0	1	4	2	1	-1	31
6122.3	1.178	2,u	1	0	0	1	1	3	1	35	6452.1	1.159	1,g	0	3	0	1	1	0	0	100
		2,u	1	0	0	1	-1	3	3	59	6452.6	1.168	2,u	0	1	1	2	2	0	0	64
6133.3	1.194	1,g	0	0	0	3	3	6	-2	28			2,u	0	2	0	3	1	1	1	34
		1,g	0	0	0	3	-3	6	4	27	6455.5	1.193	2,u	0	0	0	7	5	3	-3	58
6136.7	1.182	2,u	0	1	0	1	1	5	1	30	6456.8	1.192	0,g	0	0	0	8	-2	2	2	81
		2,u	0	1	0	1	-1	5	3	40	6459.5	1.193	0,u	0	0	0	7	3	3	-3	39
6140.9	1.180	0,u	0	0	1	0	0	4	0	73			0,u	0	0	0	7	1	3	-1	39
6151.0	1.195	1,u	0	0	0	2	2	7	-1	26	6462.2	1.192	2,g	0	0	0	8	0	2	2	44
		1,u	0	0	0	2	0	7	1	40	6468.2	1.180	1,g	0	0	1	4	0	1	1	40
		1,u	0	0	0	2	-2	7	3	28	6471.1	1.180	1,u								<25
6155.0	1.180	2,u	0	0	1	0	0	4	2	71	6487.2	1.193	0,u	0	0	0	7	-3	3	3	76
6166.6	1.183	0,u	0	1	0	1	-1	5	1	95	6492.6	1.193	2,u	0	0	0	7	-1	3	3	26
6178.9	1.183	2,u	0	1	0	1	1	5	1	47	6492.7	1.182	1,g	0	1	0	5	1	2	0	47
		2,u	0	1	0	1	-1	5	3	48			1,g	0	1	0	5	-1	2	2	32
6191.4	1.194	1,u	0	0	0	3	1	6	0	28	6495.9	1.180	1,u	0	1	0	4	4	3	-3	37
		1,g	0	0	0	3	-1	6	2	25	6499.1	1.194	2,g	0	0	0	6	6	4	-4	70
6205.2	1.195	1,u	0	0	0	2	2	7	-1	43	6502.9	1.166	0,g	0	0	2	0	0	0	0	47
		1,u	0	0	0	2	-2	7	3	47			0,g	1	1	0	2	0	0	0	26
6209.5	1.179	0,g	1	0	0	0	0	4	0	95	6503.5	1.180	1,u	1	0	0	4	2	1	-1	35
6223.6	1.179	2,g	1	0	0	0	0	4	2	95	6506.5	1.194	1,g	0	0	0	6	4	4	-4	30
6259.9	1.184	0,g	0	1	0	0	0	6	0	94			0,g	0	0	0	6	2	4	-2	27
6262.5	1.196	1,g	0	0	0	1	1	8	0	45	6509.6	1.192	0,g	0	0	0	8	0	2	0	36
		1,g	0	0	0	1	-1	8	2	48	6514.1	1.166	0,g	0	0	2	0	0	0	0	44
6267.5	1.195	1,u	0	0	0	2	0	7	1	38			0,g	1	1	0	2	0	0	0	35
6273.4	1.184	2,g	0	1	0	0	0	6	2	95	6514.7	1.167	2,g	1	1	0	2	2	0	0	89
6316.0	1.180	1,g	0	1	0	7	1	0	0	90	6515.1	1.192	2,g	0	0	0	8	2	2	0	29
6327.8	1.170	0,g	0	2	0	4	0	0	0	98			2,g	0	0	0	8	0	2	2	29
6330.7	1.170	2,g	0	2	0	4	2	0	0	98	6523.5	1.180	1,u	1	0	0	4	0	1	1	37
6335.4	1.196	1,g	0	0	0	1	1	8	0	43			1,u	0	1	0	4	4	3	-3	29
		1,g	0	0	0	1	-1	8	2	42	6523.5	1.170	0,g	1	1	0	2	0	0	0	33
6348.0	1.179	1,u	0	0	1	5	1	0	0	52			0,g	0	2	0	2	2	2	-2	44
		1,u	0	1	0	6	2	1	-1	25	6524.8	1.179	1,u	1	0	0	4	2	1	-1	31
6377.9	1.179	1,u	0	0	1	5	1	0	0	45			1,u	0	0	1	3	3	2	-2	31
		1,u	0	1	0	6	2	1	-1	37	6527.8	1.194	0,g	0	0	0	6	-4	4	4	71
6385.4	1.197	1,u	0	0	0	0	0	9	1	90	6529.5	1.193	0,u	0	0	0	7	3	3	-3	48

cont.

$G(v)$	B_v	$k,u/g$	v_1	v_2	v_3	v_4	l_4	v_5	l_5	%
6532.5	1.194	2,g								<25
6532.8	1.171	0,g	0	2	0	2	-2	2	2	85
6535.8	1.193	2,u	0	0	0	7	-1	3	3	41
6538.1	1.181	1,u	0	0	1	3	1	2	0	30
		1,u	0	1	0	4	-2	3	3	33
6541.1	1.171	2,g	0	2	0	2	2	2	0	54
		2,g	0	2	0	2	0	2	2	30
6544.4	1.179	1,u	1	0	0	4	0	1	1	27
		1,u	0	0	1	3	-1	2	2	26
6555.9	1.164	0,u	1	0	1	0	0	0	0	99
6559.8	1.180	1,g								<25
6561.2	1.171	0,g	0	2	0	2	0	2	0	67
6561.4	1.194	0,g	0	0	0	6	4	4	-4	58
6562.0	1.169	0,g	0	1	1	1	1	1	-1	63
		0,g	0	2	0	2	2	2	-2	29
6563.3	1.181	1,u	0	0	1	3	1	2	0	27
6565.7	1.194	0,u	0	0	0	5	5	5	-5	30
		0,u	0	0	0	5	3	5	-3	31
		0,u	0	0	0	5	1	5	-1	28
6568.4	1.194	2,g	0	0	0	6	4	4	-2	26
		2,g	0	0	0	6	-2	4	4	38
6569.0	1.169	0,g	0	1	1	1	-1	1	1	85
6569.0	1.171	2,g	0	2	0	2	2	2	0	42
		2,g	0	2	0	2	0	2	2	47
6578.5	1.169	2,g	0	1	1	1	1	1	1	76
6579.8	1.195	0,u	0	0	0	5	-5	5	5	72
6581.1	1.179	1,g	1	0	0	3	3	2	-2	50
6581.4	1.194	2,u	0	0	0	5	5	5	-3	33
		2,u	0	0	0	5	3	5	-1	25
6585.8	1.193	0,u	0	0	0	7	-1	3	1	42
6589.5	1.183	1,u	0	1	0	4	0	3	1	47
6592.3	1.193	2,u	0	0	0	7	1	3	1	28
6596.3	1.181	1,g								<25
6602.6	1.195	0,u	0	0	0	5	5	5	-5	63
6603.7	1.181	1,g	0	1	0	3	-3	4	4	30
6604.0	1.160	1,u	0	3	0	0	0	1	1	100
6606.9	1.194	0,g	0	0	0	6	-2	4	2	45
6610.9	1.195	2,u	0	0	0	5	5	5	-3	32
		2,u	0	0	0	5	-3	5	5	28
6614.3	1.194	2,g	0	0	0	6	-2	4	4	30
6620.7	1.180	1,g	1	0	0	3	-1	2	2	34
6622.7	1.168	0,u	1	1	0	1	1	1	-1	87
6623.1	1.180	1,g	1	0	0	3	1	2	0	30
6636.8	1.167	0,u	1	1	0	1	-1	1	1	98
6636.9	1.182	1,g								<25
6638.3	1.195	0,u	0	0	0	5	-5	5	5	26
		0,u	0	0	0	5	-3	5	3	44
6638.6	1.180	1,g	0	0	1	2	-2	3	3	29
6641.6	1.168	2,u	1	1	0	1	1	1	1	88
6644.9	1.195	0,g	0	0	0	4	4	6	-4	26
		0,g	0	0	0	4	2	6	-2	40
6647.1	1.195	2,u	0	0	0	5	-3	5	5	41
6654.8	1.172	0,u	0	2	0	1	1	3	-1	82
6660.1	1.181	2,g	0	0	1	2	0	3	1	36
6661.3	1.195	2,g								<25
6663.8	1.180	1,u	1	0	0	2	0	3	1	28
6665.8	1.194	0,g								<25

$G(v)$	B_v	$k,u/g$	v_1	v_2	v_3	v_4	l_4	v_5	l_5	%
6669.4	1.172	2,u	0	2	0	1	1	3	1	33
		2,u	0	2	0	1	-1	3	3	50
6673.2	1.194	2,g								<25
6675.7	1.195	0,g	0	0	0	4	-4	6	4	69
6682.6	1.172	0,u	0	2	0	1	-1	3	1	98
6685.9	1.195	0,u	0	0	0	5	3	5	-3	43
6687.6	1.183	1,g	0	1	0	3	1	4	0	40
		1,g	0	1	0	3	-1	4	2	35
6690.3	1.179	1,u	1	0	0	2	2	3	-1	35
		1,u	1	0	0	2	-2	3	3	55
6690.3	1.195	2,g	0	0	0	4	-4	6	6	49
6690.5	1.169	0,u	0	1	1	0	0	2	0	88
6693.0	1.172	2,u	0	2	0	1	1	3	1	62
		2,u	0	2	0	1	-1	3	3	30
6694.5	1.195	2,u								<25
6701.7	1.182	1,u	0	1	0	2	0	5	1	25
6705.7	1.170	2,u	0	1	1	0	0	2	2	80
6709.2	1.163	0,g	2	0	0	0	0	0	0	92
6713.6	1.195	0,g	0	0	0	4	4	6	-4	62
6719.4	1.180	1,u	1	0	0	2	0	3	1	30
6724.7	1.195	2,g	0	0	0	4	4	6	-2	28
		2,g	0	0	0	4	-4	6	6	33
6730.4	1.181	1,u	1	0	0	2	0	3	1	31
		1,u	0	0	1	1	1	4	0	31
6733.7	1.183	1,u	0	1	0	2	2	5	-1	26
		1,u	0	1	0	2	-2	5	3	35
6734.7	1.196	0,u	0	0	0	3	3	7	-3	34
		0,u	0	0	0	3	1	7	-1	56
6746.9	1.195	0,u	0	0	0	3	-1	7	1	26
		0,u	0	0	0	5	-1	5	1	32
6750.1	1.196	2,u	0	0	0	3	1	7	1	34
6755.0	1.182	1,u	0	0	1	1	1	4	0	30
		1,u	0	0	1	1	-1	4	2	34
6755.3	1.195	2,u								<25
6759.2	1.168	0,g	1	1	0	0	0	2	0	94
6763.7	1.195	0,g	0	0	0	4	-2	6	2	43
6773.2	1.168	2,g	1	1	0	0	0	2	2	96
6773.4	1.195	2,g	0	0	0	4	-2	6	4	29
6781.1	1.181	1,g	1	0	0	1	1	4	0	41
		1,g	1	0	0	1	-1	4	2	39
6782.6	1.196	0,u	0	0	0	3	-3	7	3	77
6784.5	1.184	1,u	0	1	0	2	0	5	1	50
6795.7	1.196	2,u	0	0	0	3	3	7	-1	31
		2,u	0	0	0	3	-3	7	5	46
6803.4	1.173	0,u	0	2	0	0	0	4	0	95
6816.6	1.173	2,g	0	2	0	0	0	4	0	96
6819.6	1.180	1,g	1	0	0	1	1	4	0	36
		1,g	1	0	0	1	-1	4	2	56
6825.1	1.184	1,g	0	1	0	1	1	6	0	31
		1,g	0	1	0	1	-1	6	2	37
6826.6	1.195	0,g								<25
6835.8	1.197	0,g	0	0	0	2	2	8	-2	58
		0,g	0	0	0	2	0	8	0	29
6835.8	1.195	2,g								<25
6836.7	1.196	0,u	0	0	0	3	3	7	-3	51
6844.5	1.183	1,g	0	0	1	0	0	5	1	74
6847.9	1.196	2,u	0	0	0	3	3	7	-1	26

$G(v)$	B_v	$k,u/g$	v_1	v_2	v_3	v_4	l_4	v_5	l_5	%
		2,u	0	0	0	3	-3	7	5	25
6850.3	1.197	2,g	0	0	0	2	2	8	0	26
		2,g	0	0	0	2	0	8	2	35
		2,g	0	0	0	2	-2	8	4	29
6877.3	1.185	1,g	0	1	0	1	1	6	0	47
		1,g	0	1	0	1	-1	6	2	46
6901.8	1.197	0,g	0	0	0	2	-2	8	2	83
6903.2	1.196	0,u	0	0	0	3	-1	7	1	39
6910.9	1.181	1,u	1	0	0	0	0	5	1	93
6913.2	1.196	2,u								<25
6914.2	1.197	2,g	0	0	0	2	2	8	0	38
		2,g	0	0	0	2	-2	8	4	46
6946.0	1.171	1,g	0	2	0	5	1	0	0	96
6947.7	1.198	0,u	0	0	0	1	1	9	-1	89
6958.7	1.181	0,g	0	1	0	8	0	0	0	82
6961.6	1.198	2,u	0	0	0	1	1	9	1	41
		2,u	0	0	0	1	-1	9	3	48
6961.8	1.181	2,g	0	1	0	8	2	0	0	83
6962.3	1.186	1,u	0	1	0	0	0	7	1	92
6973.1	1.197	0,g	0	0	0	2	2	8	-2	29
		0,g	0	0	0	2	0	8	0	29
6977.6	1.180	0,u	0	0	1	6	0	0	0	51
		0,u	0	1	0	7	1	1	-1	37
6980.3	1.180	2,u	0	0	1	6	2	0	0	51
		2,u	0	1	0	7	3	1	-1	27
6984.1	1.197	2,g	0	0	0	2	0	8	2	31
7012.6	1.181	0,u	0	0	1	6	0	0	0	44
		0,u	0	1	0	7	1	1	-1	52
7013.7	1.181	2,u	0	0	1	6	2	0	0	44
		2,u	0	1	0	7	3	1	-1	42
7016.7	1.171	1,u	0	2	0	4	2	1	-1	51
		1,u	0	2	0	4	0	1	1	26
7021.8	1.181	0,g	0	0	1	5	1	1	-1	52
7023.6	1.181	2,g	0	0	1	5	3	1	-1	37
7033.5	1.161	0,g	0	3	0	2	0	0	0	100
7034.4	1.198	0,u	0	0	0	1	-1	9	1	76
7035.0	1.182	0,u	0	1	0	7	-1	1	1	84
7036.4	1.161	2,g	0	3	0	2	2	0	0	100
7040.5	1.182	2,u	0	1	0	7	1	1	1	65
7046.2	1.198	2,u	0	0	0	1	1	9	1	40
		2,u	0	0	0	1	-1	9	3	38
7046.7	1.193	1,g	0	0	0	9	1	2	0	37
		1,g	0	0	0	11	1	0	0	42
7048.4	1.193	1,u	0	0	0	10	2	1	-1	38
		1,u	0	0	0	10	0	1	1	25
7052.2	1.171	1,u	0	2	0	4	2	1	-1	47
		1,u	0	2	0	4	0	1	1	30
7055.5	1.181	2,g	1	0	0	6	2	0	0	29
		2,g	0	1	0	6	4	2	-2	40
7055.8	1.170	1,u	0	1	1	3	1	0	0	57
		1,u	0	2	0	4	0	1	1	41
7056.5	1.181	0,g	1	0	0	6	0	0	0	33
		0,g	0	1	0	6	2	2	-2	29
7062.7	1.181	0,g	0	0	1	5	-1	1	1	47
		0,g	0	1	0	6	-2	2	2	43
7068.8	1.181	2,g	0	0	1	5	3	1	-1	31
		2,g	0	1	0	6	0	2	2	25
7071.6	1.199	0,g	0	0	0	0	0	10	0	84
7073.6	1.193	1,g	0	0	0	9	3	2	-2	30
		1,g	0	0	0	11	1	0	0	38
7080.1	1.180	2,g	1	0	0	6	2	0	0	48
7080.4	1.181	0,u	0	0	1	4	2	2	-2	29
7081.9	1.181	2,u	0	0	1	4	4	2	-2	32
7082.2	1.179	0,g	1	0	0	6	0	0	0	48
		0,g	0	0	1	5	1	1	-1	33
7084.8	1.199	2,g	0	0	0	0	0	10	2	85
7092.6	1.168	1,g	0	0	2	1	1	0	0	55
7094.1	1.182	0,g	0	0	1	5	-1	1	1	47
		0,g	0	1	0	6	-2	2	2	48
7095.9	1.193	1,u	0	0	0	10	0	1	1	42
7097.8	1.194	1,u	0	0	0	8	4	3	-3	28
		1,u	0	0	0	10	2	1	-1	26
7100.6	1.181	2,g	0	0	1	5	1	1	1	43
7104.6	1.194	1,g	0	0	0	9	3	2	-2	28
		1,g	0	0	0	9	-1	2	2	39
7108.0	1.170	1,g	0	0	2	1	1	0	0	31
		1,g	0	2	0	3	3	2	-2	28
7109.3	1.182	2,u	0	1	0	5	5	3	-3	40
7111.5	1.182	0,u	0	0	1	4	-2	2	2	50
		0,u	0	1	0	5	-3	3	3	30
7117.1	1.182	1,u	1	0	0	5	1	1	-1	38
		0,u	0	1	0	5	1	3	-1	32
7119.2	1.169	1,g	1	1	0	3	1	0	0	63
		1,g	0	2	0	3	3	2	-2	26
7120.8	1.182	2,u								<25
7121.7	1.183	0,u	0	1	0	6	2	2	-2	34
		0,g	0	1	0	6	0	2	0	46
7128.4	1.183	2,g	0	1	0	6	2	2	0	36
		2,g	0	1	0	6	0	2	2	41
7129.0	1.194	1,u	0	0	0	8	4	3	-3	28
		1,u	0	0	0	8	-2	3	3	33
7131.2	1.172	1,g	0	2	0	3	-1	2	2	46
7133.9	1.195	1,g	0	0	0	7	5	4	-4	38
7137.8	1.181	2,u	1	0	0	5	3	1	-1	29
		2,u	0	0	1	4	4	2	-2	33
7142.5	1.165	1,u	1	0	1	1	1	0	0	94
7142.7	1.182	0,u	1	0	0	5	-1	1	1	37
		0,u	0	1	0	5	-3	3	3	41
7145.8	1.181	1,u	1	0	0	5	1	1	-1	36
		0,u	0	0	1	4	2	2	-2	33
7146.2	1.181	2,u	1	0	0	5	1	1	1	26
7153.8	1.182	0,g	0	0	1	3	1	3	-1	26
7154.3	1.182	0,u	0	1	0	4	0	2	0	28
		0,u	0	1	0	5	3	3	-3	30
7155.7	1.170	1,g	0	1	1	2	2	1	-1	68
7160.2	1.162	0,u	0	3	0	1	1	1	-1	95
7160.6	1.195	1,g	0	0	0	7	5	4	-4	27
7162.5	1.182	2,u	0	0	1	4	2	2	0	38
		2,u	0	1	0	5	-1	3	3	34
7162.7	1.194	1,g								<25
7164.6	1.180	0,u	1	0	0	5	-1	1	1	45
		0,u	0	0	1	4	-2	2	2	37
7165.2	1.181	2,g	1	0	0	4	4	2	-2	30
7165.4	1.172	1,g	0	2	0	3	1	2	0	57

cont.

$G(v)$	B_v	$k,u/g$	v_1	v_2	v_3	v_4	l_4	v_5	l_5	%
7169.8	1.181	2,u	1	0	0	5	1	1	1	38
7170.3	1.170	1,g	0	1	1	2	0	1	1	58
		1,g	0	2	0	3	-1	2	2	28
7173.4	1.195	1,u	0	0	0	8	-2	3	3	29
7174.3	1.162	0,u	0	3	0	1	-1	1	1	100
7176.4	1.182	0,g	0	0	1	3	-3	3	3	42
7181.6	1.196	1,u	0	0	0	6	6	5	-5	49
7182.1	1.162	2,u	0	3	0	1	1	1	1	100
7183.4	1.182	0,u	0	0	1	4	0	2	0	31
		0,u	1	0	0	5	3	3	-3	25
7189.4	1.181	2,g	1	0	0	4	4	2	-2	52
7190.1	1.183	0,g	0	1	0	4	4	4	-4	29
7191.7	1.182	2,u	0	0	1	4	0	2	2	35
7198.7	1.195	1,g	0	0	0	7	-3	4	4	33
7203.1	1.196	1,u	0	0	0	6	6	5	-5	28
7205.4	1.183	2,g	0	1	0	4	4	4	-2	34
7206.0	1.183	0,g	0	0	1	3	3	3	-3	37
7206.3	1.182	0,g	1	0	0	4	-2	2	2	48
		0,g	0	1	0	4	-4	4	4	39
7206.9	1.169	1,u	1	1	0	2	2	1	-1	39
7213.1	1.159	0,u	0	2	1	0	0	0	0	95
7213.6	1.184	0,u	0	1	0	5	-1	3	1	66
7219.3	1.168	1,u	0	0	2	0	0	1	1	57
		1,u	1	1	0	2	2	1	-1	29
7219.6	1.182	0,g	0	0	1	3	1	3	-1	31
7219.7	1.182	2,g								<25
7221.3	1.184	2,u	0	1	0	5	1	3	1	42
7229.2	1.170	1,u	0	0	2	0	0	1	1	26
7229.6	1.182	0,g	1	0	0	4	-2	2	2	26
		0,g	0	0	1	3	-3	3	3	50
7232.4	1.196	1,u	0	0	0	6	-4	5	5	33
7234.6	1.182	2,g	0	0	1	3	3	3	-1	30
7234.7	1.182	0,g	0	1	0	4	4	4	-4	26
7235.1	1.195	1,u								<25
7235.7	1.171	1,u	1	1	0	2	0	1	1	56
7240.9	1.182	2,g								<25
7245.0	1.182	0,u	1	0	0	3	1	3	-1	31
7247.8	1.195	1,g								<25
7248.1	1.173	1,u	0	2	0	2	2	3	-1	35
		1,u	0	2	0	2	-2	3	3	45
7248.6	1.183	0,g	0	0	1	3	-1	3	1	35
		0,g	0	1	0	4	-2	4	2	38
7253.1	1.181	0,g	1	0	0	4	0	2	0	32
7253.9	1.197	1,g								<25
7258.5	1.183	2,g	0	0	1	3	1	3	1	32
		2,g	0	1	0	4	-2	4	4	33
7261.7	1.182	2,u								<25
7262.3	1.182	2,g	1	0	0	4	0	2	2	26
7264.9	1.166	1,g	1	0	1	0	0	1	1	91
7269.4	1.181	0,u	1	0	0	3	-3	3	3	69
7272.9	1.196	1,u								<25
7274.7	1.171	1,u	0	1	1	1	1	2	0	48
		1,u	0	1	1	1	-1	2	2	30
7275.6	1.183	0,g	0	0	1	3	-1	3	1	47
		0,g	0	1	0	4	-2	4	2	30
7277.0	1.197	1,g	0	0	0	5	-5	6	6	41
7281.5	1.183	0,u	1	0	0	3	3	3	-3	40
		0,u	0	1	0	3	1	5	-1	28
7281.5	1.173	1,u	0	2	0	2	2	3	-1	30
		1,u	0	2	0	2	0	3	1	50
7282.5	1.158	0,g	1	2	0	0	0	0	0	97
7285.6	1.183	2,g	0	0	1	3	1	3	1	27
7287.1	1.171	1,u	0	1	1	1	1	2	0	30
		1,u	0	1	1	1	-1	2	2	42
7291.1	1.182	2,u	1	0	0	3	3	3	-1	41
7295.9	1.182	0,u	1	0	0	3	3	3	-3	36
		0,u	1	0	0	3	1	3	-1	29
7297.6	1.164	1,g	2	0	0	1	1	0	0	86
7303.2	1.184	0,u	0	0	1	2	-2	4	2	29
		0,u	0	1	0	3	-3	5	3	42
7303.7	1.184	2,u								<25
7305.4	1.197	1,g	0	0	0	5	5	6	-4	29
		1,g	0	0	0	5	-5	6	6	34
7307.7	1.185	0,g	0	1	0	4	2	4	-2	36
		0,g	0	1	0	4	0	4	0	38
7312.1	1.195	1,g								<25
7315.6	1.182	0,u	0	0	1	2	2	4	-2	40
7316.6	1.185	2,g	0	1	0	4	2	4	0	25
		2,g	0	1	0	4	0	4	2	39
7318.7	1.183	2,u	0	1	0	3	-3	5	5	33
7319.1	1.162	0,g	0	3	0	0	0	2	0	97
7324.7	1.196	1,u								<25
7328.1	1.183	0,u	1	0	0	3	-1	3	1	29
		0,u	0	1	0	3	-3	5	3	30
7330.1	1.183	2,u	0	0	1	2	2	4	0	25
7330.5	1.170	1,g	1	1	0	1	1	2	0	43
		1,g	1	1	0	1	-1	2	2	38
7331.9	1.163	2,g	0	3	0	0	0	2	2	100
7333.2	1.197	1,u								<25
7337.7	1.183	2,u	1	0	0	3	1	3	1	28
7342.6	1.184	0,u	0	1	0	3	3	5	-3	36
7342.7	1.182	0,u	1	0	0	3	-1	3	1	51
		0,u	0	0	1	2	-2	4	2	33
7347.6	1.197	1,g								<25
7351.1	1.170	1,g	1	1	0	1	1	2	0	41
		1,g	1	1	0	1	-1	2	2	54
7351.1	1.182	0,g	1	0	0	2	2	4	-2	36
		0,g	1	0	0	2	0	4	0	27
7353.3	1.183	2,u	0	0	1	2	2	4	0	31
7355.8	1.184	2,u								<25
7364.8	1.174	1,g	0	2	0	1	1	4	0	36
		1,g	0	2	0	1	-1	4	2	43
7367.1	1.182	2,g	1	0	0	2	0	4	2	27
7367.5	1.184	0,u	0	0	1	2	0	4	0	34
7373.4	1.197	1,u	0	0	0	4	4	7	-3	28
		1,u	0	0	0	4	-4	7	5	37
7379.1	1.183	2,u	0	0	1	2	0	4	2	34
7387.4	1.182	0,g	1	0	0	2	-2	4	2	86
7391.0	1.196	1,u								<25
7393.3	1.185	0,g	0	1	0	2	2	6	-2	35
		0,g	0	1	0	2	0	6	0	27
7398.1	1.174	1,g	0	2	0	1	1	4	0	55
		1,g	0	2	0	1	-1	4	2	27
7400.6	1.182	2,g	1	0	0	2	2	4	0	26

G(v)	B_v	k,u/g	v_1	v_2	v_3	v_4	l_4	v_5	l_5	%
		2,g	1	0	0	2	-2	4	4	60
7401.4	1.197	1,g								<25
7401.7	1.186	0,u	0	1	0	3	-1	5	1	67
7402.7	1.172	1,g	0	1	1	0	0	3	1	73
7408.5	1.185	2,g	0	1	0	2	0	6	2	26
7411.7	1.186	2,u	0	1	0	3	1	5	1	37
7417.4	1.184	0,g	0	0	1	1	1	5	-1	55
7418.4	1.165	1,u	2	0	0	0	0	1	1	91
7418.6	1.197	1,u	0	0	0	4	4	7	-3	28
		1,u	0	0	0	4	-4	7	5	40
7423.3	1.198	1,g								<25
7430.2	1.183	2,g	0	0	1	1	-1	5	3	35
7431.1	1.182	0,g	1	0	0	2	0	4	0	56
7433.3	1.185	0,g	0	0	1	1	-1	5	1	26
		0,g	0	1	0	2	-2	6	2	62
7441.0	1.183	2,g	1	0	0	2	0	4	2	41
		2,g	0	0	1	1	1	5	1	33
7447.7	1.185	2,g	0	1	0	2	-2	6	4	34
7455.8	1.184	0,u	0	0	1	1	-1	5	1	64
		0,g	0	1	0	2	-2	6	2	30
7468.0	1.171	1,u	1	1	0	0	0	3	1	90
7469.0	1.184	2,g	0	0	1	1	1	5	1	29
		2,g	0	0	1	1	-1	5	3	34
7469.3	1.197	1,g								<25
7470.0	1.183	0,u	1	0	0	1	1	5	-1	77
7475.6	1.198	1,u								<25
7482.3	1.198	1,g	0	0	0	3	3	8	-2	35
		1,g	0	0	0	3	-3	8	4	40
7484.8	1.183	2,u	1	0	0	1	1	5	1	40
		2,u	1	0	0	1	-1	5	3	37
7493.0	1.186	0,g	0	1	0	2	2	6	-2	38
		0,g	0	1	0	2	0	6	0	46
7504.1	1.186	2,g	0	1	0	2	0	6	2	46
7513.3	1.175	1,u	0	2	0	0	0	5	1	92
7516.4	1.186	0,u	0	1	0	1	1	7	-1	68
7519.1	1.182	0,u	1	0	0	1	-1	5	1	93
7523.9	1.199	1,u	0	0	0	2	0	9	1	42
7528.7	1.184	2,u	1	0	0	1	-1	5	3	47
7532.7	1.184	2,u	1	0	0	1	1	5	1	44
		2,u	0	1	0	1	-1	7	3	26
7536.6	1.185	0,u	0	0	1	0	0	6	0	74
7543.4	1.198	1,g								<25
7546.6	1.198	1,u								<25
7550.9	1.185	2,u	0	0	1	0	0	6	2	72
7569.4	1.172	0,g	0	2	0	6	0	0	0	93
7572.5	1.172	2,g	0	2	0	6	2	0	0	94
7578.6	1.187	0,u	0	1	0	1	-1	7	1	90
7590.8	1.187	2,u	0	1	0	1	1	7	1	46
		2,u	0	1	0	1	-1	7	3	45
7600.9	1.183	0,g	1	0	0	0	0	6	0	91
7602.9	1.199	1,u	0	0	0	2	2	9	-1	36
		1,u	0	0	0	2	-2	9	3	40
7605.7	1.182	1,g	0	1	0	9	1	0	0	67
7612.8	1.182	1,u	0	0	1	7	1	0	0	48
7614.8	1.183	2,g	1	0	0	0	0	6	2	91
7617.8	1.198	1,g								<25
7625.1	1.162	1,g	0	3	0	3	1	0	0	99
7625.6	1.172	0,u	0	1	1	4	0	0	0	25
		0,u	0	2	0	5	1	1	-1	69
7627.8	1.172	2,u	0	2	0	5	3	1	-1	55
7637.2	1.200	1,g	0	0	0	1	1	10	0	37
		1,g	0	0	0	1	-1	10	2	45
7649.4	1.182	1,g	0	0	1	6	2	1	-1	29
7652.9	1.182	1,u	0	0	1	7	1	0	0	45
		1,u	0	1	0	8	2	1	-1	31
7653.8	1.188	0,g	0	1	0	0	0	8	0	90
7666.2	1.171	0,u	0	1	1	4	0	0	0	68
		0,u	0	2	0	5	1	1	-1	27
7666.9	1.171	2,u	0	1	1	4	2	0	0	58
		2,u	0	2	0	5	3	1	-1	35
7667.3	1.188	2,g	0	1	0	0	0	8	2	90
7667.8	1.173	0,u	0	2	0	5	-1	1	1	94
7674.7	1.173	2,u	0	2	0	5	1	1	1	75
7676.3	1.183	1,u	0	1	0	8	2	1	-1	27
		1,u	0	1	0	8	0	1	1	47
7681.3	1.199	1,u	0	0	0	0	0	11	1	26
7686.3	1.169	0,g	0	0	2	2	0	0	0	55
7687.1	1.183	1,g	0	1	0	7	3	2	-2	32
7690.0	1.169	2,g	0	0	2	2	2	0	0	56
7692.9	1.183	1,g	0	0	1	6	2	1	-1	28
		1,g	0	1	0	7	-1	2	2	28
7696.6	1.183	1,u								<25
7697.8	1.195	0,u	0	0	0	9	1	3	-1	45
		0,u	0	0	0	11	1	1	-1	38
7697.8	1.195	0,g	0	0	0	10	0	2	0	37
7700.7	1.195	2,u	0	0	0	9	3	3	-1	30
		2,u	0	0	0	11	3	1	-1	27
7700.8	1.195	2,g	0	0	0	10	2	2	0	35
7706.9	1.172	0,g	0	0	2	2	0	0	0	27
		0,g	0	2	0	4	2	2	-2	32
7707.0	1.172	2,g	0	0	2	2	2	0	0	26
		2,g	0	2	0	4	4	2	-2	45
7718.0	1.181	1,g	1	0	0	7	1	0	0	52
7725.8	1.170	2,g	1	1	0	4	2	0	0	61
7727.3	1.170	0,g	1	1	0	4	0	0	0	72
7729.2	1.183	1,g	0	0	1	6	0	1	1	36
7730.9	1.183	1,u	0	1	0	6	4	3	-3	26
7731.2	1.173	0,g	0	2	0	4	-2	2	2	76
7733.7	1.167	0,u	1	0	1	2	0	0	0	82
7733.9	1.194	2,g	0	0	0	12	2	0	0	51
7734.0	1.194	0,g	0	0	0	12	0	0	0	57
7735.5	1.163	1,u	0	3	0	2	2	1	-1	70
7737.4	1.167	2,u	1	0	1	2	2	0	0	89
7738.2	1.183	1,u								<25
7738.5	1.200	1,g	0	0	0	1	1	10	0	33
		1,g	0	0	0	1	-1	10	2	32
7739.2	1.173	2,g	0	2	0	4	2	2	0	33
		2,g	0	2	0	4	0	2	2	33
7741.1	1.195	2,u	0	0	0	9	5	3	-3	44
7744.5	1.195	0,u	0	0	0	9	3	3	-3	29
		0,u	0	0	0	11	1	1	-1	41
7748.1	1.195	0,u	0	0	0	9	-1	3	1	35
		0,u	0	0	0	11	-1	1	1	33

cont.

G(v)	B_v	k,u/g	v_1	v_2	v_3	v_4	l_4	v_5	l_5	%
7750.0	1.196	0,g	0	0	0	8	-2	4	2	37
		0,g	0	0	0	10	-2	2	2	53
7753.9	1.195	2,u	0	0	0	9	1	3	1	32
		2,u	0	0	0	11	3	1	-1	26
7756.7	1.171	2,g	0	1	1	3	3	1	-1	70
7757.2	1.195	2,g	0	0	0	10	0	2	2	30
7758.0	1.184	1,g								<25
7758.3	1.163	1,u	0	3	0	2	2	1	-1	26
		1,u	0	3	0	2	0	1	1	73
7759.3	1.184	1,g								<25
7759.7	1.171	0,g	0	1	1	3	1	1	-1	67
7763.0	1.201	1,u	0	0	0	0	0	11	1	72
7765.9	1.196	2,g	0	0	0	8	6	4	-4	53
7766.4	1.183	1,u	1	0	0	6	0	1	1	29
		1,u	0	1	0	6	4	3	-3	29
7770.5	1.182	1,u	1	0	0	6	2	1	-1	41
7771.9	1.172	0,g	0	1	1	3	-1	1	1	76
7772.5	1.196	0,g	0	0	0	8	4	4	-4	27
		0,g	0	0	0	10	2	2	-2	26
7773.4	1.195	0,u	0	0	0	9	-3	3	3	45
		0,u	0	0	0	11	-1	1	1	29
7773.8	1.174	0,g	0	2	0	4	2	2	-2	37
		0,g	0	2	0	4	0	2	0	56
7776.7	1.172	2,g	0	1	1	3	1	1	1	45
		2,g	0	2	0	4	2	2	0	38
7779.6	1.195	2,u	0	0	0	11	1	1	1	33
7779.8	1.184	1,u	0	0	1	5	1	2	0	27
		1,u	1	0	0	6	-2	3	3	27
7783.5	1.173	2,g	0	1	1	3	1	1	1	26
		2,g	0	2	0	4	0	2	2	54
7784.7	1.184	1,g								<25
7786.4	1.171	0,u								<25
7791.3	1.160	1,u	0	2	1	1	1	0	0	93
7794.9	1.182	1,u	1	0	0	6	0	1	1	40
7795.9	1.196	0,g	0	0	0	8	-4	4	4	58
7799.4	1.184	1,g	1	0	0	5	3	2	-2	28
7800.5	1.197	2,u	0	0	0	7	7	5	-5	70
7800.8	1.171	2,u	1	1	0	3	3	1	-1	38
7801.5	1.170	0,u	0	0	2	1	-1	1	1	56
7802.4	1.196	2,g								<25
7804.5	1.170	0,u	0	0	2	1	1	1	-1	61
7808.0	1.152	0,g	0	4	0	0	0	0	0	100
7809.6	1.197	0,u	0	0	0	7	1	5	-1	26
7813.0	1.184	1,u								<25
7813.3	1.196	0,g	0	0	0	8	0	4	0	26
7814.7	1.169	2,u	0	0	2	1	1	1	1	51
		2,u	1	1	0	3	3	1	-1	33
7816.3	1.172	0,u	1	1	0	3	1	1	-1	59
7817.0	1.172	0,u	0	0	2	1	-1	1	1	30
		0,u	0	2	0	3	-3	3	3	51
7817.3	1.196	0,u								<25
7819.1	1.184	1,g								<25
7819.4	1.196	2,g								<25
7822.5	1.196	2,u	0	0	0	9	-1	3	3	29
7824.0	1.184	1,g	0	0	1	4	4	3	-3	
7826.1	1.172	2,u								<25
7829.4	1.183	1,g	1	0	0	5	3	2	-2	37
7829.8	1.197	0,u	0	0	0	7	-5	5	5	61
7834.1	1.171	0,u	1	1	0	3	-1	1	1	80
7835.0	1.168	0,g	1	0	1	1	1	1	-1	85
7835.5	1.197	2,u								<25
7836.9	1.171	2,u	1	1	0	3	1	1	1	59
7837.9	1.196	0,g	0	0	0	8	4	4	-4	40
7840.4	1.184	1,u								<25
7841.3	1.174	0,u	0	2	0	3	3	3	-3	36
		0,u	0	2	0	3	1	3	-1	42
7844.0	1.196	2,g	0	0	0	8	-2	4	4	27
7846.1	1.185	1,u	0	1	0	6	0	3	1	33
7850.1	1.183	1,g	1	0	0	5	-1	2	2	29
7852.4	1.174	2,u	0	2	0	3	-1	3	3	44
7852.4	1.168	0,g	1	0	1	1	-1	1	1	96
7853.4	1.168	2,g	1	0	1	1	1	1	1	82
7855.1	1.159	1,g	1	2	0	1	1	0	0	91
7856.0	1.184	1,g								<25
7858.1	1.172	0,u	0	1	1	2	2	2	-2	47
7861.1	1.198	0,g	0	0	0	6	6	6	-6	27
		0,g	0	0	0	6	4	6	-4	26
7862.7	1.197	0,u	0	0	0	7	5	5	-5	53
7865.4	1.183	1,u	1	0	0	4	4	3	-3	39
7866.4	1.172	0,u	0	1	1	2	-2	2	2	74
7869.7	1.197	2,u	0	0	0	7	-3	5	5	27
7870.4	1.185	1,g								<25
7872.9	1.172	2,u	0	1	1	2	2	2	0	64
7875.0	1.198	0,g	0	0	0	6	-6	6	6	65
7876.5	1.198	2,g	0	0	0	6	6	6	-4	26
7877.0	1.163	1,g	0	3	0	1	1	2	0	37
		1,g	0	3	0	1	-1	2	2	52
7878.9	1.182	1,g	1	0	0	5	1	2	0	28
7881.6	1.185	1,u								<25
7882.0	1.172	0,u	0	1	1	2	2	2	-2	26
		0,u	0	1	1	2	0	2	0	54
7882.7	1.196	0,u	0	0	0	7	-1	5	1	35
		0,u	0	0	0	11	-1	1	1	25
7883.8	1.175	0,u	0	2	0	3	-1	3	1	83
7888.0	1.197	0,g	0	0	0	6	-2	6	2	29
7889.0	1.174	2,u	0	1	1	2	0	2	2	33
		2,u	0	2	0	3	1	3	1	37
7889.2	1.184	1,u								<25
7889.2	1.196	2,u								<25
7890.3	1.166	0,g	2	0	0	2	0	0	0	87
7892.5	1.166	2,g	2	0	0	2	2	0	0	84
7894.4	1.197	2,g								<25
7896.0	1.174	2,u	0	1	1	2	0	2	2	30
		2,u	0	2	0	3	-1	3	3	36
7896.2	1.164	1,g	0	3	0	1	1	2	0	56
		1,g	0	3	0	1	-1	2	2	42
7897.2	1.198	0,g	0	0	0	6	6	6	-6	62
7899.6	1.172	0,g	1	1	0	2	2	2	-2	33
7901.3	1.184	1,g	0	0	1	4	0	3	1	27
7906.1	1.198	2,g	0	0	0	6	6	6	-4	28
7907.9	1.197	0,u								<25
7909.6	1.185	1,u	1	0	0	4	-2	3	3	27
		1,u	0	1	0	4	4	5	-3	25
7912.8	1.184	1,g	0	0	1	3	3	4	-2	31

G(v)	B_v	k,u/g	v_1	v_2	v_3	v_4	l_4	v_5	l_5	%	G(v)	B_v	k,u/g	v_1	v_2	v_3	v_4	l_4	v_5	l_5	%
7914.8	1.197	2,u	0	0	0	7	-3	5	5	27	8010.8	1.200	0,g	0	0	0	4	4	8	-4	27
7917.0	1.172	2,g	1	1	0	2	2	2	0	34			0,g	0	0	0	4	2	8	-2	33
7919.2	1.171	0,g	1	1	0	2	-2	2	2	88	8012.2	1.199	2,u	0	0	0	5	-5	7	7	34
7920.1	1.184	1,u								<25	8013.6	1.184	1,g								<25
7921.5	1.170	0,g	0	0	2	0	0	2	0	67	8013.7	1.167	2,u	2	0	0	1	1	1	1	82
7930.6	1.174	0,g	0	2	0	2	2	4	-2	34	8025.7	1.200	2,g								<25
		0,g	0	2	0	2	0	4	0	25	8026.0	1.172	0,u	1	1	0	1	1	3	-1	74
7930.6	1.199	0,u	0	0	0	5	1	7	-1	30	8028.4	1.187	1,u	0	1	0	4	0	5	1	30
7932.6	1.161	1,g	0	2	1	0	0	1	1	93	8033.1	1.198	0,u	0	0	0	5	-1	7	1	35
7933.1	1.198	0,g	0	0	0	6	-6	6	6	28	8033.9	1.185	1,g	0	0	1	2	-2	5	3	27
		0,g	0	0	0	6	-4	6	4	28	8036.5	1.165	1,u	0	3	0	0	0	3	1	95
7933.9	1.184	1,u								<25	8036.8	1.198	0,g	0	0	0	4	-2	8	2	35
7935.1	1.184	1,g								<25	8041.0	1.198	2,u								<25
7935.2	1.170	2,g	0	0	2	0	0	2	2	58	8041.4	1.172	2,u	1	1	0	1	1	3	1	41
7937.0	1.186	1,g	0	1	0	5	1	4	0	29			2,u	1	1	0	1	-1	3	3	34
7941.4	1.198	2,g	0	0	0	6	-4	6	6	36	8042.7	1.185	1,u	1	0	0	2	0	5	1	26
7944.5	1.171	0,g	1	1	0	2	2	2	-2	28	8046.0	1.198	2,g								<25
		0,g	1	1	0	2	0	2	0	60	8050.4	1.185	1,g								<25
7945.1	1.173	2,g								<25	8051.7	1.199	0,u								<25
7945.8	1.185	1,u								<25	8053.9	1.185	1,g	1	0	0	3	1	4	0	26
7947.1	1.199	2,u								<25	8056.1	1.172	0,u	1	1	0	1	-1	3	1	93
7951.9	1.172	2,g	1	1	0	2	0	2	2	61	8060.9	1.199	2,u								<25
7953.4	1.175	0,g	0	2	0	2	-2	4	2	79	8063.2	1.199	0,g	0	0	0	4	-4	8	4	64
7956.8	1.197	0,g								<25	8063.6	1.176	0,u	0	2	0	1	1	5	-1	74
7961.7	1.185	1,u								<25	8067.5	1.172	2,u	1	1	0	1	1	3	1	33
7961.9	1.168	0,u	1	0	1	0	0	2	0	82			2,u	1	1	0	1	-1	3	3	59
7962.2	1.198	0,u								<25	8076.1	1.200	2,g	0	0	0	4	-4	8	6	39
7964.0	1.199	0,u	0	0	0	5	-5	7	5	58	8077.9	1.186	1,g	0	0	1	2	0	5	1	31
7964.0	1.197	2,g								<25	8077.9	1.176	2,u	0	2	0	1	1	5	1	30
7966.1	1.183	1,u	1	0	0	4	0	3	1	31			2,u	0	2	0	1	-1	5	3	44
7966.5	1.175	2,g	0	2	0	2	2	4	0	29	8087.3	1.185	1,u	1	0	0	2	-2	5	3	49
		2,g	0	2	0	2	-2	4	4	42	8088.7	1.186	1,u	1	0	0	2	2	5	-1	40
7968.3	1.183	1,g	1	0	0	3	3	4	-2	28	8099.0	1.200	0,u	0	0	0	3	1	9	-1	62
		1,g	1	0	0	3	-3	4	4	36	8101.7	1.174	0,u	0	1	1	0	0	4	0	87
7969.2	1.198	2,u								<25	8105.6	1.177	0,u	0	2	0	1	-1	5	1	94
7975.0	1.173	0,g	0	1	1	1	1	3	-1	76	8107.9	1.199	0,g	0	0	0	4	0	8	0	26
7976.6	1.168	2,u	1	0	1	0	0	2	2	85	8113.2	1.199	0,u	0	0	0	3	3	9	-3	39
7977.5	1.186	1,g								<25	8113.3	1.175	2,u	0	1	1	0	0	4	2	54
7979.5	1.199	2,u	0	0	0	5	-5	7	7	42			2,u	0	2	0	1	1	5	1	39
7979.9	1.198	0,g	0	0	0	6	4	6	-4	35	8113.6	1.186	1,u	0	0	1	1	1	6	0	27
7988.4	1.198	2,g								<25			1,u	0	0	1	1	-1	6	2	36
7990.2	1.173	2,g	0	1	1	1	1	3	1	51	8113.7	1.200	2,u	0	0	0	3	-1	9	3	37
		2,g	0	1	1	1	-1	3	3	26	8115.6	1.168	0,g	2	0	0	0	0	2	0	89
7990.6	1.185	1,u								<25	8117.5	1.199	2,g								<25
7993.0	1.173	0,g	0	1	1	1	-1	3	1	84	8118.4	1.188	1,g	0	1	0	3	1	6	0	29
7994.7	1.167	0,u	2	0	0	1	1	1	-1	79			1,g	0	1	0	3	-1	6	2	27
7995.2	1.176	0,g	0	2	0	2	2	4	-2	40	8118.8	1.199	0,g	0	0	0	4	4	8	-4	38
		0,g	0	2	0	2	0	4	0	54	8119.9	1.175	2,u	0	1	1	0	0	4	2	32
7997.8	1.184	1,g	1	0	0	3	-3	4	4	31			2,u	0	2	0	1	-1	5	3	44
7998.5	1.160	1,u	1	2	0	0	0	1	1	94	8122.6	1.199	2,u	0	0	0	3	3	9	-1	25
8001.3	1.199	0,u	0	0	0	5	5	7	-5	60	8128.8	1.199	2,g								<25
8001.6	1.167	0,u	2	0	0	1	-1	1	1	90	8129.9	1.168	2,g	2	0	0	0	0	2	2	90
8002.1	1.175	2,g	0	2	0	2	2	4	0	29	8135.4	1.186	1,u								<25
		2,g	0	2	0	2	0	4	2	30	8139.3	1.185	1,u	1	0	0	2	0	5	1	39
8006.8	1.186	1,g	0	1	0	3	-3	6	4	29	8159.9	1.186	1,u	0	0	1	1	1	6	0	30
8009.5	1.174	2,g	0	1	1	1	-1	3	3	25			1,u	0	0	1	1	-1	6	2	32
		2,g	0	2	0	2	0	4	2	27	8162.2	1.185	1,g	1	0	0	1	1	6	0	37

cont.

G(v)	Bv	k,u/g	v1	v2	v3	v4	l4	v5	l5	%	G(v)	Bv	k,u/g	v1	v2	v3	v4	l4	v5	l5	%
		1,g	1	0	0	1	-1	6	2	37	8319.6	1.184	0,u	0	0	1	6	2	2	-2	26
8165.3	1.173	0,g	1	1	0	0	0	4	0	86			0,u	0	1	0	9	1	1	-1	26
8171.4	1.200	0,u	0	0	0	3	-3	9	3	48	8320.4	1.164	2,u	0	3	0	3	3	1	-1	78
8179.2	1.173	2,g	1	1	0	0	0	4	2	87	8320.4	1.184	2,u	0	0	1	6	4	2	-2	26
8180.5	1.200	0,g	0	0	0	2	-2	10	2	31	8322.2	1.185	2,u	0	1	0	9	1	1	1	44
8184.1	1.200	0,u	0	0	0	3	-3	9	5	32	8322.5	1.184	0,g	0	0	1	7	-1	1	1	41
8188.5	1.199	0,u	0	0	0	7	-1	5	1	27			0,g	0	1	0	8	-2	2	2	34
8190.4	1.200	2,g								<25	8322.7	1.184	2,g	0	1	0	8	4	2	-2	35
8197.1	1.199	2,u								<25	8323.2	1.184	0,g								<25
8200.4	1.174	1,g	0	2	0	7	1	0	0	88	8325.2	1.201	0,u	0	0	0	1	-1	11	1	41
8203.1	1.201	0,g	0	0	0	2	2	10	-2	41			0,u	0	0	0	5	-1	7	1	25
		0,g	0	0	0	2	0	10	0	36	8329.5	1.184	2,g	0	0	1	7	3	1	-1	30
8204.2	1.189	1,u	0	1	0	2	0	7	1	41	8330.5	1.202	2,u	0	0	0	1	1	11	1	53
8210.8	1.188	1,g	0	1	0	1	1	8	0	32	8333.3	1.169	1,u	1	0	1	3	1	0	0	66
		1,g	0	1	0	1	-1	8	2	33	8336.9	1.201	2,u	0	0	0	1	-1	11	3	44
8212.4	1.177	0,g	0	2	0	0	0	6	0	89	8339.2	1.173	1,g	1	1	0	5	1	0	0	26
8217.5	1.201	2,g	0	0	0	2	0	10	2	35			1,g	0	2	0	5	3	2	-2	32
		2,g	0	0	0	2	-2	10	4	28	8343.9	1.197	1,g								<25
8220.9	1.185	1,g	1	0	0	1	1	6	0	44	8343.9	1.197	1,u	0	0	0	10	2	3	-1	27
		1,g	1	0	0	1	-1	6	2	43	8344.1	1.173	1,g	1	1	0	5	1	0	0	40
8223.6	1.163	0,g	0	3	0	4	0	0	0	98			1,g	0	2	0	5	-1	2	2	26
8225.7	1.177	2,g	0	2	0	0	0	6	2	90	8347.1	1.164	0,u	0	3	0	3	-1	1	1	99
8226.6	1.163	2,g	0	3	0	4	2	0	0	98	8348.1	1.190	1,u	0	1	0	0	0	9	1	87
8231.7	1.187	1,g	0	0	1	0	0	7	1	72	8353.0	1.164	2,u	0	3	0	3	1	1	1	80
8241.4	1.201	0,u	0	0	0	1	1	11	-1	28	8355.9	1.185	0,u	0	0	1	6	-2	2	2	42
		0,u	0	0	0	3	3	9	-3	30			0,u	0	1	0	7	-3	3	3	27
8241.5	1.173	1,u	0	1	1	5	1	0	0	29	8357.6	1.185	2,u	0	1	0	7	5	3	-3	33
		1,u	0	2	0	6	2	1	-1	38	8359.2	1.183	2,g	1	0	0	8	2	0	0	42
8249.3	1.183	0,u	0	0	1	8	0	0	0	40	8363.0	1.182	0,g	1	0	0	8	0	0	0	58
		0,u	0	1	0	9	1	1	-1	29	8363.4	1.185	0,u	1	0	0	7	1	1	-1	28
8249.5	1.183	0,g	0	1	0	10	0	0	0	38			0,u	0	1	0	7	1	3	-1	32
8252.7	1.183	2,g	0	0	1	8	2	0	0	41	8364.1	1.184	0,g	0	0	1	7	-1	1	1	46
8252.9	1.201	2,u								<25			0,g	0	1	0	8	-2	2	2	39
8253.2	1.183	2,g	0	1	0	10	2	0	0	39	8365.7	1.185	2,u								<25
8259.1	1.200	0,g								<25	8367.8	1.153	1,g	0	4	0	1	1	0	0	100
8268.4	1.200	2,g								<25	8368.6	1.172	1,g	0	1	1	4	2	1	-1	49
8283.0	1.189	1,g	0	1	0	1	1	8	0	43	8369.6	1.185	0,g	0	0	1	5	1	3	-1	25
		1,g	0	1	0	1	-1	8	2	43	8370.7	1.185	2,g	0	0	1	5	5	3	-3	29
8284.0	1.172	1,u	0	1	1	5	1	0	0	54	8373.2	1.183	2,g	0	0	1	7	1	1	1	36
		1,u	0	2	0	6	2	1	-1	32	8376.2	1.161	0,u	0	2	1	2	0	0	0	90
8284.5	1.184	0,g	0	0	1	7	1	1	-1	31	8378.1	1.173	1,u								<25
		0,g	0	1	0	10	0	0	0	34	8379.1	1.161	2,u	0	2	1	2	2	0	0	91
8286.3	1.184	2,g	0	1	0	10	2	0	0	32	8379.5	1.202	0,g	0	0	0	0	0	12	0	45
8287.8	1.170	1,g	0	0	2	3	1	0	0	54	8384.4	1.173	1,g	0	1	1	4	0	1	1	50
8291.8	1.174	1,u	0	2	0	6	0	1	1	64	8390.9	1.202	2,g	0	0	0	0	0	12	2	43
8293.7	1.186	1,u	1	0	0	0	0	7	1	88	8391.7	1.175	1,g	0	2	0	5	1	2	0	42
8296.2	1.183	0,u	0	0	1	8	0	0	0	51			1,g	0	2	0	5	-1	2	2	37
		0,u	0	1	0	9	1	1	-1	29	8392.5	1.197	1,g	0	0	0	9	3	4	-2	25
8297.5	1.201	0,g	0	0	0	2	-2	10	2	64	8392.8	1.197	1,u	0	0	0	10	4	3	-3	27
8297.8	1.183	2,u	0	0	1	8	2	0	0	49	8394.4	1.184	2,u								<25
		2,u	0	1	0	9	3	1	-1	26	8395.0	1.186	0,g	0	1	0	8	0	2	0	28
8309.4	1.201	2,g	0	0	0	2	2	10	0	27	8395.9	1.185	0,u	0	1	0	7	-3	3	3	38
		2,g	0	0	0	2	-2	10	4	37	8396.5	1.171	1,u	0	0	2	2	2	1	-1	49
8312.1	1.173	1,g	0	2	0	5	3	2	-2	28	8398.6	1.197	1,u	0	0	0	12	2	1	-1	29
8316.5	1.185	0,u	0	1	0	9	-1	1	1	57	8399.1	1.197	1,g	0	0	0	9	-1	4	2	29
8318.8	1.202	0,u	0	0	0	1	1	11	-1	68	8399.2	1.185	0,g	0	0	1	5	-3	3	3	43
8319.4	1.164	0,u	0	3	0	3	1	1	-1	90	8399.7	1.172	1,u	0	0	2	2	0	1	1	32

$G(v)$	B_v	$k,u/g$	v_1	v_2	v_3	v_4	l_4	v_5	l_5	%
		1,u	0	2	0	4	4	3	-3	27
8400.2	1.185	2,g								<25
8401.6	1.185	2,g	0	1	0	8	2	2	0	26
8402.2	1.183	0,u	1	0	0	7	1	1	-1	51
8403.1	1.184	2,u	1	0	0	7	3	1	-1	31
8405.1	1.185	0,u	0	0	1	6	2	2	-2	26
8412.5	1.185	2,u	0	0	1	6	2	2	0	27
		2,u	0	1	0	7	-1	3	3	29
8414.0	1.172	1,u	0	0	2	2	0	1	1	26
		1,u	0	2	0	4	4	3	-3	25
8414.6	1.185	2,u	1	0	0	6	4	2	-2	26
8416.2	1.185	0,g								<25
8418.0	1.197	1,g	0	0	0	11	3	2	-2	29
8420.0	1.173	1,u	1	1	0	4	2	1	-1	29
8421.2	1.169	1,g	1	0	1	2	2	1	-1	51
		1,g	1	0	1	2	0	1	1	25
8427.8	1.163	0,g	1	2	0	2	0	0	0	37
		0,g	0	3	0	2	2	2	-2	27
8428.2	1.197	1,u	0	0	0	12	0	1	1	30
8428.9	1.183	0,u	1	0	0	7	-1	1	1	58
8432.3	1.183	2,u	1	0	0	7	1	1	1	44
8434.9	1.185	0,u								<25
8437.5	1.161	2,g	1	2	0	2	2	0	0	84
8437.8	1.201	0,u	0	0	0	1	-1	11	1	51
		0,u	0	0	0	3	-1	9	1	32
8439.4	1.186	0,g	0	0	1	5	3	3	-3	33
8440.1	1.185	2,g	0	1	0	6	6	4	-4	29
8440.2	1.198	1,g	0	0	0	11	-1	2	2	30
8440.8	1.198	1,u	0	0	0	8	6	5	-5	35
8440.9	1.186	0,g	1	0	0	6	-2	2	2	30
		0,g	0	1	0	6	-4	4	4	38
8441.1	1.185	2,g								<25
8441.3	1.173	1,u	1	1	0	4	0	1	1	43
8442.4	1.162	0,g	1	2	0	2	0	0	0	54
		0,g	0	3	0	2	2	2	-2	33
8443.0	1.185	0,u								<25
8445.6	1.169	1,g	1	0	1	2	2	1	-1	37
		1,g	1	0	1	2	0	1	1	48
8447.6	1.185	2,u								<25
8447.7	1.175	1,u	0	2	0	4	-2	3	3	34
8448.7	1.165	0,g	0	3	0	2	-2	2	2	92
8448.8	1.201	0,u	0	0	0	1	1	11	1	27
		2,u	0	0	0	1	-1	11	3	25
8449.8	1.202	0,g	0	0	0	0	0	12	0	50
8451.3	1.185	2,u	0	0	1	6	0	2	2	28
8452.0	1.185	2,g								<25
8453.2	1.184	0,g	1	0	0	6	2	2	-2	27
8454.2	1.173	1,u	0	1	1	3	3	2	-2	49
8457.6	1.164	2,g	0	3	0	2	2	2	0	52
		2,g	0	3	0	2	0	2	2	30
8458.1	1.186	0,u	0	0	1	4	-4	4	4	37
8461.5	1.198	1,u	0	0	0	10	-2	3	3	26
8461.9	1.202	2,g	0	0	0	0	0	12	2	53
8462.8	1.198	1,g								<25
8467.5	1.198	1,u								<25
8467.9	1.159	0,g	0	1	2	0	0	0	0	80
8471.4	1.173	1,u	0	1	1	3	-1	2	2	30

$G(v)$	B_v	$k,u/g$	v_1	v_2	v_3	v_4	l_4	v_5	l_5	%
8473.1	1.184	0,g	1	0	0	6	-2	2	2	48
		0,g	0	0	1	5	-3	3	3	29
8474.3	1.184	2,u	1	0	0	5	5	3	-3	51
8476.3	1.184	2,g	1	0	0	6	0	2	2	28
8476.6	1.165	0,g	0	3	0	2	2	2	-2	33
		0,g	0	3	0	2	0	2	0	64
8477.9	1.186	0,g	0	1	0	6	4	4	-4	29
8478.7	1.186	0,u	0	1	0	7	-1	3	1	40
8478.9	1.199	1,g	0	0	0	7	7	6	-6	47
8479.5	1.186	0,u	0	1	0	5	5	5	-5	33
8481.7	1.173	1,g								<25
8484.8	1.165	2,g	0	3	0	2	2	2	0	37
		2,g	0	3	0	2	0	2	2	61
8485.4	1.185	2,g								<25
8485.7	1.186	2,u	0	1	0	7	1	3	1	26
8487.3	1.186	0,u	0	0	1	4	4	4	-4	33
8489.2	1.174	1,u	0	1	1	3	1	2	0	34
		1,u	0	1	1	3	-1	2	2	32
8489.9	1.197	1,g								<25
8490.1	1.168	1,g	2	0	0	3	1	0	0	72
8490.8	1.186	2,u	0	1	0	5	5	5	-3	29
8491.9	1.186	0,g	0	0	1	5	-1	3	1	28
8496.6	1.176	1,u	0	2	0	4	0	3	1	46
8497.1	1.186	0,u	1	0	0	5	-3	3	3	36
		0,u	0	1	0	5	-5	5	5	38
8499.9	1.186	2,g	0	1	0	6	-2	4	4	26
8500.9	1.199	1,g								<25
8501.7	1.172	1,g	1	1	0	3	3	2	-2	36
8502.1	1.162	0,g	0	2	1	1	1	1	-1	80
8502.4	1.198	1,u								<25
8504.9	1.186	2,u								<25
8506.7	1.183	0,g	1	0	0	6	0	2	0	35
8506.9	1.162	0,g	0	2	1	1	-1	1	1	92
8509.7	1.172	1,g	0	0	2	1	1	2	0	52
8511.6	1.158	0,u	1	1	1	0	0	0	0	95
8512.8	1.184	2,g	1	0	0	6	0	2	2	30
8513.1	1.173	1,g	0	0	2	1	-1	2	2	32
8514.9	1.185	0,u								<25
8515.7	1.162	2,g	0	2	1	1	1	1	1	90
8517.7	1.186	0,g								<25
8520.4	1.173	1,g	1	1	0	3	-1	2	2	27
8523.6	1.186	0,u	0	0	1	4	-4	4	4	50
8524.6	1.186	0,u	0	1	0	5	5	5	-5	32
8526.8	1.186	0,g	0	0	1	5	-1	3	1	36
8527.4	1.175	1,g	0	2	0	3	-3	4	4	36
8527.6	1.186	2,u	0	1	0	5	5	5	-3	26
8528.9	1.199	1,g								<25
8529.1	1.185	2,u	0	0	1	4	4	4	-2	35
8529.3	1.186	0,u	1	0	0	5	-3	3	3	25
		0,u	0	0	1	4	-2	4	2	33
8531.1	1.198	1,u								<25
8532.1	1.154	1,u	0	4	0	0	0	1	1	100
8532.8	1.199	1,g								<25
8533.2	1.170	1,u	1	0	1	1	1	2	0	43
		1,u	1	0	1	1	-1	2	2	33
8533.9	1.186	2,g								<25
8535.9	1.186	2,g								<25

cont.

G(v)	B_v	k,u/g	v₁	v₂	v₃	v₄	l₄	v₅	l₅	%
8541.5	1.186	2,u	0	1	0	5	-3	5	5	29
8543.6	1.200	1,u								<25
8547.5	1.173	1,g	1	1	0	3	1	2	0	30
		1,g	1	1	0	3	-1	2	2	35
8547.9	1.185	0,g	1	0	0	4	-4	4	4	45
8551.3	1.185	0,u								<25
8553.9	1.198	1,u								<25
8554.7	1.176	1,g								<25
8556.1	1.162	0,u	1	2	0	1	1	1	-1	78
8557.9	1.170	1,u	1	0	1	1	1	2	0	34
		1,u	1	0	1	1	-1	2	2	54
8559.2	1.185	2,u	1	0	0	5	-1	3	3	30
8560.8	1.186	0,g	1	0	0	4	4	4	-4	36
8562.4	1.174	1,g	0	1	1	2	2	3	-1	26
		1,g	0	1	1	2	0	3	1	27
8563.4	1.186	0,u								<25
8566.0	1.187	0,g								<25
8566.9	1.186	2,g	1	0	0	4	4	4	-2	28
8568.0	1.200	1,u	0	0	0	6	-6	7	7	40
8569.1	1.199	1,g								<25
8569.7	1.161	0,u	1	2	0	1	-1	1	1	96
8571.9	1.186	2,u								<25
8574.0	1.187	2,g								<25
8574.2	1.162	2,u	1	2	0	1	1	1	1	82
8576.3	1.174	1,g	0	1	1	2	2	3	-1	31
		1,g	0	1	1	2	-2	3	3	40
8576.7	1.186	0,g	1	0	0	4	4	4	-4	41
8580.0	1.187	0,g	1	0	0	4	-4	4	4	35
		0,g	0	1	0	4	-4	6	4	29
8580.3	1.187	0,u								<25
8581.8	1.187	2,g								<25
8581.8	1.168	1,u	2	0	0	2	2	1	-1	67
8583.0	1.165	0,u	0	3	0	1	1	3	-1	80
8588.8	1.186	2,u								<25
8589.4	1.184	0,u	1	0	0	5	-1	3	1	50
8594.8	1.200	1,u	0	0	0	6	6	7	-5	28
		1,u	0	0	0	6	-6	7	7	28
8596.1	1.175	1,g	0	1	1	2	0	3	1	45
8596.6	1.186	2,g	1	0	0	4	4	4	-2	29
		2,g	0	1	0	4	-4	6	6	27
8597.1	1.165	2,u	0	3	0	1	1	3	1	30
		2,u	0	3	0	1	-1	3	3	52
8597.3	1.174	1,u								<25
8597.7	1.185	2,u	1	0	0	5	1	3	1	26
8599.3	1.168	1,u	2	0	0	2	0	1	1	72
8600.1	1.186	0,g	0	0	1	3	3	5	-3	36
8603.6	1.199	1,g								<25
8603.7	1.177	1,g	0	2	0	3	1	4	0	34
		1,g	0	2	0	3	-1	4	2	38
8604.9	1.199	1,u								<25
8609.2	1.187	0,g	0	1	0	4	-4	6	4	27
8609.4	1.166	0,u	0	3	0	1	-1	3	1	96
8612.6	1.187	0,u	0	0	1	4	0	4	0	26
8613.0	1.201	1,g								<25
8614.2	1.186	0,u								<25
8614.3	1.186	2,g								<25
8615.6	1.187	0,g	0	1	0	4	4	6	-4	34
8617.0	1.187	2,g								<25
8619.9	1.166	2,u	0	3	0	1	1	3	1	60
		2,u	0	3	0	1	-1	3	3	37
8622.7	1.186	2,u								<25
8623.8	1.199	1,g								<25
8624.7	1.173	1,u	1	1	0	2	2	3	-1	39
		1,u	1	1	0	2	-2	3	3	40
8625.4	1.172	1,u	0	0	2	0	0	3	1	58
8630.0	1.186	2,u								<25
8630.4	1.185	0,g	1	0	0	4	-2	4	2	44
		0,g	0	0	1	3	-3	5	3	32
8631.4	1.187	2,g	0	1	0	4	-4	6	6	32
8633.3	1.176	1,u								<25
8638.0	1.200	1,u								<25
8640.1	1.163	0,u	0	2	1	0	0	2	0	93
8641.7	1.186	2,g	1	0	0	4	-2	4	4	25
8648.6	1.187	0,g	0	1	0	4	4	6	-4	28
8653.8	1.188	0,u								<25
8654.6	1.163	2,u	0	2	1	0	0	2	2	91
8655.5	1.174	1,u	1	1	0	2	0	3	1	47
8657.3	1.201	1,g	0	0	0	5	-5	8	6	31
8657.7	1.186	0,u	1	0	0	3	-3	5	3	52
8659.1	1.187	2,g								<25
8660.9	1.157	0,g	2	1	0	0	0	0	0	89
8660.9	1.188	0,u	0	1	0	3	1	7	-1	35
8661.6	1.170	1,g	1	0	1	0	0	3	1	75
8662.0	1.177	1,u	0	2	0	2	2	5	-1	30
		1,u	0	2	0	2	-2	5	3	39
8663.1	1.188	2,u								<25
8668.1	1.188	0,g	0	0	1	3	-1	5	1	26
		0,g	0	1	0	4	-2	6	2	26
8671.5	1.186	2,u	1	0	0	3	-3	5	5	37
8673.7	1.185	0,g	1	0	0	4	2	4	-2	26
		0,g	1	0	0	4	0	4	0	34
8676.4	1.188	2,u								<25
8676.5	1.187	2,g								<25
8676.5	1.200	1,u	0	0	0	4	0	9	1	26
8677.7	1.200	1,g								<25
8678.4	1.175	1,u	0	1	1	1	1	4	0	42
		1,u	0	1	1	1	-1	4	2	32
8684.1	1.186	2,g								<25
8688.4	1.187	0,u	1	0	0	3	3	5	-3	26
8689.7	1.201	1,u								<25
8696.8	1.169	1,g	2	0	0	1	1	2	0	30
		1,g	2	0	0	1	-1	2	2	45
8698.0	1.187	0,g	0	0	1	3	-1	5	1	38
8699.0	1.188	0,u	0	1	0	3	-3	7	3	44
8699.4	1.200	1,u								<25
8700.9	1.187	2,u	1	0	0	3	-3	5	5	26
8702.0	1.175	1,u	0	1	1	1	1	4	0	37
		1,u	0	1	1	1	-1	4	2	41
8702.3	1.201	1,g	0	0	0	5	-5	8	6	25
8703.1	1.162	0,g	1	2	0	0	0	2	0	88
8705.7	1.186	0,u	1	0	0	3	3	5	-3	29
		0,u	1	0	0	3	1	5	-1	38
8707.6	1.169	1,g	2	0	0	1	1	2	0	50
		1,g	2	0	0	1	-1	2	2	38

$G(v)$	B_v	$k,u/g$	v_1	v_2	v_3	v_4	l_4	v_5	l_5	%
8709.0	1.187	2,g								<25
8710.6	1.178	1,u	0	2	0	2	0	5	1	48
8712.5	1.188	2,u	0	0	1	2	2	6	0	26
		2,u	0	1	0	3	-3	7	5	25
8716.8	1.162	2,g	1	2	0	0	0	2	2	92
8718.2	1.186	2,u								<25
8722.5	1.187	0,g	1	0	0	2	2	6	-2	31
		0,g	1	0	0	2	0	6	0	26
8724.9	1.175	1,g	1	1	0	1	1	4	0	35
		1,g	1	1	0	1	-1	4	2	33
8728.9	1.188	0,u	0	0	1	2	-2	6	2	50
		0,u	0	1	0	3	-3	7	3	28
8737.8	1.187	2,g	1	0	0	2	0	6	2	25
8741.8	1.188	2,u	0	0	1	2	-2	6	4	34
8742.7	1.189	0,g								<25
8743.3	1.167	0,g	0	3	0	0	0	4	0	90
8747.4	1.201	1,u								<25
8749.6	1.201	1,g								<25
8752.1	1.189	2,g								<25
8752.2	1.189	0,u	0	1	0	3	3	7	-3	33
8756.2	1.167	2,g	0	3	0	0	0	4	2	93
8757.5	1.185	0,u	1	0	0	3	-1	5	1	62
8761.2	1.187	2,u								<25
8762.7	1.201	1,u	0	0	0	4	4	9	-3	29
		1,u	0	0	0	4	-4	9	5	28
8762.9	1.174	1,g	1	1	0	1	1	4	0	26
		1,g	1	1	0	1	-1	4	2	57
8765.6	1.177	1,g	1	1	0	1	1	4	0	28
		1,g	0	2	0	1	1	6	0	26
		1,g	0	2	0	1	-1	6	2	39
8766.0	1.200	1,g								<25
8769.0	1.187	2,u	1	0	0	3	1	5	1	25
8771.5	1.189	0,g	0	1	0	2	2	8	-2	35
8779.0	1.186	0,g	1	0	0	2	-2	6	2	80
8779.5	1.188	0,u	0	0	1	2	0	6	0	28
8783.3	1.202	1,g	0	0	0	3	-1	10	2	27
8785.8	1.189	2,g								<25
8791.8	1.188	2,u	0	0	1	2	0	6	2	28
8792.1	1.186	2,g	1	0	0	2	2	6	0	39
		2,g	1	0	0	2	-2	6	4	36
8797.9	1.188	0,g	0	0	1	1	1	7	-1	65
8803.7	1.176	1,g	0	1	1	0	0	5	1	84
8812.2	1.202	1,u								<25
8812.5	1.188	2,g	0	0	1	1	-1	7	3	40
8814.9	1.179	1,g	0	2	0	1	1	6	0	42
		1,g	0	2	0	1	-1	6	2	47
8815.3	1.170	1,u	2	0	0	0	0	3	1	87
8824.5	1.201	1,g								<25
8827.7	1.190	0,u	0	1	0	3	-1	7	1	45
8830.1	1.189	0,g	0	1	0	2	-2	8	2	56
8831.6	1.165	1,g	0	3	0	5	1	0	0	96
8834.5	1.175	0,g	0	2	0	8	0	0	0	79
8836.9	1.201	1,u								<25
8837.9	1.190	2,u								<25
8838.1	1.175	2,g	0	2	0	8	2	0	0	81
8838.8	1.186	0,g	1	0	0	2	2	6	-2	33
		0,g	1	0	0	2	0	6	0	46
8841.1	1.189	2,g								<25
8843.3	1.187	0,u	1	0	0	1	1	7	-1	71
8850.0	1.187	2,g	1	0	0	2	0	6	2	35
8854.2	1.189	0,g	0	0	1	1	-1	7	1	60
		0,g	0	1	0	2	-2	8	2	27
8857.8	1.187	2,u	1	0	0	1	1	7	1	35
		2,u	1	0	0	1	-1	7	3	36
8860.5	1.174	0,u	0	1	1	6	0	0	0	31
		0,u	0	2	0	7	1	1	-1	50
8863.9	1.174	2,u	0	1	1	6	2	0	0	31
		2,u	0	2	0	7	3	1	-1	36
8865.1	1.175	1,u	1	1	0	0	0	5	1	81
8867.8	1.188	2,g	0	0	1	1	1	7	1	27
		2,g	0	0	1	1	-1	7	3	31
8869.1	1.202	1,g	0	0	0	3	-3	10	4	28
8886.3	1.203	1,u	0	0	0	2	2	11	-1	35
8888.0	1.185	1,u	0	0	1	9	1	0	0	27
8889.3	1.185	1,g								<25
8890.7	1.203	1,u	0	0	0	2	-2	11	3	39
8893.1	1.171	0,g	0	0	2	4	0	0	0	52
8894.4	1.190	0,u	0	1	0	1	1	9	-1	64
8896.9	1.172	2,g	0	0	2	4	2	0	0	53
8905.5	1.173	0,u	0	1	1	6	0	0	0	34
		0,u	0	2	0	7	1	1	-1	38
8907.2	1.191	0,g	0	1	0	2	2	8	-2	32
		0,g	0	1	0	2	0	8	0	33
8907.4	1.201	1,g								<25
8907.5	1.173	2,u	0	1	1	6	2	0	0	37
		2,u	0	2	0	7	3	1	-1	36
8908.1	1.190	2,u	0	1	0	1	1	9	1	29
		2,u	0	1	0	1	-1	9	3	33
8911.8	1.165	1,u	0	3	0	4	2	1	-1	57
		1,u	0	3	0	4	0	1	1	29
8913.9	1.179	1,u	0	2	0	0	0	7	1	86
8914.1	1.187	0,u	1	0	0	1	-1	7	1	86
8915.3	1.189	0,u	0	0	1	0	0	8	0	73
8917.5	1.176	0,u	0	2	0	7	-1	1	1	83
8918.4	1.191	2,g	0	1	0	2	0	8	2	34
8923.2	1.175	0,g	0	2	0	6	2	2	-2	32
8923.5	1.176	2,u	0	2	0	7	1	1	1	65
8924.0	1.188	2,u	1	0	0	1	-1	7	3	34
8924.1	1.175	2,g	0	2	0	6	4	2	-2	36
8930.3	1.185	1,g	0	1	0	11	1	0	0	44
8931.6	1.188	2,u	0	0	1	0	0	8	2	48
8936.1	1.155	0,g	0	4	0	2	0	0	0	100
8938.3	1.185	1,u	0	0	1	9	1	0	0	51
8938.6	1.171	0,u	1	0	1	4	0	0	0	34
		0,u	0	1	1	6	0	0	0	30
8939.2	1.155	2,g	0	4	0	2	2	0	0	100
8940.5	1.203	1,g	0	0	0	1	-1	12	2	26
8942.2	1.171	2,u	1	0	1	4	2	0	0	40
		2,u	0	1	1	6	2	0	0	27
8947.0	1.166	1,u	0	3	0	4	2	1	-1	36
		1,u	0	3	0	4	0	1	1	61
8953.6	1.176	0,g	0	1	1	5	-1	1	1	27
		0,g	0	2	0	6	-2	2	2	60
8955.3	1.186	1,u	0	1	0	10	2	1	-1	37

cont.

$G(v)$	B_v	$k,u/g$	v_1	v_2	v_3	v_4	l_4	v_5	l_5	%
8957.1	1.175	2,g	0	2	0	6	4	2	-2	27
8957.2	1.186	1,g								<25
8957.6	1.186	1,u								<25
8962.5	1.173	0,g	1	1	0	6	0	0	0	52
8962.6	1.186	1,g	0	0	1	8	2	1	-1	28
8966.3	1.174	2,g	1	1	0	6	2	0	0	41
8970.4	1.163	1,u	0	2	1	3	1	0	0	85
8973.6	1.202	1,u								<25
8975.7	1.188	0,g	1	0	0	0	0	8	0	85
8977.2	1.174	0,u								<25
8977.6	1.174	2,u								<25
8978.8	1.191	0,u	0	1	0	1	-1	9	1	80
8982.9	1.199	0,u	0	0	0	9	1	5	-1	34
		0,u	0	0	0	11	1	3	-1	33
8982.9	1.199	0,g								<25
8984.3	1.173	2,g	0	1	1	5	3	1	-1	44
8985.9	1.173	0,g	1	1	0	6	0	0	0	26
		0,g	0	1	1	5	1	1	-1	39
8987.7	1.199	2,u								<25
8987.7	1.199	2,g								<25
8988.7	1.186	1,g								<25
8989.3	1.186	1,u								<25
8989.5	1.188	2,g	1	0	0	0	0	8	2	85
8990.9	1.191	2,u	0	1	0	1	1	9	1	41
		2,u	0	1	0	1	-1	9	3	40
8994.4	1.186	1,u								<25
8996.0	1.173	0,u	0	0	2	3	-1	1	1	51
8997.7	1.173	2,u	0	0	2	3	3	1	-1	52
8998.7	1.174	0,g	0	1	1	5	-1	1	1	54
		0,g	0	2	0	6	-2	2	2	30
8999.1	1.203	1,u	0	0	0	2	-2	11	3	28
9001.9	1.173	0,u	0	0	2	3	1	1	-1	42
9003.3	1.173	2,u	0	0	2	3	1	1	1	30
9003.7	1.185	1,g								<25
9004.9	1.174	2,g	0	1	1	5	1	1	1	43
9006.8	1.165	1,g	0	3	0	3	3	2	-2	34
9009.4	1.203	1,g	0	0	0	1	1	12	0	26
9010.9	1.174	0,g	0	2	0	6	0	2	0	28
9011.2	1.173	0,g	1	0	1	3	1	1	-1	31
9014.7	1.184	1,g	1	0	0	9	1	0	0	48
9016.6	1.171	2,g	1	0	1	3	3	1	-1	48
9018.5	1.176	2,g	0	2	0	6	2	2	0	28
		2,g	0	2	0	6	0	2	2	43
9019.7	1.175	0,u	0	0	2	3	-1	1	1	27
		0,u	0	2	0	5	-3	3	3	48
9021.5	1.173	2,u	0	0	2	3	1	1	1	30
		2,u	1	1	0	5	3	1	-1	27
9021.8	1.187	1,g								<25
9026.1	1.174	0,u	1	1	0	5	1	1	-1	63
9026.6	1.163	1,g	1	2	0	3	1	0	0	51
		1,g	0	3	0	3	3	2	-2	35
9028.5	1.186	1,u	0	1	0	8	4	3	-3	26
9029.9	1.175	2,u	1	1	0	5	3	1	-1	30
9032.0	1.192	0,g	0	1	0	0	0	10	0	83
9035.4	1.203	1,g	0	0	0	1	-1	12	2	27
9035.7	1.200	0,u	0	0	0	9	-3	5	3	29
9035.7	1.200	0,g	0	0	0	10	-2	4	2	37
9035.8	1.166	1,g	0	3	0	3	-1	2	2	49
9035.8	1.187	1,g								<25
9036.3	1.186	1,u								<25
9037.1	1.187	1,g								<25
9037.1	1.199	2,u	0	0	0	9	5	5	-3	27
9037.2	1.199	2,g								<25
9040.6	1.185	1,u	1	0	0	8	2	1	-1	34
9042.7	1.171	0,g	1	0	1	3	-1	1	1	60
9042.8	1.187	1,u								<25
9045.2	1.199	0,u	0	0	0	13	1	1	-1	32
9045.3	1.199	0,g	0	0	0	12	0	2	0	31
9045.3	1.192	2,g	0	1	0	0	0	10	2	83
9047.6	1.171	2,g	1	0	1	3	3	1	-1	28
		2,g	1	0	1	3	1	1	1	39
9047.9	1.161	1,g	0	1	2	1	1	0	0	69
9050.4	1.199	2,g								<25
9050.7	1.199	2,u								<25
9052.8	1.177	0,u	0	2	0	5	3	3	-3	37
		0,u	0	2	0	5	1	3	-1	28
9054.5	1.175	2,u	1	1	0	5	1	1	1	27
9055.1	1.174	0,u	1	1	0	5	-1	1	1	62
9056.7	1.175	2,u	0	1	1	4	4	2	-2	41
9062.0	1.175	0,g								<25
9063.3	1.174	0,u	0	1	1	4	2	2	-2	35
9064.2	1.187	1,g	0	1	0	7	5	4	-4	26
9065.2	1.175	2,u	0	2	0	5	-1	3	3	28
9067.4	1.187	1,u								<25
9068.4	1.187	1,g								<25
9069.3	1.167	1,g	0	3	0	3	1	2	0	55
		1,g	0	3	0	3	-1	2	2	33
9070.4	1.185	1,u	1	0	0	8	0	1	1	37
9074.1	1.198	2,g								<25
9074.5	1.155	0,u	0	4	0	1	1	1	-1	97
9076.2	1.204	1,u	0	0	0	0	0	13	1	63
9076.3	1.199	2,u	0	0	0	9	7	5	-5	47
9076.9	1.175	2,g								<25
9077.7	1.185	1,g	1	0	0	7	3	2	-2	35
9078.8	1.174	0,u	0	1	1	4	-2	2	2	54
9080.3	1.198	0,g	0	0	0	12	2	2	-2	33
		0,g	0	0	0	14	0	0	0	27
9080.3	1.175	0,g	0	2	0	4	-4	4	4	26
9081.3	1.199	0,u	0	0	0	11	-1	3	1	28
		0,u	0	0	0	13	-1	1	1	27
9081.3	1.186	1,u								<25
9083.3	1.199	0,g	0	0	0	8	-4	6	4	27
		0,g	0	0	0	12	-2	2	2	34
9083.9	1.175	2,u	0	1	1	4	2	2	0	36
9084.2	1.163	1,g	0	2	1	2	2	1	-1	70
9085.7	1.159	1,u	1	1	1	1	1	0	0	89
9086.2	1.172	0,g	0	0	2	2	2	2	-2	35
9086.5	1.199	2,u								<25
9086.7	1.199	0,u	0	0	0	11	3	3	-3	46
9088.4	1.156	0,u	0	4	0	1	-1	1	1	100
9090.0	1.187	1,u								<25
9093.8	1.199	2,g	0	0	0	12	0	2	2	29
9094.2	1.171	2,g	2	0	0	4	2	0	0	39
9095.2	1.164	1,g	0	2	1	2	0	1	1	76

G(v)	B_v	k,u/g	v_1	v_2	v_3	v_4	l_4	v_5	l_5	%
9096.2	1.156	2,u	0	4	0	1	1	1	1	100
9097.8	1.175	0,u	0	1	1	4	2	2	-2	25
		0,u	0	1	1	4	0	2	0	36
9099.8	1.200	2,g	0	0	0	8	8	6	-6	69
9099.8	1.173	0,g								<25
9100.9	1.174	0,g	0	0	2	2	-2	2	2	51
		0,g	0	2	0	4	-4	4	4	26
9101.4	1.200	0,g	0	0	0	10	4	4	-4	34
9102.0	1.172	0,g	2	0	0	4	0	0	0	37
9102.7	1.173	2,g	2	0	0	4	2	0	0	25
		2,g	1	1	0	4	4	2	-2	38
9102.7	1.188	1,u								<25
9103.4	1.200	0,u								<25
9103.7	1.186	1,g	1	0	0	7	-1	2	2	34
9104.4	1.172	0,u	1	0	1	2	2	2	-2	41
9105.4	1.199	0,u	0	0	0	9	-5	5	5	30
		0,u	0	0	0	11	-3	3	3	36
9105.9	1.175	2,u	0	1	1	4	0	2	2	40
9106.8	1.173	0,g	1	1	0	4	2	2	-2	30
9107.5	1.200	2,u								<25
9108.8	1.187	1,g								<25
9109.1	1.173	2,g	0	0	2	2	2	2	0	45
9110.2	1.178	0,u	0	2	0	5	-1	3	1	67
9111.8	1.200	2,g								<25
9112.8	1.175	2,g	0	2	0	4	4	4	-2	28
9114.3	1.188	1,u								<25
9115.0	1.187	1,g								<25
9115.8	1.200	0,g	0	0	0	8	6	6	-6	25
9116.9	1.188	1,u								<25
9118.7	1.177	2,u	0	2	0	5	1	3	1	39
		2,u	0	2	0	5	-1	3	3	27
9119.2	1.199	2,u								<25
9120.8	1.187	1,g								<25
9121.2	1.172	2,u	1	0	1	2	2	2	0	42
9121.4	1.176	0,g	0	2	0	4	4	4	-4	28
9122.3	1.174	0,g	1	1	0	4	-2	2	2	72
9124.8	1.186	1,u	1	0	0	6	4	3	-3	30
9125.2	1.173	2,g	0	0	2	2	0	2	2	37
9125.9	1.171	0,u	1	0	1	2	-2	2	2	84
9127.8	1.164	1,u	1	2	0	2	2	1	-1	30
9129.7	1.201	0,g	0	0	0	8	-6	6	6	49
		0,g	0	0	0	10	-4	4	4	27
9134.7	1.176	2,g								<25
9138.0	1.200	2,g								<25
9143.0	1.200	0,u	0	0	0	9	5	5	-5	26
9143.2	1.185	1,g	1	0	0	7	1	2	0	25
9143.6	1.187	1,g								<25
9144.0	1.164	1,u	1	2	0	2	2	1	-1	53
9144.7	1.188	1,u								<25
9145.5	1.203	1,g								<25
9146.6	1.199	0,g	0	0	0	14	0	0	0	32
9147.9	1.187	1,u								<25
9148.6	1.175	0,g	0	1	1	3	3	3	-3	33
		0,g	0	1	1	3	1	3	-1	30
9149.7	1.204	1,u	0	0	0	0	0	13	1	25
9150.3	1.200	2,u								<25
9151.3	1.152	0,u	0	3	1	0	0	0	0	97
9151.7	1.172	0,u	1	0	1	2	2	2	-2	49
		0,u	1	0	1	2	0	2	0	35
9152.4	1.187	1,u	0	1	0	6	-4	5	5	25
9152.9	1.199	2,g	0	0	0	14	2	0	0	33
9154.4	1.164	1,u	1	2	0	2	0	1	1	62
9154.5	1.174	0,g	1	1	0	4	2	2	-2	28
		0,g	1	1	0	4	0	2	0	40
9154.5	1.171	2,u	1	0	1	2	2	2	0	27
		2,u	1	0	1	2	0	2	2	41
9154.8	1.177	0,g	0	2	0	4	-2	4	2	57
9157.2	1.175	2,g	1	1	0	4	0	2	2	31
9158.0	1.201	0,u	0	0	0	7	7	7	-7	34
9158.4	1.176	0,g	0	1	1	3	-3	3	3	41
9161.1	1.187	1,g								<25
9161.4	1.187	1,g								<25
9162.5	1.176	2,g	0	1	1	3	3	3	-1	48
9163.0	1.166	1,u	0	3	0	2	2	3	-1	31
		1,u	0	3	0	2	-2	3	3	46
9165.8	1.201	0,u	0	0	0	7	-7	7	7	38
		0,u	0	0	0	9	-5	5	5	26
9166.0	1.176	0,u								<25
9166.4	1.200	0,g								<25
9168.6	1.176	2,g	0	2	0	4	-2	4	4	28
9170.2	1.201	0,g								<25
9170.3	1.188	1,g								<25
9170.8	1.201	2,u								<25
9170.8	1.200	2,g								<25
9171.9	1.201	0,u	0	0	0	7	-7	7	7	29
9175.5	1.170	2,u	2	0	0	3	3	1	-1	67
9176.3	1.170	0,u	2	0	0	3	1	1	-1	69
9176.8	1.175	0,g	0	1	1	3	3	3	-3	30
		0,g	0	1	1	3	1	3	-1	30
9178.4	1.161	1,u	0	1	2	0	0	1	1	75
9178.9	1.186	1,u	1	0	0	6	-2	3	3	26
9179.4	1.200	2,u								<25
9179.7	1.200	2,g								<25
9183.0	1.176	2,u								<25
9186.2	1.175	2,g	0	1	1	3	-1	3	3	28
9187.8	1.188	1,g								<25
9189.6	1.201	0,u	0	0	0	7	7	7	-7	50
9189.7	1.188	1,u								<25
9192.7	1.175	0,u	1	1	0	3	-3	3	3	50
9195.3	1.188	1,g								<25
9195.7	1.167	1,u	0	3	0	2	2	3	-1	25
		1,u	0	3	0	2	0	3	1	58
9197.4	1.170	2,u	2	0	0	3	-1	1	1	83
9198.1	1.201	2,u	0	0	0	7	7	7	-5	26
9198.4	1.173	0,u	0	0	2	1	1	3	-1	32
		0,u	1	1	0	3	3	3	-3	26
9199.9	1.176	0,g	0	1	1	3	-1	3	1	62
9200.0	1.188	1,u								<25
9200.9	1.171	2,u	2	0	0	3	1	1	1	47
9201.8	1.189	1,g								<25
9204.6	1.200	0,u	0	0	0	13	-1	1	1	35
9204.6	1.176	0,u	0	2	0	3	1	5	-1	27
9205.4	1.187	1,g								<25
9206.9	1.188	1,u								<25

cont.

$G(v)$	B_v	$k,u/g$	v_1	v_2	v_3	v_4	l_4	v_5	l_5	%
9208.0	1.200	0,g	0	0	0	6	-4	8	4	26
9209.3	1.176	2,g	0	1	1	3	1	3	1	31
		2,g	0	1	1	3	-1	3	3	28
9210.7	1.174	2,u	1	1	0	3	3	3	-1	27
9211.4	1.200	2,u								<25
9212.4	1.179	0,g	0	2	0	4	2	4	-2	37
		0,g	0	2	0	4	0	4	0	40
9212.6	1.174	0,u	0	0	2	1	-1	3	1	52
9212.7	1.201	2,g								<25
9214.0	1.164	1,u	0	2	1	1	1	2	0	53
9214.9	1.151	0,g	1	3	0	0	0	0	0	95
9215.8	1.202	0,g								<25
9216.5	1.160	1,g	1	1	1	0	0	1	1	76
9219.0	1.172	0,g	1	0	1	1	1	3	-1	68
9219.2	1.173	2,u	0	0	2	1	1	3	1	40
9221.4	1.175	0,u	1	1	0	3	3	3	-3	40
		0,u	1	1	0	3	1	3	-1	38
9222.1	1.186	1,u	1	0	0	6	0	3	1	27
9222.1	1.178	2,g	0	2	0	4	0	4	2	40
9222.2	1.176	2,u								<25
9222.2	1.187	1,g								<25
9223.0	1.164	1,u	0	2	1	1	1	2	0	32
		1,u	0	2	1	1	-1	2	2	57
9224.8	1.177	0,u	0	2	0	3	-3	5	3	46
9227.7	1.188	1,g								<25
9227.9	1.201	0,u								<25
9230.2	1.174	2,u	0	0	2	1	-1	3	3	32
9233.9	1.202	2,g								<25
9234.0	1.201	2,u								<25
9235.1	1.172	2,g	1	0	1	1	1	3	1	40
		2,g	1	0	1	1	-1	3	3	29
9236.3	1.189	1,g								<25
9239.2	1.159	1,g	2	1	0	1	1	0	0	76
9239.4	1.187	1,u								<25
9240.1	1.177	2,u	0	2	0	3	-3	5	5	35
9240.9	1.201	0,u								<25
9241.6	1.201	0,g								<25
9244.3	1.188	1,u								<25
9246.2	1.156	0,g	0	4	0	0	0	2	0	95
9249.1	1.201	2,g								<25
9249.1	1.189	1,u								<25
9250.4	1.201	2,u								<25
9251.8	1.202	0,g	0	0	0	6	-6	8	6	52
9253.3	1.172	0,g	1	0	1	1	-1	3	1	74
9253.5	1.177	0,u	0	1	1	2	2	4	-2	49
9256.1	1.187	1,g								<25
9256.9	1.175	0,u	1	1	0	3	-1	3	1	64
9258.5	1.156	2,g	0	4	0	0	0	2	2	100
9259.0	1.178	0,u	0	2	0	3	3	5	-3	49
9261.8	1.176	2,u	1	1	0	3	-1	3	3	28
9262.1	1.164	1,g	1	2	0	1	1	2	0	37
		1,g	1	2	0	1	-1	2	2	31
9265.6	1.202	2,g	0	0	0	6	-6	8	8	36
9265.9	1.172	2,g	1	0	1	1	1	3	1	27
		2,g	1	0	1	1	-1	3	3	55
9269.0	1.177	2,u	0	1	1	2	2	4	0	33
9270.2	1.171	0,g	2	0	0	2	2	2	-2	58
9270.6	1.202	0,u								<25
9270.9	1.200	0,g								<25
9270.9	1.188	1,u	1	0	0	4	-4	5	5	39
9272.7	1.189	1,g								<25
9273.1	1.177	2,u	1	1	0	3	1	3	1	29
9275.5	1.176	0,u	0	1	1	2	-2	4	2	66
9277.2	1.189	1,u	1	0	0	4	4	5	-3	25
9278.6	1.200	2,g								<25
9280.6	1.171	0,g	2	0	0	2	-2	2	2	75
9280.8	1.201	2,u								<25
9282.3	1.163	1,g	1	2	0	1	1	2	0	38
		1,g	1	2	0	1	-1	2	2	54
9282.9	1.177	0,g	1	1	0	2	2	4	-2	26
9287.7	1.189	1,u								<25
9288.9	1.202	0,g	0	0	0	6	6	8	-6	45
9289.1	1.176	2,u	0	1	1	2	2	4	0	30
		2,u	0	1	1	2	-2	4	4	38
9289.3	1.171	2,g	2	0	0	2	2	2	0	53
9289.8	1.202	0,u	0	0	0	5	5	9	-5	31
9291.4	1.188	1,u								<25
9291.7	1.189	1,g								<25
9292.2	1.167	1,g	0	3	0	1	1	4	0	32
		1,g	0	3	0	1	-1	4	2	41
9297.8	1.202	2,g								<25
9298.4	1.188	1,g								<25
9298.8	1.177	2,g								<25
9300.0	1.170	0,g	2	0	0	2	2	2	-2	26
		0,g	2	0	0	2	0	2	0	60
9301.1	1.203	2,u								<25
9302.0	1.177	0,u	0	1	1	2	2	4	-2	31
		0,u	0	1	1	2	0	4	0	44
9304.0	1.186	1,g								<25
9308.1	1.189	1,u								<25
9311.9	1.170	2,g	2	0	0	2	0	2	2	61
9312.0	1.202	0,u								<25
9312.6	1.202	0,g								<25
9312.8	1.177	2,u	0	1	1	2	0	4	2	40
9315.3	1.179	0,u	0	2	0	3	-1	5	1	69
9317.4	1.174	0,g	0	0	2	0	0	4	0	53
9318.5	1.189	1,u								<25
9319.9	1.176	0,g	1	1	0	2	-2	4	2	78
9322.7	1.202	2,g								<25
9322.9	1.202	2,u								<25
9324.1	1.168	1,g	0	3	0	1	1	4	0	51
		1,g	0	3	0	1	-1	4	2	42
9324.3	1.177	0,g	0	2	0	2	2	6	-2	28
9326.2	1.179	2,u	0	2	0	3	1	5	1	33
		2,u	0	2	0	3	-1	5	3	33
9327.5	1.189	1,g								<25
9331.9	1.174	2,g	0	0	2	0	0	4	2	55
9332.3	1.187	1,u	1	0	0	4	-2	5	3	26
9333.3	1.176	2,g	1	1	0	2	-2	4	4	56
9337.8	1.201	0,u								<25
9339.2	1.177	2,g								<25
9341.2	1.202	0,g								<25
9347.1	1.201	2,u								<25
9348.0	1.190	1,g								<25

$G(v)$	B_v	$k,u/g$	v_1	v_2	v_3	v_4	l_4	v_5	l_5	%
9348.4	1.202	2,g								<25
9349.1	1.202	0,u	0	0	0	5	-5	9	5	44
9349.4	1.173	0,u	1	0	1	0	0	4	0	68
9350.0	1.188	1,g	1	0	0	3	3	6	-2	29
9350.2	1.165	1,g	0	2	1	0	0	3	1	91
9353.1	1.189	1,u								<25
9358.2	1.202	2,u	0	0	0	5	-5	9	7	27
9359.6	1.176	0,g	1	1	0	2	2	4	-2	37
		0,g	1	1	0	2	0	4	0	38
9359.8	1.179	0,g	0	2	0	2	-2	6	2	72
9361.8	1.204	0,g	0	0	0	4	2	10	-2	25
9364.0	1.173	2,u	1	0	1	0	0	4	2	70
9367.0	1.177	2,g	1	1	0	2	0	4	2	25
9367.9	1.159	1,u	2	1	0	0	0	1	1	87
9370.3	1.177	0,g	0	1	1	1	1	5	-1	66
9374.9	1.178	2,g	0	2	0	2	-2	6	4	30
9374.9	1.190	1,g								<25
9376.4	1.190	1,u								<25
9377.5	1.189	1,g								<25
9377.7	1.204	2,g								<25
9378.6	1.203	0,u	0	0	0	3	1	11	-1	26
9384.9	1.177	2,g	0	1	1	1	1	5	1	45
		2,g	0	1	1	1	-1	5	3	26
9385.9	1.187	1,u	1	0	0	4	0	5	1	26
9387.0	1.172	0,u	2	0	0	1	1	3	-1	68
		0,u	1	0	1	0	0	4	0	26
9388.3	1.202	0,g	0	0	0	4	4	10	-4	31
9391.0	1.203	2,u								<25
9395.4	1.190	1,g								<25
9397.4	1.203	2,g								<25
9401.5	1.178	0,g	0	1	1	1	-1	5	1	81
9401.5	1.171	2,u	2	0	0	1	-1	3	3	48
9403.3	1.171	0,u	2	0	0	1	-1	3	1	87
9403.5	1.202	0,u								<25
9405.3	1.188	1,g								<25
9406.8	1.189	1,u								<25
9408.8	1.189	1,u								<25
9410.2	1.165	1,u	1	2	0	0	0	3	1	82
9411.8	1.202	2,u								<25
9412.3	1.177	0,u	1	1	0	1	1	5	-1	62
9413.6	1.178	2,g	0	1	1	1	1	5	1	34
		2,g	0	1	1	1	-1	5	3	40
9413.9	1.201	0,g								<25
9416.1	1.171	2,u	2	0	0	1	1	3	1	51
		2,u	2	0	0	1	-1	3	3	35
9416.7	1.180	0,g	0	2	0	2	2	6	-2	39
		0,g	0	2	0	2	0	6	0	47
9420.7	1.201	2,g								<25
9426.7	1.190	1,g	0	0	1	2	-2	7	3	30
9427.0	1.177	1,u	1	1	0	1	1	5	1	32
		2,u	1	1	0	1	-1	5	3	31
9428.6	1.180	2,g	0	2	0	2	0	6	2	45
9439.6	1.204	0,g	0	0	0	4	-2	10	2	31
9445.6	1.166	0,g	0	3	0	6	0	0	0	93
9448.9	1.166	2,g	0	3	0	6	2	0	0	94
9450.1	1.203	0,u	0	0	0	3	-3	11	3	25
9452.4	1.169	1,u	0	3	0	0	0	5	1	87
9452.7	1.204	2,g								<25
9455.4	1.191	1,g								<25
9455.9	1.180	0,u	1	1	0	1	1	5	-1	26
		0,u	0	2	0	1	1	7	-1	67
9456.9	1.191	1,u	0	1	0	2	0	9	1	40
9458.9	1.204	2,u								<25
9459.3	1.204	0,u	0	0	0	3	1	11	-1	37
9460.6	1.176	0,u	1	1	0	1	-1	5	1	87
9460.8	1.191	1,u								<25
9466.2	1.188	1,g	1	0	0	3	1	6	0	27
9467.8	1.203	0,g	0	0	0	4	-4	10	4	35
9469.1	1.179	2,u	1	1	0	1	-1	5	3	36
		2,u	0	2	0	1	1	7	1	32
9472.5	1.176	1,g	0	2	0	9	1	0	0	60
9473.4	1.177	2,u	1	1	0	1	1	5	1	53
9474.1	1.188	1,u	1	0	0	2	2	7	-1	34
		1,u	1	0	0	2	-2	7	3	41
9475.2	1.204	2,u	0	0	0	3	-1	11	3	26
9476.1	1.203	2,g								<25
9483.6	1.175	1,u	0	1	1	7	1	0	0	30
9484.1	1.202	0,u								<25
9484.9	1.190	1,g								<25
9485.4	1.190	1,u	0	0	1	1	1	8	0	27
		1,u	0	0	1	1	-1	8	2	36
9491.6	1.202	2,u								<25
9494.7	1.178	0,u	0	1	1	0	0	6	0	85
9501.2	1.204	0,g	0	0	0	2	0	12	0	28
9503.2	1.172	0,g	2	0	0	0	0	4	0	84
9507.1	1.173	1,g	0	0	2	5	1	0	0	46
9508.2	1.178	2,u	0	1	1	0	0	6	2	82
9510.8	1.166	0,u	0	3	0	5	1	1	-1	80
9513.0	1.167	2,u	0	3	0	5	3	1	-1	62
9514.2	1.181	0,u	0	2	0	1	-1	7	1	88
9514.4	1.204	2,g								<25
9515.6	1.156	1,g	0	4	0	3	1	0	0	99
9517.4	1.172	2,g	2	0	0	0	0	4	2	85
9522.1	1.186	0,u								<25
9522.4	1.187	0,g	0	0	1	9	1	1	-1	26
9526.7	1.186	2,u								<25
9526.8	1.181	2,u	0	2	0	1	1	7	1	41
		2,u	0	2	0	1	-1	7	3	45
9527.0	1.186	2,g								<25
9527.3	1.191	1,u	0	1	0	2	-2	9	3	27
9528.3	1.190	1,g	1	0	0	1	1	8	0	35
		1,g	1	0	0	1	-1	8	2	33
9528.5	1.174	1,u	1	0	1	5	1	0	0	33
		1,u	0	2	0	8	2	1	-1	27
9529.3	1.203	0,u	0	0	0	3	3	11	-3	36
9538.6	1.203	2,u								<25
9539.7	1.192	1,g								<25
9540.3	1.176	1,g								<25
9542.0	1.188	1,u	1	0	0	2	0	7	1	36
9550.2	1.177	1,u	0	2	0	8	2	1	-1	28
		1,u	0	2	0	8	0	1	1	40
9551.6	1.167	0,u	0	3	0	5	-1	1	1	94
9552.9	1.203	0,g								<25
9553.0	1.191	1,u	0	0	1	1	1	8	0	25

cont.

G(v)	B_v	k,u/g	v_1	v_2	v_3	v_4	l_4	v_5	l_5	%
		1,u	0	0	1	1	-1	8	2	28
9553.7	1.177	0,g	1	1	0	0	0	6	0	76
9555.1	1.174	1,u	0	1	1	7	1	0	0	47
9556.5	1.167	2,u	0	3	0	5	3	1	-1	30
		2,u	0	3	0	5	1	1	1	62
9560.7	1.203	2,g								<25
9561.6	1.204	0,u	0	0	0	3	-3	11	3	36
9567.5	1.177	2,g	1	1	0	0	0	6	2	78
9570.3	1.164	0,u	0	2	1	4	0	0	0	79
9570.9	1.205	0,g								<25
9572.5	1.177	1,g								<25
9573.6	1.204	2,u	0	0	0	3	-3	11	5	28
9573.8	1.164	2,u	0	2	1	4	2	0	0	78
9575.9	1.186	0,u	0	0	1	10	0	0	0	43
9577.6	1.186	0,u	0	0	1	10	2	0	0	39
9577.9	1.186	0,g	1	0	0	10	0	0	0	28
		0,g	0	1	0	12	0	0	0	30
9579.8	1.186	2,g	1	0	0	10	2	0	0	25
		2,g	0	1	0	12	2	0	0	27
9581.2	1.192	1,g	0	1	0	1	1	10	0	28
		1,g	0	1	0	1	-1	10	2	32
9581.7	1.205	2,g								<25
9585.5	1.175	1,u	1	0	1	5	1	0	0	25
9588.8	1.204	0,g	0	0	0	2	-2	12	2	50
9588.8	1.175	1,g	1	1	0	7	1	0	0	27
9589.2	1.187	0,g								<25
9590.1	1.187	0,u								<25
9591.1	1.166	0,g	0	3	0	4	2	2	-2	33
9593.3	1.167	2,g	0	3	0	4	4	2	-2	47
9596.0	1.187	2,g								<25
9596.6	1.187	2,u								<25
9601.2	1.204	2,g	0	0	0	2	-2	12	4	36
9601.9	1.191	1,g	0	0	1	0	0	9	1	66
9602.8	1.187	0,g	0	0	1	9	1	1	-1	46
9602.9	1.174	1,u	0	0	2	4	2	1	-1	28
9603.0	1.173	1,g	1	1	0	7	1	0	0	26
9603.0	1.187	0,u	0	1	0	11	1	1	-1	43
9604.5	1.181	0,g	0	2	0	0	0	8	0	83
9604.6	1.187	2,u	0	1	0	11	3	1	-1	31
9605.0	1.187	2,g	0	0	1	9	3	1	-1	33
9610.4	1.174	1,u								<25
9611.5	1.189	1,g	1	0	0	1	1	8	0	42
		1,g	1	0	0	1	-1	8	2	34
9612.8	1.193	1,u								<25
9615.2	1.175	1,g	0	1	1	6	2	1	-1	37
9617.7	1.181	2,g	0	2	0	0	0	8	2	84
9618.0	1.188	2,g								<25
9618.3	1.164	0,g	0	1	2	2	0	0	0	27
		0,g	1	2	0	4	0	0	0	29
9618.3	1.188	0,g	0	1	0	10	2	2	-2	30
9618.6	1.164	2,g	1	2	0	4	2	0	0	34
		2,g	0	3	0	4	4	2	-2	31
9619.3	1.201	1,u								<25
9619.3	1.201	1,g								<25
9619.8	1.204	0,u								<25
9620.6	1.175	1,g	0	1	1	6	0	1	1	34
9624.4	1.168	0,g	0	3	0	4	-2	2	2	85
9626.5	1.205	0,u	0	0	0	1	1	13	-1	64
9627.2	1.204	2,u	0	0	0	1	1	13	1	26
9627.6	1.188	2,u								<25
9627.9	1.187	0,u								<25
9629.5	1.188	0,u	0	1	0	11	-1	1	1	36
9631.2	1.176	1,u								<25
9631.7	1.167	2,g	0	3	0	4	2	2	0	29
		2,g	0	3	0	4	0	2	2	35
9635.8	1.187	2,u	0	1	0	11	1	1	1	29
9636.8	1.157	1,u	0	4	0	2	2	1	-1	72
9637.8	1.162	0,g	0	1	2	2	0	0	0	39
		0,g	1	2	0	4	0	0	0	50
9638.9	1.178	1,g	0	2	0	7	1	2	0	36
9639.2	1.187	0,g	0	0	1	9	-1	1	1	41
9640.2	1.159	0,u	2	0	1	0	0	0	0	28
		0,u	0	0	3	0	0	0	0	62
9640.5	1.205	2,u	0	0	0	1	-1	13	3	43
9640.8	1.163	2,g	0	1	2	2	2	0	0	41
		2,g	1	2	0	4	2	0	0	49
9641.5	1.175	1,u	1	1	0	6	2	1	-1	36
9643.2	1.187	2,g	0	0	1	9	1	1	1	33
9649.1	1.174	1,g								<25
9655.4	1.188	0,g	0	1	0	10	-2	2	2	44
9656.2	1.188	0,u								<25
9656.4	1.176	1,g								<25
9656.6	1.189	2,u	0	0	1	6	6	4	-4	27
9657.1	1.188	2,g								<25
9659.3	1.157	1,u	0	4	0	2	2	1	-1	25
		1,u	0	4	0	2	0	1	1	74
9661.8	1.190	1,u	1	0	0	0	0	9	1	80
9662.0	1.188	0,g								<25
9662.1	1.188	2,g								<25
9662.4	1.158	0,g	1	0	2	0	0	0	0	82
9663.6	1.188	0,u	0	0	1	8	-2	2	2	35
9665.2	1.188	0,g								<25
9665.3	1.188	2,u								<25
9665.9	1.168	0,g	0	3	0	4	2	2	-2	38
		0,g	0	3	0	4	0	2	0	55
9666.5	1.188	0,u								<25
9667.1	1.178	1,u								<25
9667.8	1.161	0,u	1	1	1	2	0	0	0	68
9669.1	1.161	2,u	1	1	1	2	2	0	0	83
9671.1	1.187	2,g								<25
9672.1	1.167	2,g	0	3	0	4	2	2	0	33
		2,g	0	3	0	4	0	2	2	36
9672.8	1.202	1,g								<25
9672.8	1.202	1,u								<25
9673.5	1.176	1,u	0	1	1	5	3	2	-2	38
9674.1	1.188	2,u								<25
9674.6	1.165	2,g	0	2	1	3	3	1	-1	59
9675.1	1.186	0,g								<25
9676.6	1.164	0,g	0	2	1	3	1	1	-1	68
9677.1	1.175	1,u	1	1	0	6	2	1	-1	33
9678.6	1.205	0,g								<25
9678.9	1.194	1,g	0	1	0	1	1	10	0	37
		1,g	0	1	0	1	-1	10	2	36
9679.7	1.187	2,g	0	1	0	12	2	0	0	25

$G(v)$	B_v	$k,u/g$	v_1	v_2	v_3	v_4	l_4	v_5	l_5	%
9679.7	1.176	1,g								<25
9683.6	1.186	2,u	1	0	0	9	3	1	-1	32
9684.0	1.175	1,g	0	0	2	3	3	2	-2	30
9684.5	1.189	0,u	0	0	1	6	-4	4	4	32
9684.6	1.186	0,u	1	0	0	9	1	1	-1	46
9685.7	1.165	0,g	0	2	1	3	-1	1	1	84
9687.5	1.188	2,u								<25
9688.1	1.205	2,g								<25
9690.5	1.201	1,g								<25
9690.5	1.201	1,u								<25
9690.5	1.173	1,u								<25
9692.9	1.189	0,g	0	1	0	8	-4	4	4	28
9693.8	1.165	2,g	0	2	1	3	1	1	1	75
9695.1	1.188	2,g								<25
9695.6	1.167	0,u	0	3	0	3	1	3	-1	32
9696.4	1.175	1,u								<25
9698.2	1.204	0,g	0	0	0	2	-2	12	2	33
		0,g	0	0	0	4	-2	10	2	31
9700.1	1.188	0,g								<25
9701.5	1.205	0,u								<25
9702.0	1.175	1,g								<25
9703.6	1.176	1,g								<25
9705.4	1.189	2,u								<25
9707.2	1.188	2,g								<25
9708.2	1.187	2,g	1	0	0	8	4	2	-2	42
9709.1	1.176	1,g								<25
9709.2	1.205	2,g								<25
9709.5	1.189	0,u								<25
9710.3	1.166	2,u								<25
9711.9	1.205	2,u								<25
9712.6	1.189	0,g								<25
9713.0	1.168	0,u	0	3	0	3	-3	3	3	61
9714.0	1.188	0,u	0	0	1	8	0	2	0	26
9714.6	1.187	0,g	1	0	0	8	2	2	-2	25
9714.7	1.176	1,u								<25
9714.7	1.187	2,u								<25
9715.7	1.186	0,u	1	0	0	9	-1	1	1	52
9715.9	1.154	1,u	0	3	1	1	1	0	0	96
9717.4	1.171	1,g	2	0	0	5	1	0	0	64
9718.0	1.174	1,u	1	0	1	3	3	2	-2	41
9719.5	1.194	1,u	0	1	0	0	0	11	1	77
9722.7	1.146	0,g	0	5	0	0	0	0	0	100
9725.1	1.187	2,u								<25
9725.3	1.165	0,u	1	2	0	3	1	1	-1	61
9725.9	1.189	2,g								<25
9726.6	1.189	0,u								<25
9727.3	1.177	1,g								<25
9727.4	1.201	1,g								<25
9727.5	1.201	1,u								<25
9729.0	1.165	2,u	1	2	0	3	3	1	-1	61
9731.6	1.188	2,u								<25
9732.9	1.201	1,u								<25
9732.9	1.189	0,u	0	1	0	7	-5	5	5	31
9733.4	1.179	1,u	0	2	0	6	0	3	1	34
9733.5	1.176	1,g	1	1	0	5	-1	2	2	27
9733.8	1.202	1,g								<25
9734.8	1.189	0,g	0	0	1	5	-5	5	5	27
9735.4	1.188	2,u	0	0	1	6	6	4	-4	30
9737.3	1.189	0,g								<25
9738.4	1.188	2,g								<25
9738.7	1.205	0,u	0	0	0	1	-1	13	1	52
9739.2	1.164	0,u	1	2	0	3	-1	1	1	57
9739.3	1.187	0,g	1	0	0	8	-2	2	2	46
9741.7	1.165	0,u	0	1	2	1	1	1	-1	47
		0,u	0	3	0	3	1	3	-1	27
9742.3	1.166	2,u	1	2	0	3	1	1	1	44
		2,u	0	3	0	3	3	3	-1	34
9742.4	1.188	2,u	1	0	0	7	5	3	-3	34
9745.0	1.167	0,u	0	3	0	3	3	3	-3	36
9745.2	1.201	1,u								<25
9746.6	1.177	1,g	0	1	1	4	4	3	-3	40
9747.2	1.201	1,g								<25
9748.6	1.189	0,g								<25
9749.1	1.177	1,u								<25
9749.3	1.205	2,u	0	0	0	1	-1	13	3	28
9749.6	1.189	0,u								<25
9749.6	1.187	2,g								<25
9752.4	1.174	1,u	1	0	1	3	1	2	0	31
9752.5	1.188	2,g								<25
9753.2	1.188	0,u								<25
9753.7	1.201	1,g	0	0	0	9	7	6	-6	25
9755.8	1.167	2,u	0	3	0	3	-1	3	3	45
9756.3	1.163	0,u	0	1	2	1	-1	1	1	60
		0,u	1	2	0	3	-1	1	1	29
9756.6	1.189	2,u								<25
9757.5	1.188	2,g								<25
9760.9	1.190	0,g	0	0	1	5	5	5	-5	27
		0,g	0	1	0	6	6	6	-6	35
9761.3	1.201	1,u								<25
9761.5	1.206	0,g	0	0	0	0	0	14	0	63
9764.1	1.163	2,u	0	1	2	1	1	1	1	61
9765.0	1.178	1,g								<25
9766.0	1.177	1,g								<25
9767.5	1.190	0,g								<25
9768.0	1.189	0,u	0	1	0	7	5	5	-5	27
9770.9	1.189	2,g								<25
9771.3	1.177	1,g	1	1	0	5	1	2	0	28
9771.7	1.174	1,u								<25
9771.8	1.189	2,u								25
9772.8	1.188	0,u	1	0	0	7	-3	3	3	42
9773.3	1.206	2,g	0	0	0	0	0	14	2	64
9773.8	1.153	1,g	1	3	0	1	1	0	0	83
9774.1	1.201	1,g								<25
9774.2	1.162	0,g	1	1	1	1	1	1	-1	75
9775.2	1.202	1,u	0	0	0	8	8	7	-7	55
9775.2	1.176	1,u								<25
9776.6	1.189	0,u								<25
9780.0	1.188	2,u								<25
9781.9	1.189	0,g	1	0	0	6	-4	4	4	28
		0,g	0	1	0	6	-6	6	6	37
9782.8	1.187	0,g	1	0	0	8	0	2	0	26
9784.4	1.189	2,u								<25
9785.3	1.169	0,u	0	3	0	3	-1	3	1	84
9785.6	1.190	0,u								<25

cont.

$G(v)$	B_v	$k,u/g$	v_1	v_2	v_3	v_4	l_4	v_5	l_5	%
9785.6	1.187	2,g								<25
9787.0	1.189	2,g								<25
9787.0	1.165	0,u	0	2	1	2	2	2	-2	47
9789.1	1.176	1,g								<25
9789.3	1.176	1,u	1	1	0	4	4	3	-3	39
9790.6	1.157	1,g	0	4	0	1	1	2	0	32
		1,g	0	4	0	1	-1	2	2	52
9790.9	1.202	1,u								<25
9791.7	1.173	1,u	2	0	0	4	2	1	-1	33
9792.7	1.162	2,g	1	1	1	1	1	1	1	70
9792.7	1.174	1,g								<25
9792.8	1.189	0,g	0	0	1	7	-1	3	1	28
9793.1	1.169	2,u	0	3	0	3	1	3	1	51
9793.3	1.162	0,g	1	1	1	1	-1	1	1	91
9793.9	1.165	0,u	0	2	1	2	-2	2	2	74
9795.6	1.202	1,g								<25
9798.1	1.176	1,u								<25
9800.5	1.165	2,u	0	2	1	2	2	2	0	62
9801.9	1.202	1,u								<25
9802.0	1.188	2,g								<25
9802.2	1.189	2,u								<25
9802.5	1.176	1,u								<25
9804.0	1.190	0,g								<25
9804.3	1.189	0,g								<25
9805.8	1.166	0,u	0	2	1	2	2	2	-2	29
		0,u	0	2	1	2	0	2	0	58
9808.3	1.187	0,u	1	0	0	7	3	3	-3	26
9808.4	1.203	1,g								<25
9808.7	1.190	2,g								<25
9808.9	1.190	0,g	0	1	0	6	6	6	-6	35
9808.9	1.158	1,g	0	4	0	1	1	2	0	56
		1,g	0	4	0	1	-1	2	2	41
9809.7	1.174	1,u	2	0	0	4	0	1	1	27
9811.8	1.175	1,u								<25
9812.3	1.202	1,u								<25
9812.6	1.188	2,u	1	0	0	7	-1	3	3	31
9813.3	1.177	1,g	0	1	1	4	0	3	1	31
9815.0	1.189	0,u								<25
9816.3	1.189	0,g	1	0	0	6	-4	4	4	31
9816.5	1.189	2,g								<25
9817.1	1.166	2,u	0	2	1	2	0	2	2	67
9818.6	1.165	0,g								<25
9820.3	1.161	0,g	2	1	0	2	0	0	0	62
9821.3	1.189	0,u	1	0	0	5	-5	5	5	29
9822.3	1.174	1,g	1	0	1	2	2	3	-1	26
		1,g	1	0	1	2	-2	3	3	47
9823.3	1.160	2,g	2	1	0	2	2	0	0	77
9824.0	1.189	2,g								<25
9825.0	1.189	2,u								<25
9825.5	1.177	1,u								<25
9827.5	1.200	1,g	0	0	0	15	1	0	0	42
9829.9	1.202	1,u								<25
9830.3	1.177	1,u								<25
9830.6	1.189	0,u								<25
9831.6	1.180	1,g	0	2	0	5	1	4	0	28
		1,g	0	2	0	5	-1	4	2	26
9832.7	1.190	0,g								<25
9835.4	1.190	2,u								<25
9835.6	1.158	0,u	2	0	1	0	0	0	0	68
		0,u	0	0	3	0	0	0	0	30
9836.0	1.166	2,g								<25
9836.2	1.190	0,u	1	0	0	5	5	5	-5	27
9837.4	1.202	1,g								<25
9839.6	1.165	0,g	1	2	0	2	-2	2	2	76
9840.7	1.190	2,g								<25
9842.5	1.188	0,g								<25
9842.6	1.189	2,u								<25
9844.6	1.177	1,u								<25
9845.6	1.167	0,g	1	2	0	2	2	2	-2	44
9846.9	1.190	0,g								<25
9848.8	1.189	2,g	1	0	0	6	-2	4	4	28
9851.3	1.175	1,g								<25
9852.2	1.189	0,u	1	0	0	5	5	5	-5	39
9853.2	1.205	0,u	0	0	0	3	-1	11	1	34
9853.3	1.202	1,u								<25
9853.6	1.190	0,u	1	0	0	5	-5	5	5	42
9853.8	1.202	1,g								<25
9854.1	1.205	0,g								<25
9855.0	1.190	2,u								<25
9855.8	1.190	0,g								<25
9855.9	1.189	2,g								<25
9856.5	1.177	1,g								<25
9856.6	1.187	0,u	1	0	0	7	-1	3	1	34
9858.0	1.167	2,g	1	2	0	2	2	2	0	48
		2,g	0	3	0	2	-2	4	4	29
9859.1	1.203	1,g	0	0	0	7	-7	8	8	43
9859.9	1.173	1,g								<25
9859.9	1.178	1,u	0	1	1	3	3	4	-2	26
		1,u	0	1	1	3	-3	4	4	33
9861.7	1.188	2,u								<25
9861.9	1.165	0,g	1	2	0	2	0	2	0	54
9862.9	1.205	2,u								<25
9864.0	1.205	2,g								<25
9864.2	1.178	1,u								<25
9866.1	1.190	2,g								<25
9867.1	1.169	0,g	0	3	0	2	-2	4	2	77
9869.1	1.155	1,g	0	3	1	0	0	1	1	96
9869.2	1.166	2,g	1	2	0	2	0	2	2	55
		2,g	0	3	0	2	2	4	0	28
9870.0	1.178	1,u								<25
9870.0	1.190	2,u	1	0	0	5	5	5	-3	26
9872.6	1.190	0,g								<25
9873.0	1.190	0,u								<25
9874.1	1.202	1,u								<25
9875.5	1.173	1,g	2	0	0	3	-1	2	2	31
9877.8	1.164	0,g	0	1	2	0	0	2	0	66
9878.7	1.203	1,g								<25
9879.0	1.190	0,u	0	0	1	4	4	6	-4	28
9879.5	1.203	1,u								<25
9880.6	1.168	2,g	0	3	0	2	-2	4	4	40
9881.9	1.188	0,g	1	0	0	6	-2	4	2	36
9883.1	1.189	2,u								<25
9884.8	1.177	1,u								<25
9885.7	1.190	0,u								<25

$G(v)$	B_v	$k,u/g$	v_1	v_2	v_3	v_4	l_4	v_5	l_5	%
9886.4	1.203	1,g								<25
9886.6	1.190	2,g								<25
9887.2	1.190	0,u	0	1	0	5	5	7	-5	26
9889.7	1.178	1,g								<25
9889.8	1.189	2,g								<25
9890.1	1.190	2,u								<25
9891.9	1.164	2,g	0	1	2	0	0	2	2	71
9893.0	1.190	2,u								<25
9894.2	1.176	1,g								<25
9895.5	1.190	0,g								<25
9896.6	1.204	1,u								<25
9899.6	1.177	1,g								<25
9905.2	1.190	2,g								<25
9905.4	1.172	1,g	2	0	0	3	1	2	0	46
9906.4	1.190	2,u	0	1	0	5	-5	7	7	28
9907.2	1.170	0,g	0	3	0	2	2	4	-2	37
		0,g	0	3	0	2	0	4	0	56
9908.8	1.175	1,u	1	0	1	1	1	4	0	34
		1,u	1	0	1	1	-1	4	2	26
9909.5	1.162	0,u	1	1	1	0	0	2	0	60
9910.8	1.189	0,u	0	0	1	4	-4	6	4	35
9912.6	1.178	1,u	0	1	1	3	1	4	0	28
9913.3	1.190	0,g								<25
9914.0	1.166	0,g	0	2	1	1	1	3	-1	72
9914.2	1.176	1,g								<25
9916.6	1.169	2,g	0	3	0	2	2	4	0	32
		2,g	0	3	0	2	0	4	2	50
9919.1	1.190	0,u								<25
9919.4	1.203	1,u								<25
9920.3	1.203	1,g								<25
9922.8	1.190	2,u								<25
9923.6	1.177	1,g	1	1	0	3	-3	4	4	26
9923.7	1.189	0,g	1	0	0	4	-4	6	4	38
9923.8	1.190	2,g								<25
9924.7	1.191	0,g								<25
9925.0	1.191	0,u	0	1	0	5	5	7	-5	28
9925.1	1.162	2,u	1	1	1	0	0	2	2	68
9926.4	1.191	2,u								<25
9927.4	1.178	1,g	0	2	0	3	-3	6	4	26
9927.7	1.167	0,g	0	2	1	1	-1	3	1	90
9928.4	1.166	2,g	0	2	1	1	1	3	1	53
9928.8	1.154	1,u	1	3	0	0	0	1	1	91
9931.2	1.181	1,u	0	2	0	4	0	5	1	32
9934.6	1.188	0,g								<25
9935.4	1.190	2,u								<25
9936.2	1.189	2,g								<25
9936.5	1.161	0,u	2	1	0	1	1	1	-1	65
		0,u	1	1	1	0	0	2	0	30
9937.4	1.191	0,u								<25
9938.4	1.202	1,g								<25
9939.6	1.161	0,u	2	1	0	1	-1	1	1	87
9940.0	1.190	2,g								<25
9941.8	1.167	2,g	0	2	1	1	1	3	1	29
		2,g	0	2	1	1	-1	3	3	55
9942.3	1.189	2,g								<25
9943.6	1.204	1,u								<25
9945.1	1.204	1,g								<25

$G(v)$	B_v	$k,u/g$	v_1	v_2	v_3	v_4	l_4	v_5	l_5	%
9947.7	1.191	2,u								<25
9949.0	1.178	1,g								<25
9949.5	1.204	1,u								<25
9951.5	1.174	1,u	1	0	1	1	1	4	0	29
		1,u	1	0	1	1	-1	4	2	36
9954.1	1.161	2,u	2	1	0	1	1	1	1	70
9954.7	1.190	0,g	1	0	0	4	4	6	-4	27
9954.9	1.190	0,g								<25
9956.1	1.166	0,u	1	2	0	1	1	3	-1	58
		0,u	0	3	0	1	1	5	-1	25
9956.3	1.188	0,u	1	0	0	5	3	5	-3	29
9962.6	1.158	1,u	0	4	0	0	0	3	1	91
9963.3	1.179	1,g								<25
9963.4	1.191	0,g	0	1	0	4	-4	8	4	34
9963.5	1.173	1,u	2	0	0	2	-2	3	3	26
9965.1	1.190	2,g								<25
9965.1	1.189	2,u								<25
9965.3	1.204	1,g								<25
9967.9	1.190	2,g	1	0	0	4	-4	6	6	29
9968.7	1.179	1,g								<25
9969.5	1.191	0,u	1	0	0	3	1	7	-1	26
9971.3	1.167	2,u	1	2	0	1	1	3	1	35
		2,u	1	2	0	1	-1	3	3	27
9972.8	1.179	1,u								<25
9974.4	1.189	0,g								<25
9974.6	1.191	0,u								<25
9977.2	1.191	2,g								<25
9978.9	1.178	1,g	0	1	1	2	2	5	-1	28
		1,g	0	1	1	2	-2	5	3	26
9980.1	1.173	1,u	2	0	0	2	2	3	-1	28
		1,u	2	0	0	2	-2	3	3	41
9980.6	1.204	1,u								<25
9985.2	1.191	2,u								<25
9985.3	1.191	2,u								<25
9985.4	1.166	0,u	1	2	0	1	-1	3	1	87
9987.3	1.190	2,g								<25
9990.4	1.169	0,u	1	2	0	1	1	3	-1	30
		0,u	0	3	0	1	1	5	-1	66
9991.1	1.203	1,g								<25
9992.3	1.156	0,g	3	0	0	0	0	0	0	92
9997.0	1.166	2,u	1	2	0	1	1	3	1	27
		2,u	1	2	0	1	-1	3	3	60
9997.9	1.191	0,g	0	0	1	3	-3	7	3	32
9999.4	1.191	0,u								<25

AUTHOR INDEX

Numbers in parentheses are reference numbers and indicate that the author's work is referred to although his name is not mentioned in the text. Numbers in *italic* show the pages on which the complete references are listed.

389

Heinze, J., 56(397), *304*
Helgaker, T., 42(275), *301*
Heller, D. F., 283(1131), *327*
Helminger, P., 255(782), *316*
Henderson, J. R., 16(94), 45(94),
   51(331–333), *295, 302–303*
Henri, H., 54(364–365), 98(364–365), *303*
Henry, A., 246(729), 248(729), *314*
Henry, B. R., 29(161), 51(334), 148(599),
   221(334,703–705,710–714), 281(1085),
   283(714,1085), *297, 303, 310,* 314, *326*
Hepp, M., 259(867), 260(867,894),
   261(884a,893,893a,894,894a,895),
   319–320
Herbelin, J. M., 266(950), *322*
Herlemont, F., 256(801), *317*
Herman, M. 195(655,676), 233(721),
   237(721), 243(722), 257(862), 258(864),
   259(866–867), 260(867,894), 261(721,
   880,884a,886,889–891,893,893a,894,
   894a,895–896,906–907),
   278(1055), 279(1055), 288(1055),
   312–314, 319–321, 325 37(237),
   56(386,394–395,414), 70–71(444),
   77(450–451,453), 78(450), 82(450),
   83–86(444), 88(450,468–469), 90(450),
   93(444), 112(444,504–505), 119(516),
   122(516), 125(444,450,468–469,525),
   126–128(444), 135(505), 137(450),
   138(505), 141(451), 147(575,586,589),
   148(444,450,525,586,592–594,
   596,610), 149(596), 150(450,596),
   151(596), 155(614,617),
   156(468,614,619),
   157(450,453,468,525), 158(592),
   160(589), 161(469,610), 163(610),
   172(394), 175(394), 182(610,655),
   183(525,610), 185(589), 187(516),
   188(594), 189(469,504,655), 190(655),
   192(655), 193(450,504,655), 194(655),
   195(451,505), 196(451),
   197(450,468,516,525,610,614),
   198(610), 221(444,468, 596),
   222(468,596), 245(596), 248–250(451),
   252(504), 255(505), 256(453), 261(586),
   269(596), 278(516,525,596,614),
   283(516), 288(516,614), 291(469),
   364(450), *300, 304–306, 308, 310–311*
Herman, R., 246(724), *314*
Hermans, C., 255(787), *316*

Hermina, W., 282(1103), *327*
Herzberg, G., 6(3), 56(3,385), 62(3),
   119(515), 122(515), 182(515), 184–
   185(515), 197(515), 208(697), 224(697),
   255(748–751), 262(748), *293, 304, 308,*
   315, 340(697)
Herzberg, L., 255(749–750), *315*
Hess, P., 281(1083,1092–1093), *326*
Heyl, A., 26(130), *297*
Hidalgo, A., 32(194), 47(194,315), 50(194),
   298, 302
Hietanen, J., 189(669), *312*
Hilico, J.-C., 147–148(583–584,586),
   261(586), *310*
Hillman, J. J., 189(671), 195(675), *313*
Hindmann, J. C., 278(1051), *325*
Hineman, M. F., 284(1146), *328*
Hinkley, E. D., 264(930), *321*
Hippler, H., 56(409), 290(1199–1201), *305,*
   330
Hirano, T., 26(130), *297*
Hiraoka, S., 291(1227), *330*
Hiris, H., 264(927), *321*
Hirsch, G., 6(29,37), 53(37), *293–294*
Ho, J., 163(628), *311*
Ho, T. S., 40(257), *300*
Hobbs, P. C. D., 264(936–937), *321*
Hodges, T., 267(956–957), *322*
Holland, J. K., 147(574–575), 261(896), *310,*
   320
Hollenstein, H., 147–148(588),
   261(874,876,884b,897,915), 284(1145),
   310, 319–321, 328
Holme, T. A., 62(438), 70–71(438), 73(438),
   92(438,475), 148(438), *305, 307*
Homann, K., 198(680), *313*
Horn, D., 257(858), *319*
Horn, H., 116(510), *308*
Horn, T. R., 12(61), *294*
Hornberger, C., 147(561), 282(1096),
   283(1096,1127), *309, 327*
Horneman, V.-M., 257(860), *319*
Hornos, J. E. M., 94(489), 147(572),
   148(489), *307, 310*
Hornos, Y. M. M., 147(572), *310*
Hougen, J. T., 12(59), 53(59), 56(393),
   98(494–496), 100(498), 140(541),
   173(540), 182(59), 213(700), 227(393),
   294, 307, 309, 314
Houston, P. L., 291(1221), *330*

SUBJECT INDEX